Springer Atmospheric Sciences

The Springer Atmospheric Sciences series seeks to publish a broad portfolio of scientific books, aiming at researchers, students, and everyone interested in this interdisciplinary field. The series includes peer-reviewed monographs, edited volumes, textbooks, and conference proceedings. It covers the entire area of atmospheric sciences including, but not limited to, Meteorology, Climatology, Atmospheric Chemistry and Physics, Aeronomy, Planetary Science, and related subjects.

Vladimir L. Bychkov

Natural and Artificial Ball Lightning in the Earth's Atmosphere

Springer

Vladimir L. Bychkov
Faculty of Physics
Lomonosov Moscow State University
Moscow, Russia

ISSN 2194-5217 ISSN 2194-5225 (electronic)
Springer Atmospheric Sciences
ISBN 978-3-031-07860-6 ISBN 978-3-031-07861-3 (eBook)
https://doi.org/10.1007/978-3-031-07861-3

Translation from the Russian language edition: "Natural and Artificial Ball Lightning in the Earth's Atmosphere" by Vladimir L. Bychkov, © MAKS Press 2021. Published by MAKS Press. All Rights Reserved.

This Springer imprint is published by the registered company Springer Nature Switzerland AG
The registered company address is: Gewerbestrasse 11, 6330 Cham, Switzerland

Preface

The book is called "Natural and artificial ball lightning in the Earth's atmosphere" because it is devoted to long-lived luminous formations of both natural and artificial origin. The first is generally accepted name—ball lightning (BL); the second is long-lived luminous objects (LLO), usually obtained in laboratories in experiments with various kinds of discharges or combustible substances, and also more rarely in devices used in everyday life.

The term LLO—more corresponds to the object than the frequently used term "plasmoid", or artificial BL. It contains at least two properties that distinguish this object from a number of natural and artificial objects: a long lifetime and glow. It separates these objects from "plasmoids", the properties of which are not clearly defined, and the name contains the word plasma, which makes this object supposedly known.

Natural BL has a large number of observable properties, and it is obviously unclear which of them is the main. Using the term LLO for them highlights glowing objects. The term "long-lived object" also requires decoding. It is an object with an existence time greater or comparable to the minimum time of object perception by observers.

The words about atmospheric pressure emphasize the conditions for the appearance of these objects and separate them from the multitude of objects realized in experiments in gas discharge and beam plasma at reduced pressure and in various gases other than air.

The joint consideration of objects under the term LLO emphasizes the need for a comprehensive study of complex objects of artificial and natural origin. It was the study of the properties of LLO in the twentieth-twenty-first centuries that allowed a better understanding of the possibility of experimentally modeling BL analogues.

The problem of BL exists for quite a long time. For decades, it has aroused great interest among people. There was always a small group of brave scientists who, based on the latest scientific achievements was trying to explain this rather rare and "rather simple" compared, for example, with the linear lightning phenomenon, so far, all attempts by this group to unravel this phenomenon are considered unsuccessful, since everyone expects from them a recipe or implementation of this object with all its anomaly parameters.

It has even become commonplace to use in the media the phrase " BL not solved by science." That is a strange problem that humanity for some reason cannot or does not want to solve, while the nature of space and time, black holes, and dark energy is already generally clear. At the same time, the most unpleasant thing is that, unlike these super-fundamental and generally safe for ordinary people phenomena, BL often commits real destruction of surrounding objects, is the cause of injuries and deaths of people and animals. Maybe it would be better to postpone the study of this object for later, until better times. Is it possible to stop taking this phenomenon seriously at all, in connection with declaring it a physical or physiological illusion, and to put cases of death and injuries under the cloth? However, this is actually how it happens. Instead of looking at this situation as nature's indication of the need to search for a new paradigm, or return to the paradigm of the nineteenth century, and involve additional researchers and finance in the study of BL, support for research on this phenomenon has practically ceased.

Nevertheless, enthusiasts are conducting research in the following areas.

Collection and statistical analysis of observations showed that BL or LLO is not only phenomena associated with thunderstorm conditions but also with geophysical anomalous phenomena in the atmosphere and ocean. They are a much broader phenomenon than is commonly believed.

Experimental modeling of LLO and similar objects. Analysis of the results of earlier experimental studies that are not described within the framework of existing approaches and paradigms.

Creation of theoretical models that allow to explain many properties of LLO or BL and present them as objects of physical vacuum realized in anomalous conditions.

This book is devoted to the coverage of these topics. The material is presented concisely, with the idea that a particular researcher will find in this book the right direction of research or references to literature and will avoid trivial mistakes in research.

The book consists of five chapters. The first chapter is devoted to the historical facts of BL research. Its purpose is to present to the public scientists who have made a significant contribution to research on LLO and BL. I would like to show real heroes of this complex science.

The second chapter presents data on the natural LLO and BL. It contains observations describing the usual and unusual properties of BL in different physical, geographical, and geophysical conditions. These events are usually not presented in a single collection. However, such a collection of data shows that BL is not such a rare phenomenon, as it is commonly believed.

The third chapter presents statistical data on the BL properties and the results of modern statistical analysis of a number of BL properties.

The fourth chapter is devoted to the description of numerous studies on the laboratory production of LLO. The presentation shows that currently there are a number of experimental methods leading to the creation of LLO with properties close to those of natural BL, i.e., the problem of experimental production of artificial BL analogues of several types is practically solved.

In the fifth chapter, we briefly describe the BL theories, allowing us to look at this phenomenon from different positions, but the main properties of anomalous BL are their high energy and long lifetime.

I want to thank my wonderful friends and comrades, enthusiasts and colleagues, research partners, and co-authors of the book. I grieve for the departed researchers who were close friends over life.

I thank Prof. B. M. Smirnov for introducing me to this problem challenging the scientific community, and instilling in me a persistent interest in it. Being an outstanding scientist, he constantly oversees and amicably supports BL research.

I thank Prof. A. F. Alexandrov, who for more than 30 years stimulated the BL research at the Department of Physical Electronics of the Physical Faculty of Lomonosov Moscow State University. Looking historically, Moscow University should have been actively engaged in these studies, since M. V. Lomonosov was the world's first researcher of BL and the university's duty to him is to unravel the nature of this phenomenon.

I sincerely thank Academician of the Russian Academy of Natural Sciences, DS A. I. Nikitin, and Corresponding Member of this Academy Dr. T. F. Nikitina. Communication with them and joint research brought me a lot of knowledge, sincere, and friendly warmth. Without discussion with A. I. Nikitin, many sections of this book would not have appeared, for which I thank him very much.

I thank the Academician of the Russian Academy of Natural Sciences, prof. A. I. Klimov, with whom we have traveled many roads together in the analysis of complex phenomena, obtaining analogues of BL, LLO, low-temperature plasma, and plasma aerodynamics.

I remember with special warmth the untimely passing of the remarkable experimental scientist S. E. Emelin, without whose work a new direction of experimental BL research would not have taken place.

I thank Dr. G. C. Dijkhuis (the Netherlands). Being a good theoretician and experimentalist, he is always interested in BL research. He helps scientists in every way to establish international scientific contacts and finance their participation in international forums.

I thank Dr. A. G. Keul (Austria) for his hard work in the collection and analysis of ball lightning observations, friendly attitude to Eastern colleagues, and the provided help during the most difficult period of the 90s of the twentieth century.

I thank Prof. I. I. Esakov and Prof. A. L. Kuranov for their constant assistance in finding funds for BL research and all friendly support during research. I also thank Dr. J. Tishkov (USA) and Dr. Gill Won Suh (Korea) for their support of BL and LLO research and the holding of international symposia in Russia.

I am sincerely grateful to my family members, my wife Nataliya, and my sons Andrey and Dmitry, who for more than 30 years have constantly, selflessly, and uncomplainingly helped me both in research and in writing the book; otherwise, it would not have been born.

Moscow, Russia
January 2022

Vladimir L. Bychkov

Contents

Chapter 1
Long-Lived Luminous Formations (LLO), Ball Lightning (BL), and Their Researches

Abstract This chapter has a historical character. It is devoted to the contribution of the main ball lightning researchers. From it, one can learn about the role of M. Faraday, M.V. Lomonosov, F. Arago, N. Tesla, P.L. Kapitsa, Ya.I. Frenkel, S. Singer, etc., in the formation of research on Ball lightning (BL) and modeling of long-lived luminous formations (LLO)—artificial BL. It provides data on groups of modern researchers of BL and LLO. Four areas of research are identified: the collection of observational data on BL, statistical analysis of observations, experimental studies on the modeling of LLO and BL, theoretical models of BL.

Keywords Ball lightning · Long-lived luminescent objects · History · Main investigators · Directions of investigation

There is a natural phenomenon in the world that arouses both burning interest and strong dislike. It is ball lightning (BL). Without going into its description in detail, one can say that this is a phenomenon that has many different manifestations. A common feature of observations is the visibility of this, as a rule, the spherical shape of an object against the background of the sky or a room, not necessarily due to luminescence and autonomy (Fig. 1.1).

A burning interest is aroused by this object among those who want to understand the nature of the phenomenon, even realizing that in their entire life one may not get an answer about its essence. There are not many such people among professional naturalists, because the majority believe that it is necessary to deal with tasks that are promising from the point of view of success and applications, and not with a problem with an unclear result in advance.

The strongest dislike of BL is aroused by some professionals, who perceive it as an object for some kind of competition: who will win whom? Nature—with the preservation of the mystery, or a given ambitious researcher who must not lose and must come up with an explanation for some feature, which he will then call the main one. But, having competed, and recognizing the futility of personal attempts in this competition, they either say that this object does not exist as such, and its existence is an idle invention of a sick imagination or the action of a complex of natural events leading to the appearance of a visual image of this object. The futility of searching for an answer to the nature of BL is declared, for this reason, a ban on its research or

V. L. Bychkov, *Natural and Artificial Ball Lightning in the Earth's Atmosphere*, Springer Atmospheric Sciences, https://doi.org/10.1007/978-3-031-07861-3_1

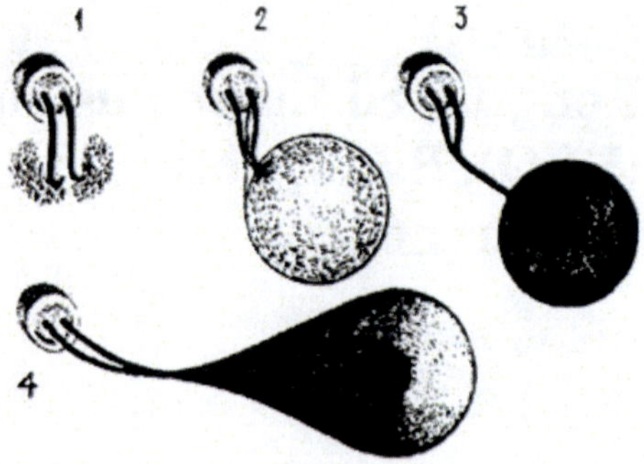

Fig. 1.1 The emergence of BL from the socket. The ball of 10 cm. Observer Evdokimov 1983

even discussion is announced. Those who nevertheless do research, they treat with undisguised irony, lack of respect, and ostracism.

There are also those who intuitively understand that many years of fruitless attempts to understand this phenomenon speak of fundamental problems in the modern description of nature, and if this is so, then his research should be attributed to later when natural science grows, but for now, it should not be considered, not discussed and do not waste energy and forces on the search for an essence of incomprehensible objects. They are scared that answering questions about the nature of this phenomenon will require a restructuring of the entire system of physical views, the overthrow of established agreed concepts, and the destruction of the property status of the established scientific elite. They, like the blind, do not see that only the solution of just such problems leads to an understanding of nature at a new higher level.

Modern society, which has grown up on the principles of competition, sometimes biased, in which contractual representations in natural science have already won out, stands on the position of neglecting and ignoring the existence of this phenomenon and the like, humiliates, and marginalizes its researchers. In this regard, the statement of Russian Academician V.L. Ginzburg: "When a discussion of BL starts at my seminar, I immediately stop it" is typical. This attitude only increases the interest of ordinary people and media representatives in such phenomena and their researchers.

Based on the state of the scientific community described above, the heroes among BL researchers in Russia and the world remain unknown to the public, and I would like to partially fill his gap. I will confine myself to facts well known among BL researchers and the experience of personal communication with some BL Russian and Western researchers.

1.1 A Bit of History

Of course, we must start with M.V. Lomonosov (Singer 1971; Stenhoff 1999). He was interested in the phenomenon of ball lightning. He described the death of Professor G. Richman when he tried to measure the charge of a cloud in his laboratory. Having

Fig. 1.2 M.V. Lomonosov

launched the kite towards the cloud, he expected that a current would arise, which would remove the charge from the cloud and transfer it along a waxed rope from the cloud to the electroscope. As can be seen from Sokolov's engraving, a luminous ball separated from one of the metal rods, which hit the professor in the chest, after which the latter died (Fig. 1.2).

It was the lack of confidence in the appearance of the ball that caused a discussion, the ball or just the discharge separated from the bar. Even the presence of an engraving by the artist observing this event did not affect the discussion.

Such a mistrustful attitude towards incomprehensible luminous balls in the atmosphere persists to this day (Fig. 1.3).

However, M.V. Lomonosov took this phenomenon seriously. He knew about this phenomenon, because in his time, of course, he still remembered that the peasants called the BL by the name of Rodia, meaning either red or the fruit of a pomegranate. The scientist described it and came up with the first hypothesis about its chemical nature in a form of fatty vapors. At that time, the scientist was not afraid that his colleagues would condemn or ridicule for this.

It should be noted that the result of G. Richman's experiment suspended the study of atmospheric electricity for a whole century. We must not forget the religiosity of scientists of that time, who could consider the appearance of the ball as a divine warning.

The history of the work of scientists on BL is described in the books by S. Singer and Mark Stenhof (Singer 1971; Stenhoff 1999).

Michael Faraday is the first well-known scientist who left a significant mark on physics. M. Faraday—the discoverer of the electromagnetic field, experimental

Fig. 1.3 Death of Professor G. Richman 1753, engraver Sokolov

physicist, and chemist. Creator of the first model of an electric motor and transformer. Many saw in him the lord of lightning and the king of physicists (Fig. 1.4).

In the work, "Faraday M. Experimental researches on electricity V.1. Bernard Quaritch. London 1839, P. 523." He argued that the speed of movement and the lifetime of BL are incomparable with all the known properties of ordinary electric discharges, and from this, he concluded that any connection between BL and ordinary lightning or atmospheric electricity is more than doubtful.

Fig. 1.4 Michael Faraday (1791–1867)

Fig. 1.5 Dominique Francois Jean Arago (1786–1853)

The first in-depth analysis of materials on BL was made in 1838 in the work of astronomer and physicist D.F. J. Arago. He published a review of twenty reports of BL observation (Fig. 1.5).

Arago made notable contributions to physics. At the age of only 23, he was elected a member of the Paris Academy of Sciences, later he became its permanent secretary. He was appointed manager of the Paris Bureau of Geographical Longitudes. He was a professor of analytical geometry at the Ecole Polytechnique from 1809 to 1830 and director of the Paris Observatory. He was engaged in research on magnetism and established a connection between the aurora borealis and changes in the Earth's magnetic field. His most recent research is in the field of astronomy, and he also worked with Fresnel to study the polarization of light. He listed some of the difficult questions of science of that time and tried to find out if there is BL or is it just an optical illusion and came to the conclusion about its real existence.

C. Flammarion was a French astronomer who was admitted to the Paris Observatory in 1858 and founded the Juvisy observatory in 1883 (Fig. 1.6).

He was a great popularizer of science, writing many books on astronomy, ballooning, and psychology. In 1874, he published a collection of 50 reports on the observation of ball lightning in his book L'Atmosphere (Atmosphere), written in a very flamboyant magazine style.

R. Plante is a French physicist. Since 1854, he has been a preparator at the Conservatoire national des arts et métiers, and since 1860, he is a professor of physics at the Association Polytechnique in Paris. His works on electricity were published mainly in the Comptes rendus. In 1859–1860, he invented the first lead-acid battery. In 1979, the International Astronomical Union named the crater on the far side of the moon after Gaston Planté (Fig. 1.7).

There is his report (Plante 1891) the main conclusion of which: BL is a special, primary form of a stable state of electrical energy. Linear lightning is simply BL combination. There were two types of current sources that made it possible to do

Fig. 1.6 Camille Flammarion (1842–1925)

Fig. 1.7 Raymond Louis Gaston Plante (1834–1889)

such experiments. Some were intended to produce very strong, in kA currents at a voltage of 4–7 V. Others gave high impulse voltages up to 10,000 V at currents of tens of amperes. In any case, Plante received a unique combination of amperage, voltage, and time of their action. According to Plante, with an EMF of a battery of 2000–4000 V, tiny BL were obtained. Plante's book states that BL can be seen in almost every thunderstorm. You just need to look carefully ... Note that today there are no ready-made sources of electricity with the characteristics of the Plante's battery. They can be created on the basis of special electric generators with flywheels or ion-capacitor banks.

Fig. 1.8 Nikola Tesla (Serbian Nikola Tesla; born 07/10/1856, Smilyan, Gospic)

Nikola Tesla was the next prominent scientist who took the BL phenomenon extremely seriously. N. Tesla was a scientist, engineer, and physicist. Born in the Austrian Empire and raised in Austria-Hungary, in the following years, he mainly worked in France and the USA. In 1891, he received US citizenship (Fig. 1.8).

He created alternating current devices, polyphase systems, a synchronous generator, and an asynchronous electric motor, which allowed for the so-called second stage of the industrial revolution.

He is also known as a supporter of the theory of ether existence—thanks to his numerous experiments and experiments aimed at showing the presence of ether as a special form of matter that can be used in technology.

A unit for measuring the density of magnetic flux (magnetic induction) is named after the inventor. Among the many awards of the scientist are the medals of E. Cresson, J. Scott, and T. Edison.

In experiments in a laboratory in Colorado Springs using two coils in an electrical circuit under resonant conditions, he obtained luminous balls in 1899 (Tesla 1989, pp. 368–372, 431–443).

Almost 100 years later, in 1988, American scientists, K.L. Corum and J.A. Corum (1988) in the USA, and V.V. Kopeikin from IZMIR AN in Russia (Kopeikin 2011) in 2018 repeated the experiments of N. Tesla (thus obtaining luminous balls 3–5 cm in size, having properties similar to those observed in real BL. Unfortunately, the physical community did not pay close attention to these remarkable works. Most likely, this was caused primarily by the hostile attitude towards N. Tesla in scientific circles in America and Western Europe. This follows from the history of Lord Rayleigh's attitude to the results of N. Tesla with distrust, and who did not admit the erroneousness of Hertz's conclusions. It should be noted that N. Tesla's views on nature run counter to the dominant interpretation of physical phenomena in the twentieth century, based on conflicting theories and fitting the results of obscure theories to experiments. He argued that all his discoveries were based on ideas about the ether (now called the physical vacuum), rightly criticized the unfounded interpretations by Hertz of experiments on the propagation of electric waves, and even more so, he did not believe in the general theory of relativity and publicly sarcastically over it.

In Russia, N. Tesla and his discoveries were also disliked by the scientific community, most likely due to the Western worship of scientists dealing with gas discharges where German researchers reigned before the start of the Second World War. Therefore, there he was known only as an electrical engineer. The general attitude towards N. Tesla in Russia changed only after 1990 when N. Tesla's books and patents appeared in Russian, and the opportunity arose to test his ideas and concepts.

W. Brand (1923) in 1923 published a monograph entitled Der Kugelblitz (BL) (Brand 1923), which at that time was the most comprehensive review of BL research. Brand collected about 600 BL reports from the university libraries of Berlin, Marburg, and Göttingham, as well as the Marine Observatory in Hamburg. Of these, he selected 215 of the most reliable messages that appeared in the 100 years before the publication of his book and presented the full original text of 108 of them. Based on statistical processing of 215 messages, he compiled a table of the manifestation frequency of various properties of the BL. He also reviewed the theories proposed to explain this phenomenon, from Muschenbrock to Tepler. Brand noted that there were no BL photographs at the time.

B.L. Goodlet (1937) published not in a scientific journal "An often-described observation is the so-called "keg of water case." It was in a letter to the editorial office of the English newspaper The Daily Mail. This event made it possible to estimate the BL energy, which turned out to be more than 1 MJ. It had a significant impact on the course of a discussion about possible models of BL.

In the 1960s of the twentieth century, a number of BL reviews appeared: (McNally 1966; Rayle 1966; Dewan 1964; Barry 1966, 1980). They substantiated the reliability of the BL phenomenon, described their properties and provided information on obtaining long-lived objects (LLO).

Dr. Singer was born in North Adams, Massachusetts on October 2, 1925. From the University of California at Los Angeles (UCLA), he received his BS (1946) and his doctorate (1950) in chemistry. His professional career spanned a range of interdisciplinary fields, usually related to chemistry, physics, and engineering. From 1951 to 1958, he worked at the US Navy Test Station (China Lake, California), the last 6 years of which he was head of their Liquid Fuel and Combustion Division. From 1958 to 1963, he served as Division Chief of the Aviation Division of the Hughes Tool Company and Maremont Corporation and from 1964 to 1965 was Director of Physics at Dynamic Science Corporation (Fig. 1.9).

Dr. Singer was best known as one of the world's leading experts on BL. His book The Nature of Ball Lightning (1971) was a pioneering monograph on this topic. He was the founder and first president of the International Committee for Ball Lightning (ICBL). The ICBL holds symposia twice a year, the first of which was held in Tokyo in 1988. At the time of his death, he was working on the opening speech at the 2008 Kaliningrad Symposium. Perhaps his most important professional legacy left behind a scientific organization that actively studies ball lightning, which has been almost ignored until now.

Fig. 1.9 Dr. Stanley Singer (1925–2008)

Formation of the international BL committee in 1988 significantly improved communication between scientists from different countries. Its backbone included and includes Dr. S. Singer (USA) (S. Singer), prof. Y.-H. Ohtsuki (Japan), Dr. Dijhuis G.C., (Netherlands) Prof. Smirnov B.M. (Russia), Dr. H. Ofuruton (Japan), Dr. Keul A.G. (Austria), Dr. Handel P.H. (USA), Prof. Jennison R.S. (Great Britain), Dr. Egeli J. (Hungary), DS Bychkov V.L. and DS Nikitin A.I. as well as Prof. A.I. Grigoriev (Russia) and Strand E. (Strand E). (Norway). They participated in the organization and holding of international forums dedicated to BL in the period 1988–2020.

Yoshihiko Otsuki was born on June 18, 1936, in the city of Kakuda, Miyagi, Japan. Son of Otsuki Sato Sazo and Otsuki Taki. Assistant at the University of Tokyo, 1963–1967, lecturer, 1967–1968. Associate Professor, Waseda University, Tokyo, 1968–1973, professor, since 1973. Visiting Professor, Nagoya (Japan) University, 1975–1977, University of Munich (Germany), 1979–1980. Director of the Japan Physical Society, Tokyo, 1983–1984. Book Interaction of a charged particle beam with solids, 1983 (Fig. 1.10).

Dr. Ofuruton H. together with Prof. Yo-X Otsuki carried out the collection and analysis of data (more than 2060 cases) of BL observations in Japan; conducted experiments on obtaining BL in discharges with the addition of organic components, conducted experiments on obtaining BL in microwave discharges (Fig. 1.11).

Dr. G.C. Dijkhuis, b. 1940 April, Dutch physicist, researcher. M.Sc. from Delft U. of Technology (1968), Ph.D. in Applied Physics, Stanford U., California (1973). Retired from Mathematics Department, Eindhoven U. of Technology (2008). Co-founder, past Secretary and current President of the International Committee on Ball Lightning. Leader of ongoing test program for laboratory reproduction of ball lightning. Collaborator at Laboratoire de Recherche sur la Foudre, France. Studied hydrodynamics and astrophysics at Stanford, Ph.D. Received glowing balls in experiments with powerful electric batteries (Fig. 1.12).

Fig. 1.10 Prof. Yo-X Otsuki

Fig. 1.11 Dr. H. Ofuruton

Fig. 1.12 Dr. Geert Cornelius Dijkhuis

He was engaged in the analysis of BL observations. Created a theory of powerful BL. Deals with mathematical issues of modeling the transition of linear lightning to BL.

Dr. Alexander G. Keul conducts research in applied psychology, environmental psychology, and health psychology.

The term "BL" has been used since the time of Arago (1837) to denote an inexplicable class of metastable light phenomena in atmospheric electricity. BL seems to be a random object in time and space, lasts only a few seconds, disappearing with or without traces. BL causes epistemological problems in case classification, residue analysis and laboratory experiments.

He conducts collection and analysis of BL observations. He created a bank of observational data of BL (more than 300 cases), taking into account the psychological state of the observer and the quality of his vision. Received the observation frequency of BL $9 \bullet 10^{-10}$ km^{-2} s^{-1}, which is 20 times higher than the previously known value. He held the Vizotum Symposium in Salzburg (Fig. 1.13).

Dr. G. Egeli. A remarkable and courageous Hungarian scientist. He collected and processed 278 cases of BL observations (Egeli 1987). He presented information about the anomalous energy of the BL and its ability to lift heavy objects. Born in 1950, he studied at the Budapest University of Technology, Faculty of Mechanical Engineering, Department of Engines. He graduated from it in 1974. At the university, he was mainly engaged in the transfer of heat and energy. He continued his research in this area of science at the Department of Nuclear Technology of the Central Research Institute of Physics of the Hungarian Academy of Sciences (Fig. 1.14).

In 1982, he received his Ph.D. in technical sciences in heat engineering from the Budapest University of Technology.

Fig. 1.13 Dr. Alexander G. Keul

Fig. 1.14 Dr. Georges Egeli

Dr. P.H. Handel is a theoretical physicist in the field of atmospheric phenomena in the University of Missouri—St. Louis (USA). He created the theory of BL as an atmospheric maser. He held a BL Symposium at the University of Missouri—St. Louis (Fig. 1.15).

Prof. R.S. Jennison is an astrophysicist at the University of Canterbury. He observed BL on the plane. The description of the observation in Nature contributed to the growth of interest in BL research. He held a BL symposium in Canterbury (UK) (Fig. 1.16).

We present information about scientists who have made a significant contribution to the physics of BL, and who participate in the work of the International Committee on BL.

Fig. 1.15 Dr. Peter H. Handel

Fig. 1.16 Prof. R.C. Jennison

M. Stenhoff. In 1999, he published the book (Stenhoff 1999) Stenhoff M. Ball lightning: an unsolved problem in atmospheric electricity appeared. Kluwer Academic/Plenum Publishers. New York. 1999. Mr. Stenhoff, former BL Research Director of the Tornado and Storm Research Organization (TORRO), Mark Stenhoff has been researching BL for over two decades. Since the publication of a world-wide report on BL in 1976 in Nature, Stenhoff quickly collected over 200 first-hand accounts of personal experiences, and the collection has continued to grow since then.

Fig. 1.17 R. Golka

R. Golka (USA) Engineer inventor. Participated in the creation of large Tesla coils. Investigated the creation of glowing balls in a liquid. He invented the well-known lamp the "Plasma Ball" (Fig. 1.17).

Dr. R. Lowke (Australia). He is a gas discharge expert. He is the author of the theory that BL is a complex discharge that appears when linear lightning strikes the ground or airplanes.

Dr. John Abrahamson (New Zealand) is an Engineer Chemist at the University of Canterbury in Christchurch. After many years of searching for an answer about the nature of BL, he proposed a model in which, when lightning strikes the Earth, BL is formed from small particles—the products of lightning impact the soil. He collected observational data, in which there is a BL with a diameter of 100 m. After the work of Abrahamson et al. (2002), many scientists began to take seriously the pioneering work of B.M. Smirnov (1989) and V.L. Bychkov (1994) on the heterogeneous nature of BL, and papers (Stephan and Massey 2008) and (Paiva et al. 2007) appeared on the effect of laser excitation and thermal heating plasma on solids such as silicon plates or various kinds of chips (Fig. 1.18).

Raymond Picolli (France). The head at Laboratoire de Recherche sur la Foudre, France. Last decade, he became the leader of Conferences devoted to linear and ball lightning and environment processes in Europe.

Apparently, the first experimenter who received luminous balls in Russia in 1893 was N.A. Gesechus in a discharge realized over water with the addition of salt.

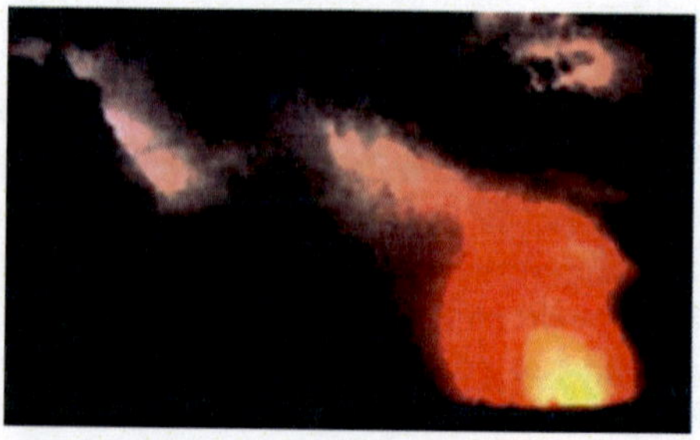

Fig. 1.18 Ball lightning with a diameter of 100 m

Fig. 1.19 Ya. I. Frenkel -Corresponding member of USSR Academy of Sciences

His work received unexpected continuation in the research team led by an engineer G.D. Shabanov from the Institute of Nuclear Physics in Gatchina near St. Petersburg when at the beginning of 2000 charged luminous structures were obtained in discharges above liquid (water) 10–20 cm in size and with a long lifetime. The main difference between the G.D. Shabanov from the discharge of N.A. Gesechus was that it existed up to 1 s after disconnecting the galvanic connection with the electrodes.

In the period before the Second World War in Russia, there were practically no works on BL, information about such a phenomenon leaked only to books on lightning discharges and atmospheric electricity by I.S. Stekolnikov and P.N. Tverskoy and in the form of an article by Corresponding Member of the USSR Academy of Sciences Ya.I. Frenkel in 1940. But because of the Second World War, this article was unnoticed. In 1949, he published a book (Frenkel 1949) devoted to atmospheric electricity in which there was a chapter on BL1.19.

A breakthrough in publications about BL after the war was an article by Academician of the USSR Academy of Sciences Nobel Prizewinner P.L. Kapitsa in 1955, who argued that BL is born in a beam of radio waves and is a gas discharge. Thus, he introduced BL as such into Russian scientific circulation. By that time, active studies of plasma showed that luminous plasma regions could serve as a model for BL. The first book on BL after the war by R.A. Leonov appeared in 1965 in a popular science series. After that, for some time, the taboo on studies and articles on BL was taken away solely thanks to the authority of Academician P.L. Kapitsa, who was one of the first to experimentally simulate BL. Since the mid-1970s, books by both domestic and foreign authors devoted to BL began to be published, and the first foreign publication in 1973 was S. Singer's book "The Nature of Ball Lightning."

But this fertile time for the BL ended in the beginning of the 90 s of the twentieth century, and the possibilities of publishing books on BL ended. Not knowing the

Fig. 1.20 P.L. Kapitsa—Nobel Prize winner, Academician of the USSR Academy of Sciences

anomalous properties of BL, but focusing only on average data, Ginzburg V.L. began to preach that BL is not a fundamental problem. His authority and the expressed negative opinion on the BL problem have so far closed the regular academic studies of BL in Russia, so the Russian Academy of Science (RAS) does not support research, publications, and conferences on this topic (Fig. 1.20).

1.1.1 Collection and Analysis of Observational Data on BL in Russia

To begin with, the most important aspect of BL research is the collection of observations of this object. Without such a collection and clarification of the main properties, it is impossible either to hypothesize about its nature or to think about setting up experiments.

Until the mid-60 s, foreign literature on BL practically did not appear in Russia. Neither the work of W. Brandt (1923) nor the reviews of the American Laboratory at Oakridge (McNally 1966) were known. Therefore, Russian scientists started practically from scratch. For the study of BL, scientists from the areas in which this phenomenon was observed, or similar processes in manifestation, were detected. In the 80 s, they were primarily prof. I.M. Imyanitov, a specialist in atmospheric electricity, dealing with the tasks of protecting aircraft from lightning strikes, radiologist prof. M.T. Dmitriev, who became interested in this problem in connection with the injury of people in contact with BL and prof. I.P. Stakhanov, a specialist in plasma physics.

At this time, their works appeared almost simultaneously. From the articles of prof. I.T. Dmitriev followed that BL is a high-energy formation, most likely plasma with a high surface temperature of up to 6000 K. BL produces electromagnetic oscillations, and NO_2 and ozone molecules appear near its surface.

From the book of prof. I.M. Imyanitov (Imyanitov and Tikhyi 1980) also followed that BL is a high-energy formation with an energy of $2.7 \cdot 10^9$ J/kg, which is three orders of magnitude greater than the energy density of explosives, and it can leave the room through the window glass without leaving a trace. In these works, the properties of the observed BL were discussed, but no statistical data on their properties were made.

Prof. I.P. Stakhanov started a work on these properties. Through the journal Nauka I Zhizn (Science and Life), he asked people who watched BL to send answers to questions about their properties and a request for a description of the observed events. In response, he received over fifteen hundred responses. Based on the observational data from these letters, he constructed histograms of the BL distribution over the parameter intervals. Let us note his conclusion: "Perhaps the most striking example of the connectivity of the BL substance is that it can penetrate through narrow holes and even slots, being very deformed at the same time and again restoring its spherical shape after entering free space." In addition, in prof. I.P. Stakhanov's book, the observational material is quite well separated from the hypotheses of the BL structure. Of course, many of the data were not covered by these histograms. Data on the anomalously high energy of the BL and special manifestations in the interaction of BL with some objects have absolutely dropped out of his book (Fig. 1.21).

Prof. I.P. Stakhanov placed them in the book (Stakhanov 1985) either without discussion, or very subjectively. If there were cases with anomalous energy release, then he attributed them to the action of linear lightning during the existence of the BL. It should be noted that the book of prof. I.P. Stakhanov, with some incorrect conclusions, was published three times. Since no new books have been published

Fig. 1.21 Prof. I.P. Stakhanov

Fig. 1.22 Prof. B.M. Smirnov

available to the general reader, special attention should be paid to this fact and the comments from these books should be treated with great caution. The results of I.P. Stakhanov have served many experimenters and theorists as a false benchmark for research.

A significant contribution to the processing and analysis of BL observations was made by prof. B.M. Smirnov from the I.V. Kurchatov institute for nuclear energy. He developed kinetic and hydrodynamic approaches to the analysis of BL observations published in the book (Smirnov 1989). He presented the BL model as a fractal aggregate that accumulates chemical energy on the surface. His team made pioneer experiments on the combustion of fractal aggregates and their analogues. Analyzing the properties of the BL based on the data of prof. I.P. Stakhanov and the data sent to him by the Austrian researcher Dr. Köhl (further Keul), B.M. Smirnov also received low average values of the BL energy "by the value of the energy of six matches." Therefore, just like I.P. Stakhanov, he did not take seriously the observations of BL with anomalously high energy (Fig. 1.22).

In 1980–1990, a professor from Yaroslavl University A.I. Grigoriev collected data based on surveys and correspondence with observers. He published a small print run not comparable with the print run of prof. I.P. Stakhanov books. His books (Grigoriev 2010; Grigoriev and Shiryaeva 2019) appeared, to be useful for Russian researchers, but inaccessible. In them, on the basis of a larger number of observations than Prof. I.P. Stakhanov described the anomalous properties of BL: the possibility of changing its shape, high energy, passing through glass without leaving traces, and data on the death of observers upon contact with BL. In addition, he clarified and expanded the statistical data on the properties of BL in comparison with the simple analysis of I. P. Stakhanov. These observations were in agreement with the observations of Prof. I.M. Imyanitov and Prof. M.T. Dmitriev and showed the real anomalousness of the BL phenomenon, which must be studied! However, the book clearly expressed a negative attitude towards the construction of theoretical models of the BL. To the merit of Prof. A.I. Grigoriev should include the holding of the first all-Union seminars on

Fig. 1.23 Prof. A.I. Grigoriev

the BL problem with a wide range of participants, including foreign scientist, no information on BL experimental modeling was presented (Fig. 1.23).

In the 90s of the twentieth century, Dr. V.L. Bychkov in collaboration with A.Kh. Amirov performed an extended computer analysis based on a package of modern statistical programs using the data of prof. I.P. Stakhanov, Dr. A. Keul, and own 200 cases, collected on the basis of interviews and materials of the USSR Air Force Main Meteorological Center. The analysis showed the existence of high-energy BL with an energy density of 10^6–10^{10} J/m^3. There are not many such cases, less than 1% of all observations in the world. However, they are highly reliable. The lifetime of BL increases with its size, which may lead to the conclusion that some of the UFOs are large BL. Computer analysis of a large array of data based on modern statistical programs led to an unexpected result: it is impossible to determine the factors influencing the color of the BL since the anomalies in observations are primarily associated with the observation conditions and physiological characteristics of the observers. The scarcity of new results in comparison with the statistical data of Prof. I.P. Stakhanov, Prof. A.I. Grigoriev, and Prof. B.M. Smirnov, nevertheless, was brought in by DS (Doctor of science) V.L. Bychkov together with DS A.I. Nikitin to the conclusion that it is the study of the anomalous properties of BL that can answer the question of its nature.

Since the end of the 80 s of the twentieth century, academician of the Russian Academy of Natural Science (RANS), DS, A.I. Nikitin, and Corresponding Member RANS Ph.D. T.F. Nikitina from INEPKhF RAS have been involved in active studies of the BL observational data. They set out to analyze photo and video evidence of the existence of BL. After analyzing the existing photographs, taking into account the specific properties of the cameras used, they came to the conclusion about the sphericity of the BL, and not ovality, as in the photo. In addition, analyzing the videos, they, after conducting a comprehensive study of the observation site, determined the

Fig. 1.24 Dr. Nikitina T.F., DS Nikitin A.I., and DS Bychkov V.L. at the stand with BL artifacts

real size of ball lightning, which in the case of BL in Dolgoprudny led to its size of about 3 m. This group collects artifacts of BL interaction with objects. This team executed a number of experiments on modeling BL interaction with glasses. The most recent important studies of observational data on BL include the analysis of DS V.L. Bychkov together with DS A.I. Nikitin of an almost imperceptible hole left after the passage of the BL through the glass, showing the material (and not purely electromagnetic) nature of the BL (Fig. 1.24).

Observational data collection work in Russia is proceeding slowly. Perhaps only Prof. A.I. Grigoriev and DS V.L. Bychkov continue to systematically collect and analyze observations. Recently, people, as a rule, do not send letters with descriptions but communicate over the Internet. The information received requires special care in processing, since Internet communication turns out to be much less effective than personal interviewing of observers. Research of photo and video materials comes across the analysis of a large number of forgeries and false images.

Experimental research in Russia. It should be noted that the difficulty of analyzing the experimental modeling of BL in Russia lies in the fact that there are no traces of the pre-Second World war data, and in the post-war period the materials are scattered over scarcely available sources, most of which were lost in the hard times of the 90 s of the twentieth century. Years when many technical libraries were closed and research reports in applied and departmental institutes were destroyed. Therefore, the information presented here from accidentally survived works is extremely fragmentary and incomplete. The author apologizes to the readers whose works are not mentioned in this review.

After the work of Academician P.L. Kapitsa on modeling luminous formations in the form of spheres in a discharge in hydrogen at low pressures and the emergence in the 60s of interest in plasma in connection with its applications, research began in various laboratories for the study of plasma formations, which were carried out until the 90 s of the twentieth century. From that time to the present, experimental work in the field of BL in Russia has been of an accidental nature, carried out by a small number of groups, which is caused by an almost complete lack of funding.

In the postwar years, many institutes, primarily applied ones, such as the Scientific Institute of Radio Technical Industry (NIRP), State optical institute (GOI), Radiotechnical Institute (RTI), not to mention military and completely closed institutes and laboratories, began to develop various types of plasma, electron beam, and applied optical devices. They also attracted civilian universities and laboratories to work.

In the course of their work, various types of plasma formations appeared, so-called LLOs in properties similar to those of BL. They could not present their results in the open press, but they informed the public at various small seminars and conferences with legal and neutral names about them. Therefore, some information that was not translated into foreign languages got into the scientific circulation, with which interested scientists could familiarize themselves.

Let's move on to describing specific results, clearly understanding that experimental research is carried out almost blindly, without the support of developed theoretical concepts.

The concept of LLO or artificial BL appeared in the second half of the twentieth century in the course of various studies of plasma discharges in air at atmospheric pressure by teams of DS R.F. Avramenko from NIRP and prof. A.F. Alexandrov from M.V. Lomonosov Moscow State University (MSU), the latter carried out research for NIRP. This includes the work of the team of DS A.I. Klimov at that time from the Moscow Radiotechnical Institute (MRTI) RAS, who worked closely with DS R.F. Avramenko (Figs. 1.25. 1.26, 1.27, and 1.28).

A feature of these objects is the anomalously long lifetime (luminescence time) of these objects relative to the time of their realization by the discharge source or the typical time of plasma disappearance in atmospheric conditions (extinction of plasma in recombination processes in air). Autonomy is another feature of LLO, that is, their ability to move in space, retaining their shape, size, and color in time compared to their lifetime. In terms of these properties, they are similar to natural BL. LLO differs from St. Elmo's lights in this very property since such lights are

Fig. 1.25. DS R.F Avramenko

Fig. 1.26 Prof. A.F. Alexandrov

Fig. 1.27 Prof. A.I. Klimov

usually attached to charged surfaces or water droplets. This feature is absent in the case of plasma objects generated by microwave radiation or the field of unoriented contours.

In the works of successfully interacting groups DS R.F. Avramenko, Prof. A.F. Alexandrov, and Prof. A.I. Klimov in 1990–2000 in the twentieth century, with capillary discharges in polymer channels, when creating a plasma, LLO appeared with many properties similar to those of natural BL. They could pass through holes in dielectrics and have a selective effect on objects. The structures of large LLO are observed during the ejection of plasma flows from plasma generators. In the works of the group of Prof. A.F. Alexandrov formations were associated with the

Fig. 1.28 Radio Engineer S.E. Emelin

development of the Kelvin–Helmholtz instability. The main difference between these objects andnatural BL is their pure vortex shape and decay without explosion.

An important step in the creation of the LLO was the experiments with the Gatchina discharge, carried out by the group of G.D. Shabanov. and later S.I. Stepanov, from the Institute of Nuclear Physics in Gatchina, and S.E. Emelin (since 2002) at the Department of Radiophysics of St. Petersburg State University at the beginning of 2000.

The essence of the experiments is as follows. A ring-shaped copper electrode is located at the bottom of a vessel filled with water and is connected by an insulated bar to the positive pole of the battery. The negative pole of the battery is connected to an electrode located in the center of the vessel near the surface of the water and pointing upward towards the air. A quartz tube surrounds this cylindrical electrode made of coal, copper, or aluminum; it is 3–5 mm above it and above the surface of the water. The battery capacity is 0.6 μF and charges up to 5.5 kV. The central electrode is wetted with a drop of water to generate LLO. When the spark gap is quickly closed, a brightly luminous spherical formation 10–20 cm in diameter with a lifetime of several units of a second and more appears. Moreover, it has a charged shell, which plays a major role in the behavior of this object. It is this fact that raised the question of taking the shell into account in theoretical models (Fig. 1.29).

Somewhat later, the experimental studies of S.I. Stepanov. and calculations of the theoretical group of DS V.L. Bychkov and DS N.P. Savenkova from MSU showed that this is a complex plasma-gas-dynamic formation, changing its shape from spherical to vortex in different stages of existence.

It should be noted that these experiments led to a new method of creating vortex plasma-gas-dynamic luminous structures, which are of undoubted interest from the point of view of fundamental physics. It would seem that there is an excellent

Fig. 1.29 Engineer G. D. Shabanov near the spherical LLO created by him

analogue of the BL with the parameters of the average statistical BL from the books of B.M. Smirnov and I.P. Stakhanov, whose research should be continued and funded!

In the experiments of DS B.N. Shvilkin and Prof. R.N. Kuzmin from MSU carried out at the same time, the appearance of luminous spheres up to 10 cm in diameter was observed during the explosion of a thin metal wire (made of Cu, Ni, Ti) in a miniature plastic bath filled with a combustible mixture. There was a hole in the bath for the outlet of the products of the electrical explosion. The LLO size reached 5–6 cm, and the lifetime was 2–3 s. Objects transformed into vortex rings at the last stage of their existence are similar to those observed in the experiments of G.D. Shabanov.

In the works of the team from the I.V. Kurchatov Nuclear Energy Institute under the guidance of DS L.I. Urutskoev, luminous balls 10–15 cm in size and with a lifetime of up to 5 ms appeared during the explosion of titanium foil in cuvettes with water. They were fixed over vacuum-sealed cuvettes. The energy applied to 8 cells with titanium foil ($50 \times 1 \times 5$ cm) was 50 kJ, and the pulse duration was 60 μs. During the lifetime of the object, the glow changed from purely diffuse to fill with luminous particles. This made it possible to speak of condensation processes occurring in a metal plasma that penetrated through the seal.

Experiments of Prof. V.N. Kunin and Prof. L.V. Furov from the Vladimir State University in 1990–2000 can be attributed to this type of experiment. In their laboratory, a pulsed plasma generator was created, which, when discharged, evaporated a metal membrane. A brightly glowing region of erosive plasma with a diameter of about 10–35 cm and a lifetime of about 2 s appeared. The properties of this object fit into the range of parameters of the average BL from the works of Prof. B.M. Smirnov and Prof. I.P. Stakhanov. It can be stated that in this case, too, a laboratory analogue of BL was obtained, the properties of which required the continuation of the study with the help of modern equipment, but due to the lack of funding, the research was stopped (Figs. 1.30, 1.31 and 1.32).

The works on the creation of an erosive LLO include the experiments of Prof. A.I. Vlasov from Ryazan State Technical University in 2000–2010. He was able to create a plasma generator and explode the coiled bundles of wires. In this case, large

Fig. 1.30 Prof. V.N. Kunin

Fig. 1.31 Prof. L.V. Furov

formations up to 5 cm in diameter appeared with a lifetime of up to 1 s. Unlike real fireballs, these LLO did not fly or explode on their own but were metal oxidation processes.

The next direction includes studies of structure generation under conditions of evaporation of materials in the presence of an electric discharge field.

One of the first in this direction was the research in the 80 s of the twentieth century by a team from GOI under the leadership of DS I.V. Podmoshensky, who passed the school of Arzamas. In their experiments, the formation of filamentary-type conglomerates was observed during the explosion of thin wires or tungsten foils at a pressure of 100–760 Torr and a pulsed supply of energy up to 50 J. The thickness

Fig. 1.32 Prof. A.I. Vlasov

of the conglomerate microfilaments was 20–30 nm. The results of these experiments were associated with the possibility of BL formation from aerosol particles. These studies served as the basis for the work of Prof. B.M. Smirnov, in which a hypothesis was put forward that the frame of BL can be a fractal aggregate consisting of solid particles filled with a combustible substance.

A similar group includes experiments on the creation of LLO in 1990 and 2000 with the introduction of metallic or organic components into the flow. In the works of Radio Engineer S.E. Emelin, Prof. A.I. Klimov, and DS V.L. Bychkov plexiglass, caprolon, wax, paraffin, and rosin with natural fillers from wood and lignin were used as plasma-forming materials. The discharge products were injected into air, liquid nitrogen, and air saturated with vapors of organic materials. The explosive creation of luminous regions with dimensions of 10–30 cm and a lifetime of about a fraction of a second was realized at a distance of 30–100 cm from the injector in experiments on plasma generation in a channel with metal particles. The next type of experiments by these authors at the same time consisted in the creation of closed areas with conditions similar to those that are realized when linear lightning strikes the soil or during an electric explosion in a closed space. The discharge or plasma was created in a closed volume. Then the generated luminous LLO, up to 1 cm in size and more, consisting of a mixture of steam plasma, particles of molten electrode material, and ionized air, came out through a specially prepared hole. The lifetime of such LLO reached 5–7 s. They exploded when hitting obstacles, leaving traces like a multilayer object with a molten core. Sometimes it was possible to find their remains, which resembled oxidized metal shells. This was one of the few types of experiments when the LLO explosion occurred.

In the experimental series carried out by the group of DS V.L. Bychkov, using a 1–2 mm capillary made of a material of a complex organic mixture consisting of paraffin, rosin, and crushed wood, luminous objects with a typical visible size of 1–2 cm and a lifetime of 1–2 s were obtained. These LLO almost completely copied the observed BL movements. They suddenly appeared from the channel of the plasma

Fig. 1.33 Prof. R.N. Kuzmin

generator, 30 ms after the plasma exit. Visually, they suddenly disappeared as in the observations of real BL, although the video showed their gradual extinction within 0.3–0.4 s. But they did not explode and did not exhibit electrical properties.

Let us dwell on the experiments continuing the experiments of S.E. Emelin. They were performed by a group of DS V.L. Bychkov and Assoc. Prof. V.A. Chernikov. In the experiments, plasma from capillary plasmatrons was directed to metal wires and foils. At the same time, it was possible to obtain luminous balls up to 5 mm in diameter, which lived up to 4–5 s, and exploded when they hit obstacles. Studies of their structure have shown that they consist of a molten core and a quasi-solid shell. Moreover, the density of their energy was comparable to the energy of real fireballs. Similar experiments were carried out at St. Petersburg University after the untimely death of Engineer S.E. Emelin, his follower, Ph.D. A.L. Pirozerskiy A. in discharges with eroding tin electrodes. They showed the existence of LLO similar to real BL.

It is necessary to mention the researchers of BL, who carried out experiments in difficult experimental conditions. One of them is professor R.N. Kuzmin from MSU, a specialist in the field of solids and cold transformation of the nuclei of chemical elements. Together with DS B.N. Shvilkin from MSU, he created an object based on the combustion of a combustible gas and its excitation by a gas discharge.

Prof. R.N. Kuzmin in 2004 together with DS Bychkov created a Moscow-wide seminar "Physics of environments," which in 2006 became known as "Seminar on BL," then it was renamed "Physics of high-current discharges, long-lived plasma formations LLO, BL and electrical phenomena in the atmosphere."

Since there is still no generally accepted hypothesis or theory of BL, it makes little sense to describe them due to their large number and variety of approaches. One can only casually touch on several groups of researchers on the theory of BL.

The most obvious idea of the plasma nature of BL has attracted a large number of researchers. These include the corresponding member USSR Academy of Sciences Ya. I. Frenkel, Prof. A.A. Vlasov from MSU, Corresponding Member USSR

Academy of Sciences L.M. Biberman from the Institute for high-temperature RAS, Prof. S.I. Yakovlenko from Physical Institute of AS USSR, Academician B.B. Kadomtsev from the I.V. Kurchatov Nuclear Energy Institute, DS A.I. Nikitin from N.N. Semenov Chemical Physics institute, Prof. I.P. Stakhanov from IZMIRAN USSR, Prof. O.A. Sinkevich from Moscow Power Institute, Prof. A.I. Vlasov from RRTU, DS M. Shmatov from the I. Ioffe Physical Technical Institute RAS, Prof. M. Ya. Ivanov from CIAM, and many others. The trouble is that the experimental verification of the hypotheses of these scientists did not lead to objects with a long lifetime, high energy, the ability to pass through holes, and other important properties of a real BL.

Another direction, considering BL as an aggregate of solid particles, polymer fibers, or condensed matter, non-ideal plasma was developed in the works of Prof. B.M. Smirnov, DS V.L. Bychkov from MSU, Prof. E.A. Manykin from the I.V. Kurchatov Nuclear Energy Institute, and Prof. G.E. Norman from the Institute for High-temperature Academy of Sciences of the USSR. This direction was verified experimentally. LLO was obtained that have many properties of real BL, but with low energy, small size, and short lifetime compared to those of real BL.

The third area includes the work of Prof. A.I. Grigoriev, Prof. V.A. Saranin, DS A.I. Nikitin, and DS V.L. Bychkov, who considered BL as objects with a core and a shell. In this case, the core could be liquid, vapor or gaseous, and the shell could be solid or liquid. The presence of a solid phase or a hard shell allows such a BL to have significantly more energy with the possibility of radiation. The latest model of BL as an object with a vaporous core and a solid shell (V.L. Bychkov) made it possible to experimentally obtain luminous, exploding uncharged LLO of small size, but with a high supply of internal energy. These studies also showed that levitation requires the creation of a charged object, but not in a plasma, which is extremely difficult in a gas discharge or plasma experiment.

The direction developed in the works of Prof. Atsukovsky V.A. from TsAGI, Prof. Zaitsev, and DS V.L. Bychkov from MSU goes back to the works of N. Tesla, who considered the BL as a vortex object of the physical vacuum and was confirmed by the experiments of N. Tesla himself, as well as K.L. and J.A. Corum from the USA and Ph.D. V.V. Kopeikin from IZMIR AN.

In general, we can conclude that BL is most likely a heterogeneous formation or an object of physical vacuum.

From the description of the theoretical direction, it is clear that the level of research and scientists is extremely high and wide. The lack of a complete solution to the BL problem is primarily due to the obstructionist policy of the Academy of Sciences and the Ministry of Science in relation to the development of this promising direction and the complete lack of financial support.

1.2 Conclusion to the Chapter

In the twentieth and twenty-firstcenturies, BL was studied as an object with many unusual properties.

Works have been completed and continue to study BL observational data. Investigations have been carried out to obtain LLO-analogues of BL, a number of them with a heterogeneous phase can be considered successful. The existing theories cannot explain the whole phenomenon of BL.

Much work on the establishment and functioning of the International BL Committee has improved the exchange of ideas and information among scientists from different countries.

The BL problem is fundamental and requires unconventional joint approaches in the field of theory and experiment.'

References

Abrahamson J, Bychkov AV, Bychkov VL (2002) Recently reported sightings of ball lightning. Phil Trans Roy Soc 360(1790):11–36

Arago DF (1838, 1859) Sur Ie Tonnerre, p. 221, Annuaire au Roi par le Bureau des Longitudes, Notices Scientifiques. Arago F (1859) Thunder and Lightning, Paris.

Barry JD (1966) Ball lightning—a natural phenomenon in atmospheric physics, M. S. Thesis, Calif. State College, Los Angeles, California.

Barry JD (1980) Ball lightning and bead lightning: extreme forms of atmospheric electricity, Plenum Press, NY

Brand W (1923) Der Kugelblitz. Hamburg Verlag von Henry Grand, Hamburg

Bychkov VL (1994) Polymer ball lightning model. Phys Scr 50:591–599

Corum KL, Corum JF (1989) Tesla's production of electric fireballs. Tesla Coil Builder's Association News. 8(3):13–18

Dewan EM (1964) Attempted explanations of ball lightning, paper No. 67, AFCRL-64–927, U.S. Air Force Cambridge Research Laboratory, Cambridge

Egeli G (1987) Hungarian ball lightning observations. Institute for Physics Preprint KFKI-1987–10/D. Budapest

Frenkel Y (1949) Theory of atmospheric electricity phenomena, Gos izd tekh –teor lit Publishers, Moscow–Leningrad

Goodlet BL (1937) Lightning. IEEJ, London

Grigoriev AI (2010) Ball Lightning. Yaroslavl state university publishers, Yaroslavl

Grigoriev AI, Shiryaeva SO (2019) Ball lightning by the witness's eyes. Directmedia Publishers, Russia (In Russian)

Imyanitov I, Tikhy D (1980) Beyond the laws of science. Atomizdat Publishers, Moscow

Kopeikin V (2011) Fireball-Ball Lightning. Trovant, Troitsk

McNally JR (1966) Preliminary Report on Ball Lightning, Report No. ORNL-3938, Oak Ridge National Laboratory 1964, 1966. Oak Ridge, Tennessee

Paiva GS, Pavão AC, de Vasconcelos EA, Mendes OJr, da Silva EF (2007) Production of ball-lightning-like luminous balls by electrical discharge in silicon. Phys Rev Lett, 98: 048501-1–048501-4

Plante G (1891) Report electrical phenomena in the atmosphere. Academy of Sciences, St. Petersburg

Rayle WD (1966) Ball lightning characteristics, NASA Tech. Note, NASA-TN-D-3188, National Aeronautics and Space Administration, Washington, DC
Singer S (1971) The nature of ball lightning. Plenum press, New York
Smirnov BM (1989) The problem of ball lightning. Nauka Publishers, Moscow
Stakhanov IP (1985) On the physical nature of ball lightning. Energoatomizdat, Moscow
Stenhoff M (1999) Ball lightning: an unsolved problem in atmospheric electricity. Kluwer Academic/Plenum Publishers, New York
Stephan KD, Massey N (2008) Burning molten metallic spheres: one class of ball lightning? J Atmos Solar-Terr Phys 70:1589–1596
Tesla N (1989) Colorado Springs Notes 1899-1900. Nolit, Beograd, pp. 368-720, 431-443

Chapter 2
Observation Data on BL Events

Abstract This chapter is devoted to the presentation of observational data on events related to ball lightning. Data on the energy and external features of the BL—its size, color, and shape are given. The electrical properties of the BL and its interaction with various objects are described. The meetings of BL with aircraft are also described, data on the observation of BL at sea and during geotectonic events are given. Data on BL in Japan, artifacts, and traces of BL are presented. The anomalous properties of BL and their connection with the appearance of crop circles and deep holes in the ground are discussed. A discussion of photographs and video materials of the BL observation is given.

Keywords Observation data · Ball lightning · Properties and features · Meetings with aircrafts · Ball lightning in geography and geophysics · Similar phenomena

This chapter is devoted to the presentation of cases of observation of BL is a compilation consisting of observational data collected by the author through surveys and questionnaires of observers, as well as cases of observations from various sources, literary, and media. We present data on rare properties of BL, as a rule, not known to a wide circle of people interested in BL—these are observational data, which, in the author's opinion, are of interest from the point of view of the properties of BL and their interaction. Our collection includes not all cases of available BL observations; those that are of purely statistical interest did not make it into this chapter. As our experience shows, at present, the possibilities of statistical analysis of observations are very limited for a number of reasons, see this chapter; therefore, the presentation of observations is a necessary material for setting up new experiments and creating new models of BL. It should be noted that, when citing observational data, we often provide comments highlighting the interesting properties of BL and their interactions with objects, and, where possible, a brief interpretation of the phenomenon.

V. L. Bychkov, *Natural and Artificial Ball Lightning in the Earth's Atmosphere*, Springer Atmospheric Sciences, https://doi.org/10.1007/978-3-031-07861-3_2

2.1 BL Energy

The most important and difficult to explain property of BL is its ability to accumulate large amounts of energy in a limited volume. The works of I. Imyanitov (Imyanitov and Tikhy 1980), M. T. Dmitriev (Dmitriev et al. 1981), G. Egely (Egely 1987), B. N. Kozlov (Kozlov 1978), and S. I. Stepanov (Stepanov 1990) explicitly state about the high energy of this object. In the book by Singer S (1971) observations of the destructive effect of BL are given, showing its high energy. However, as a rule, estimates in these works were given without linking energy to the volume of the BL, which did not allow comparing the efficiency of energy input into the volume of the BL. This defect was later corrected in the work of V. L. Bychkov and J. Abrahamson (Bychkov et al. 2002). The question of BL as a high-energy object was examined in detail in the works of A. I. Nikitin and V. L. Bychkov (Bychkov et al. 2010; Bychkov and Nikitin 2014; Nikitin et al. 2014), who showed that the energy of this object can be great and why.

At the same time, in the works of I. P. Stakhanov (Stakhanov 1985), B. M. Smirnov (Smirnov 1988), and M. Stenhoff (Stenhoff 1999), the low energy and insignificant hazard of BL and most of the destruction attributed to the action of ordinary linear lightning.

This showed that BL are often high-energy objects of both scientific and practical interest. In this section, we present observational data and estimates of the energy of high- and medium-energy BL.

Observations

(Imyanitov and Tikhy 1980), (Barry JD 1980) 10/05/1936. The first documented case of observation of high-energy BL is considered to be the publication in the newspaper The Daily Mail (London) on November 5, 1936, a letter to the editor entitled "The Mystery of a Thunderstorm": "Sir, -During a thunderstorm I saw a large, red-hot ball come down from the sky. It struck hit our house, cut the telephone wires, burnt the window frame, and then buried itself in a tube of water which was underneath.

The water boiled for some minutes afterwards, but when it was cool enough for me to search I could find nothing in it." Dorstone, Hereford. W. Morris.

The description of the case, published in the newspaper, became known to the scientific community thanks to the comments of the eminent physicist and lightning researcher Sir Charles Vernon Boyes. This happened during a discussion at the Institute of Electrical Engineers in London on January 7, 1937, following the presentation of Professor BJ Goodlet's review article "Lightning." Boyes said: "The importance of this observation is that for the first time, as far as I know, the energy in a hot red ball was roughly measured. I found out that the ball was about the size of a large orange, and for 20 min, the water was too hot for Mr. Morris to dip his hands in it. The amount of water was about 4 gallons." During the discussion, Professor Goodlett, based on these data, determined the energy content in the ball. He accepted that Mr. Morris's description of heating water corresponded to an increase in water temperature to 60 °C. He carried out two calculations, in the first he considered that the initial temperature was 200 °C and obtained an estimate for the energy density of the sphere as $0.6 \cdot 10^{10}$ J/m^3. After that, he carried out another calculation, based on the assumption that all the water was heated to 100 °C, and the mass of 1.8 kg had evaporated. For this variant, the energy density of BL turned out to be equal to $1.9 \cdot 10^{10}$ J/m^3 (this is tens of times higher than the energy density of explosives).

Due to the impossibility of giving an acceptable explanation of the mechanism of accumulation of such a high-energy density, most authors of the theories of ball lightning preferred to declare the results of estimates made by Goodlet unreliable. For example, I. P. Stakhanov (Stakhanov 1985) believed that "such energy cannot be accumulated in the electron shell at all, and the possibility of accumulating the energy of ball lightning in nuclear form seems unlikely. For these reasons, it was suspected that Goodlet's estimate was wrong. Indeed, it is possible that the 'boiling' of water in the barrel was associated not with the high temperature of the entire mass of water, but with the formation of steam bubbles in places of local heating at a low average water temperature. True, an eyewitness said that even after a few minutes the water was hot to the touch. But these sensations could be exaggerated, and the actual temperature of the water was not measured." Literature.

1. (Covington AE 1970) On a summer evening in 1928 or 1929, during a severe thunderstorm, a luminous ball fell from the clouds, which hit a wooden pile of a small pier on the lake. This blow was accompanied by a loud sound and caused the column to split into long splinters.

Comment. Based on this observation, an assessment of the energy that was spent on the destruction of the pile was carried out. It was assumed that the water originally contained in the volume was heated by the BL energy to a boil. This volume expanded to the size of the diameter of the log, and the pressure in it increased to the size of the wood rupture. The estimated energy density was equal to $1.9 \cdot 10^9$ J/m^3.

2. (Imyanitov and Tikhy 1980) Frogs. August 1962 Ukraine, Transcarpathia near the town of Perechin.

A BL the size of a tennis ball fell into a trough with water for watering cattle. The water from the trough had almost completely boiled away, there were boiled frogs at the bottom.

Comment. The size of the trough was 0.3×2.5 m, the depth of the water layer was 15 cm. The energy density of this BL was 2.38×10^{12} J/m^3. This density is about 100 times greater than the energy density of the "English" BL (see above). The reason for this could be an error in the observer's determination of the size of the BL. This case removes Stakhanov's objection [2] about the absence of heating water to a high temperature: boiled frogs became an "indicator" of high water temperature.

3. 1964. The story of engineer N. F. Ignatov. July 1964 on the Ucha River near the Mamontovka platform in the Pushkin District of M/O. The narrator was 20 years old at the time. "At that time I was a student of the College of Communication. At 4 o'clock in the morning, I was fishing on the Ucha River. The sun had not yet risen, the weather was clear, there was no rain this morning and the night

before. At the fishing spot, the width of the river was 20–30 m. I had a simple fishing rod with a reed rod, with an ordinary fishing line and an ordinary hook. Before the described event, I managed to catch 5–6 roaches, which I kept in a galvanized iron bucket with a capacity of 10 L. The bucket was half-full of water. I felt that the fish had eaten the worm off the hook. I took the fishing line with an empty hook out of the water and when I carried the hook hanging on the fishing line over the edge of the shore (it was at a height of about 1 m from the ground), a glowing ball with a diameter of about 5 cm suddenly 'clung' to it. I did not notice where the ball came from. The weight of the sinker with the hook, when the ball stuck to it, did not increase. I do not remember the color of the ball, it glowed a little brighter than a fluorescent lamp, and its light did not blind my eyes. The outline of the ball was clear. I was confused and slowly lowered the hook with the ball hanging on it into the bucket with the caught fish. The water immediately bubbled and hissed, as it happens when a red-hot stone is thrown into it. The water boiled for a few more minutes. All the fish was cooked, and the way it happens if you cook it for too long: the skin with scales separated from it and the 'giblets' fell out. When boiling, thick steam came out of the water. The hook attached to the fishing line disappeared. The fishing line remained intact, only slightly shortened. Out of fear of the unknown of what happened, I threw out the remains of the water with the fish on the ground and came to the hostel with an empty bucket. When I told my comrades about the incident, they did not believe me: they decided that I had not caught anything and was making excuses. I had to go show them the discarded cooked fish."

Comment. After making the estimates, we find that the energy density in the BL was not less than 3×10^{10} J/m^3. As you can see, this figure is consistent with the value of the energy density of the "English" 1.9×10^{10} J/m^3 BL. Apparently the fishhook disappeared due to the influence of the electromagnetic field emitted by the BL on it. People who have met with BL have repeatedly reported the disappearance of metal objects (rings, bracelets, etc.) Perhaps the heating of the water also occurred due to the effect of this electromagnetic radiation on it.

Ignatov said that behind him, at a distance of 215 m, there was a track of an electrified electric railway, along which freight and passenger trains passed at intervals of 5–10 min. According to an eyewitness, there was no thunderstorm at that time, so it is possible that the BL appeared due to the occurrence of an electric arc on the current collector of the locomotive. Since this happened behind the narrator's back, he could not notice the process of formation and the approach of the glowing ball to him. In the literature, there are descriptions of cases of the appearance of BL when the contacts of power plants are broken, including the contact network of trolleybuses.

On 18.06.1981, a resident of Odessa, T. V. Borodina (female), observed an unusual picture: "At the trolleybus that turned into the intersection, the bugle jumped off. At the same time, it hit the wires several times, causing short-term short circuits, there was a crack, and sparks fell. Suddenly, a round bright ball with a diameter of 15–20 cm, resembling an electric welding arc in color and brightness, separated from the total mass of sparks. The balloon crossed the roadway at the height of the trolleybus wires, landed on the wires of the opposite trolleybus line, and began to move along them at a speed slightly higher than before. The wires were mounted on poles with lighting lamps. As the ball passed under the lamps, they flashed in turn. So the ball sailed past 3–4 lamps." Trolleybuses are powered by a direct current of 200 A at a voltage of 500–750 V. Thus, if we assume that the formation of BL lasted about 1 s, then its energy could range from 100 to 140 kJ, and the energy density of $1.4–4.6 \cdot 10^7$ J/m^3.

4. (Nikitin et al. 2014) on 27.05.2013, at 15 o'clock local time, a BL destroyed a residential building located on the outskirts of the village of Mogsokhon in the Kizhinginsky District of Buryatia. According to eyewitnesses of the event, "nothing foreshadowed a tragedy that day. The usual rain began, but suddenly thunder struck with such force that passers-by and cattle who were on the street at that time ducked to the ground in fright and ran in different directions. A minute after that, a bright glowing ball descended from the sky, which penetrated through the roof into the house of Bayan Sandanov's family. There was a deafening explosion in the house. At the time of the explosion, the hostess, Bayan's wife, was in the house, cleaning the kitchen after lunch. She ended up under the rubble of the wall and received a serious contusion, partially losing her hearing. The victim was taken to the hospital, and her health recovered. After the explosion, the house was in critical condition: doors and windows were knocked out and thrown a long distance, part of the wall fell, and the roof was partially destroyed (see Figs. 2.1 and 2.2). Many residents of neighboring houses had household appliances out of order from the explosion of BL."

The energy density of the "Buryat" ball lightning can be estimated from 2.1×10^9 J/m^3 to 9.3×10^9 J/m^3 (Nikitin et al. 2014). These figures are of the same order as the energy density values of the high-energy BL mentioned above.

5. (Bychkov et al. 1996) 15.08. 1994 Eduard Kalaijyan (husband), 69 years old, architect. The message was received 3 months after the observation, an aluminum saucepan was attached to the message. The event took place on Sunday at 7 p.m. local time, at a dacha near Sergiev Posad, 67 km north of

Fig. 2.1 Photo of a house destroyed by a ball lightning explosion

Fig. 2.2 The wall of the house thrown out by the explosion

Moscow. The day was very sultry and hot, windless. The event took place at the beginning of a thunderstorm with lightning, hail, and rain. Linear lightning struck the roof of a neighboring house 30 m from the observer's house. Both of these houses had a common electrical circuit. The lightning strike was so strong that a boy standing on the porch of a neighboring house was knocked to the ground by the force of the explosion. Lightning burned through the roofing material on the roof, broke several boards, passed through the attic, split the rafters, and "smashed to pieces" electric machines, but there was no fire. Sparks flew 100–150 m away from it. The observer was in the next house in a room near the kitchen. A few seconds after the explosion in the neighboring house, he went into his kitchen. There he found that the metal sink was split and holes appeared, traces of melting and soot on the metal saucepan and its lid, which had not been there before. Photos of the saucepan and its lid are shown in Figs. 2.3a, b. The saucepan was on an electric stove that was turned off. The stove was next to the sink. The observer did not remember an explosion or a bang being heard in the kitchen. The observer did not find any traces that would indicate how the results of the exposure appeared. The windows and doors in the kitchen were closed.

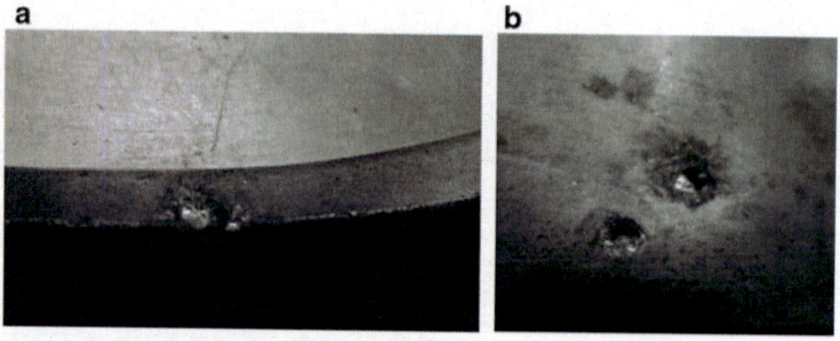

Fig. 2.3 **a**, **b** Traces of the impact of BL on the surface of the aluminum saucepan

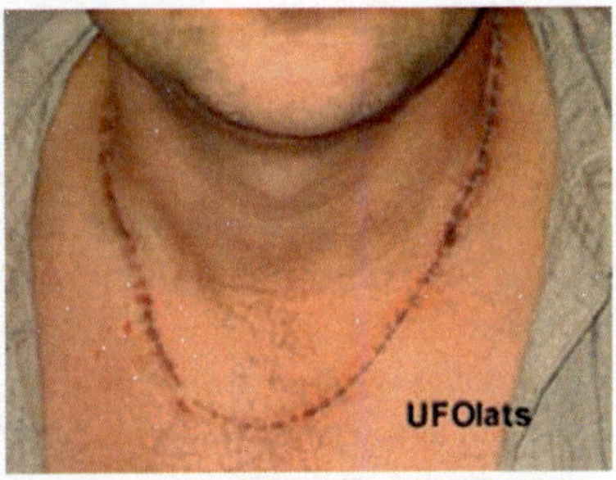

Fig. 2.4 Trace of the missing gold chain

The parameters of the saucepan with lid were as follows: saucepan, material—aluminum, diameter 140 mm, height 95 mm, average thickness 1.5 mm; lid, material—aluminum, diameter 155 mm, average thickness 1.0 mm. The largest soot trace had a shape close to circular, with a diameter of 60 mm. It was located near the bottom of the saucepan, on its side. There were also two holes.

Observation analysis. The observer did not directly observe the BL, since he was not in the kitchen at that time. Traces of soot on the surface of the saucepan, especially a large, round trace, make it possible to assume that the collision of some spherical object with the saucepan led to the appearance of such a trace. The interaction of any current with metal is unlikely, since there are no traces left at the bottom of the saucepan, and the circuit of the electric circuit of the electric stove on which it stood was not destroyed. A linear lightning strike first into the saucepan, and then into the sink is also unlikely. A metal sink, due to its greater mass than that of a saucepan, would inductively attract the discharge of linear lightning more strongly, and its impact would fall directly into the sink. In addition, the observation time coincides with the seasonal and daily time of the appearance of ball lightning.

Therefore, there is every reason to assume that the traces of melting on the saucepan were left by a BL that appeared in the kitchen during a thunderstorm, and from the traces of destruction of the metal surface of the saucepan to estimate the possible energy of BL.

Energy assessment. Traces of destruction on the surface of the saucepan and its lid allow us to estimate the necessary energy for such an impact. Metal was vaporized on the lid in the form of a rounded cone, with a volume of 3.4×10^{-9} m^3.

Holes appeared on the side surface in the form of spherical segments with a volume of 3.15×10^{-8} m^3, so the total volume of evaporated aluminum was assumed to be of the order of $V \sim 3.5 \times 10^{-8}$ m^3. Using data on the heat capacity and the heats of melting and evaporation of aluminum, it is possible to obtain an estimate of the energy $E_0 \approx 1050$ J. The need to take into account the evaporation of metal is based on the fact that the traces of destruction were typical for the boiling of metal, and there were no traces of melting.

If we assume that the energy carriers (for example, charges) of the BL are located near the surface of the BL, then the total energy of the BL E_{tot} can be estimated by the formula $E_{tot} \approx E_0 \cdot R_s/R_h$, where R_s and R_h are, respectively, the surface area of the sphere that left a trace on the surface of the saucepan and the surface area above the holes. Whence it follows that $E_{tot} \approx 5.4 \cdot 10^4$ J, or the energy density of the sphere $\approx 4.8\ 10^8$ J/m^3.

6. (Wittman 1971). Outdoor surveillance during thunderstorms. A thunderstorm with heavy rain occurred in the Neustadt area near Coburg, Germany. I was sitting with some other people right outside the ground-floor window, and I could look at the weather. The field of view faced the street and was limited by the window frame. Suddenly, at a height of about 16 m above the ground and at a short distance of about 24 m, I saw a spherical plasma ball painted in a bright-yellow-white color. This object had a diameter of 50–100 cm. It was moving vertically downwards at a speed of about 4 m/s, and its path ended at the top of the tree branches, at a height of about 9 m. Upon touching the tree, the ball instantly disintegrated into 8–12 smaller spheres. They were the same color as the big one, and each had a diameter of 12–15 cm. They fell to the ground, guided by the outer contour of the tree, and moved vertically for the last few meters in the absence of branches. Upon reaching the ground (the asphalt road and the adjacent pedestrian path), the spheres instantly disappeared. There was no noise (other than the noise of rain and lightning) associated with the primary plasma-like sphere. Three to five minutes after that, the same phenomenon occurred again as before, indicating that conditions were still maintained or restored during the elapsed time. Immediately after the rain stopped, I went outside to find additional evidence. There were round spots of molten asphalt pavement on the roadbed, on which the interference colors of thin layers are visible. Their diameters ranged from 12 to 15 cm, and they obviously marked the impact zones of smaller spheres. This event was described in a similar way by all the witnesses.

Comment. Since road asphalt, as a rule, contains bitumen B-80—a thermoplastic from which liquid components disintegrate at a temperature of about 170 °C, it is possible to calculate very approximately the minimum amount of energy required to obtain the observed areas. Assuming that the 0.5-mm-thick water layer evaporated at 20 °C and that the 1-mm-thick asphalt layer was then heated to 170 °C, and assuming that the average density of 1.0 g $\times$ cm^{-3} with an average specific heat c_p = 0.46 cal g^{-1} for B-80, it is easy to calculate the energy density of plasma spheres, and it was at least 1.9×10^7 J $\times$ m^{-3}.

7. (Egely 1987) Saliy Janos, Pech. "It happened in clear, sunny weather in July 1972. At noon, most of the workers of our factory had lunch, but some of them stayed in the factory building. Suddenly, they saw a bright sphere the size of a football rolling down a lightning rod. This was accompanied by a hissing sound and was colored with a color transitional between pink and yellow. It passed through a long glass window on the roof. It then disappeared into a water pit in the outer corner of the building next to a lightning rod with a loud sound. After that, the air felt rich in ozone. The water disappeared from the pit, leaving no trace. Although the pit was already completely dry, the pit was still floating for half an hour. About 30 people saw this phenomenon for about 8–10 s, the

speed of the bright sphere was about 3–5 m/s. The sound it made was similar to the sound of a supersonic airplane, that is, it was a hissing, whistling sound. Initially, the pit contained about 120 L of water, since it was only 2/3 full. My colleagues examined it carefully, no water splashed on the walls, all the water evaporated. I am sure it evaporated even for a long time after the event. None of my colleagues felt any radiated heat or felt an impulse. The windows did not vibrate, although some people had a strong ringing in their ears due to the explosion. One of my colleagues by the name of Bela Trisher was just looking out the window when a fireball fell into a pit with water, and he saw not only steam, but also that some grass leaves around the pit caught fire and eventually burned down. The grounding cable was not damaged at all. In the corner of the building, next to the pit, the foreman and several other colleagues were sitting. They didn't feel anything."

Comment. When water with a volume of 120 L evaporates, about 200 m^3 of water vapor is formed, that is, a cube with sides 6 m long. About 2.8×10^8 J of energy is needed to turn this amount of water into steam. If we consider the diameter of the BL to be 25 cm, then the energy density is $3.4 \times {}^{10}10$ J/m^3.

8. Rudkovsky M. date of birth 4.06. 1940. Village Utchanka Kurgan region. 1952. Mid-July. After a thunderstorm and rain, noon. At a distance of 2 m. The color of the ball is bright red.

The dark cloud grew and approached. A squally wind blew in, lifted the dry ground by the wheels into the air, played with paper garbage, and quickly turned into a dust storm. Windows and panes slammed in the houses. A third of the firmament was quickly covered with a black, torn on the sides, heavy cloud, through some gaps of which the sun's rays hardly broke through. The cloud swelled, tore, and merged again into a gloomy mass. Lightning flared at the horizon, creating the impression of a distant fire. Lightning with a deafening crack tore the air very close to the house, and such an avalanche of a rolling chord fell on it that the windows in the frames trembled. The first large drops rattled on the windows and roofs. The long-awaited rain was gaining strength and soon fast streams rushed along the roads and ditches, dirty at first, then lighter. The rain stopped. We were about to go home when we suddenly saw a fireball the size of a rubber hand ball two meters away from us. It was hanging, swaying from side to side.

What's it?—Yurka whispered for some reason.

- It burns like red-hot—Sashka wanted to get closer, but the ball bounced aside.
- Wow!—I waved my hand, and the ball of fire flew even further away. …

Suddenly our stranger staggered, sank a little lower and, picking up speed, rushed towards the railway track that passed not far from our street. We, as if enchanted, slowly trotted after him. The ball disappeared from our eyes very quickly. Suddenly, a deafening blow and explosion shocked us. Out of fear, we, who grew up on war films, plopped down on the earth. A second later, we jumped up, climbed onto the railway embankment and were stunned by what they saw.

One rail for thirty centimeters was a molten cake, from which steam was coming.

Comment. Let us estimate the energy density of the observed BL, which had a radius of 2.5–4 cm (it is unknown what kind of rubber hand ball, most likely 2.5 cm). The length of the destruction of the rail is 0.3 m.

The rail, the most common for conventional railways (in the USSR) of the mark 50 or P 50, has a linear mass of 51.67 kg/m, so the mass of the melted rail was $M =$ 15.5 kg according to [Gost 51,685–2000.]. Specific heat of melting: $\lambda = 84$ kJ/kg. Specific heat $Cp = 685$ J/(kg × °C) in the range $T = 1300$–1400 °C. Melting point of carbon steel is $T1 = 1535$ °C. The air temperature in summer heat $T2 = 35$ °C, $\Delta T = T1 - T2 = 1500$ °C. Then $E1 = Cp \times M \times \Delta T = 1.59 \times 10^7$ J. $E2 = M \times \lambda = 1.3\ \ 10^6$ J.

The total energy used to heat and melt the rail is equal to $E = 1.72 \times 10^7$ J. Hence, the energy density in the case of a ball with a radius of 4 cm $W = 6.6 \times 10^{10}$ J × m^{-3}, and in the case of a ball with a radius of 2.5 cm $W = 2.6 \times 10^{11}$ J × m^{-3}.

9. (Data 2006) 10.07.2006 The Gulf of Riga near Jaunkemeri. About 4 p.m.

Oleg, his wife, and son went swimming. His wife and son were already on the shore when Oleg also went ashore. Suddenly (according to his wife) a bright-yellow-orange ball, the size of a tennis ball (approximately 5 cm in diameter), flew nearby and there was a strong rumble.

Oleg fell face down on the sand. "I did not lose consciousness, but I was very inhibited, my consciousness was scattered." When they turned him over, his face was covered by soot.

There was a trace on the observer's body. People swimming in the bay at this time felt a noticeable electric shock. An ambulance was called, which took Oleg to the hospital.

The gold chain around Oleg's neck evaporated, leaving a noticeable burn around the neck in the form of a dotted line (see Fig. 2.4). On the stomach and on the heels there was also a trace of lightning (a current channel).

Comment. Estimates of the BL energy. Let us estimate what was the energy of the BL that vaporized a chain of gold weighing 5 g. The energy that could go into heating, melting, boiling, and evaporation of the gold chain can be estimated as

$$\begin{aligned}\Delta Q &= CP1 \times m \times (Tpl - To) + m \times \lambda pl + CP2 \times m \times (Tkip - Tpl) \\ &\quad + m \times Lisp \\ &= m \times [CP1 \times (Tpl - To) + \lambda pl + CP2 \times (Tkip - Tpl) + Lis].\end{aligned}$$

Here $Tpl = 1064$ °C is the melting point of gold; $Tkip = 2807$ °C is the boiling point of gold; $To = 20$ °C is the ambient air temperature; gold density $\rho_{\text{gold}} =$ 19,320 kg/m^3; $m_{\text{gold}} = 5 \times 10^{-3}$ kg is the estimated value of the chain mass; $CP1 =$ 0.13×10^3 J/(kg × K) is the heat capacity of gold; $CP2 = 0.13 \times 10^3$ J/(kg × K) is the heat capacity of gold in the molten state; $\lambda pl = 66.2 \times 10^3$ J/kg is the specific heat of melting of gold; and $Lis = 1650 \times 10^3$ J/kg is the specific heat of evaporation of gold.

Based on the equation, we obtain the following value of the BL energy ΔQ, which went to heating, melting, boiling, and evaporation of the chain: $\Delta Q = 1.05 \times 10^4$ J, and the power (with a possible explosion duration of 10–2 s) is 0.1 MW. At the same time, the specific energy per 1 kg of the chain substance was $\Delta Q/m \approx 2.1 \times 10^6$ J/kg, and the volumetric energy density of BL $W = \Delta Q/m \times \rho_{gold} \approx 4.1 \times 10^{10}$ J/m^3.

If we assume that evaporation occurred as a result of the effect of the electric field of the charge present in the BL, then the electric field of the BL can be estimated by the following formula: $E^2 = 2 \times \Delta Q/(\varepsilon_0 \varepsilon) \approx 9.26 \times 10^{21}$ (V/m)^2; $E \approx 9.6 \times 10^{10}$ V/m. $\varepsilon_0 = 8.85 \times 10^{-12}$ F/m is the dielectric constant; and $\varepsilon = 1$ F/m is the permittivity of air. This value is about 3000 times greater than the breakdown field of air. A possibility of implementing such fields will be discussed in other chapters.

10. (Imyanitov and Tikhy 1980) Evaporation of the gold bracelet by BL.

"There was a thunderstorm. A woman was sleeping in a house whose doors and windows were securely locked from the inside. She was awakened by the sound of a thunderstorm discharge, similar to a shot. Going to the window, she was surprised to find a small hole in it, about fifty kopecks (diameter 24 mm). Suddenly she noticed that a gold bracelet had disappeared from her hand. Only a dark stripe remained on the wrist. (The special analysis showed that it was gold oxide.) The most thorough searches yielded nothing. Only a round piece of glass with melted edges was found on the floor. It exactly matched the hole in the window glass. A dog killed by lightning was found in a booth near the house."

Comment. Let us estimate the energy spent by the BL on the evaporation of the bracelet. The total evaporation energy of the bracelet is 10^5 J, and the evaporation of one gram of gold requires an energy of 2.11×10^3 J. If we take the evaporation time of the bracelet to be 10^{-2} s, then the power of the energy source should be at least $P = 10$ MW.

11. (Likhosherstnykh 1994) observer Petrov A. A., City Tver.

Two men and a woman were sitting on the hay on the threshing floor. Suddenly, a red-hot body larger than an apple zigzagged up to them at a distance of one meter. The body of the BL sparkled, at the same time the hair of the woman (their length was 20–30 cm) stood on end (fan), and the man's hair moved under the cap.

The ball swayed from side to side in front of them, like a cobra, and left the threshing floor through a crack in the roof. Astonished observers, coming to their senses, discovered strange things.

The rings from the fingers of the woman disappeared, the chain from the folding knife, the metal tip from the fountain pen, and the metal rims on the shoes securing the edges of the holes from the laces disappeared from the man.

Comment. The effect of BL on a woman's hair. The disappearance of rings and metal objects. If the reason for the disappearance of these items was their evaporation by the energy of BL, then for a gold ring weighing 14 g, this energy was 30 kJ, and for all missing items—several times more (about 100 kJ). For apple of 8 cm diameter, the energy density is about 3.7×10^8 J/m^3. Note that the BL spent only part of its energy on performing this action.

12. (Stakhanov 1985) Observation of the mechanical engineer Ya. V. Berezovsky. Germany in the spring of 1945, where he was stationed while serving in the army. At 0 a.m. he lay down to rest in a room on the 2nd floor with an open window, placing his carbine at the headboard near the window with the butt on the floor. I woke up from a strong crackling near my head ("... it seemed that a carbine had fired"). Opening his eyes, he saw a glowing white "ball" moving, stretched out in the direction of movement, which smoothly slid along the walls parallel to the floor. Skirting the stove and walking along the walls around the perimeter of the room, the ball flew out the window. In the morning, when examining the carbine, it turned out that the upper head of the ramrod was melted, but there was no melting or smudge of metal. The metal simply disappeared (evaporated), leaving a hole with a diameter of 5 mm and a depth of 3 mm on the ramrod. No traces of heat exposure were found on the walls. He did not feel the warmth on his face either, although he was half a meter away from the carbine.

Comment. It is of interest to determine the energy of the BL that vaporized the steel ramrod head and left a hole with a diameter of 5 mm and a depth of 3 mm.

To assess the energy that could go on heating, melting, boiling, and evaporation of a piece of steel rod, we write the following equation:

$$\begin{aligned}\Delta Q &= C_{P1} \times m \times (T_{pl} - T_0) + m \times \lambda_{pl} + {}_2 \times m \times (T_{kip} T_{pl}) + m \times L_{isp} \\ &= m \times [C_{P1} \times (T_{pl} - T_0) + \lambda_{pl} + C_{P2} \times (T_{kip} - T_{pl}) + L_{isp}].\end{aligned}$$

Here $T_{pl} = 1400$ °C is the melting point of steel; $T_{kip} = 3000$ °C is the boiling point of steel; $To = 20$ °C is the ambient temperature; $\rho_{stali} = 7800$ kg/m^3 is the density of steel; $C_{P1} = 0.426 \times 10^3$ J/(kg × K) is the heat capacity of steel; and $C_{P2} = 0.54 \times 10^3$ J/(kg × K) is the heat capacity of steel in the molten state.

Since the type of steel from which the carbine was made is unknown and there is no exact value for λ_{pl} and L_{isp}, then take the values for stainless steel, which consists of 93% iron and 7% chromium, then the values of λ_{pl} and L_{isp} will be close to the values of λ_{pl} and L_{isp} for iron. $\lambda_{pl} = 270 \times 10^3$ J/kg—specific heat of melting of iron $L_{isp} = 6.120 \times 10^6$ J/kg—specific heat of iron evaporation. In order to find the mass m of the evaporated steel, imagine the tip of the ramrod in the form of a truncated cone. The volume of a truncated cone with bases R and r : $1/3 \times \pi \times h \times (R^2 + R \times r + r^2) \approx 30.6 \times 10^{-9}$ m^3, and its mass: $m = \rho_{stali} \times V \approx 0.24 \times 10^{-3}$ kg.

Based on the above equation, we obtain the following value of the BL energy, which went to heating, melting, boiling, and evaporation of a piece of ramrod: $\Delta Q = 1882$ J. At the same time, the specific energy per 1 kg of ramrod substance was $\Delta Q/m \approx 7.8 \times 10^6$ J/kg, and the volumetric energy density of the BL was $\Delta Q/m \times \rho_{stali} \approx 6.1 \times 10^{10}$ J/m^3. If we assume that evaporation occurred as a result of the electric field of the charge accumulated by the BL, then its electric field estimated by the formula: $E^2 = 2 \times \Delta Q/(\varepsilon \cdot \varepsilon_0) \approx 1.37 \times 10^{22}$ (V/m)^2. $E \approx 1.17 \times 10^{11}$ V/m. We will return to the electric field's value later.

13. (Bychkov and Vaulin 2013) The "young man" left the house early in the morning and saw a glowing ball moving parallel to the ground. The BL flew past a bucket of water, after the flight the ball stopped glowing, and the water from the bucket evaporated.

Comment. Let's estimate what was the energy of the BL that evaporated water from a bucket of 10 L. To estimate the energy that could go to heating, boiling, and evaporation of water, we write the following equation:

$$\Delta Q = C_{P1} \times m \times (T_{kip} - T_0) + m \times L_{isp} = m \times [_P \times (T_{kip} - T_0) + L_{isp}].$$

Here $V \approx 10\,\mathrm{l} = 10 \times 10^{-3}$ m^3 is the volume of evaporated water; $T_{kip} = 100$ °C is the boiling point of water; $T_0 \approx 10$ °C is the ambient temperature; $\rho_v = 1000$ kg/m^3 is the density of water; $C_P = 4.187 \times 10^3$ J/(kg × K) is the heat capacity of water; $L_{isp} = 2256 \times 10^3$ J/kg is the specific heat of water evaporation; and m of water = $\rho_v \times V = 10$ kg—the mass of water in the bucket.

Based on the above equation, we obtain the following value of the BL energy, which went to heating, boiling, and evaporation of water from a bucket: $\Delta Q = 2.6 \times 10^7$ J. Thus, the specific energy of 1 kg of water was $\Delta Q/m \approx 2.6$ J/kg and a volumetric energy density of BL: $W = \Delta Q/m \times \rho_v$ is $2.6 \cdot 10^9$ J/m^3. Assuming that evaporation occurred as a result of the influence of the electric field of the charge accumulated BL, the BL electric field can be estimated by the equation $E^2 = 2 \times \Delta Q/(\varepsilon \cdot \varepsilon_0) \approx \approx 6.32 \cdot 10^{20}$ (V/m)^2; $E \approx 2.5 \times 10^{10}$ V/m.

14. (Olkhovatov 2008) 6.08.1868 County Donegal Ireland.

A bright red round fireball with a diameter of about 30 cm at the beginning and 8 cm at the end of the observation floated in the air for about 20 min, "lazily" bouncing off the ground, covering a distance of about 1.6 km. When examining its trajectory, it was found that the object had dug a trench in a peat bog approximately 100 m long and more than 1.2 m deep. During its journey, the ball decreased in diameter (especially after each encounter with the earth) and was about 75 mm across before finally sinking into the peat ridge. The phenomenon was observed by engineer Michael Fitzgerald. In his report to the Royal Society, he also said that he found a hole in the ground with an area of about 2 m^2 in the place where the ball first touched the ground. At the same time, the ground was as if cut out with a giant knife. Also, severe destruction was made on the bank of the stream. Traces of these injuries could be observed in 2006 in the expedition.

Comment. The assessment of the energy required to commit such destruction, done on the basis of observational data and engineering physics, turned out to be 4×10^7 J. If we assume that the average diameter of the BL was 19 cm, then the energy density is 1.4×10^9 J/m^3, which indicates a high energy of the BL.

15. (Dmitriev et al. 1981) on 24.08.1978 at 23.20.00. Khabarovsk city, Khasanovskaya Street.

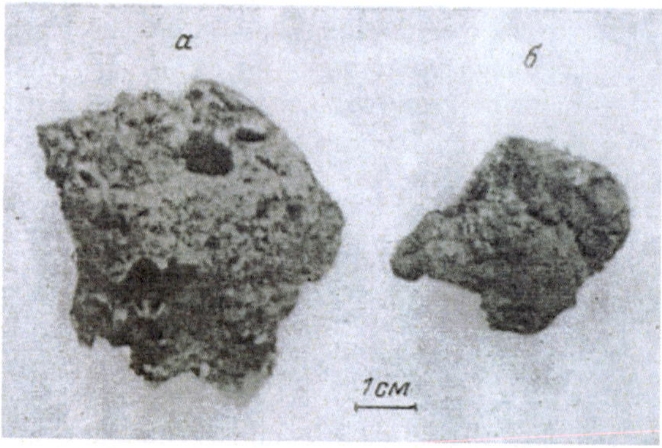

Fig. 2.5 Pieces of slag that appeared after the explosion of the BL above the ground

During a heavy rain, the audience, who came out of the Zarya cinema, heard a sharp whistle, reminiscent of the operation of a jet engine. The whistle was accompanied by a crackle. It became as bright as day. Then a BL with a diameter of about 1.5 m appeared over the cinema building. The color of the BL was orange, sparks poured out of it. The BL began to descend. It descended to the surface of the earth through the branches of trees, flashed above the ground for a moment and rose up again. A strong explosion took place; it became dark and quiet. BL was observed for about a minute. At a distance of up to 100 m, the electrical wiring was disabled. Despite the torrential rain and soaked soil, its section with a diameter of 1.5 m and a depth of 20–25 cm was charred, slag formed in it. The volume of the zone filled with slag was about 0.4 m^3. The slag did not consist of a solid crust, but of many irregular pieces with an average size of 5–6 cm. Some pieces were fused together. In total, about a thousand such pieces were found. The initial soil was a sandy loam based on quartz sand (Fig. 2.5).

Comment. The authors of the article conducted experiments to determine the methods of energy impact on the soil, the result of which is the formation of slag similar to the slag that appeared during the explosion of the BL. They found that similar effects on the soil were obtained when it was heated by the high-frequency radiation at a frequency of 13.56 MHz with a power of 60 kW. The authors estimated the energy density released during the BL explosion 6×10^8 J/m^3. If one takes the duration of the BL flash over the ground equal to 0.01 s, then the power of electromagnetic radiation is $P = 10^9$ W.

16. Astana, 18.08.2014-IA Novosti-Kazakhstan, TV channel "TDK-42."

The BL practically destroyed a house in the village of Akzhayyk in the Terekta District of the West Kazakhstan region. No one was injured. Rain. A huge fireball penetrated into the Kashkarovs' house through the chimney, smashed the walls, disabled all electrical appliances, and smashed furniture and dishes. Jadiger Koshkarov (male): Everyone at home was very scared, I told them to calm down that it was lightning. The eldest daughter could not get up from fear. Then we noticed that all the veins appeared on the body, but in the morning, everything returned to normal. BL that night deprived the entire village of electricity, 20 families were left without appliances and electrical appliances.

Comment. BL destroyed the walls of the house.

17. http://don-news.net/mestnoe/11840-v-artemovske-videli-sharovuju-molniju.html 23.06.2015 16.45. Artemovsk.

Lightning of an unusual kind was visible to residents of house No. 3 Komsomolskaya Street. It happened on Saturday, when there was a strong thunderstorm with a downpour in the city. According to an eyewitness, after the start of the downpour, she went to the window in order to cover it. At that moment, a bright fireball that did not exceed the size of a soccer ball was flying at the level of the third floor. A few seconds later, there was a loud crash, followed by the sounds of bricks and fragments of slate falling on the asphalt. As it turned out, the BL flew into a brick chimney, which was located above the entrance of the house. It smashed its head, damaged the slate roof. All residents of the entrance marked by lightning complain about electrical appliances that have failed—televisions, computers, boilers, receivers, and other equipment. One of the Internet service providers was also surprised, a cable ran through the attic, four meters of which simply burned out and evaporated.

Comment. BL destroyed the entrance of the house; electrical appliances failed: televisions, computers, boilers, receivers, and other equipment.

18. (Olkhovatov 2008) J. Reed. Message 2001. The event of 1954. The northern side of the suburb of Montreal. "A hot, stuffy day turned into a humid evening. Somewhere in the vicinity, there was a faint rumble. The observer and his friend were walking through an open field. Almost reaching the chain of trees on the border of the field, they suddenly saw a white ball, no bigger than a football, which was flying at a pedestrian speed. It was about a meter above the ground at a distance of 30 m. The ball hit the first tree in its path and cut it down, then did the same with all the trees in its path. The observers were frightened by the ease with which the ball cut down trees. They saw how 6 out of 7 trees were cut down and hurried home. The next morning it turned out that the trees were cut at a height of 1 m, i.e. approximately at the height of the ball's flight, and they fell in the direction of impact. The trees had been cut so cleanly that no signs of charring or burning could be seen. Despite the discoloration, it was still possible to count the annual rings on them. His friend counted more than 200 on one, after which he abandoned the idea."

Comment. The ball cut 6 large trees on impact. If we assume that the destruction energy of 1 tree is of the order of the BL energy in the case of Sect. 2.1, then the energy density estimate gives 1.1×10^{10} J/m^3.

19. (Olkhovatov 2008) English Mechanic magazine. 1872. Bunbury. England. Thunderstorm.

A giant fireball, accompanied by buzzing and roaring, moved at a height of 2–3 m above the ground. The phenomenon was accompanied by a kind of smoke that "whistled" and rose high up, and the accompanying strong gust of wind carried a cloud of branches and destroyed everything in its path. Large trees were uprooted or broken off at a height of about 3 m above the ground. According to eyewitnesses, the smell of sulfur accompanied the ball. The ball disappeared soundlessly.

Comment. The movement of the ball during a hurricane.

20. (Bychkov and Vaulin 2014) 2002. Settlement Mostovsky Krasnodar Territory.

BL appeared at six o'clock in the morning after heavy rain, accompanied by thunderclaps. A fireball descended from the sky onto the territory of the new school of the village and moved towards the building, and then, having reached the foundation, exploded with great force. The explosion was so powerful that finishing stones flew away from the foundation of the school by more than 20 m, glass was broken on the opposite side of the building. In the school itself, there was a short circuit of the electrical wiring, all phones stopped working. According to local residents, one of the four lightning rods located on its territory apparently saved the school from great destruction, near which the BL exploded. Residents of neighboring houses also suffered from this BL. Many of them burned down televisions, refrigerators, and computers.

Comment. The explosion of the BL destroyed the foundation of the school.

21. The newspaper "Premier" 4.09.2000 3.00 p.m. Vologda region. The village of Pelpakhta, Babaevsky District.

The BL destroyed an apartment building when it crashed into it in the midst of a severe thunderstorm. According to eyewitnesses, "it crashed so hard that everyone almost went deaf." The newly renovated building almost completely collapsed in front of the villagers—the front wall collapsed, the doors flew out. Fortunately, there was no one in the house at that moment. After this explosion, electricity and telephone service were cut off throughout the village. Alarmed villagers ran to the destroyed building, expecting to see a fire started there. But the wooden house wasn't even smoking. After taking turns until midnight, the locals went to bed, but about two o'clock in the morning the village was woken up by a terrible crash. People ran out into the street and were speechless with surprise—the building was engulfed in flames. By morning, there were only smoking remains of it.

Comment. The explosion of the CMM destroyed the walls of the house.

22. (Olkhovatov 2008) Snegirev P. E., Komi-Permyak Autonomous Okrug. June 1956. The village of the Golden Cape of the Kudymkarsky District. A village on the elevated bank of a forest river, with an average width of about 3 m. Evening about 6 p.m.

The observer first heard a terrible hum, and then saw a grayish ball the size of a football flying along the river with a terrible hum. The flight height was 1.5–2 m. The ball repeated all the bends of the river and flew at a speed of about 100 m/min. There was a glow around the ball like a halo around the sun, and the water under it in the river was foaming. There was a bathhouse on the shore, 50 m from the river.

When the ball flew past the bathhouse, the roof was knocked down from it, which slid off the log house at the speed of the ball's flight (the weight of the roof was about 700 kg). After the flight of the ball, the observer found that at a distance of 2–10 m from the river about a dozen trees (the diameter of the trunks 15–20 cm) were as if knocked down by a blunt object, smooth stumps about 1.5 m high stuck out. The willows remained bent, all the grass along the river lay down. The air was clean at the same time, there was a feeling of lightness. After some time (about 2–3 h), a huge black cloud came to the village from the "wet corner" (from the west) and a terrible thunderstorm of unprecedented strength began.

Comment. BL tore off the roof of the house.

23. (Likhosherstnykh 1994) Khvoshchev N. A., G. Tarusa. Summer 1943.

"On the observation tower of the forestry at an altitude of 20 m there was an air defense observation post equipped with a telephone. A girl with a rifle was on duty at that time. Although it had recently rained with a thunderstorm, it was hot. Suddenly, the tower lit up brightly, as if a searchlight had been pointed at it. There was a slight explosion. The tower swayed. People who ran up to the tower saw a girl lying at the foot of it in a greatcoat and kirza boots, without a cap. She was dead.

There was a halo of white fluff around her body, which turned out to be, as it turned out later, crushed underwear and soldiers' footcloths, also brought to a state of fluff. At the same time, the tunic and skirt were intact and there were no traces of burns on the body. Only on the finger where the ring was, a circular burn was noticeable, the ring itself disappeared. Later, the melted metal parts of the rifle were found on the side."

Comment. The disappearance of the ring. Destruction of the linen material. The death of a woman. Melted metal parts of the rifle.

24. (Batygin and Mosin 1989) 1989, Khabarovsk, Russia.

BL heated 7 cubic meters of water to a boil, on the basis of which its energy was estimated as 3000 MJ.

Comment to the section

So, we have estimated the energy of BL, which went to heating, melting, evaporation, and boiling of various substances, the maximum values of which are in the range of 10^{10}–10^{11} J/m^3. The oxidation energy of the components of boiling materials (combustion) could additionally be released in the air; however, the corresponding calculations show that this energy does not exceed ~25% of the energy of the BL itself, so the values obtained are an estimate from below.

In Fig. 2.6, we present the dependence of the energy density on the size of the BL, constructed on the basis of observations. Cases are presented when an energy estimate was made for a known size of the BL.

In order to understand whether the energy of BL is large or small, and since in observations during the explosion of BL their action is often compared with the action of an explosive substance—TNT, we give the value of its enthalpy H = 15 kcal/mol,

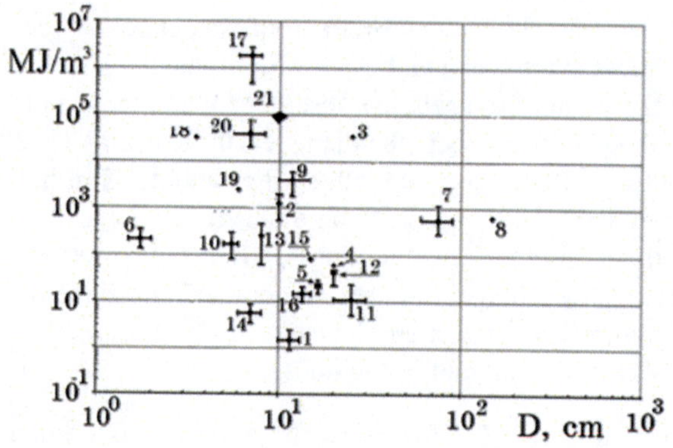

Fig. 2.6 Dependence of the BL energy density on the diameter. Events: № 1 (Grigoriev 2006, № 22); № 2 (Egely 1987, No.270c); № 3 (Egely 1987, № 222); Egely 1987, № 4 (Egely 1987, № 58); № 5 (Egely 1987, № 4); № 6 (Stenhoff 1999, case 1981, p. 65); № 7 (Kozlov BN 1978); № 8 (Dmitriev et al. 1981); № 9 (Singer 1971); № 10 (Stakhanov 1985, № 30); № 11 (Stakhanov 1985, № 33); № 12 (Stakhanov 1985, № 34); № 13 (Stakhanov 1985, № 35); № 14 (Imyanitov and Tikhy 1980). № 15 (Singer 1971, Ch. 4.2, case 2); № 16 (Wittman 1971); № 17 (Imyanitov and Tikhy 1980, Transcarpathia 1962); №18 Ignatov's observation, see above; № 20; Rudkovskyi observation, see above case № 8; № 21 Observations in the Gulf of Riga № 9, see above

which, with its composition $C_7H_5N_3O_6$, gives the value of the specific energy released during the explosion equal to 2.8×10^5 J/kg. However, see above, the energy density of the BL reaches 7.8×10^6 J/kg. This means that the energy released by the BL is at least ten times greater than the energy of common explosives.

One should note that the least reliable parameter in the observations is the size of the BL. The error can be two to five times.

All of the above suggests that BL is a dangerous natural phenomenon. This conclusion does not agree with the opinion of some researchers about the insignificant danger of BL. For example, (Stenhoff 1999) considers it not very dangerous and explains most of the destruction attributed to it by the action of ordinary linear lightning.

The main question remains how the BL energy can find applications.

2.2 The Appearance of BL

This section is important because it shows the appearance of BL in various conditions. At the same time, the unexpected appearance or appearance of BL from "nothing" poses a problem for BL researchers: how to conduct experimental or theoretical modeling. Often in nature and laboratory research, some time passes between the moment of energy input and the appearance of an object, so that the appearance of "nothing" requires a careful analysis of the surrounding conditions, which is often not reflected in the descriptions of observations. The same applies to the appearance of BL in good weather, in principle, information is needed about the distribution of thunderstorm activity around the CMM observation area (lightning can fly from the area of thunderstorm activity), which is also not reflected in the descriptions.

1. (Bychkov et al. 2003) V. A. Rantsev-Kartinov, Ph.D. physicist, born in 1939, survey 2000.

1944. Summer, June–July, day 13–16 h. Kupino Village near Novosibirsk. A strong thunderstorm was coming to an end, the rain had stopped. There was a strong wind. Observation of the BL formation in a room with a Russian stove in a wooden rural house from a distance of 7–10 m from the object. The shortest distance to the BL during its movement is 0.5 m. The window in the room on the first floor is clogged with a rusty iron sheet.

After a linear lightning strike into a birch tree in the yard, a ball appeared on a sheet of iron. At first, a glowing dot appeared, which turned into a ball with a diameter of 7–9 cm, then it did not change. The ball grew and trembled, there was a feeling that it was moving, its center was fluctuating. Its border was diffuse, not sharp. It resembled a cotton ball with fluffed edges, a wool ball, only colored. The ball flew to the stove at a height of 1.5 m, then turned 90° and flew into the canopy. The ball moved horizontally, smoothly, without jerks. Slowly at first, but then accelerated.

There was an explosion, and the corner of the hall was torn apart, as if by an explosion, but without charring. In the house, the doors did not fit tightly and there were cracks, the BL flew into the corner not through a draft.

Life time 7–9 s (growth time 2–3 s, 3 s flew around the room, 1.5–3 s flew in the hall). The glow was bright, white–pink with a blue periphery. It could not be compared with an incandescent lamp, because it was possible to look at it. The glow can be compared with a 40-W fluorescent lamp. From this light, the dark corners of the room were not illuminated any brighter. The color did not change. The BL flew 6 m for 3–4 s. The shortest distance to the observer was 0.5 m. The number of observers was 2.

Comment. Pay attention: The ball appeared on a sheet of iron. The appearance of BL on a sheet of iron hinders the development of plasma theories of BL, since plasma immediately disappears on a metal.

2. (Bychkov et al. 2003) A. S. Timoshuk, Associate Professor, Chemist. Describes in 1990 his observation of BL as a child, in 1946 in the city of Belaya Tserkov, Kiev region at 10–11 a.m. at the end of April. The distance from the observer was about 25 m. The beginning of a thunderstorm, a thundercloud is rapidly approaching, thunder is heard. A branching linear lightning crossed and closed the wires running across the street from a wooden telegraph pole over a previously broken branch of an old poplar tree.

At the moment of a short circuit on the telegraph pole, a yellow-green flash of electric welding brightness was generated at the upper roller. A white ball with a diameter of about 15 cm grew in it. It slowly rolled along the inclined wire, increasing its speed, acquires the color of red-hot metal, and when rotating it abundantly scattered sparks. The ball was perceived as a light, but not hollow formation. In the place of the greatest sagging, it jumped to the lower wire 5 cm from the pole. In the place of sagging of the lower wire, it fallen on a poplar branch from a height of 0.5–0.8 m from it. At the moment of contact, a strong thunder or crack was heard. Breaking, the branch tilts. The main ball decreased, and six to seven red balls with a diameter of 4–6 cm ran along the branches. Then they did not go out at the same time.

After about 3 s, a new ball with a diameter of about 12 cm appeared "out of nothing." The place of its formation was a meter to the right of the place of formation of the first ball. It moved along the leaning thick branch, accelerating the movement, smoothly and elastically jumping on irregularities. It decreased to 9–10 cm, scattering sparks abundantly. Jumping off onto the pavement, it smoothly jumped up like an air balloon. The heights of the jumps were 20, 12, and 5 cm, while the amplitudes decreased and the frequencies of the jumps increased. Then the ball split into three large shell-shaped pieces and several small ones. The destruction was viscous, like a rolled snowball, the structure was layered. The pieces did not melt at the same time. The duration of this process was 18–20 s.

Comment. Independent survey of the observer by the author. Pay attention: at the moment of a short circuit on the telegraph pole, a yellow-green flash of electric welding brightness was generated at the upper roller. A white ball grew in it.

3. (Bychkov and Vaulin 2014) erytryec, Igo 28.04.2011, 14:33.

I saw the BL 2 years ago (2009) when I went to the peak of Cho Oyu. We got into a thunderstorm at an altitude of more than 5000 m, and there were a lot of BL. Mostly bluish-white and of different sizes. They just appeared out of nowhere and from them lightning struck the mountainside opposite. I can say for sure that I have seen several times how around an ordinary lightning that appeared from a ball, the effect of heated air was observed as from a fire when it rises and a distorted image is visible through it. That is, the temperature, most likely, was very big.

Comment. Pay attention. BL appeared out of nowhere and lightning struck the mountainside from them.

4. Prof. Kuzmin R. N. DS. The end of July 1997, Moscow.

Before the storm. Time-14.30 LT. Leninsky Prospekt, the bridge over the district railway, next to the monument to Yuri Gagarin. The BL suddenly passes smoothly, from tche observers' faces at a distance of 15 cm. Two observers see a globular blue jellyfish with a slight yellowish tinge 10–12 cm in diameter. It is as if it is alive. The film is moving along the surface of the BL, the shades of its colors change, as on a soap bubble or a thin film of gasoline in a stream of water after rain on the sidewalk. The heat from the BL is not felt, the cosmonaut's pedestal is barely visible through the film. The BL peacefully floats over the railing of the bridge and descends to the railway track, drawn by the breeze, goes under the bridge from the eyes of observers. It smelled of ozone. There is a pothole on the bridge, the trolleybus rod momentarily breaks away from the wires carrying at least 500 V. The birth of another BL takes place five steps away from the observers. The discharge between the trolley rod shoe and the current-carrying wire generated a white ball that broke away from the place of birth, grew a little and repeated the previous observation. The BL smoothly rolled over the railing of the bridge and, descending to the railway tracks, exploded almost without noise, about one and a half to two meters from the ground surface, scattering a shower of sparks. The size of the salute ball was more than a meter in diameter. Before the explosion, the BL slightly decreased in size, turned red, and a darkening appeared inside (core).

Comment. Pay attention. The BL suddenly passed smoothly, from the observers' faces at a distance of 15 cm. Two observers saw a globular blue jellyfish.

5. (Olkhovatov 2008) Australia. 1956. Coast. End of July (winter). Without thunderstorms, lightning, and rain.

For about 2 h, about 1.000 very bright fireballs of a meter in diameter fell from the clouds into the sea.

Comment. Pay attention. Without thunderstorms, lightning, and rain. BL of a meter in diameter fell from the clouds into the sea.

6. (Bychkov et al. 2003) Observers Pletneva V. A. (female) and Pletnev Yu. A. (male) (born in 1923 and 1924) place of residence Moscow. Observation 1978, night 2 a.m., September, Abkhazia, New Athos.

During a thunderstorm, without rain. Strong gusts of wind from the Black Sea. With gusts of wind, glowing balls of 40–50 pieces appeared on the tree. They fell down and went out before reaching the ground. The phenomenon was repeated about 10 times for 20 min.

The parameters of the balls: lifetime 2 s, diameter 5–6 cm, distance to objects 10 m, distance traveled by the object 5–10 m, shape—sphere, color—blue, radiation power compared to an incandescent lamp 500 W.

Comment. Pay attention. With gusts of wind, glowing balls appeared on the tree.

7. (Bychkov et al. 2003) A. I. Romanov, Philologist, 47 y.o. at the time of observation, 1994 (survey of 1994). The reaction is calm. Mid-September–early October. 4–5 p.m. local time. A village of the Ivanovo region on the riverbank. Weather: temperature 8–10 °C, dry, windless.

After lunch, the observer nailed the slate on the roof of the house. He needed to put the second layer of slate on the first layer, which lay on the iron roof. Between the old roof and the slate, a large amount of compressed poplar fluff and buds from a large poplar standing nearby have accumulated. He started throwing this stuff down. Suddenly, two tangles appeared from under the iron roof and old slate at a distance of 1 m from the observer, initially resembling tangles of poplar fluff with a diameter of 3–3.5 cm. There was no heat from the balls. They began to descend, soaring flying like soap bubbles. The balls differed from the balls of fluff in that when they hit the trunks of trees standing next to the house, they made a crackling sound like a distant salute. The balls crackled, emitted a faint glow, like a dying sparkler, the color of the glow (spark) of both balls is white-pink. Each ball resembled a dense lump of poplar fluff, and everything was tied up in it, such a gray lump. The glow appeared only at the moment of impact. In flight, the surface of the ball vibrated like jelly and BL resembled a transparent gelatinous substance. After flying 10 m, the balls went out with the sound of fireworks. Observation time 15–20 s.

Comment. Pay attention: suddenly, two tangles appeared from under the iron roof and old slate at a distance of 1 m from the observer, initially resembling tangles of poplar fluff.

8. (Bychkov et al. 2003) A. V. Bogdanov, Artist, 43 years old at the time of observation (1978–1979), 65 years old at the time of the survey in 2003. Place of observation Tikhvin, Pionerskaya str. Mid-August, 10 p.m.

"During a thunderstorm with rain, linear lightning streamers fell into a telegraph pole with an electric light bulb. The observation took place from a distance of 20–30 m from the objects with good visibility. Balls of 10–15 cm in diameter appeared near the light bulb, when they fell they decreased to 5 cm (walnut). It seemed that the drop was coming off and falling. They fell like soap bubbles. The balls lived 2–3 s. Their color was blue. Falling almost to the ground, they disappeared before reaching it. The whole process of the appearance and disappearance of the balls lasted about 15 min."

Comment. Pay attention: balls appeared near the light bulb.

9. (Bychkov et al. 2003) Yu. L. Serov Ph.D., Physicist, survey 1990 about his observation in July 1958, when he was 13 years old.

A thunderstorm was coming; clouds were low over the ground. The first drops of rain were falling. The air temperature was 20–23 °C. At a distance of less than a kilometer from the observer from the ground, under a cloud, a ball appeared in the grass. Instantly, in about 0.5 s, it soared to a height of tens of meters and exploded. The energy of the explosion was two to three times inferior to the energy of ordinary lightning. Then up to five similar phenomena were observed. The time between the phenomena was 3–5 s. The balls flew up and exploded. The place of occurrence of the BL moved synchronously with the movement of the cloud.

Comment. Pay attention. Under a cloud, a ball appeared in the grass. It flew up to a height of tens of meters. The place of origin of the BL moved synchronously with the movement of the cloud.

10. (Olkhovatov 2008) Summer 2000–1999, Murom District, Vladimir region. Near Lake Visha.

Eyewitnesses saw a heavy rain cloud float out from behind the forest, which soon covered the area. After a bright flash, dazzling white balls suddenly rained down from the sky, which, twisting and spinning, slowly approached the ground and burst soundlessly at the level of the tops of the trees. The balls looked like giant fireflies. Some sank very low, it seemed that you could even touch them with your hand. Then, right in the middle of the lake, 3 m from the surface, a huge red ball with black veins hovered. Ten seconds later, it split in two with a bang, and its two halves, twisting and leaving a tracer trail behind them, flew in different directions.

Comment. Pay attention: after a bright flash, dazzling white balls suddenly rained down from the sky.

11. (Bychkov et al. 2003) Observer Troinov E. A., Engineer (male), born in 1973, survey 2003. Observation 1984. July, Vadyb Village, Sysolsky District, Komi ASSR. 19 h. On a hill, in a meadow. Thunderstorm. 3 observers. Observation from 15 m.

A lightning bolt hit one of the poles of the power line. A ball about 10 cm in size appeared. The ball was slowly flying 50 m. Moved parallel to the slope at a height of 2 m from the ground. The color was yellowish-reddish.

Comment. Pay attention: a lightning bolt hit one of the poles of the power line. A ball appeared.

12. Observer Bychkov V. L., (male), Physicist, Ph.D. 54 years old at the time of the survey in 2005. Observation in Thousand Oaks, California, 10/25/2002, 9.30 p.m.

"On this day, a strong Santa Anna wind was blowing from the desert towards the ocean. The city is separated from the ocean by a ridge of mountains, upon meeting with which condensation and smog precipitation occur. Smog hung over the city, lying on a mountain ridge. The planes that usually often flew over the city were not visible, the sky resembled dark milk. It was dark all around.

I was returning from work along a dimly lit street going downhill. I went down the mountain and looked down at the houses. Suddenly, at ~3–4 m above a palm tree located 50 m in front of me, and at a height of ~20–30 m, a silently bright object flashed. It resembled a uniformly glowing circle. The power in comparison with an incandescent lamp was 500–1000 W. It was a little lower than me in height. It fell down for 2–3 s, after which it silently went out. When falling, it passed ~2–3 m. It was a bright flash of white. The approximate size of the object was 30–40 cm.

The reaction was calm, interest and the need to remember everything that is possible. The observation took place from top to bottom. The palm tree stood near the house. There were no people nearby, at this time in this poorly lit part of the city no one was in the streets, so I can't assume that someone fired a firecracker. I also can't assume that the object fell from the plane, because the hum of the engines was not heard."

Comment. Pay attention: Suddenly … a bright object flashed silently.

13. Tom Kammer. Born in 1958 (male) Teacher of Physics and Natural Sciences. 32 years old at the time of observation in 1990–1991. Survey of 1995. Place of observation Onalaska, Wisconsin, USA. Summer. June–July. 21 h. Observation from a restaurant-bar in the city. The object was outdoors. Heavy rain. The wind is not strong.

"I was standing at the counter and watched a strong thunderstorm. Linear lightning struck the top of a telephone pole located 40–45 m away from me. An object more similar to a sphere with a diameter of 30–35 cm appeared on the top. It was very bright. Quickly in 1 s, it decreased and disappeared. It was bright with an orange center, and when it decreased, it turned orange. It shone like a welding flame of 300 W. A flame appeared at the top of the pillar. Heavy rain extinguished it after a minute or two.

All the lights in the restaurant went out and the music stopped playing. It turned out that the electric fuse at the top of the pole was destroyed."

Comment. Pay attention: linear lightning struck the top of the telephone pole. An object more like a sphere appeared at the top.

14. (Olkhovatov 2008) M. V. de Visser (male) in his 1913 book "Dragons in China and Japan."

The evidence of an eyewitness of the phenomenon, which most likely occurred in 1295, is given. In the sky near the lake, "two dragons" of vague shapes suddenly appeared, which turned around each other, as if struggling, and then suddenly unexpectedly fell into the lake. During the following short period of time, a strong wind appeared, which raised a wave 3 m high.

Then more than a dozen fireballs the size of a large house fell from the sky. At the same time, these two dragons immediately rose into the sky. After a while, torrential rain poured down.

Comment. Pay attention. In the sky near the lake, "two dragons" of vague shapes suddenly appeared, spinning around each other. Then more than a dozen fireballs the size of a large house fell from the sky.

15. (Olkhovatov 2008) 6.07.1976, Sichuan Province, China. 10.55 p.m. local time.

More than 20 people saw something like a fireball rise from one of the wastelands and, having soared to a certain height, then gradually disappeared. A few minutes later, this phenomenon was repeated, and the fireball rose to the height of the cornice of the house.

Comment. The appearance of two fireballs.

16. (Keul AG 2021) Walther Gerlach, Physics Professor at the University of Tuebingen, wrote a report to Naturwissenschaften on May 20, 1927:

"On May 9, 1927, 8:00 [LT] I observed ball lightning [at Tuebingen].. I happened to stand at the window.. In the northeast, linear lightning went down with strong ramifications. From it (apparently out of a sharp kink) came out a bright luminous yellowish-white ball in a considerable height and flew to the southwest. The time between lightning and the ball's flyover of our institute was roughly one second. It could be observed another second in the same appearance on its straight trajectory, soundless. Two seconds later, a not very strong thunder started and 1.5 s after its start, there was an extraordinarily violent detonation, like an explosion bang." Gerlach did his calculations: with a speed of sound of 330 ms^{-1} and observed times of the lightning's thunder and the explosion, he gets a ball lightning trajectory of 1 300 + 1 150 = 2 450 m. With 2 s flighttime, "ball lightning had a mean speed of about 1200 ms^{-1}." "I learned at noon that the ball hit a small, barn-like house on the edge of Tuebingen which is 1 100 m distant.. The house was not damaged, but the top of an electrical mast was smashed nearby." Gerlach also mentioned bluish "spraying" of electrical wires and blown fuses in "a whole row of houses" (Gerlach 1927). His uncommented speed calculation of 1 225 ms^{-1} is amazing, because that is Mach number 3.6 without effects ("soundless"). This is one of the rare occasions where ball lightning met a gifted experimental physicist. Gerlach is renowned for his 1922 experiment about spin quantization in a magnetic field.

Comment. For 2 s of flight, "the BL had an average speed of about 1200 m/s." This is an unusually high speed for average BL.

17. (Keul AG 2021) In August 1968, Graduate Engineer Alfred Geiswinkler stayed with his wife at a Maria Woerth villa, Austria, near Lake Woerther. Around 15:00 LT, during a heavy thunderstorm.

A bright shining ball of about 20 cm with a discharge corona crossed the room 1.5 m away. Its outline looked sharp and the color was white-blue with a reddish touch. Hissing like a Xmas sparkler, it went out on the balcony, along a balcony pillar to the ground and a further 300 m out onto the lake, disappearing with thunder and lightning. It followed a linear track above the floor and rotated, singed the wooden balcony pillar, and caused a sulfur-like smell in the room. Both witnesses suffered a shock. Alfred Geiswinkler, who made his report to the author 1979, was then Technical Director of BEWAG, the power company of Burgenland Federal Province (Keul 1980, 1981).

Comment. Both witnesses suffered a shock.

18. (Keul AG 2021) Coroner Leopold Breitenecker, later Forensic Medicine Professor at Vienna University, encountered ball lightning at Ternitz, Austria, in the summer of 1955, 16–17:00 LT. He carried out an autopsy in an old mortuary during a heavy thunderstorm, when lightning struck and threw the electricity meter out of the wall. From the hole in the wall, a fist-sized, sharply outlined ball descended to the floor, radiating a bluish-white light like a welding arc. It moved 5 cm above the floor to the open door, but disappeared into the stone wall beneath the door frame. The witness threw down his instruments and leapt to the door just in time to see the ball (apparently after passing through a crack in the wall) move across the vestibule and out of the mortuary's front door. It bounced down the steps like a toy ball and disappeared into a grave-mound after traveling 8–10 m in about 5 s (Keul 1980, 1981). Breitenecker saw ball lightning a second time from Muehlbach/Attersee, Austria, in the summer of 1975, late afternoon, looking towards Mt.Schafberg in a thunderstorm with many lightning flashes. Suddenly, a distant fireball traveled horizontally along a mountain wall, then burst sparking, like a spray candle.

Comment. From the hole in the wall, a fist-sized, sharply outlined ball descended to the floor, radiating a bluish-white light like a welding arc.

19. (Tar 2006a) 1954. male, Physicist-Student, Budapest, Hungary, Summer day, temperature 25–27 °C; very maggy, humidity about 100% because of the rain; about 10 am; thunderstorm.

D. Tar witnessed the formation of BL when terrible flash of linear lightning struck the ground 50 m away from him. Two seconds after that, a sphere (ball) with a diameter of 30–40 cm appeared at a distance of 2.5 m to the left of the impact site at a height of about 1.2 m. The ball rotated counterclockwise around a horizontal axis perpendicular to the line connecting the ball to the place of the lightning strike. The ball shone brightly, one or two small "tails" were visible on its surface, which were drawn into the sphere after 0.3 s. The ball was moving to the left at the same height at a constant slow speed of 1.17 m/s, no change in its diameter. After 3 s, it disappeared soundlessly, as if a soap bubble burst.

Comment. Detailed description and discussion is given in (Tar 2006b).

General comment to section

It was not possible to find an explanation for all the observations of the appearance of BL from a single position.

2.3 External Features of the BL

2.3.1 Size, Color, and Shape of the BL

This section shows the possibility of the BL to have not only a spherical shape. The BL substance allows this object to transform from a ball into a ribbon and vice versa. When BL appears from sockets in the form of streams, they behave not like plasma, but like currents that are attracted to each other according to the Ampere law. This clearly indicates that the BL is not a plasma object.

1. Arsenyev VK (1951) Observation of the BL in the Summer of 1909.

After about half an hour, the solid forest ended, and I went out on a hillock… I stopped for a minute and at that time. I saw a strange light ahead of me. Someone was walking towards me with a lantern.

"What a weirdo, I thought." On such a bright night, someone is walking with fire." After a few steps, I saw that the lantern was round and opaque. "What a miracle, I thought again.—Who could think of the idea of walking through the taiga by moonlight with a paper lantern."

At this time, I noticed that the light object was quite high above the ground, significantly exceeding the height of a person…

A strange light was approaching. Since the terrain was uneven and the path then rose a little, then fell into a pothole, then the lantern, consistent, as it seemed to me, with the movements of the mysterious pedestrian, then lowered to the ground, then rose up. I stopped and listened...

However, the silence was complete... Suddenly I saw that the lantern was not moving along the path, but to the side, to the left of me, a shrubby overgrowth.

I was scared because I could not explain who or what I was dealing with. It was some kind of glowing ball the size of two fists, matte in color. It slowly floated through the air, adjusting to the topography of the place. Sometimes descending where there were depressions on the ground and where vegetation was lower. Then rising up where the soil rose and shrubs grew higher, and at the same time it avoided contact with tree branches; grass in every possible way; and carefully avoided every twig, every twig, and blade of grass.

When the glowing ball caught up with me, it was no more than ten paces away from me, and therefore I could see it well. Once or twice, its outer shell seemed to burst, and then a bright-white-blue light was visible inside it....A thread-thin fiery tail stretched from the lightning ball, which at times gave tiny flashes in different places.

I realized that I was dealing with a BL, with absolutely clear skies and with complete calm.... I stood as if chained, and did not dare to move. The glowing ball was steadily moving all in one direction. It crossed my path obliquely and began to climb the hillock. Along the way, it climbed quite high and passed over a bush, then began to descend to the ground and then disappeared behind a hill.

Comment. An ideal description of the observation. BL with a shell and a tail, there is no thunderstorm. Once or twice, its outer shell seemed to burst, and then a bright-white-blue light was visible inside it.

2. (Bychkov et al. 2003) Toloknova E. A. (female born 1918) and Barinova M. V. (female) cardiologists, observation 1952. Moscow (survey 2000). Spring, late May–early June. In the afternoon, 12–13.00 local time. The weather is clear, sunny day, and there is no thunderstorm.

The doctors were in a large room with fluorography and other measuring electrical equipment on the second floor of a two-storey building. A ball entered the open window 1.5 × 2.5 m. When it entered, it seemed larger than the window. The longitudinal and transverse dimensions did not differ. The impression was that an object similar to the sun entered, golden, orange, inside (inner diameter 70 cm) the color is more intense. The edges are less bright. Along the edges, many objects similar to small snowflakes were observed. The color did not change. The border of the ball is blurry. The glow is bright, more than 100 W compared with an incandescent lamp. When the ball approached, the observer was afraid to burn out, had a state close to shock. The ball passed at a distance of 3–4 m from the observers. There was no warmth from it.

The ball moved slowly, smoothly, in a straight line from the window passed through the room, went out through the open door into the corridor, and with a loud sound flew into yellow-orange snowflakes, hitting the wall of the corridor (5–5.5 m from the observers). The sound is strong, like thunder. In 1.5–2 min, it passed 12 m. There were no thermal traces or traces of the explosion left. BL did not affect the operation of the appliances in the room.

Comment. A ball with a diameter larger than the window 1.8 × 2 m. The color is similar to the sun, golden, orange, inside (inner diameter 70 cm) the color is more intense.

3. (Bychkov et al. 2003) E. A. Megera (female) observation 1948. Engineer born in 1943, survey 1992. Specialty-Software Engineer. The observer's reaction is fear. Observation time: summer, July, 16.00 local time. Location: Moscow region, village Rozhdestvenno.

On one of the hot July days (with a strong wind) at the beginning of a thunderstorm, all the household (three people) were sitting at a table in the dining room of an old hut. Suddenly, a spherical glow appeared in the right corner, in relation to the observer, low above the floor (0.5 m). The BL was round; its border was blurry. The diameter was 20–25 cm. The color was white-bluish, did not change during observation. The radiation power in comparison with an incandescent lamp was 40–50 W. The ball rose to the ceiling, to a height of 2.5–3 m, and then moved to the center of the room to an electric lamp hanging on an electric cord. On the fabric, white cotton braid of the electric lamp cord, the BL left a black thread-like trace along the length of the cord. Then it returned to the wall, where it appeared and went into the crack in the window flattened and stretched. The BL flowed out through a crack in the window, flattened into a flat stream, then disappeared into the crack.

The ball moved smoothly, at a speed of 0.5 m/s, the maximum distance it traveled was 10 m, the minimum distance to the observer was 3.5–4 m. There was no heat.

Comment. The BL was round. It went into a crack in the window, the BL flowed out into a crack in the window, flattened, and stretched into a flat stream.

4. (Bychkov et al. 2003) I. A. Zemnukhova (female) Mathematics Teacher born in 1957, survey 1993. Observation 1964–1967. Place of observation: Tambov region, Zherdevsky District. Summer, August 20, 19–20.00 local time. Strong wind. The observer first conducted surveillance through the window of a village house.

Under the influence of the wind, tree branches flew, buckets and barrels left in the yard, and thatched roofs flew off from various buildings. It was dark because of a large cloud that had moved in. Suddenly everything calmed down, the dust settled, and a huge ball (with a diameter of more than 1 m) became visible in the steppe, which smoothly rolled slightly bouncing. It was approaching the village. The ball looked like a fiery bun. The color was a mixture of yellow-orange to brown. There was a halo around it, consisting of swirling dust; its color was like dust—gray-yellow. The ball from the steppe rolled into the village street, smoothly moving along the middle of the street. It sometimes stopped, as if determining where to go next.

The ball turned smoothly from the street to our house. There were poles at the edges of the street. Supports were attached to the pillars at an angle. Such a support turned out to be in the way of the ball. However, the ball somehow passed through the support and continued its movement.

Then it rose above the ground, above the house, above the trees. It seemed that it was flying up. All this was happening very quickly. Nevertheless, the ball landed at the window and penetrated through the window pane into the room in which the observers were. The window was closed, but without any locks. Maybe the ball pushed the window door into the room, maybe the wind did it. Now the ball was already without a ring, as if crawling through the window, the ball mixed with the ring in color, brightness, and density. Climbing through the window, the ball stretched out like a rubber ball, acquiring an ellipsoidal shape, and then restored the shape of the ball. The distance from the ball to the floor was ~20 cm. People (five people) stood pressed against the walls. There was no heat coming from the ball. The ball did not move around the room, it was not particularly attracted to anything, to anyone.

Grandma, after taking a few steps smoothly, opened the door to a long corridor, and the ball probably with a draft (a noise similar to a strong gust of wind was heard) instantly flew out into the street. When the ball moved quickly along the corridor, it stretched out and resembled a huge melon. After flying ~15 m from the house to the garden, the ball came across, caught on a cherry. It disintegrated into fiery shreds, patches of 15–20 cm in size, which hung and glowed on the branches for a short time, and then the tree caught fire. There was no explosion. Right after that, there was a linear lightning, it started to rain, and the tree was still burning and hissing.

The ball was similar in density to orange smoke balls, different in density, brighter in the center—more yellow. It also looked like a glowing balloon filled with smoke. It also resembled balls of poplar fluff of different density and a different (not poplar fluff) color. It was as if the glowing ball was filled with smoke. It seemed that rays were coming from the dust covering the ball. Large particles entered the ball and moved up and down in it. The border of the ball is blurry. The intensity of the glow is similar to the intensity of the glow of the fire. The center of the ball is orange-yellow, more than 200 W compared to an incandescent lamp.

Comment. A ball with a diameter of more than 1 m. The color is a mixture from yellow-orange to brown. There was a halo around it, consisting of swirling dust; its color was like dust—gray-yellow. The ball inexplicably passed through the support. It entered the room through the window. The window was closed, but without any locks. Climbing through the window, the ball stretched out like a rubber ball, acquiring an ellipsoidal shape, and then restored the shape of the ball. Disintegrated into fiery shreds, patches of 15–20 cm in size.

5. A. V. Elinek (male). Chemist, Ph.D., born in 1953. 20.03.1994 (early spring). Local time was 16–17.00, Moscow. Survey 22.03.1994. 1 observer, reaction-interest at the absence of fear. On this day, there was a thunderstorm with strong gusts of wind and wind with snow. Partly cloudy, but at the time of the appearance of the BL it was clear.

The observer was under the car in the garage and noticed an unusual glow. The garage door was not closed tightly; there was a gap of 3–5 cm. The edges of the door were covered with iron. The observer got out from under the car and saw a ball of not strict dimensions floating in the air at a height of 1.5 m. The size of the ball was 15–20 cm. The color was yellow-green, shone as a low-intensity fluorescent lamp or as a 40-W incandescent lamp. Towards the edges, the color intensity was less than in the center. The ball had a translucent halo. The BL hovered and wavered in the air.

When the observer went to the BL, it began to move away from him, he stopped-it stopped. The nearest distance to the object was 2.5–3 m. At that, there was a thunderstorm in the street. The observation lasted about 1 min. Smoothly floating up to the crack in the door, the BL passed through the crack. It decreased within a fraction of a second. When the observer went out for the BL, there was a blue sky in the sky in the gaps, thunder rang out, and there were discharges. BL was nowhere to be found. The BL passed about 2 m during the observation. It moved horizontally, as if swinging on a swell.

The starter in the Lada VAZ-21-011 car, which worked well before, did not work for 2–3 h after this event, then worked well again. There were no sound effects. The color of the BL did not change.

Comment. The ball had a translucent halo. Smoothly floating up to the crack in the door, the BL passed through the crack. It decreased within a fraction of a second. Had an impact on the car starter.

6. V. P. Frolov (male). Chemist. in 1949. Summer. description of 2003.

"Spooked from the river by an impending thunderstorm, driven by almost continuous thunderclaps, the boys were already near the houses when the cloud covered them. Lightning flashed from right and left at the same time. The lightning was unusual—like shiny upside-down bushes without leaves. The thick channel branched almost immediately from the cloud into thinning twigs, most of which ended right in the air. Suddenly, a large yellow spark 'bounced' from one of the branches. The boys stopped dead in their tracks: the lightning had long gone out, and it flew in silence, turning into a large red ball, which gradually darkened and 'went out' after 5–7 s, moving away from the place of its appearance by about a quarter of the width of the lightning bush.

Then heavy and cold raindrops splashed on the shorn heads. Suddenly there was a loud and sharp bang like the click of a shepherd's whip right over the roof; at the same time, sparks fell from the power line pole standing in front of the window, as if from electric welding. Sparks rained down on the ground; one of them, the largest and brightest, smoothly changed its flight to horizontal, not reaching the ground about a meter. The amazing spark almost disappeared behind the semi-basement boiler room adjacent to the barracks.

Then the BL (and it was it) turned in the direction of the boy. …the BL was slightly larger than a three-kopeck coin (with diameter 2.1–2.2 cm), light yellow, like a small sun and fluffy. Luminous trickles-hairs flew out of it by 2–10 cm, the longest of which bent towards the BL, similar to small fireworks falling to the ground, in that they had a sparkling droplet at the end.

The BL hit the glass 20 cm from the observer's nose with the sound of a pistol shot and a blue-white flash of light. Immediately the rain began to pour.

It turned out that a spark gap was installed on that pole—a special device to protect transformer substations from overvoltage, which can be caused by lightning hitting the wires."

Comment. The BL was slightly larger than the coin (with diameter 2.1–2.2 cm), light yellow, and fluffy. Luminous trickles-hairs flew out of it by 2–10 cm, the longest of which bent towards the BL, similar to small fireworks falling to the ground, in that they had a sparkling droplet at the end.

7. (Bychkov et al. 2003) Datum from 2.02.1996 (female) The event occurred on 5.09.1995 in the afternoon.

"I was doing something at the table at a distance of about two meters from the window, ajar by about 10 cm. Suddenly something made me raise my head; and I saw at a distance of about 3.5 m from the window a shining, and maybe sparkling, but not blinding, ball, as if motionless hovering in the space of the room. After a time of about 1 s, it disappeared, immediately followed by a deafening thunderclap, and immediately after this blow, a downpour poured down like a wall.

The boundaries of the ball were quite definite, but not smooth, but as if slightly blurred. Its entire surface seemed to be boiling—it consisted of randomly arranged 'hills' with pronounced peaks, and with depressions in the intervals between them. At the same time, the fluctuations were either noticeable, or were drawn by imagination. Jets were shooting out of the ball in all directions, mainly in directions extending its radii. The whole system had different shades of steel in a state of pale pink and white heat. The ball as a whole was intensely pink, but not dark pink. The 'peaks' were darker pink, and the gaps between them were rather white, as if clouded or shaded. The diameter of the ball was about 35 cm, and the distance between the 'peaks' was about 3 cm. The jets were shaped like tubes, more precisely cylinders, of different lengths, up to 15–20 cm, and with a diameter of about 5 mm (the observer doubted the reliability of this size). They were rather white at the origins, dark pink at the ends, especially the longest ones. These emissions behaved like fountains in the gravity field. The few, the longest, having reached their greatest height, they abruptly changed direction, and in the form of drops fell down to the surface of the ball. See Fig. 2.7. I managed to notice this only for the part of the ball facing the ceiling, so I do not know if it was a consequence of gravity or the influence of the ball itself. The distance between the exit points of the jets from the ball was about 5–10 cm."

Comment. Jets were shooting out of the ball in all directions, mainly in directions extending its radii. The few, the longest, having reached their greatest height, they abruptly changed direction, and in the form of drops fell down. This suggests that this type of BL is a shell filled with some melt. In this case, when defects appear in the shell from the BL, jets will beat.

8. (Bychkov et al. 2004) I. M. Shakhparonov survey of 1991 (male), Phys.-Chemist, Experimenter. 53 years old at the time of the description of events.

8.1 "In the summer of 1950, I was in the town of Nikolina Gora near Moscow. At that time, it was quite deaf. At this point, the Moscow river makes a big

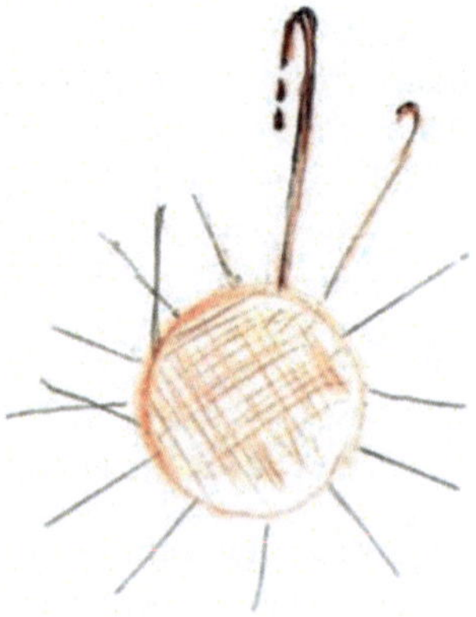

Fig. 2.7 BL with jets from the surface

loop, inside of which there is a huge meadow, and the opposite bank is quite steep. In those places, there are strong thunderstorms coming from the south, i.e. from the meadow.

One day in July after a severe thunderstorm, at about 14–15.00 local time, when all the surrounding space was covered with a post-storm fog, I noticed two transparent balls interlocked with each other. They first moved along the ridge of the iron roof of the house, then along the drainpipe to the ground. Each of the balls was about 50 mm in diameter, had a glass luster and a thick double shell, see Fig. 2.8. Later, the balls became opaque. The thickness of each layer of the double shell was about 1 mm, and stood out against the general background in a gray-bluish color. The tandem of balls moved silently, and silently disappeared, disappearing into the ground at the end of the drainpipe. The whole phenomenon lasted about 30 s. The total length of the path of the balls was 18–20 m.

Comment. Tandem of balls having a double shell.

8.2 The observation took place in the same place in 1954 in the summer, in July. From 15 to 16.00 local time, there was a strong thunderstorm. Linear lightning hit a step-down transformer located on a pole. Smashed the transformer housing and dropped it to the ground. The village left without light. At 17.00 o'clock,

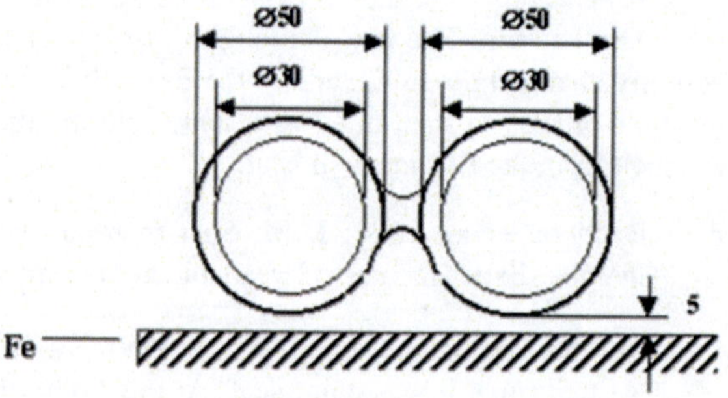

Fig. 2.8 Tandem of BL (dimensions, mm)

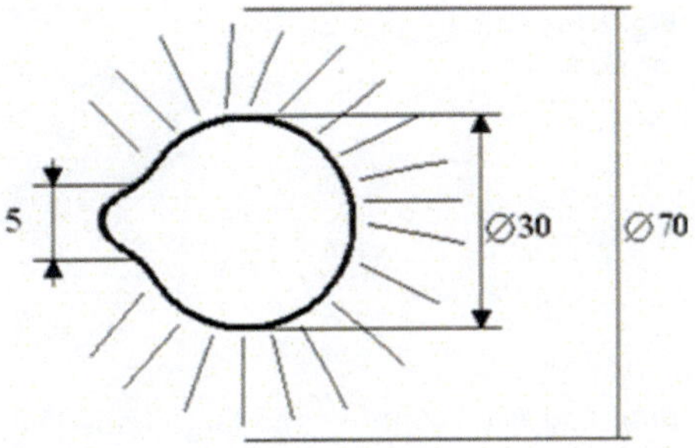

Fig. 2.9 Pear-shaped ball lightning (dimensions in mm)

I went to see what had happened. At the scene, I saw a twisted and slightly melted transformer. I was struck by the fact that its windings seemed to be unwound in layers and twisted in a completely chaotic way. In some places, there were breaks of individual conductors. There was steam in the air.

Suddenly there was a hissing and a faint crackling in the transformer, and a small "ball" with a diameter of about 30 mm, uneven in color and pear-shaped, floated out of the open casing with tangled windings, see Fig. 2.9. The bulk of the BL had a thick blue color, thinning-red with a bluish, and transparent shell. A soft white glow in the form of arrows spread around the ball lightning.

The BL hung for about 10 s over the broken transformer, and then, gaining a height of about ~3 m, flew into the thick of the spruce forest, skirting the trees.

Comment. BL appeared from the broken transformer. A small "ball" with a diameter of about 30 mm, uneven in color and pear-shaped, floated out of the open casing with tangled windings.

8.3 This incident occurred in 1956 on Nikolina Gora, in the summer at the beginning of July. It was evening. A thunderstorm was coming from the same meadow. A thick fog hung over the river and the shore. It was warm. There was no wind. I was walking home, it was dark, and the shore was illuminated by the flashes of an approaching thunderstorm. I had a pocket flashlight that gave a bright beam of light. Lighting up the place in front of me, I suddenly saw two green lights, very similar to the eyes of a cat. Only this "cat" was moving along the path in a strange way, sideways, climbing up the path of the steep bank (over an arc of 15 m), and moving down with the same speed. When I got closer, I saw two interlocked transparent balls, the surface of which shone with a glassy sheen, reflecting the green part of the spectrum. These balls moved evenly over a dry sandy path going up a steep slope to a pine forest. Suddenly, about 3 m from the path, I noticed a third transparent ball of much larger size. If two interlocked balls had a diameter of ~100 mm, then the ball to the side had a diameter of ~500 mm and was gray-transparent. The grass was visible through it, as through a slightly smoky glass.

The light reflected from a kind of glass surface was white. The ball lay motionless on the grass, having crushed it. I took a long dry pine stick, strong enough, ~1.5 m

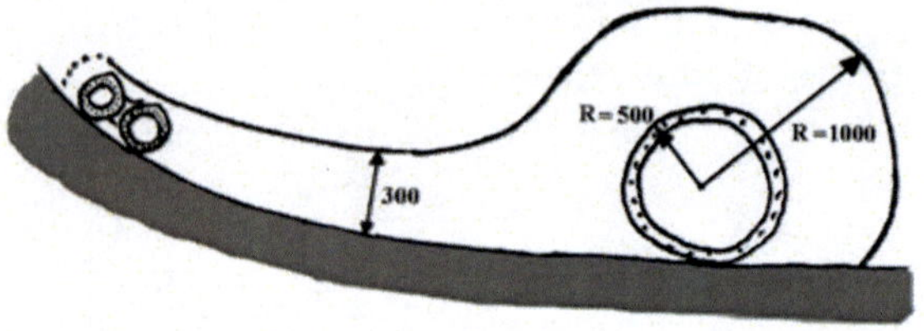

Fig. 2.10 Ball lightning on the slope

long and tried to move the large ball. The feeling was that the stick rested on hard rubber. At first I pressed lightly, trying to push the ball out of place, when it failed, I applied all my strength. The ball stood as if filled with mercury. Then I took off my jacket, along with keys, small coins in pockets, etc., it weighed about ~2 kg. I tried to shield the big ball from the small ones by throwing a jacket over it. However, the action did not bring the expected success, because the jacket straightened out and hung almost flat pancake over the ball at a height of ~120–150 mm. All attempts to move the jacket by a stick to the side were unsuccessful, but the jacket rose up easily. At that I noticed that the storm was already close. By removing the jacket from the ball by the stick, it was possible to very roughly determine the radius of action of this field. It represented a "bird's beak," at the end of which there were two interlocked balls, see Fig. 2.10.

The storm was already very close. Suddenly, the inner space of the big ball lit up with purple light, and the small ones with blue. Then the whole trio solemnly rose into the air and, growing brighter and gaining altitude, sailed towards the impending thunderstorm.

Comment. Let's consider the questions about the parameters of a large BL and leave aside the question about the parameters of two small ones, since there is not enough information for them to make estimates. At the same time, we will make the simplest estimates, realizing that due to the limited observation data, it makes no sense to make detailed calculations. Take into account that after the interaction of the big BL with the observer's jacket, the latter hung over the BL. The simplest explanation of this phenomenon can be obtained by assuming that the BL has a unipolar charge. Then it is clear that when the jacket came into contact with the BL, part of the charge passed from the BL to the inner surface of the jacket. After that, there was a repulsion of the charges on the jacket and the BL with the same sign. During the interaction, the jacket did not catch fire or melt, i.e. the BL was low temperature.

Now one can estimate the charge of the BL. We introduce the surface density of the q_s charges located on the jacket and the BL. We assume that the main charge is located on their surface, with equal charge density. This is the second simplification of the task. Now consider the jacket as some effective ball, with a charge $q_c = q_s \cdot L^2$. L is the linear dimensions of the jacket, for estimates we take $L = 1$ m. We accept its mass $M_c = 2\,kg$. The distance between the centers of the BL and the effective ball accept $R_{bl} + d$, where $R_{bl} = 0.5$ m, $d = 0.1$ m. Then we can get an estimate for the surface charge density on the BL and the effective ball under the conditions of the jacket hovering over the BL:

$$M_c \cdot g = \frac{q_s \cdot L^2 \cdot q_s \cdot 4\pi \cdot R_{bl}^2}{4\pi \cdot (R_{bl} + d)^2 \cdot \varepsilon \cdot \varepsilon_0}.$$

Here L^2 is the area of the charged surface of the jacket, $\varepsilon \cdot \varepsilon_0$—the dielectric permittivity of the air. Substituting the known values into the equation, we obtain an estimate for the surface charge density of q_s ~ 1.6 · 10^{-6} C/m^2. The charge of the BL turns out to be $q_{bl} \approx 5 \cdot 10^{-5}, C/m^2$, and the electric field on the surface is E_s ~ 1.8 · 10^6 V/m. Considering that the breakdown field in the air is E ~ 3.0 × 10^6 V/m, it becomes clear that there are practically no discharges on the surface of the BL, therefore the surface is translucent and cold.

Let us estimate the mass of such a BL. Note that when the cloud approached, the BL flew towards it. Consequently, the charge of the BL is opposite in sign to the charge of the cloud, which is realized after a linear lightning strike, when the negative charge flows down to the ground, and the cloud remains positively charged. In stationary conditions, when the force of the cloud on the BL is balanced by its weight, we obtain the following equation: $M_c \cdot g = q_{bl} \cdot E_{cl}$, where E_{cl} is the electric field strength under the cloud, the average value of which according to research is $E_{cl} \approx 3.5 \cdot 10^3$ V/m. Then from this equation we get the proper mass of the BL, which turns out to be of the order $M_c \cdot \approx 2.5 \cdot 10^{-2} kg$, i.e. quite light.

Then the question arises why the observer could not raise it. For the answer, let us recall the power of the image, which acts on the BL:

$$F_{pol} = \frac{q_{bl}^2}{16\pi \cdot (x)^2 \cdot \varepsilon \cdot \varepsilon_0}.$$

Here x is the distance from the center of the BL to the surface of the Earth, q_{bl}—the charge of the BL. According to the description, the BL was on the surface of the Earth, therefore $x = R_{bl}$. Substituting the known values into the equation, we get that the force of attraction turns out to be about 200 N, i.e. it effectively corresponded to an object with a weight of ~20 kg. This explains the difficulty with lifting the BL.

The conclusions of the assumption of the unipolarity of the BL and a number of simplifications allowed us to explain the observational characteristics of the BL and its behavior.

9. (Bychkov et al. 2004) V. S. Shcherbak, (male) Power Engineer, born in 1946, survey in 2003, observation 07.1989, 14.00 local time. Shakhtersk, village. Oleksiy Orlovka, in the main lift room (a machine lifting coal from a mine located on the surface of the earth).

There was a very strong thunderstorm. At the moment of the linear lightning strike, a strong sound was heard, the high-voltage input switches (6 kV) were switched off with the engine running (300–400 kW, asynchronous with a co-phase rotor). At this time, I was looking at the rotating drum of the engine. Simultaneously with a lightning strike, a ball appeared 30–35 cm from the born head (battery outlet), then it disappeared. The engine bore included three cable cores, see Figs. 2.11 and 2.12.

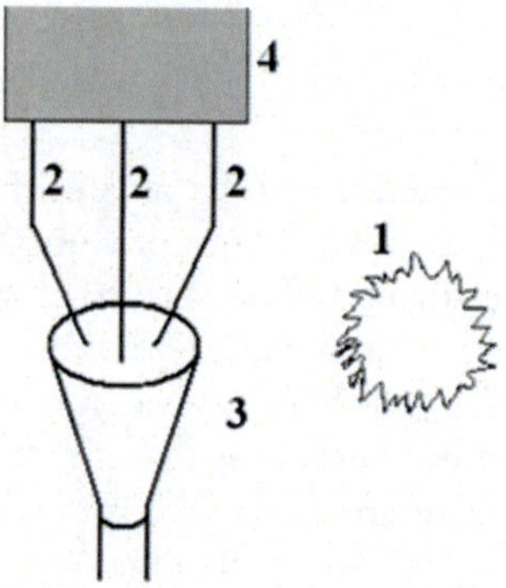

Fig. 2.11 The scheme of the appearance of the object (side view): (1) ball, (2) cables, (3) coupling, born head

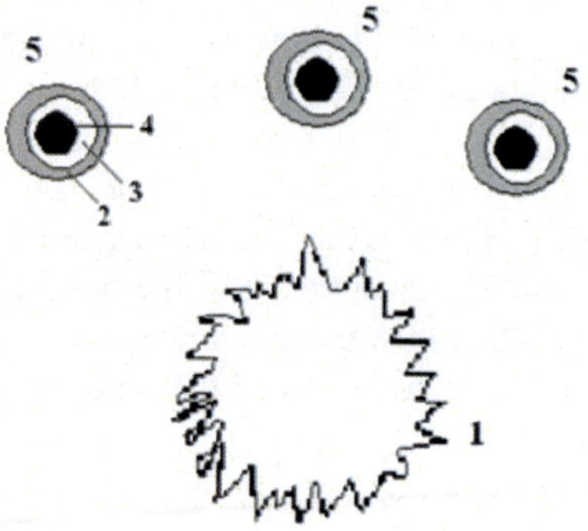

Fig. 2.12 The scheme of the appearance of the object (top view): (1) ball, cable structure: (2) cambrick, (3) PVC insulation, (4) copper wire, (5) cables

They were coming out of a coupling filled with resin. The distance between the veins was ~15 cm, but they were not in the same plane, see Fig. 2.12.

Suddenly, a ball appeared in the area near the veins, see Figs. 2.12 and 2.13. The distance from me to the ball was 3 m. The ball was the size of a table tennis ball, 4–5 cm in diameter. It was round, and it was boiling all over. On the surface there were

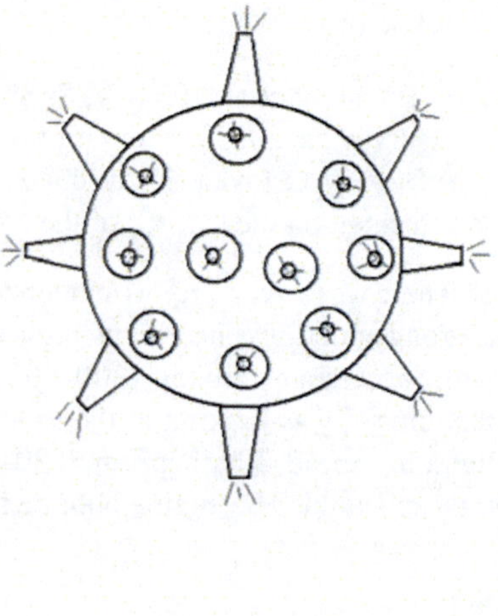

Fig. 2.13 The scheme of the "boiling" ball

protrusions of "volcanoes," see Fig. 2.13, height ~ 1 cm. The top of the "volcano" was, as it was torn, it was boiling, sparks were coming from it, as if the material was flying away. The inside of the ball was indistinguishable. The color of the glow was blue with a yellowish tinge. In comparison with an incandescent bulb, its power was 100 W. Without changing the brightness, it went out.

Nothing happened to the engine. There was no soot or burnt PVC insulation.

Comment. Pay attention: The appearance of the ball. It was round, and it was boiling all over. On the surface there were protrusions of "volcanoes" with a height of ~1 cm. The top of the "volcano" was, as it was, torn, boiling, sparks were coming out of it, as if the material was flying away.

10. (Bychkov and Bychkov 2008) Z. L. Shulimanova (female), Physicist, DS, 53 y. o. at the time of the survey in 2006. 13 y. o. at the time of observation, Kargapolye Station, Kargapolsky District, Kurgan region, May, 14–15.00 local time. The day is cloudy, warm, there is no rain.

"I came home from school. I was standing at the window of our one-story house, looking out into the courtyard. It suddenly darkened quickly and became dark as in the evening, but there was no rain. Suddenly I saw a glowing (fiery) ball descending into the yard and slowly floating to the window and almost flew into it. I was scared and hid under the table. After a while, I looked out the window. The ball was already floating above the ground. Burst with a strong crack. At this point, a white glow appeared larger than the size of the ball."

The color of the ball was white-yellow, like ivory. The diameter of the ball was 5–6 cm. The observation time is difficult to estimate. The shortest distance to the ball was 50 cm. The maximum distance to the ball was 8 m. The surface of the ball resembled a hedgehog, needles emanated from it, the ball was fluffy.

Comment. Pay attention. The surface of the ball resembled a hedgehog, needles came out of it, the ball was fluffy.

11. (Bychkov and Bychkov 2008) A. N. Kabanov (male), Chemist, DS. 9 y.o. during the observation in 1954. The Kazakh Republic of Chymkent. Observation in the afternoon after 12.00 local time, survey 2000.

"There was a strong thunderstorm; then the sun came out. I was in the garden, where the planting is low. There was a tree on the right, an explosion sounded above it. The explosion was sharp. I raised my head and saw at the level of 6 m a clot of orange-yellow color 30–40 cm in diameter, from which tentacles like an octopus with a length of 50–60 cm emanated. It disintegrated in a split second.

Comment. Pay attention. A clot of orange-yellow color 30–40 cm in diameter, from which tentacles like an octopus with a length of 50–60 cm emanated.

12. (Bychkov and Bychkov 2008) T. D. Zolotareva (female), Teacher, born in 1914. Observation of 1931, survey of 1995, July 15–16. Cloudy day, 11–12.00 local time. 4 observers.

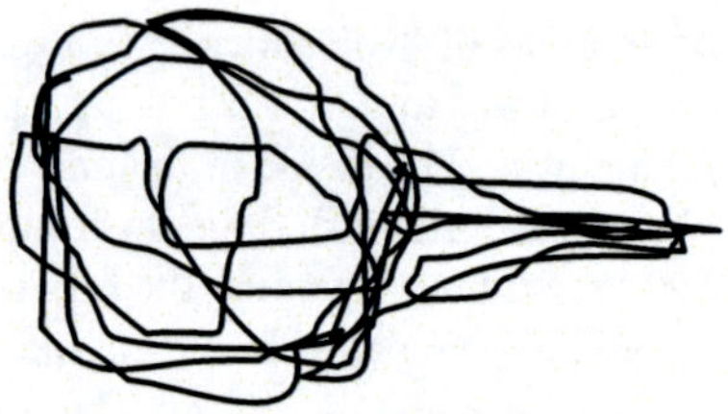

Fig. 2.14 Diagram of the ball surface

"It was sunny; a black cloud appeared in the middle of the day. There were 4 women and a child in the house. The child opened the door to the vestibule, where there was a dormer window. In the room in the window, the corner of the glass was broken off. At 12.30 a.m., a ball entered the window through the corner of the glass, went around the table, walked along the wall very slowly, went out the door into the hallway and went out through the dormer window. The surface of the ball, oscillating, vibrated like a soap bubble. The ball had an even volume of bright-yellow color. The intensity of the glow corresponded to an incandescent lamp of 100 W. The ball had a bright-yellow, lemon tail. The nearest distance to people was 1 m. I was scared, but the ball did not hit anyone. As soon as the ball came out, a thunderstorm began." See Fig. 2.14. The diameter of the ball is 8–12 cm. It walked 5 m. The observation time is 40–50 s.

Comment. Pay attention. The ball had a bright-yellow, lemon tail. Entered the window through the corner of the glass, exited through the dormer window.

13. (Bychkov and Bychkov 2008) http://seh.ru/shm 2007-02-15. Valery (male), 14.08.1981, Bratsk. Evening, thunderstorm, about 23.00 local time.

"A thunderstorm has started. I was lying on the couch in my room. Suddenly, orange reflections appeared on the walls of the room—as if a bonfire had been lit outside the window. I thought so for a few seconds. Suddenly it dawned on me—what kind of fire in such a downpour, and I live on the ninth floor? I jumped up from the sofa and went to the window. Outside the window, about half a meter from the frame, an orange ball was slowly rising. Its diameter was about 15 cm. The ball had a smooth, shiny surface; the blue halo around the ball was about 10 cm thick. The ball hissed and pulsed with a frequency of 2–3 Hz. Fountains of sparks were ejected from its surface. I quickly slammed the window shut. The ball rose up and went out of sight. There was a hiss for a few more seconds; then everything went quiet."

Comment. The ball had a smooth, shiny surface; the blue halo around the ball was about 10 cm thick. The ball hissed and pulsed with a frequency of 2–3 Hz. Fountains of sparks were ejected from its surface.

14. (Bychkov and Bychkov 2008) NLO No. 5 (52) June 1998. "BL bounces." Vladimir Sorokin (male). 6.07.1983 Altai Territory, near the village of Togul. 2 observers. Outside, by the river.

"... My son and I stopped at the confluence of two rivers-Chumysh and Uksunai. The day was very hot, and in the evening, a severe thunderstorm broke out. Lightning often flashed in the evening sky, and there was a continuous rumble in the air from the thunderclaps. …By midnight, the storm moved east, lightning flashes became less frequent. The rain stopped.

On the east side, about two hundred meters away, I saw a bright matte ball the size of a car headlight. …The ball was about two meters from the ground … the ball was moving against a quiet wind with the speed of a calmly walking person in my direction.

… Here I noticed an amazing feature: the ball and the outgoing thunderstorm were on the same line of my observation. …With each flash of lightning, the ball was thrown up by about 20 cm. Moreover, it flew higher from near discharges than from distant ones.

The ball approached me at a distance of 20 m and hovered over a small lowland overgrown with small willow bushes. It froze and slowly began to stretch out, taking the shape of a disk with a diameter of 30 cm. After a fraction of a second, the disk again took the shape of a ball, then again the shape of a disk. It burst without a sound, the glow disappeared- only the air in this place glowed with a dim blue light for several seconds."

Comment. At each flash of lightning, the ball was thrown up by about 20 cm. Moreover, it flew higher from near discharges than from distant ones. The ball froze and slowly began to stretch out, taking the shape of a disk with a diameter of 30 cm. After a fraction of a second, the disk again took the shape of a ball, then again the shape of a disk.

15. (Bychkov and Vaulin 2014) "Evening Novosibirsk." Release date: 04.08.2010 17:15. Title: A BL flew over Kargat last Monday, a hail shower hit Kargat, and soon local residents noticed a BL in the air.

An unusual phenomenon hovered over private houses located in the Northern Lane. According to eyewitnesses, the lightning was large, black, and a bright red train was trailing behind it. It was flying towards the house and hit the TV antenna. The refrigerator, TV, and computer were not damaged. Only the stereo system connected to the network burned down.

Comment. Pay attention. The BL was big, black, and a bright red train trailed behind it.

16. (Bychkov and Vaulin 2014) V. M. Petrov (male) Physicist, Ph.D. Interview May 2012. The place of observation is Moscow. The premises of the MOSENERGO SKTB VKT laboratory. July 2000. 12.15 local time. Three more observers all Ph.D.

"It was a hot day. A thunderstorm with a strong wind began, the downpour stood like a wall. Lightning flashed continuously. The rainwater was blowing under the roof.

I was looking at the wall opposite the windows. I stopped looking at a crack in the plaster on the wall near the ceiling. I was surprised that the crack began to get wet and expand. I drew the attention of my colleagues to it.

At this time, a red-pink bubble appeared in the crack and began to inflate like an air ball. When it got out, it was flat, and when it got out and reached a diameter of 15 cm, it became perfectly round, and burst.

Then several bubbles appeared in the crack, they grew and absorbed each other, and turned into one ball 10–25 cm in diameter, which burst.

The balls sizzled when they appeared, and burst without a sound.

The color is reddish-pink. It is as if the inside is empty, and only the shell glows. The glow is weak ~15 W incandescent lamp.

This happened 3–4 times and 3–5 bubbles appeared in the crack. The time of their existence from the origin to the explosion was about 10 s. Then everything was gone. There were no smells, no traces left. Only a wet spot remained.

Comment. Pay attention. A red-pink bubble appeared in the crack and began to inflate like an air ball. When it got out, it was flat, and when it reached a diameter of 15 cm, it became perfectly round, and burst. Then several bubbles appeared in the crack, they grew and absorbed each other, and turned into one ball 10–25 cm in diameter, which burst.

17. (Bychkov and Vaulin 2014) Oleg (male) and Natalia (female) Zubtsov. 2013. June. City Voronezh. Observers: he is a Police Officer of the linear Department of Internal Affairs on railway transport, she is a Primary School Teacher at school No. 33. Observation at a dacha 25 km from the city near the village of Platonovo.

At the weekend, the family was at the dacha and because of the incessant rain was going to return home. There is no electricity in the village (it was switched off). The observer reached for his raincoat and remembers nothing further. "I came to my senses on earth." His wife was bothering him. " I didn't have the strength to get up: my legs refused." Leaning on his wife, he got up and went into the house."

Natalia: "It was getting late, and it seemed that the rain had stopped. A double rainbow took up half of the sky. Then the clouds thickened, the husband was going to go outside for shovels. Mom was standing in the doorway of the inner room.

Suddenly, at a distance of about a meter from me and half a meter from my husband, a white smoke column rose directly from the floor, and on its top a jelly-like ball, like a living being, began to form, shimmering with light. It was absolutely quiet, the ball suddenly began to crackle in the way electricity usually crackles when a short circuit occurs.

Involuntarily, I began to squat, and then there was a pop, and something hit me hard in the head. I had time to think: 'Did the light bulb explode?'

When I sat up, there was a ringing in my ears, and the whole space, vibrating, floated, as if it had lost its rigidity. A moment later, I saw Oleg's legs convulsively beating, and the rest of the body on the street: threw him out. I ran to him. I cannot say he was unconscious: his eyes were rolling, blood was flowing from his nose, but he did not understand anything. Then he came to himself."

Natalia's mother told what she saw. "The ball turned red and exploded. Some kind of tentacle or sleeve from him stretched out to the metal clip on the woman's head, and the main discharge hit Oleg's chest, just where the silver cross was. He fell to the ground, and the discharge went out through his ankle, leaving two dark marks there. Lightning-like zigzags were visible on Oleg's chest and back for almost a week after the incident."

The family got into the Lada car, and the man safely drove the car home. When Oleg returned, he went to Semashko Hospital. Electrocardiograms were immediately made to both victims. Oleg was placed in intensive care. A week has passed. And the patient was discharged from the hospital in order for him to go to the hospital in a departmental hospital.

Comment. Pay attention. A white column of smoke rose straight from the floor, and on top of it a jelly-like ball, like a living being, began to form, shimmering with light. The ball turned red and exploded. Some kind of tentacle or sleeve from him stretched out to the metal clip on the woman's head, and the main discharge hit Oleg's chest, just where the silver cross was. He fell to the ground, and the discharge went out through his ankle, leaving two dark marks there.

18. (Bychkov and Vaulin 2014) Ukrinform 21:0714.08.2013. The small village of the Simferopol District of the Crimea during a thunderstorm. BL were flying along the village street.

It was already dark; it was raining. Lightning flashed. Suddenly, in the yard to Lilia Belyalova (female) "landed," scattering a sheaf of sparks, BL. It fell on the grapes, which caused the leaves of the plant to immediately wither. The "flame" engulfed the doorway of the barn. "It's like a sack of fire was thrown," Lilia Belyalova said. The woman was sitting in the kitchen at that time. The window was open, and a black ball the size of a large fly suddenly flew in. According to Lilia Belyalova, the lightning did not sparkle, but it made a buzzing sound. The mosquito net, as it turned out, does not stop the ball lightning at all: the ball entered the house, damaging the wire, and flew to the chandelier. It flared up and the lights went out in the house immediately. The black ball flew straight to the outlet behind the TV. Lightning blew apart the plug: plastic parts and wires flew in different directions. The TV had to be repaired. This was not the end of the lightning's journey through the house: from the outlet, it rushed up to the antenna. But instead, it crashed into the wall and left a deep dent there.

Comment. The window was open, and a black ball the size of a large fly suddenly flew in. The BL didn't sparkle, but it made a kind of buzzing sound. The mosquito net does not stop the BL: the ball entered the house, damaging the wire, and flew to the chandelier.

19. (Bychkov and Vaulin 2014) http://www.paralljelnyj-mir.com/products/tri-vstrechi-s-sharovymi-molniyami-/Stavropol region. Russia. August 1955. It was 30 °C hot that day. By 5 o'clock in the evening, the sky began to tighten with haze. An hour later, it was noticeably fresher, and the top of the 990-m Snake Mountain, which was less than four kilometers from our house, was covered with a huge black and blue cloud. The first thunderclaps were heard, repeatedly amplified by the mountain echo. It started to rain.

The house where I lived was adobe, with a high tiled roof and rather dilapidated. After one particularly powerful thunderclap, I got up from the table and went into the next room. Going to the window, covered with a shutter, I felt the smell of dampness coming from the windowsill. Suddenly, a deafening crash was right behind the shutter, followed immediately by a terrible thunderclap.

I involuntarily shuddered, but then a small blue ball that suddenly appeared in the upper part of the window frame attracted my attention. The unknown object was no more than half a meter away. Watching the ball, I noticed that it is quite beautiful: it is no bigger than a tennis ball, painted in a bluish-blue color, while its middle was rather bluish-pink. Meanwhile, the balloon, crackling softly, slowly descended down to the windowsill. Suddenly, following the first one, three more balls appeared in the same place, but they were no bigger than a walnut. Following each other, the balls floated in the air down to the barely noticeable gap between the window frame and the windowsill and disappeared into it.

No more than five seconds after the balls disappeared, an explosion of such force was heard from the outside of the window that the glass in the frame shattered, and the shutter was torn off the metal bar. Immediately everything went quiet, as if there was no thunderstorm.

The blue balls that I saw next to me burned all the electrical wiring going from the house to the main pole. An electric cable about 10 m long seemed to have evaporated.

Comment. Pay attention. Following the first one, three more balls appeared in the same place, but they were no bigger than a walnut. Following each other, the balls floated in the air down to the barely noticeable gap between the window frame and the windowsill and disappeared into it. The balls burned all the electrical wiring going from the house to the main pole. An electric cable about 10 m long seemed to have evaporated.

20. (Bychkov and Vaulin 2014) Internet. 10.07.2011 in the Czech city of Liberec, the BL appeared in the dispatch building of the city emergency services. A ball with a two-meter tail jumped to the ceiling directly from the window, fell to the floor, jumped to the ceiling again, flew 2–3 m, and then fell to the floor and disappeared. This scared the employees, who smelled burnt wiring and thought that a fire had started. All the computers were frozen (but not broken), the communication equipment was out of order overnight until it was repaired. In addition, one monitor was destroyed.

Comment. Pay attention. A ball with a two-meter tail jumped to the ceiling directly from the window, fell to the floor, and jumped to the ceiling again.

21. (Bychkov and Vaulin 2014) O. S. Surkont, male, physicist, 34 years old interview 21.03.2012. The age at observation is 10 y.o. The daytime is the middle of the day. Location—Moscow region, Barybino Station. The distance to the object was 3 m. The object was 4 m above the ground. The object was floating slowly. The color was gray, cloudy-transparent, and similar to a soap bubble. Diameter 5–10 cm. Lifetime: more than 30 s. The object has gone out of sight.

It was in the summer when I was at a pioneer camp. We were walking with Mom. It was raining a little (not a downpour). From afar, we noticed a flying ball that looked like a soap bubble. We came up to it by almost 3 m. It behaved like a soap bubble that was slowly moving at the same height, about three meters above the ground. (I wanted to find a stick and touch it, so I remember the distance.) Mom took me home. I do not remember if there was a thunderstorm. It was amazing that the soap bubble floats during the rain, does not fall, and does not burst.

The ball appeared in the middle of the road. On the one hand, there was a house at a distance of 10 m from the ball; on the other side, there was a tree at a distance of 5 m from the ball.

Comment. The ball like a soap bubble floats during the rain and does not fall, and does not burst.

22. (Bychkov and Vaulin 2014) O. A. Mizenina, Ph.D. in biology, born 1951. Interview 9.11.14. Observation 20.00 local time. 08. 2014. New York, Rego Rank, house on the mountain, the last 15th floor. There was no rain or thunderstorm. "The door to the loggia is closed. I am standing in the room, going to take a shower. I do not know what made me turn around. A huge white cloud, in color-ignited magnesium-piercingly white, slowly floats through the window opening into the loggia. This glow is 3–4 m in length, and more than 2.5 m in height (the view was limited by the size of the loggia). The glow is dense and opaque.

Then there is a huge roar on the roof. There was the sound of falling bricks. On the roof, there is a 3-storey water tank, like a water tower, lined with bricks. The house shook; it seemed to me that the dishes rattled. A hole appeared on the roof (see photo) measuring 4 × 3 m. There were many fallen bricks, having fallen they blocked the entrance to the underground garage.

The object is missing. The duration of the observation is 1.5 s. Nothing has gone wrong in the apartment." (Fig. 2.15).

It is known from the local newspaper that the BL was seen from 2 metro stops located 3 km away. There was also a bang.

Comment. Pay attention. A white cloud, in color-ignited magnesium-piercingly white, slowly floats through the window opening into the loggia. This glow is 3–4 m

Fig. 2.15 The photo shows a picture of the destruction of the house roof

in length, and more than 2.5 m in height (the size of the loggia limited the view). The glow is dense and opaque. It is assumed that the ball destroyed a piece of the roof.

23. (Bychkov and Vaulin 2014) V. A. Chizhov, male, Ph.D., Metallophysicist, 26 y.o. during the observation. Interview March 2015. Event in late May–early June 1980. The place of the incident is Podolsk M/O, Pleshcheyevskaya str., 17, second floor of a three-storeyed brick house.

"At night, at about 12 at night or at 1.00 o'clock, I was doing my diploma in the bathroom, which used to be a kitchen and had a chimney. There is a 25 × 25 cm grid in the ceiling with 1.5–2 cm holes, about 8–10 holes. The weather was hot and windless.

Suddenly, in the corner where this grate is embedded on the ceiling, I heard a crackling (buzzing). Turning sharply, I saw a glowing ball of bright red-yellow color, perfectly even shape, without halos and cores, with a diameter of about 7 cm, buzzing and crackling, appear from the hole. The ball from the pipe (grate) began to descend smoothly into the bathroom (cast iron with white enamel). The bathroom was empty without water and dry. The speed of the ball lowering was about 1.5–2 m/s. The ceilings in the apartment are 3.2 m high. Having reached half the height of the tub, the ball made zigzag movements along the tub, going about 2/3 of the length of the tub, as if it was looking for something. Then it rose about 1 m, and stopped in front of me. I froze, not knowing what to do. The ball about 2 s stood in front of me at chest level, buzzing and crackling, at a distance of 25–35 cm, as if it looked at me, then, without changing color, at the same distance from the floor, it swam towards the sink, i.e. further away from the chimney from where it came, along an arc-shaped trajectory from me. The crackling of the ball intensified, and it began to shrink. Burning. Then it seemed to me that the burning with sparks and crackling intensified, and the burning process was going inside the ball to the center. Then there was a very strong and very sharp clap, like the clap of a rattling mixture in a chemistry lesson, but much stronger.

I did not feel any heat throughout the observation of the ball. There was also no blinding effect of the brightness of the ball. I cannot say exactly when I crouched down from fear, when the ball began to shrink, or when a strong short pop thundered. I estimate the whole process observed by me from the appearance of the ball to the explosion by 10–15 s. I did not notice any foreign odors or any other secretions.

Surprisingly, the clap was very strong and short in time, but none of the sleeping household woke up."

Comment. Pay attention. The ball entered the bathroom through a grate in the wall. The ball began to shrink, and then exploded.

24. (Bychkov and Vaulin 2014) Niclaus Pfirsig. I was a witness of the BL. I think I was 11 or 12 y. o. I visited my grandparents who lived in an old farmhouse on the edge of the Barrens Plateau, near Beach Grove, Tennessee.

The day was very hot and humid; it started to rain a little with thunder. My grandfather, uncle, and I were sitting on the front porch, enjoying the cold mist of rainwater raised by the wind under the porch. Suddenly the rain stopped and the sun came out. Then lightning struck a tree that was about 25 feet away from the house. Not only did I hear the thunder, but also I felt it hit me like a giant hot pillow. Hearing the ringing in my ears, I quickly looked at my grandfather and uncle and saw them looking at something. I looked in this direction and saw a glowing ball, the size of a soft sick. The ball floated towards the house, quickly passing between my uncle and me. It walked through the door screen into the house, leaving a round hole on the screen that was smaller than a ball. By the time I turned around to look into the house, the ball had already reached the kitchen door and seemed to be heading for the massive cast-iron stove. The phone mounted on the wall next to the front door, it rang, and the ball rushed to the phone, which stopped ringing. The ball immediately headed back to the kitchen, but at a slower speed. When it got to the middle of the room, it disappeared. The whole event lasted a few seconds. A few days later, the master came out to fix the phone. He said the phone melted inside and he replaced it. In my case, there were three witnesses to this event, each with their own point of view. All three of them saw the ball enter the house. My grandfather also saw it touch the phone.

Comment. Pay attention. The ball passed through the door screen. Passing by the phone, it melted everything inside it.

25. (Bychkov and Vaulin 2014) Portal of Natural Sciences. Forum. 10.1.2014, 20:36. The observer is a man. "1955, I am 6 y.o. Dad and mom worked in a field geophysical party somewhere on Mangyshlak or on the Ustyurt Plateau. They lived in dugouts. One morning, for some reason, they did not let me out to the street. No one paid attention to me, and I decided to see—what kind of "storm" is this? I turned away the flannelette blanket that covered the entrance, pushed the plank door…

The picture that opened imprinted in my memory—the whistling of the wind, the burning muddy shroud, gray-yellow jets of sand.

There were many colored balls jumping up and down in this veil! A variety of sizes and colors—such bright and beautiful I have not seen before. Against the background of the monotonous howling of the wind, everyone buzzed and crackled in a special way…The first breath filled my mouth with dust, then someone pulled on the back of my shirt, the door slammed shut with a bang, and my mother dragged me down the stairs. 'No, Mom, there are a lot of balls there, I saw them. Let us go get one and come right back'. In the end, Mom gave in, we climbed the earthen steps again; she opened the door. The wind was still whistling, jets of sand were flying… However, there were no 'balls' anymore. Let me remind you that when Mom and I looked out again after 5–10 min, it was already over.

Then, I was not very surprised by another oddity-despite the heavy wind, the 'balls' were, as it were, tied each to its place. That is, they jumped up, floated in the air, slowly descended, but they hardly reacted to the wind. I can add (this is without certainty) that their color seemed to depend on the size-the small ones glowed blue or greenish, the larger ones yellow, the largest ones orange, red. One, purple, even had a kind of brown stripe along the equator, and an almost black spot on the side."

Comment. Observation of many jumping up and down colored balls of various sizes and colors.

26. (Bychkov and Vaulin 2014) yaba. 18.05.2010 17:44:16. I once saw this blue ball in a field about 100 m away from me. In general, the wind was strong, and this trouble famously carried around the field. Its size was visually at least half a meter. However, the most interesting thing is that, in the end, it did not explode, but disintegrated into several small balls, which then disappeared.

Comment. BL disintegrated into several small balls.

27. (Bychkov and Vaulin 2014) 14.02. 2014. TORRO. SM Newquay, Cornwall.

During a thunderstorm with linear lightning and wind gusts up to 34 knots, "two very large stationary pale blue lightning balls were spotted southwest of Newquay Airport. They existed simultaneously for about a second and seemed to extend at least ten meters in radius (only the upper half of the balls are visible). The phenomena seemed to have a brighter surface compared to the ball itself, like a jellyfish or a mushroom cloud. They appeared and disappeared instantly and simultaneously with a local power outage and power outages.

Comment. Observation of two BL with a radius of 10 m.

28. (Bychkov et al. 2015a, b) http://ufo-online.ru/file_3384.html. In July 1980, in Frunze, BL "visited" the local historical museum. Around midnight, the janitor noticed a blindingly bright light in the utility room. Looking into the control room, she saw a ball slowly floating through the open window, looking like a ball of yarn fluffed up on the floor. It approached the devices, and a strong crackling was heard in them, then, describing a smooth semi-arc, it slipped out into the street.

Comment. BL looks like a half-fluffed ball of yarn.

29. (Bychkov et al. 2015a, b) M. Chikalina, female, Novovyatsk. RF. 9.08.2016.

In Novovyatsk, the BL "burned" household appliances in almost 30 apartments. A thunderstorm thundered over Novovyatsk on the afternoon of August 3. As the resident of the house No. 26 on Kirova Street R. Koposova (female) told, having heard thunder, she and her granddaughter turned off all household appliances. Then the woman went out on the balcony and saw a BL with a diameter of about a meter flying from the Vyatka (city) side at about the 5th floor level to the house.

R. Koposova: I myself am from Siberia, in my childhood I often saw such BL, but they were all small. This one was about a meter in diameter.

Before reaching 30 m to the house, the BL exploded, scattering into small sparks. Then it began to rain with hail and squally wind. After a thunderstorm, it turned out that due to a BL strike in two entrances of house No. 26, the intercom broke down, and all household and computer equipment burned out in the apartments.

Comment. The BL was about a meter in diameter.

30. (Bychkov et al. 2003) T. S. Sychevskaya, female, Teacher of preschool. Born 1954 and observed in 1971. June (summer), Western Dvina. 20.00 local time. An hour after a thunderstorm in the open air near a railway. Survey of 1998. The number of observers is five people.

My friends and I were going to a dance. It was very fresh outside after the storm; there was no wind. We were walking through the Moscow-Riga railway. As soon as we crossed the rails, we suddenly saw a round BL with a diameter of 70–80 cm above our heads. It appeared as if "out of nothing." We were scared, squatted down, and held hands, forming a circle. Suddenly the BL went above us in a circle up and down. Then it "chose" my head and started jumping up and down like a ball. Jumped more than 20 times. The BL felt cold. Then it rose and disappeared suddenly.

Inside it (1/3 of the diameter), the glow was brighter than outside. The BL was as soft as an air bubble; it did not even feel like weight. The border is like a soap bubble. It resembled a white-gray jelly. It glowed like a bright 200-W bulb. It was located at a distance of 0.5 m from the ground. After interacting with the BL, everyone vomited.

Comment. BL appeared as if "out of nothing." Round BL with a diameter of 70–80 cm. The BL felt cold. The BL was as soft as an air bubble; it did not even feel like weight.

31. (Bychkov et al. 2003) Observer L. G. Romanova female, Geneticist, Ph.D. Age 14 y.o. at the time of observation. Observation of 08.08.1954 Survey 2000. 59 y.o. at the time of the survey. The place of observation is the village of Taiga, 80 km from Krasnoyarsk on the bank of the Yenisei. The middle of the day, before a very strong thunderstorm.

"My mom and I were looking out the window. Suddenly, the window lit up. Then we noticed a hole in the window - the edge of the glass was broken off. It was white, bright, like a 100-W lamp; it was hard to look at. The door in the corridor suddenly opened, and BL was pulled into the corridor and out of sight." BL walked around the room for about 18 m, moving with uneven movements. The shortest distance from the observers is about 2 m. The observation time is about 20 s.

Comment. A ball the size of a tennis ball entered the room through the window. It was white, bright, like a 100-W lamp.

32. (Alexeff et al. 2014) I. Alexeff, Professor, Physicist, description 2005.

The BL appeared after the linear lightning strike, immediately after the lightning discharge. The event occurred when a street baseball game was taking place. A thunderstorm started, it started raining, but the observers lost the ball and began to look for it. A strong explosion came from above, and everyone looked up, where they saw a bright orange ball the size of a basketball, which was 30 feet above the ground. The intensity of the glow was not very strong, the ball was clearly visible, but it did not blind the eye. It existed for a few seconds. Waves were moving on the surface of the object (like a liquid), so it was not a visual afterimage. No sound or smell associated with this object was observed. It seemed that it was slowly descending.

Since the ball was over the observer, the latter ran away as fast as he could, and did not see how the object disappeared.

Comment. Waves were moving on the surface of the object (like a liquid), so it was not a visual afterimage.

33. Observer Yu. Nikonov, male, Engineer, 35 y.o. during the survey in 1989. 12 y.o. at the time of observation in 1966. Village, Tambov region. July.

"A thunderstorm with a squally wind was coming. I was walking along the river bank along a path in the bushes. I saw it. I followed it. It exploded–scattered a sheaf of sparks." Color of the BL is yellow-white, the brightness is 60–100 W compared to an incandescent lamp. The diameter of the BL is 20–30 cm. The observation time is 10–20 s. The shortest distance to the BL is approximately 20–30 m.

Comment. A ball in vegetation. It swam and clung to the foliage. At the same time, sparks fell every time.

34. BL in Budapest, Internet. On 01.2011, the Editorial Office of the Hungarian newspaper Időkep received a photo of BL taken by the couple of Popele near Budapest. Figure 2.18 shows an enlarged fragment of this figure. The photo was taken with a flash by Samsung's KENOX S760 digital camera. The focal length of the camera lens is 10 mm, the aperture is 3.4, and the exposure time is 1/50 s. The diameter of the shining sphere is defined as 50 cm. After 15 s, a second picture of this place was taken, in which the sphere was missing. The most amazing thing is that after the disappearance of the ball, the snow surface remained untouched (Fig. 2.16).

Comment. When discussing this event, the fact that there was no melting of snow in the place where the ball lay was noted. As well as the fact that a faint glow of air was observed to the left and right of the ball. As we already know, the glow of the air near the BL can be caused by a corona discharge. The BL was surrounded by a spherical halo, glowing near the sphere, lying on the snow, distributed asymmetrically: adjacent to the bottom of the ball. The photo shows that the border of the glowing area is located

Fig. 2.16 Photo of the alleged BL

at a distance of $R_h = 50$ cm from the center of the sphere. Based on this, we can estimate the magnitude of the charge of this BL: $Q_{bl} > 4\pi\varepsilon_0 E_{bra} R_h{}^2 = 8.3 \cdot 10^{-5}$ C. The inequality is obtained due to the fact that, in principle, the magnitude of the electric field strength on the surface of the BL can, in principle, be greater than the breakdown field of the atmosphere $E_{bra} = 3 \cdot 10^6$ V/m. We will discuss this issue later. The fact that during the time the BL was on the ground, the snow did not melt under it. (In the note of Cooray et al. 2011, there is no information about the lifetime of the BL.) This suggests that the charge drains from the BL and the current does not heat the snow, because the snow is a strong dielectric. At the same time, the heat from the inside of the BL, if the core is hot, does not pass out through the shell. This can occur in the case of absorption of infrared radiation by the shell and the plasma above it, or the heated interior of the BL surfaced inside the shell and did not transfer heat to it.

35. (Keul AG 2021) Wilhelm von Haidinger, member of the Imperial Academy of Sciences at Vienna, observed ball lightning from his home at Vienna-Landstrasse (3rd district) on October 20, 1868, in a heavy thunderstorm between 17:15 and 17:30 LT.

"The electric fireball was one of the very first discharges of the thunderstorm. It stood maybe 2–3 s in front of the right pane of the window. It is indeed simple to describe the ball, of a vivid yellow light, especially on the right while on the left clearly red, with longer rays than the ball, firing from it to the right and towards the earth, of red, yellow color, up to the most dazzling white. In addition, left there were shorter rays inclined obliquely towards the earth... Several discharges followed and after [them] hail and rain." Haidinger determined the object's angular diameter as 0.83°, which would give 50 cm at the assumed distance of 30–40 m. He described the "electric meteor" in a report to his Academy (Brand 1923, 34).

Comment: BL with longer rays than the ball, firing from it to the right and towards the earth.

2.3.2 Complex Structures

1. (Bychkov et al. 2003) G. V. Donskikh, male, a resident of the village of Blagodatnoye, Donetsk region, sent to M. T. Dmitriev. The duration of observation is 2–3 s.

In the summer, in June–July 1978, a very strong thunderstorm broke out in the afternoon, accompanied by rain and gusts of strong wind. I was sitting by the window and looking out into the courtyard. Suddenly I saw a moving ball in the garden, a ball of fire. It had a rather rounded shape, but its contours were indistinct, as if ragged. This ball, like a wheel, rolled over the ground, contacting its surface. Its diameter was up to 50 cm. The dense core had a much smaller size. The large size of this formation gave a halo of glow and shaggy spiral-shaped fiery bands surrounding the core. It seemed that a ragged ball of fire was rolling on the ground and could not condense in any way. Behind it stretched a fiery plume of fire and spark strips up to 80 cm long. The color of the ball and its sparkle was yellow-reddish. The speed of the ball was very high, faster than that of a motorcyclist rushing along the highway. After rolling through the garden, the yard of the house, the ball rolled out through the open gate to the street and disappeared from sight. At the same moment, a powerful explosion rumbled.

Comment. The diameter of the BL was up to 50 cm. The large size of this formation gave a halo of glow and shaggy spiral-shaped fiery bands surrounding the core.

2. (Bychkov et al. 2004) A. T. Belomestnov, male. O. I. Mitrofanov male, Physicist DS, Chita region, Borzinsky District, Sherlovaya Gora settlement-1. (1982–1983). 2 observers.

It was a sultry day, and clouds began to thicken by lunchtime. From the western side, from behind the mountain, thunder rumbled. A cloud, swirling, suddenly rushed out from behind the mountain. It got dark. Two lightning bolts flashed in zigzags to the center of the cloud at the same time, the earth shuddered from the strong thunder. My companion was on the mountain 40 m above me. He shouted. I looked around. From his side, at a height of 5–6 m, a glowing ball was flying in my direction. It was gradually descending and seemed to be flying right at me. I stood as if hypnotized. The ball flew slowly and passed me at arm's length. Its size was 30–35 cm across. The light is not blinding, luminescent. The shell was gaseous; sometimes balls of gas or steam separated from it, and immediately disappeared. The ball was not translucent, club-shaped clumps of "plasma" were constantly moving inside it from left to right. They were constantly changing in color from pale purple to reddish. Discharges were constantly crackling in the ball, but there was no warmth or cold from it. The smell was the same as from the spark of an induction coil. There was a gnarled larch tree 10 m away from me. The ball reached it. The ball touched the larch, and slowly began to descend the trunk, leaving behind a light trail of evaporation. When it reached the ground, it disappeared after 5 s. Jumping from stone to stone, I ran to the larch tree, felt the trunk. It was cold; there were no burns on the bark. Only from under the larch, quieting down, the discharges crackled, as if someone was slowly breaking a bunch of dry thin branches. Figure 2.17 shows this BL.

Description 2. In the same summer, at midnight above the horizon, between the clouds, three lightning bolts flashed simultaneously. There was no thunder. A ball appeared, flew like number eight. When it flew up to the point of appearance (when connecting of an eight), it flashed dazzlingly.

Comment. The ball was not translucent, club-shaped clumps of "plasma" were constantly moving inside it from left to right.

3. Belokon V. A., male, Theoretical Physicist, Ph.D. 7 y.o. at the time of observation in 1941, survey 2005. July or early August, Zolotonosov, Poltava region, south of Kiev, against Cherkassy on the other side of the Dnieper River. Observation time is 14.00–16 0.00 local time. After the end of the storm. The clouds are

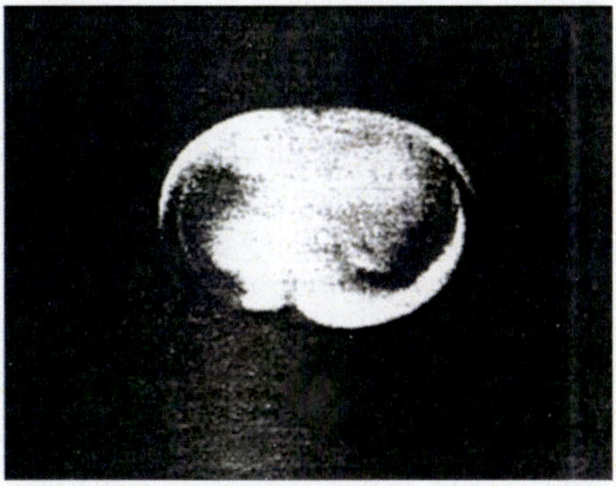

Fig. 2.17 Appearance of BL (Mitrofanov O. I.)

weak. A place of observation outside and inside an old-style Ukrainian hut with a pointed roof covered with packed hay.

"The storm was coming to an end, five lightning bolts were observed in half an hour. In the hut, the ceiling was at a height of 3 m, there was a lamp under the ceiling at a distance of 1–1.5 m from it. At the time of observation, it was turned on. There was a terrible crash, then a thunderclap. Sparks came from the light bulb, like bayonets, not straight, but saw tooth. It was a sheaf of sparks, where the length of the sparks reached 0.6–1.0 m. Everyone froze, and I jumped up and ran outside to see if the hut was on fire. There was no thunderstorm. The roof was not on fire. Returning, I began to open the door. I pull the door towards me and with my peripheral vision, I saw a bright spot behind me, I turned around and saw the light source: a melon-shaped body slightly flattened with dimensions of 20–50 cm. It was moving with a stream of air (at a speed of ~0.3 m/s), ahead of me, entered the house, passed through the hall, passed the rooms, slightly flattened, went into a smaller one, flattened again passed through a small window, came out in the form of a ball.

When the object passed by me, it was at the level of my eyes or the top of my head. I examined it carefully. It was translucent, resembled a cellular structure like a honeycomb or cells (maybe it was filled with vortices), ripples were visible inside it. The observation time was 15–20 s. The color was like a hospital-frosted lamp. The shortest distance was ~30–50 cm. The distance traveled was about 20 m. Besides me, the members of our family, sitting in a room under a lamp, were watching the BL."

Comment. The ball was translucent, resembled a cellular structure like a honeycomb or cells (maybe it was filled with vortices), ripples were visible inside it.

4. (Bychkov and Bychkov 2008) female, Newspaper "On the verge of the impossible" 0.1986. Altai Territory. The first decade of August. After midnight. City.

"Before the weekend, I was reading for a long time in the kitchen and went to bed at three o'clock in the morning, closing the window for the night (1st floor). I had just gone to bed when the light bulb under the ceiling lit up, and a glowing drop slowly flowed from it, like from an unclosed tap. A meter above the floor, the drop took the shape of a ball 8–10 cm in diameter. From the light bulb, the 'drop' did not flow at a right angle, but with a visible slope. "The ball hung motionless for 3 s, and then it moved in my direction. It moved smoothly, slowly, along a curved line, not straight. Approaching, hovered over the naked body (at a distance of 20 cm), walked very slowly along the arm.

The ball was no more than 10 cm in diameter. The color was blue-white; the surface was smooth. From the surface outwards went like the needles of a hedgehog—short rays ending in 1.0–1.5 cm from the surface. I did not notice any sound, smell, heat, or unpleasant sensations. The light coming from the surface did not illuminate anything; when the 'drop' drained from the bulb, the room became light.

After walking along the arm, the ball also slowly floated out into the corridor at a height of 1.5–1.7 m. I did not find any traces of it leaving the apartment... There was no smell of burning, no explosion. In the corridor, children's underwear was drying on a rope, nothing was damaged, and I saw that the ball floated at a height where the underwear reached."

Comment. The light bulb under the ceiling lit up, and slowly dripped from it, like a glowing drop from an unclosed tap. A meter above the floor, the drop took the shape of a ball. The structure of the ball was layered.

5. Amirov AKh., Bychkov VL (2011) I. D. Myznikov, male. Cartographer. Engineer-Surveyor. 25 y.o. at the time of observation in 1984. Place of observation Irkutsk region, Baikalsk. Mid-July. At 2–3 o'clock in the morning. Outdoors.

In the woods near the house. The shape of the BL was indeterminate, yellow-orange in color, moved quickly, and noticed at the last second. The diameter was about a meter. The vertical and horizontal dimensions of the BL differed, the diameters differed by two–three times, the horizontal diameter was larger. The border of the object was blurry. The core was bright and round. The halo has a vague shape. BL shone with high power as a good street spotlight. The smallest distance to the ball lightning is 20 m. During the observation, I walked 10–15 m. The speed of the object is ~2–3 m/s. I observed ball lightning less than 2 s at a distance of 8–10 m. The color of the object is orange-yellow. The color of the object's core is bright-yellow-white. The color of the halo is white-yellow. I did not feel the heat from the ball lightning. The size of the object did not change during the observation. The color of the ball lightning did not change. I did not see the moment of the ball lightning formation. When it contacted with the branches of a tree, it exploded, there was a powerful loud explosion, then it rang in my ears for a long time, and temporary deafness appeared. Sparks rained down from the tree during the explosion. When interacting with the BL, the branches of the tree apparently caught fire.

The electrical appliances turned on at night in the house turned out to be non-working in the morning, with the exception of the refrigerator. There are street lighting wires hanging under a tree nearby. The BL exploded at a distance of less than 20 m. It was overcast, there was a thunderstorm, and the wind was weak from the northwest.

After observing the BL, a heavy downpour began. I was the only one watching in the dead of night.

Comment. The diameter of the BL was about a meter. The vertical and horizontal dimensions of the BL differed, the diameters differed by two–three times, the horizontal diameter was larger. The BL was moving, horizontally smoothly did not rotate; against the movement of the wind. The trajectory of the BL is against the wind, almost parallel to the ground.

6. (Olkhovatov 2008) 02/13/1853, 19.00 local time. The State of Maine. USA. Winter.

The event occurred during a hurricane. The wind died down, and the strongest flashes of lightning began. Houses were shaken to the foundations. The lightning were purple in color and sometimes looked like fireballs that penetrated through windows, doors, and chimneys. They rotated and made crackling sounds. In one case, a lightning strike inside the house threw an eyewitness and lit a match. In another house, lightning lit a box of matches lying on a table in the middle of the room. In one of the sections of the yard, trees were "pulled out" from the ground along with earth and stones on the roots and scattered in different directions. One of them remained hanging, its roots caught on the tops of other trees at a great distance from the place where it grew before. Large stones were thrown out of the ground, which lay scattered on the snow; the latter in some places was completely swept away, as if it was not there. One of the lightning struck the mast for raising the flag and after it entered the ground to a depth of 1–2 m and formed a hole with a diameter of 2.5–3 m, it dispersed in 4 directions. In one of them, it made a ditch 1–1.5 m deep and 110–150 m long in an open area, scattering, turning over, and lifting blocks of frozen earth, some of which reached 3–3.3 m in length and 1.2 m in width. One of the boulders the size of a single-masted fishing vessel was thrown 23 m.

There were no wounded or killed and none of the houses were damaged.

Comment. The lightning were purple in color and sometimes looked like fireballs that penetrated through windows, doors, and chimneys. They rotated and made crackling sound.

7. (Tsarev and Tsareva 1982) 1980. S. Moiseenko, Male. Yakutia at the Indigirka River.

I had a rare stroke of luck. An extraordinary stone weighing about a kilogram, in the center of which there was a funnel with glassy edges, rushed into the eyes, despite the twilight of the thunderstorm and water flows. I realized that I had found fulgurite, a rare formation that occurs when a powerful lightning strike hits quartz sands. Looking at it enthusiastically, I felt rather than saw a flash of lightning, the bright light of which flashed down the hill above the stream about 50 m behind me. Almost instantly, a powerful thunderclap broke out—I had never heard such a force before—almost deafened me. Then the second one erupted from heaven. Suddenly… At the point where the blinding arrow struck the ground, the lightning suddenly curled into a fiery vortex and rushed up the valley of the stream. BL, creeping over the surface, I had to observe more than once. However, there was something else.

It was not a ball rolling on the ground, but a blinding fire vortex stretched upwards, then disintegrating into separate rings of flame, then reconnecting together, was rushing (fortunately not in my direction, because I would not have had time to escape from it). The fiery swirling flame burned the remains of stunted vegetation on its way, uprooted small bushes of birch elfin, and weaved them incinerating them. As it moved, the extraordinary tornado gradually lost its brightness. At the end, the fire attacked a stunted larch. The bright green of its branches instantly flared up to the top. After destroying the tree, the dying fire dragon turned into a smoky cloud, only to disappear soon, scattering across the valley. The swirling lightning streaked across the ground for almost a hundred meters.

Comment. It was not a ball rolling on the ground, but a blinding fire vortex stretched upwards, then breaking up into separate rings of flame, then joining again.

8. (Tsarev and Tsareva 1982) 10.27.1954 Italy. It all started back in Venice. Gennro Lucetti and Pietro Lastrucci were standing on the terrace of the hotel in St. Mark's Square.

They saw two "glowing spindles" flying at great speed in the sky, leaving a fiery white trail behind them. Then the object disappeared in the direction of Florence.

A few minutes later, the events continued at the Florence stadium. The phenomenon was witnessed by football players, referees, and ten thousand spectators who gathered for the match of the Fiorentina club.

Performing dizzying maneuvers, the glowing spindles swept over the football field three times, and then an unknown substance, similar to thin silver threads, began to fall from the sky. Getting into people's hands, the threads melted without a trace, like snow, although they were not cold to the touch. Student Alfrede Jacopozzi wound a ball of fibers on a stick and placed them in a glass jar.

The so-called "angel hair" came to the University of Florence to Professor Danilo Gozzi. He made tests of the substance and stated: "This is a fibrous material that has significant resistance to stretching and twisting. When exposed to heat, it darkens and evaporates, leaving a melting transparent precipitate. Analysis of the sediment showed the content of boron, silicon, calcium and magnesium in it. Hypothetically, this substance could be something like boro-silicon glass."

Comment. Two "glowing spindles" flying at great speed.

9. (Tehknika molodezhy 1982) O. I. Mitrofanov (b), male, DS Physicist. 1974. July. Ryazan region. On the left bank of the Oka River. Evening, July 23. Three observers. The sun was baking during the day, as it happens before a thunderstorm. The night was windless without a hint of a thunderstorm or rain.

A light appeared from the thickets located 70 m away from us along the shore. At first, I thought it was a lantern, but how can you see it where there is impenetrable shrubbery and 10-m trees?

I was the first to notice this strange "lantern," and according to my reaction, Viktor (male) and Olga (female) turned in the same direction. As if on command, we stood up and froze without any agreement: the BL was slowly moving right towards us, as

Fig. 2.18 Appearance of BL (Mitrofanov O. I.)

if surfacing from behind the bushes and increasing in size. It was 2 p.m. and 10 min on the clock.

All events took place in 4 min—the lifetime of the BL. It was slowly moving (floating) along a horizontal or slightly ascending line at a height of several meters in the direction slightly to the left of us (with good accuracy, its trajectory coincided with the north-south line). It moved in complete silence, as it disappeared without any sound. It was a dim, lantern-milky color, about the same as a low-pressure mercury lamp looks through a plate of frosted glass. The border of the ball was not blurred. (The traveler Arsenyev met a similar-colored BL, see above.) (Fig. 2.18).

It was not possible to see any internal structure, but some jumping light dots were noticeable against the background of the ball, quite bright, like moths at a lantern.

The lightning was approaching us, its apparent diameter was increasing, and the viewing angle was changing. However, it seemed that it was no bigger than a volley ball in size. When the lightning was already relatively close, a ring of the same color suddenly separated from it. It began to expand slowly and evenly symmetrically relative to the center of the ball. Perhaps it was the most unusual in the behavior of lightning and the most beautiful. Remotely, the phenomenon resembled the divergence of a circular wave from a stone thrown into the water, but in slow time. The ring increased in diameter, as it seemed to us, to 1–3 m and, having melted, disappeared. After that, the ball "gave out" a second ring, less bright than the first. It also blurred and began to disappear. At that moment, the lightning began to deform, took a pear-shaped shape and somehow quickly disappeared. There was no smell, no traces of BL left. The BL was bluish purple.

Comment. Pay attention. A ring suddenly separated from the BL. It began to expand slowly and evenly symmetrically relative to the center of the ball.

10. (Corliss WR 1995) J. of Meteor, U.K., 2:271, 1977. 'On June 8, 1977, an officer at the coastguard station at Dyfed West Wales observed a very large ball of electrified air. Note that the radio static and bounciness support the

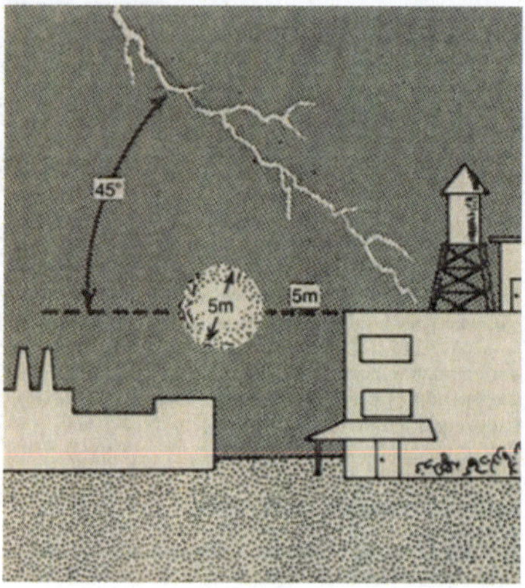

Fig. 2.19 Picture of BL. Big BL

idea that the large balls are electrostatic induction phenomena. "The ball lightning phenomenon was very large and estimated to be about the size of a bus (Fig. 2.19).

It was described as a brilliant, yellow-green, transparent ball with a fuzzy outline which descended from the base of a towering cumulus over Garn Fawr Mountain and appeared to 'float' down the hillside. Intense light was emitted for about three seconds before flickering out. Severe static was heard on the radio. The object slowly rotated around a horizontal axis and seemed to 'bounce' off projections on the ground. It was noticed that cattle and seabirds in the immediate vicinity became disturbed."

Comment. A large BL which size is greater by 25 times than those of the average BL.

11. (Corliss WR 1982) Circa, 1920. Johannesburg, South Africa.

My first experience goes back to 1920, when as a boy in Johannesburg, I observed ball of lightning rolling slowly up the slope up an incline about 50 cm above and parallel to the sloping ground. This ball of light was about 50 cm in diameter with irregular saw-tooth streaks of light around it, and it had a comet-like tail. The ball continued until it met a dry-stone terrace wall, where it exploded in a flash of flame, leaving an acrid smell. The wall was not damaged in any way. It had been possible to see that the ball of light was rolling in anti-clockwise direction (Fig. 2.20).

Comment. The ball of light was about 50 cm in diameter with irregular saw-tooth streaks of light around it, and it had a comet-like tail.

12. (Corliss WR 1982) W. J. Humphreys, meteorologist 1936. USA. (1936) Proc. Amer. Phil. Soc. 76:613.

"Mrs. Ames was standing on a rug during a thunderstorm with her hand at her waist, one finger more or less extended. I was about five feet away and noticed the

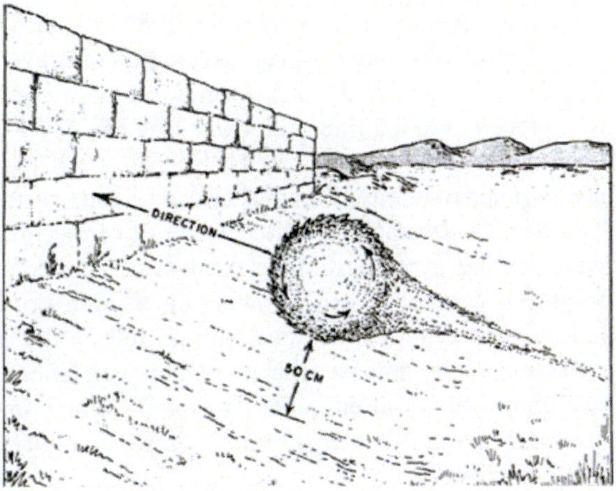

Fig. 2.20 BL with a comet-like tail rolls along ground

air between her finger and the floor was quivering so that it looked just like the hot air over a field. I noticed something rise slowly from the floor upwards her finger. It was not very bright and appeared to shine through a haze. There came a flood of lightning outside and ten the fireball disappeared." (Fig. 2.21).

Comment. The small BL.

13. (Abrahamson et al. 2002) Brett Porter, Ranger, Research & Planning, Queensland National Parks, Australia.

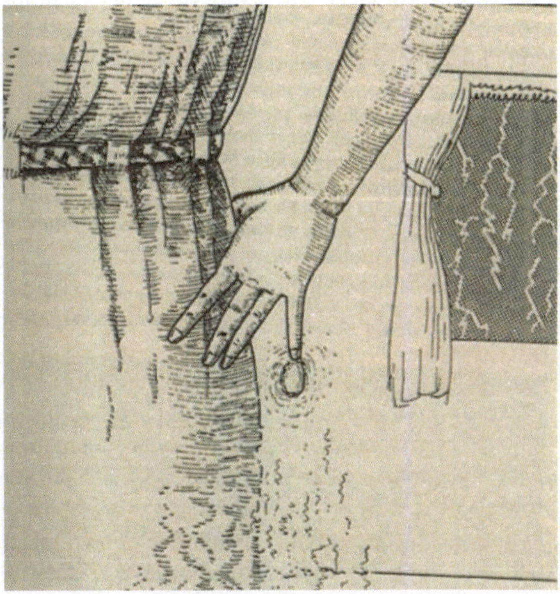

Fig. 2.21 A small BL

"On the evening of 20 November 1987 at approximately 8.05 p.m. I was in the company of Mr Gordon Maag carrying out fieldwork on stage I of the Kangaroo Population Model Validation Project II. We were working near the southern boundary of a property named Ranchall, owned by Mr Des Whittle and family. Our position was approximately 28° 09′ 30"N, 149° 39′ 18″E (northwest of Goondiwindi). The nearest main road was the Lapgate Road running north–south and forming the eastern boundary of Ranchall six kilometers east of our position. Individual storm clouds were passing over us from the southwest. The clouds did not form a single large front, but were separated, showing clear night sky between them and displaying independent electrical activity. No rain fell on our position or any of the country we subsequently traversed that night, although Goondiwindi received a severe storm that night which caused major damage to some properties. (Wind speed at ground level during the following event was zero, humidity 100% RH, with moisture in a warm relatively transparent mist, making one's skin wet.)

At 8.13 p.m. in order to observe the peculiar behavior of a conspicuous setting star, we turned our vehicle towards it, stopped the engine and turned out all vehicle lights. At 8.23 p.m., less than thirty seconds after the star had set, I noticed a pink-orange reflection on the inside of my open door. I looked behind me and saw a pink-orange glow emanating from behind the near horizon, a ridge that ran north–south about 300 m east of our position. This light was not visible only minutes before as I had been making quick scans of the entire horizon while watching the star. Consequently, I believe that I saw this light no more than one minute after it would have first been visible from our position.

From the first instant of sighting the light, it increased in intensity and size. After approximately three minutes the light reached its maximum intensity and size. This was approximately 2–3 times that shown in the accompanying slide (printed in Chap. 1) which was taken as the light was fading. At its brightest the light had a base along the horizon of approximately 100 m and the glow rose skyward in a fan shape about 200 m high. There was a distinct mistiness surrounding the glow which was possibly associated with the storm activity. (The deep red color of the early stages changed to orange yellow and then at its brightest the central emission zone appeared near-white with a brilliant red halo. The cloud formations were remarkably low and clearly reflected the light and this is apparent in the photograph.)

However, after about 40 s at its brightest the light began to fade at approximately the same rate at which it increased its intensity. As the light was fading I saw a white light beam sweep the sky about 60° further north and apparently much further away. This was most probably the lights of a tractor driven by Mr. Whittles' son, Paul, who was plowing about 7 km north by northeast of our position. Paul Whittle also observed a pink-orange light that night south of his position at the same time as our sighting. The intersection of the two independent lines of sight confirms the location of the light source as being on Ranchall and not more than three km east of our position. (The most probable position was around 300 m from us, with the light at or close to ground level. With this interpretation, the central emission zone was about 20 m diameter, and no less than 50 m at its brightest.)

After I took my photograph we drove towards the fading light. By the time we reached the top of the ridge the light was gone. There was no evidence of the source or its location. There was no fire or embers to indicate a lightning strike, although this was no surprise as the behavior and intensity of the light was inconsistent with a bush fire. The night revealed nothing but blackness illuminated occasionally by lightning. There was no sound associated with the sighting and the spectacular light show was over in not more than seven minutes. (An internal property wire fence ran approximately 50 m to the west of us, so that we were between the fence and the light. The nearest point of another fence to the estimated position of the event was >200 m away. I did not see any lightning strike the ground, but lightning frequently showed in the cloud overhead.)

A subsequent inspection of the area on horseback and by motorcycle did not reveal any evidence of the sighting (nor of any metal objects such as wire). The slope of the area where the sighting was made is approximately 1 in 30. Average slope of the property is approximately 1 in 130. (There are no submerged water pipes within several km, and the owner believes it unlikely that there are any buried metal objects.)

The photograph of the sighting was taken in the last 90 s or fading phase of the light, with Fujicolour 100 ASA slide film in a Nikon FE2 camera through a Tamron 28–135-macro zoom lens and a Hoya 67-mm Skylight lB filter. It was set on automatic wide-angle f.4. Exposure time was about 15 s."

JA has more detailed reports, discounting other people in the area, and discounting the possibility of it being a refraction phenomenon.

Comment. Pay attention. The BL size was about 100 m. One can see the photo in Fig 1.18 of Chap. 1.

2.3.3 BL in the Form of a Thread or a Harness

In the collection of observations by A. I. Grigoriev (Grigoriev AI 2006) (Grigoriev 1990), it was stated that the ability of the BL to unwind into a "ribbon" or "snake" and twist back into a ball occurs in about 2.5% of reported cases. Here we present such cases.

1. (Grigoriev 1990) 07.14.1972, 13–14. 00 local time. Researchers, Biologists A. V. Duben (male) and S.Poshevalov (male) on the southeastern slope of the second spur of the Tybga ridge, at an altitude of 2700–2500 m above sea level, got into a thunderstorm cloud.

Something unimaginable has begun. Lightning appeared on the left, on the right below and above us. Each lightning bolt was accompanied by a deafening crack, the sound resembling the explosion of a heavy bomb.

Suddenly, next to us, 20–35 m away, lightning struck the rock above us. For a fraction of a second, the lightning was motionless, and then it began to gather into a lump on the surface of the rock.

It seemed that someone was collecting a fiery roulette, winding it into a bay. As a result, a ball slightly larger than a billiard ball was formed. It jumped 30–35 cm above the rock surface and slowly swam in an easterly direction. When the ball clung to the stones, sparks rained down and a sound similar to the metallic ringing and banging of a baby rattle was heard at the same time.

Comment. Winding the ball from the thread.

2. (Grigoriev 1990) Suvorov, male, June 1969, Krasnokholm, Orenburg region. Between 14 and 16.00 local time of the day.

Returning from haymaking, two of my comrades and I were caught in heavy rain with a thunderstorm. We took shelter from the rain under a large oak tree. Suddenly, lightning struck 100–150 m away from us with a strong roar. After the discharge, the lightning channel began to descend like a thread wounding into a ball, and a fireball of bright-white color in a blue shell was formed near the ground, surrounded by a whole cloud of sparks. The size of the ball from where we were standing was about the size of a fist. There was a feeling that the ball was spinning rapidly in place. No more than ten seconds passed, and the ball exploded with a sharp, loud, dry crack, throwing a whole shower of sparks to the sides.

Comment. Pay attention. The lightning channel was wounding like a thread into a ball.

3. (Grigoriev 1990) A. F. Pryakhin, male. The end of July 1968. The village of Kuzminovka, Oktyabrsky District, Orenburg Region.

A strong thunderstorm with a downpour had already begun to pass, but it was still raining. My sister, brother and I were sitting in the room at a table by the window. Suddenly, a loud and strange sound of whistling, buzzing and hissing at the same time we heard above our heads. We looked up. The sound came from a hole with a diameter of a centimeter and a half, through which electrical wires entered the house (right above the window). After about five seconds, some kind of luminous flagellate mass began to come slowly out of the hole. It was almost cylindrical in shape and had a diameter of three centimeters. While the beginning of this tourniquet was moving deeper into the room, something similar to a body was pulling out of the hole. The movement of the 'head' was not rectilinear, but took place along a spiral trajectory, while the rest of the harness exactly repeated the movement of the front particles. It was clearly visible how bright dots, stripes, and some glowing clumps of various shades were rushing along intricate trajectories with great speed in this bundle. They turned out to be at the beginning, then at the end, then in the middle of the harness. Everything inside the harness was spinning in a frenzied whirlwind, but could not fly out of it.

The total length of the harness, when it completely came out of the hole, turned out to be 60–70 cm. Flying away from the wall by a meter and a half, the harness began to curl into a ball at high speed (it was clearly noticeable). First, the front end of the harness bent like the handle of a cane, and starting from it, the harness rolled into a ball. A rapidly rotating ball with a diameter of 10–13 cm was formed. There was some kind of fiery carousel in it, accelerating all the time. The ball crackled, whistled, hissed. A second and a half after its formation, it exploded with a strong roar. We were all stunned. The explosion occurred about one and a half meters above our heads. No sparks or splashes were noticeable during the explosion. Just a sharp, ringing clap and nothing more. All this lasted about five seconds.

Subsequently, in the army, I served as a bomber and repeatedly observed explosions of different strengths. That memorable explosion of BL was approximately equal in sound effect to the explosion of 250–300 g of TNT. Only in the case of TNT explosion, there are usually thermal effects, and the substance flies apart. In the case of the BL, there was no feeling of warmth. We did not notice any waves, the spread of the mass. That is just that our hearing was slightly impaired, and when we shared our impressions, we had to shout (at least raise our voice a lot), and our ears seemed to be blocked (there was ringing in our ears). By the evening, the hearing recovered a little, and the next morning it returned to normal.

Comment. Pay attention. Winding the harness into a ball. The explosion of the ball.

4. (Bychkov et al. 2003) V. B. Medvedev, male, Businessperson. 24 y.o. during the observation. Survey of 2001. Observation in 1985. The city of Khimki, Moscow region. The place of observation is a city apartment.

There was no thunderstorm. "The device (radio transmitter, see Fig. 2.22) worked in an abnormal mode, the oscillations were disrupted. I began to bring my right hand closer to the device. There were no sensations at this time.

When the connector was 20–25 cm away, there was a hiss. I continued to bring my palm closer to the connector and did not pay attention to the hissing. When about 15 cm remained to the connector, two yellow small jets appeared from it, similar to a viscous liquid, see Fig. 2.23.

Their diameter was equal to the diameter of the connector holes (the holes are hollow from the inside to connect the feeder line). The trickles were opaque. The tops of the streams began to diverge; the outer edge became diffuse. They parted, and I kept moving my hand.

The process of ball formation occurred suddenly. The ball was a sphere of opaque orange-yellow (red color). There was no halo. The light intensity from it compared to an incandescent lamp was 40–60 W.

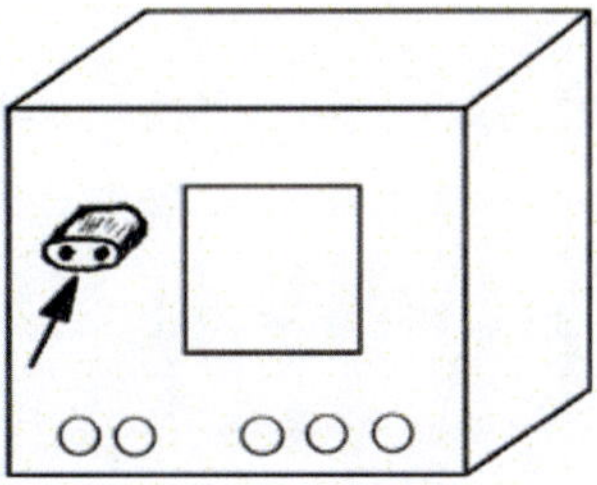

Fig. 2.22 The device with a connector. An arrow indicates the connector

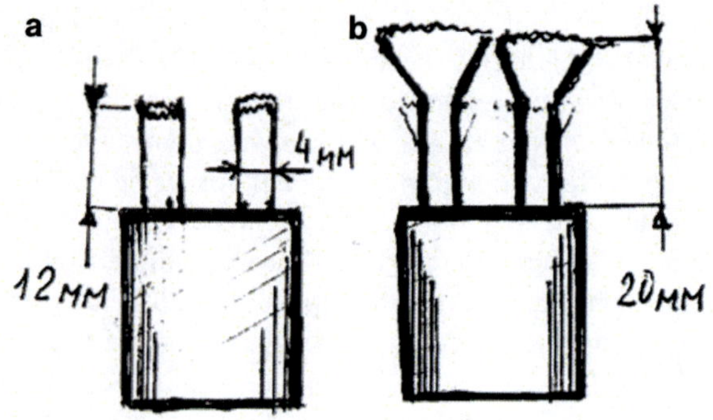

Fig. 2.23 Changing the shape of the jets coming out of the connector. **a** Initial trickles, **b** trickles before turning into a ball

I saw something incomprehensible to me, so I was not afraid, and by inertia, I continued to move my hand. The whole palm sank into the surface of the ball with a diameter of 15–18 cm. There was a feeling of heat, similar to the feeling of heat from the flame of a fire.

The surface of the ball was slightly deformed. If you look at the edges of the ball, you can see the structure visible on the top of Fig. 2.24. The density from the edge from the airside increased inside the object at a distance of 0.5–1 cm. I saw how the substance of the ball from the surface flows through my fingers. It flowed through my fingers, as if in sequins thinner than cotton threads. However, the palm did not get inside the ball. All the time of the existence of the ball, there was a hiss.

Then I got scared and jumped back about three meters.

The ball remained in place. It began to shrink in size. Having reached a diameter of about 5 cm, it disappeared (melted into the air). The hissing disappeared with it. The intensity of light decreased as the size of the ball decreased, but only slightly.

The duration of the entire observation was 12–15 s. The ball was stationary during the observation.

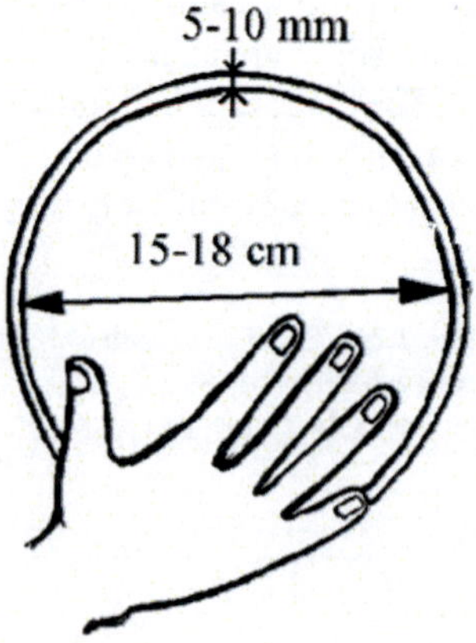

Fig. 2.24 A ball formed from jets

Contact with the ball did not leave burns and other physical consequences. Subsequent examination of the device showed that there was a slight melting of metal parts in it. The appearance of the ball did not cause subsequent disturbances in the operation of instruments and radio equipment. At the moment of observation, I was alone."

Comment. The appearance of BL in the form of small jets from a radio transmitter. The behavior of the jets resembles the attraction of currents according to Ampere's law.

5. (Bychkov et al. 2003) Observer A. L. Tseskis, male, Physicist, Ph.D. 50 y.o. at the time of the survey in 1997. Place of observation of the city of Vidnoye, Moscow region, in the premises of the All-Union Research Institute of Canning and Vegetable Drying Industry. In a room on the third floor. End of July 1983.

"The day was very hot, more than 30 °C. The room in which I stayed was very hot. In the afternoon, around 12–13.00 local time, I noticed two snakes crawling out of an electrical outlet in the wall with a dull crackling. They were bright blue in color. Brighter inside. If I were comparing it with the glow of an incandescent lamp at night, I would estimate a power of about 100 W. In the bright light of day, they were visible as bright formations. Very quickly, in about 1 s, they connected into a semicircle arc that burned for 3–4 s, and disappeared.

The arc length was 4.5–5 cm. I observed a very bright cord about 2 mm in diameter inside it. The periphery was dim diffuse, up to 5 mm thick. The cord was a smooth channel that wriggled and quivered slightly, like a small channel of linear lightning (Fig. 2.25).

After the object disappeared, there were no traces left on the electrical outlet. The electrical devices at the institute have not deteriorated.

Comment. The appearance of a jet from the outlet. This resembles the appearance of a current jet. It remains unclear how the emission of charges and the nature of the jet itself took place.

6. (Likhosherstnykh 1994) T. I. Kaznacheeva, female, city Krasnodar.

The incident happened when I was 15 y.o. I was lying in bed in the morning, reading. Suddenly, a big wind rose outside, it began to rain and it became dark.

Just when I wanted to get up and close the window, a wriggling blue 'snake' with an orange surface flew into; it was 20–25 cm long, 1 cm thick.

Fig. 2.25 Blowing the arc out of the socket

I froze and shrank into bed. Ahead of the 'snake' was something like an arrow, at the end of which shone a bright red spot, similar to a crystal eye. It was constantly in motion, making throws in one direction or the other, as if it was looking for some kind of victim, and at the same time, it clearly reacted to the surrounding objects, as if it saw them. I could not move and thought only about one thing-how would Mom not come into the room now because the 'snake' might pounce on her. After going around the whole room in a circle at a height of about 2 m, the 'snake' flew out of the window again.

Comment. The appearance of a wriggling blue 'snake' with an orange surface of 20–25 cm long.

7. V. Yu. Shemilov, male, Biologist, survey of 1994. Observation 8.07.1986, 23.30 local time. Mosolovo Village, Ryazan region, Shilovsky District. Surveillance outside the premises on the territory of the village. 3 observers.

"The day was hot and dry. The temperature was more than 30 °C. According to weather data, a cold front was approaching from the southeast. Cirrus clouds. The storm started at night. Lightning flashed all over the horizon except the northeast.

The BL appeared out of nothing in the northwest. There was no connection with the linear lightning flash. The object was located near the lower boundary of the clouds. The first feeling was that it was a lighting rocket. The color is white, like a 500 W incandescent lamp. In the center of the BL was a core of an orange-brownish hue. During the flight at an angle of 30° to the horizon, downwards, the BL changed its shape like an 'amoeba'. It flowed from place to place as a rubber ball filled with water. A smoky, non-luminous tail with a length of 50–100 m stretched behind the BL. The BL was moving, making a spiral, see Fig. 2.26.

The observation took place from an elevated place, and it was clear that the BL flew into the transformer booth. After that, the lights went out in the village. It was not possible to establish the destruction."

The distance to the object was 300 m. Observation time 5–6 s. The distance covered by the BL was 1000 m.

Comment. The BL appeared out of nothing in the northwest. BL was as "amoeba." It flowed as a rubber ball filled with water. A smoky, non-luminous tail was trailing behind the BL.

8. V. A. Kurakin, male, Physicist, Ph.D., 10 y.o. at the time of observation, 48 years at the time of the survey, event in 1997. Six more people were observed. Moscow region, Dmitrovsky District, the village of Bely Rassvet. Russia. Summer. June 1–August 20. During a thunderstorm, 14–19.00 L.T. The air temperature is 20–25 °C. The house is wooden, rustic. Survey of 2006.

Fig. 2.26 **a** BL diagram, **b**—object trajectory

"It was raining. Lightning flashed in the distance. I was in a hut. The BL appeared through a hole near the ceiling for electrical wires. It descended along the wires that went to the electrical outlet. Went into the socket.

Shape—the object was elongated like a snake. It looked like a snake thread: the central part is dense, and loose at the edges, similar to cotton wool.

The lifetime of the object was 4 s. The distance traveled was 2 m."

Comment. Changing shape of the BL (and as a snake). Penetration through the hole. It left through the outlet.

9. V. A. Kurakin, male, Physicist, Ph.D., 20 y.o. at the time of observation, 48 y.o. at the time of the survey, event occurred in 1997. Another 4–5 people observed it. M/O., Dmitrovsky District, the village of White Dawn. The house is wooden, rustic. Survey of 2006.

"The village stood on a hill, on its top there was a church. People in the village observed BL often. Now the BL disappeared after the construction of a power plant in the village with power lines and lightning rods.

BL appeared in the house. Its shape was changing."

Comment. Changing the shape of the BL.

10. (Keul 2021) The German Graduate Physicist Hans Dolezalek, working about atmospheric electricity, after 1961 for the US Navy, reported his own observation at Tuebingen, Germany (Dolezalek 1951). At June 8, 1951, about 17:00 LT, a thunderstorm started. Minutes before 18:00 p.m.

Dolezalek saw ball lightning like a car headlight 30 m away, moving downwards at an angle with about 50–100 m/s, to disappear behind trees. A violent bang was heard, like an explosion or a grenade launcher, followed by a rising blue smoke cloud. The object's impact with 10 m bright rays spreading upwards was seen by another observer in a meadow. Later, a 50 cm burned spot was found. Two isolators of a 5 kV power line pylon 5 m from the impact point were damaged. Neckar Power Station Hirschau (connected to the 5 kV line) registered an interruption at 17:54 p.m.

Comment. The object's explosion with 10 m bright rays spreading upwards.

11. (Keul 2021) Eric Dunford, Space Scientist at the Rutherford-Appleton Laboratory, Chilton, UK.

Observer with his wife and another person saw BL on April 18, 1968, at about 19:15–30 LT during a thunderstorm with heavy rain. After a (ground) flash less than a mile from the house, an intense spherical ball of light appeared, angular diameter... about half the sun's... brightness was sufficient to light up the surrounding fields... there was a tail of light (or smoke) attached to the ball. It 'did not move very much', lasted about 5 s, then disappeared slowly. The true diameter could have been 4–8 m (Stenhoff 1999, 168–169).

Comment. There was a tail of light (or smoke) attached to the ball.

12. (Keul AG 2021) On July 19, 1978, Aeronautical Meteorologist Wolf-Dietrich Wagner watched a nighttime thunderstorm from the window of his Drobollach, Austria summer resort. A few minutes before a shower, at 0:58 LT, lightning struck Lake Faaker 400 m away.

At the impact point, the witness saw "an ellipsoid, a luminous form like a discuss that moved slowly northeastwards and exploded after a short lifetime (2–3 s)." The fireball illuminated "about half of the lake's surface with pure sodium yellow light." It had a 1 m compact, dazzling core in a blurred foggy shape of 2–3 m. The cylinder moved slowly over the water surface without rising according to Wagner's 1979 report to the author.

Comment. The cylinder moved slowly over the water surface.

13. (Keul AG 2021) Friedrich Wieser, Austrian Air Force colonel and member of the Tactical Air Command, drove in his car from Atzelsdorf to Blindenmarkt, Austria, on October 10, 1977, around 19:00 LT.

Moving parallel to Federal Highway B1 in darkness and a light rainshower, a reflection in a window front made him aware of an extremely bright object on B1, a bluish-white field of car size. Wieser slowed down the car and lowered the window: "Suddenly it contracted like threads into a globe, jumped off at a high rate, turning yellow, orange and red." The diffuse red body followed a slanted trajectory and touched down at the B1 2–5 km eastwards. "It went orange, yellow, then dazzling again, was stationary for a short time, then contracted to red, flying up at an oblique angle to disappear." The phenomenon lasted less than a minute and did not flicker or make a sound. Colonel Wieser had 10 years of military radar experience when he reported his observation to the author 1979 (Keul 1980, 1981). In his interview, Friedrich Wieser told the author about his observation as a schoolboy in the summer of 1945/46 at home with his mother in Neumarkt, Austria, at 16–17:00 LT. A luminous ball the size of an apple came in through the open top of a window, bounced elastic from windowsill and table down to the floor where it exploded loud after less than five seconds. There was also a lightning strike to the neighbor's chimney and to the Western railway line.

Comment. Suddenly BL contracted like threads into a globe.

14. (Corliss WR 1982, 1995). J. Meteorology U.K. (1976) 1:310. Miss M. E. MacArthur. Argyillshire, Scotland.

I call it a bulb of lightning, it was opaque; it may or may not have lit inside. It would be 9 inches in length, slim, dome-shaped. I saw it from my chairs the fire come down in front of the kitchen window; it did not touch the ground; it hovered about 9 in. above… Attached to the bulb was a long white crinkly tail, not an unearthly white, 1 yard or more in length, 1 in. or slightly more in width. It was the old-fashioned watered silk ribbon we used to buy in the draper's shop. The tail kept to the end of the bulb, also kept slightly higher; it did not fall down as a ribbon would have done. Bulb and tail snuffed out like a candle, leaving no trace. That was the end of the thunderstorm (Fig. 2.27).

Fig. 2.27 The shape of a BL with a ribbon tail

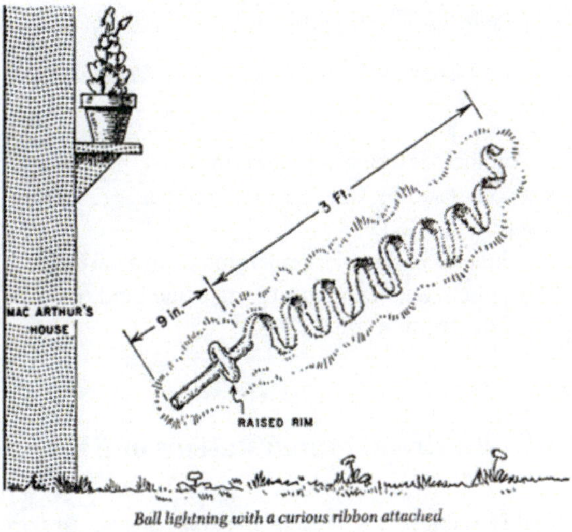

Comment. Attached to the bulb was a long white crinkly tail, not an unearthly white, 1 yard or more in length, 1 inch or slightly more in width.

15. (Corliss WR 1982) 1887. England. (1909) J. G. "Fireballs," English mechanic, 90:95.

"While walking during a heavy thunderstorm under an umbrella. Without any noise or notice, I saw what appeared to be a ball of live light fall just in front of me and slide along the road, and quietly disappear after going a few yards. The ball was about the size of a cricket ball, and appeared to be made of vibrating mass of strips of light, in rapid rolling motion. This, I suppose, was a sample of BL, and at the time I felt very thankful it had fallen just near my umbrella rather than on it. What struck me most forcibly was the extraordinary appearance of the part fibers of light wriggling like a mass of eels. Most of all, I was struck by the unusual appearance of visible light fibers on it, writhing like a mass of eels."

Comment. Unusual appearance of visible light fibers on the surface of the BL, wriggling like a mass of eels.

16. (Corliss WR 1982) (1964) Matts E. 1A Fire-Ball& Weather. 19:228 November 10, 1940, Coventry, England.

I was working at the far end of my garden, the weather was normal, no rain, no signs of thunder. Suddenly, I seemed to be in the center of intense blackness and looking down observed at my feet a ball about 2 ft across. It was of a pale blue-green color and seemed made of a mass of writhing strings of light, about ¼ in. in diameter. It remained there for about of 3 s and then rose, away from me, just missing a poplar tree about 8 ft away. It cleared the house by about 20 ft and landed at the rear of the Weavers arms on the Bell Green road, a distance of about ¼ mile. There was a loud explosion and much damage was done to the public house.

Comment. BL made of a mass of writhing strings of light.

17. (Gasper GEM and Tanner BK) Weather, 26 January, https://doi.org/10.1002/wea.4144

Authors describe an entry in a twelfth-century monastic chronicle, compiled and composed by Gervase of Canterbury (*c.* 1145–*c.* 1210), which gives a credible description of BL.

7 June 1195 an object emerging from a dark cloud, falling in the form of a rotating fiery sphere and then performing some kind of horizontal movement is very similar to the descriptions of BL.

2.4 Electrical Manifestations of BL

In this section, we present observations of BL, in which the connection of BL with electricity is evident. The described phenomena of the appearance of BL from electrical appliances make this section a continuation of Sect. 2.3.

1. (Slyusarev NM 1989) A. E. Mikelson, male, Technical DS.

It was late evening on 7.06.1967. Thunderclouds hung over the horizon. At about 23.00 local time. I went into a cottage-type house located 17 km from Riga. It wasn't raining. Flashes of distant lightning were visible from the window. For 20 min. I listened to music on the shortwave band of the "Spidola" radio receiver, whose antenna was extended by about 40 cm. "Spidola" crackled slightly, working on its own battery, it was not included in the network. At about 23.20.00, lying on my back, I raised my left hand to turn off the radio.

Now, when the fingertips did not reach the switch by 8–10 cm, a white with a blue tint ball with a diameter of about 20 cm crawled out of the front metal panel of the Spidola and stopped over the hand.

The ball was not bright, it did not dazzle the eyes; it stood in one place all the time. I froze in surprise. The hand remained in the same position. It took 3–4 s, the ball exploded, or rather, as if it slammed shut and disappeared. The slamming was accompanied by a short, dry, sharp and very loud sound. There was no smell before and after the collapse. The sound woke up wife and two children who were sleeping in the same room.

My first thought was—am I alive? Having risen from a lying position to a sitting position, I realized that I was alive. However, the left hand felt nothing. The fingers, the joints of the hand, and elbow did not move, apparently it was paralysis. The right hand was unharmed. After intensive rubbing of the hand for an hour, the numbness stopped. The arm came to life, but it was difficult to move it. During the collapse, part of the charge passed through my body through my back. This could be judged by the formation of a black spot on the skin of the back with a diameter of 3–4 cm. The second part of the charge passed through my right hand to my wife's hand, our hands touched. The wife clearly felt the characteristic "electric shock." In addition, 10–12 red circles with a diameter of 1–2 mm were noticed on the skin of my left hand, from the hand to the elbow.

The house had the following characteristic features. Made of plywood and boards. Wooden sawdust was filled in between the plywood boards. The height of the house was not more than 3 m, the area was about 15 sq.m. The input of electrical wires was made through a metal pipe. The pipe towered over the house by about 2 m and descended the wall without touching the ground at a distance of 1 m. In addition, the pipe was zeroed, i.e. the zero wire was connected to the pipe. The main wires went to the shield through two pairs of fuses—for 6 and 15 A.

The Spidola was standing on a linen dresser, the TV next to it; it was not turned on, and its cord was not plugged into the socket. Our bed (wooden with metal springs inside the mattress) stood on one side of the chest of drawers, the children's metal bed on the other.

Later it was found out that the fuses for 6 A burned out, and for 15 A remained intact, the entire electrical network and the TV were intact.

In the morning, neighbors at the dacha said that about 23.30.00 local time they heard a very strong thunderclap. After leaving our house, we made sure that our house was not on fire, we decided that lightning had hit a well; it was located 10 m from the house.

Neither before the moment of the ball collapse, nor after it, I did not hear thunderclaps. The spots on my arm did not hurt at all. There was a slight pain from the spot on my back only until the moment I went to bed, an hour after the incident. The next day I didn't feel any pain, and the marks on my back disappeared after 3–4 days." (Paralysis of the arm could occur when the current passes below 80 mA.)

Comment. Pay attention. The ball crawled out of the Spidola radio set and it collapsed. Injury of the observer.

2. V. L. Bychkov, male, Physicist, DS, 54 y.o. at the time of the survey in 2005. Observation, Moscow, 1958, spring, evening.

"I got under the desk and decided to check why it is forbidden to touch the entrances to the electrical outlet with hands. I partially submerged the electrical contacts from the plug of the electric lamp into the socket. I took a scalpel (made of durable stainless steel), and touching the blade to one contact, I began to slowly move it to the other contact. When there was a flash between the blade and the contact about 0.5–1 mm, a blue ball 3 cm in diameter appeared and rolled into the opening under the bottom of the table. The shortest distance from me to the ball was 15–20 cm. The ball resembled a blue cloud with blurred edges. The glow is milky 5–10 W. It passed 10–15 cm. The ball left no traces on the floor. The whole observation took 1–2 s. I did not feel the heat from the ball.

The reaction is interest, and fear that parents may punish me for experimenting with an electrical outlet. I examined the floor; there were no traces. The end of the scalpel has melted. I feared that parents would scold me, so I did not tell anyone about this case. I remembered about this at the age of 40 y.o., when began to collect information about BL."

Comment. The appearance of the ball during a short circuit.

3. (Bychkov et al. 2003) R. Sh. Islamov, male, Physicist, Ph.D., 36 y. at the time of the survey in 1990, observation in 1959, summer. The village of Stary Sibai Bashkir ASSR.

"At home, in the apartment, I closed the Ogonyok radio receiver with a nail. A ball with a diameter of 1.5 cm flew out of it. The glow was like a candle, like a yolk, dark yellow. Flew smoothly. At a height of 1 m 20 cm above the floor, it made a semicircle. Disappeared -as if burned out. Passed 3–4 m. The observation lasted 10–20 s. The shortest distance from the observer was 10–20 s. My father ran after it (the ball) for some reason. The reaction is surprise. I did not feel any heat. I did not notice any nail changes."

Comment. The appearance of a ball with a short circuit in the radio receiver.

4. (Bychkov et al. 2003) L. B. Ibragimova, female, Physicist. Autumn 1956–1957, being 2–3 year student. 1990 survey, Moscow. MSU.

"The ball flew out of the switching electrical panel. Diameter 15–20 cm. The color was bright white. The power in comparison with an incandescent lamp is 200–250 W. The distance to the object was 30–50 cm; the lifetime is 1–2 s. Separated from the shield and disintegrated in 1 s. The reaction is surprise."

Comment. The appearance of a ball from an electric shield.

5. (Bychkov et al. 2004) S. Evdokimov, male, winter, late 1960, Leningrad.

"There was one window in the room; to the right of the window on the wall there was a massive rosette made of clay covered with white glaze. One evening I had to remove the plug from the socket. For a while, it was not possible, as the fork sat tightly due to the misalignment. Finally, I pulled it out, put it on the table and, looking around, and froze in place.

A bubble, similar to a soap bubble, was slowly blowing out of the outlet hole. Having reached 40 mm in diameter, it separated from the socket, rose 10 cm up and hung. Its coloring was similar to that of a soap bubble film. Life seemed to be pulsating in the hanging bubble—it seemed that some kind of oscillatory process was taking place in it. The bubble did not make any sounds; it did not smell like anything. It was transparent. It has a shape of an irregular sphere. The shape was constantly changing, albeit within small limits. This bubble emitted a soft diffused light. Now the bubble made several oscillations around the vertical axis, froze for a moment, and then smoothly floated in a straight line to the middle finger of my right hand. The bubble smoothly touched my middle finger, "pierced" its spherical body to the depth of the nail and, rapidly decreasing in volume, and silently disappeared. I felt a warmth in my finger, which seemed to spread through my hand all over my body and also disappeared."

Comment. The appearance of a bubble from the outlet. The bubble touched the observer's finger, which felt warm.

6. (Bychkov et al. 2004) V. N. Inozemtsev, male, Economist. 54 y.o. at the time of the survey in 1992. Observation of 1973, 16.00 local time, July. 15 km from the city of Klin, Moscow region. Rural area. Surveillance in a one-story house. There is no thunderstorm.

"I was on the phone. Suddenly the tube lit up, a ball crawled out of the tube, passed 3 m and went into an open window, moving at a height of 1.8 m above the floor. The size of the ball was 8–10 cm; the observation time was 60 s. The color of the ball was light white with a pink tinge. The glow is even, the intensity is average—60 W compared to an incandescent lamp, the border of the ball is clear. Passed 3 m."

Comment. The appearance of the ball from the telephone tube.

7. (Bychkov et al. 2004) V. F. Kutukov male, Physicist, Ph.D., 60 y.o. during the survey in 2005, observation in 1986. Observation during night, autumn, Moscow.

"I woke up to the wild shrill meowing of a cat. I opened the door to the corridor. I saw there a ball at the electrical outlet, slightly glowing. There was a cat nearby, arching its back, and meowing loudly." The diameter of the ball was 30 cm; the color was pale white. Observation time 2–3 s. I did not observe any visible.

Comment. The appearance of a ball from an electrical outlet.

8. (Bychkov and Bychkov 2008) V. N. Panov, male V. N. born in 1971, a Physicist (male) of 11 years at the time of observation between March 25 and April 11, 1983. 15 h. Survey of 1997. Kolomna M/O. In a room on the 2nd floor of a two-storeyed house. The beginning of a thunderstorm. 2 observers.

"A thunderstorm has started. I went to look at it at the window. It began to rumble. It hasn't rained yet. Grandma screamed. Grandma then said that there was a flash, and a ball popped out of the socket. I turned around. The red-hot ball was moving from the socket under the table. It was falling almost in a parabola. Jumped under the table. Exploded with a flash. It left a diffuse black trail of burnt floor under the table, painted with red meerkat, 25 cm long. Grandmother could not wash it.

The shape was close to the ball. Size 5–6 cm. It was light, but the lamp was on. The ball did not overshadow the lamp. The radiation power was less than 60 W (compared to an incandescent lamp). The nearest distance to the ball was 3 m. The ball passed 1.5 m. The ball was falling like a ping-pong ball. Observation time 2–5 s. The color of the object is yellow, red—similar to molten metal (copper at a temperature of 1200–1300 K). The size of the ball did not change."

Comment. The ball popped out of the socket. Left a trail of burnt floor.

9. (Amirov and Bychkov 2011) Lugansk.proUA.com "A week." 2007.07.10 14:39.

Last Saturday, during a thunderstorm, a BL hit a number of houses in Alchevsk. It disabled household appliances that were connected to the power grid.

"How many years I've been living in the world, I've never seen this, and I didn't think I would! It was rattling and sparkling outside—it was creepy. I jumped out of the apartment to close the door to the entrance (my husband said that these lightning bolts (BL) like drafts), and then it appears Bluish from the neighbor's bell, looks like a hedgehog—all in 'thorns', the size of a football somewhere, and at me … it was as if someone pushed me inside the apartment. I did not understand what happened, I came to myself near the ottoman, but there was no lightning anymore," said a resident, female, of house No. 62 on Gorky Street.

Besides her, her neighbor also saw the lightning, who was talking on a radiotelephone during a thunderstorm. In her apartment, the lightning was in the form of an electric arc, which "ran" through the baguette in the corridor and "disappeared" into the wall.

In the house No. 63 on Gorky Street, it disabled the TV, which was plugged into the power grid during a thunderstorm. As a resident of this house told, lightning in the form of an arc along the antenna cable passed through the TV and disappeared into the wall. As a result, electricity was lost in two entrances of the 62nd house, televisions, a water heater burned down in some apartments, Polish antennas, and radiotelephone bases failed.

Comment. From the bell, a BL appeared, covered in "thorns." Another BL in the form of an electric arc "ran" in the corridor and "disappeared" into the wall. The third BL in the form of an arc along the antenna cable passed through the TV and disappeared into the wall.

10. (Bychkov et al. 2015a, b) Male. In 1975 and in 1979 BL visited his house.

I was born in 1974. Accordingly, I was about one and about 5 y. o. when all this happened. The first time I do not remember, and the second time I remembered a terrible explosion inside the house. Later, the parents talked about how it happened.

Both times the "lightning" visited the house in clear, windless weather. The first time in the afternoon. The second time in the evening. Both times, there were many people in the house. More precisely, the whole family… My grandmother's house, my parents, and I came to visit her on Saturdays, in the bathhouse.

First visit: Summer day. The weather is clear. A square room. There is a socket in the northwest corner of the room. My mother, grandfather, and I are sitting on the sofa by the south wall. Father and grandmother in the eastern part of the house.

We see: The BL is blown out like a soap bubble from the outlet in the corner. The diameter is about 10 cm. The color is orange, not dazzling, no sound and crackling; everything is in complete silence. The mother said she was numb with fear.

The ball rose to a level of about one and a half m and flew the letter Γ across the room, "hit" the furnace and knocked a brick out of it on the floor. I do not know how long BL lived, but the whole family saw this "thing."

Second visit: Autumn. Evening. The weather is clear. Windless. I (5-year-old) was sitting with my grandfather in the kitchen. Father and mother were sitting on the sofa in the room (against the south wall). Grandma was standing in the center of the room facing the sofa (to the south). There was a daylight lamp hanging on the western wall under the ceiling (it is still there, but it is located differently).

A ball with a diameter of about 30 cm "flew out" from the lamp starter (an aluminum barrel). The glow was brighter than the first time, but not dazzling. The ball was translucent, i.e. everything behind it could be seen through it. Everything was in complete silence again. The sphere was irregular, amorphous, changing shape from a circle to an oval. My mother said that it hung over my grandmother's head and "flattened," "like a pillow." Grandma did not feel anything. No heat and did not even hear anything. She did not move. I do not know how long it hung over my grandmother's head. After that, the "lightning" slowly floated into the far room, where the electricity meter was located in the corner, and "entered" it. The meter burned out and the wiring next to it—too. There were no casualties. I, sitting in the kitchen, heard only a very loud bang.

Comment. The BL was blown out of the socket. Knocked a brick out of the oven on the floor. Another BL "blew out" of the lamp starter. BL flattened out like a pillow. The BL "entered" the electricity meter.

11. (Bychkov and Vaulin 2013) http://otvet.mail.ru/question/25540349, male.

I saw a BL in 2000. On a stuffy summer night, a strong thunderstorm began, I was closing the balcony door—the rain was lashing into the room. At that moment, a "linear" lightning flashed (I live on the 9th floor) 300–400 m away, "ordinary" lightning struck the roof of a low house, a thunderclap was heard.

From the "bend" of the lightning line above the roof, a bright-yellow-red ball with sparks about a meter in appearance "jumped out," bounced to the right and quickly rushed, gaining speed, to the right and gaining altitude with a buzzing sound like an airplane, piercing the air. After flying about 2 km, the ball "exploded," with the sound of cannon shot, and a "salute" of sparks turned out! Everything happened in 5–6 s. I saw it alone.

Comment. The BL emerged from the "bend" of linear lightning.

12. Male Kolya. http://electrik.info/main/fakty/15-vsja-pravdv-i-vymysly-pro-sharovuju.html 4.09.2015. Kolya.

I saw BL perfectly. It flew into our house from an electric outlet. A mirror with a painted side was leaning against the outlet. It flew out of the socket and shuffled down the mirror, entered the wall, the wall was of boards and slag.

It went inside the wall, went back into the house, making a huge hole in the wall. The ball was the diameter of a tennis ball. The hole in the wall was 10 times bigger, and the hole was also where it went out of the wall, but smaller. This event was seen by 5 people.

Comment. The BL came out of the socket. Entered the wall. Came out of the wall. The hole in the wall was 10 times bigger than the BL.

13. https://ru.wikipedia.org/wiki Male.

I have observed a BL or something similar. A thunderstorm was coming. I was sitting at a table in the kitchen of a village house. My wife was standing behind the stove. Not far away, next to the window, there was a refrigerator powered from an outlet on the wall.

Suddenly I heard a hiss. Since I am a physicist who has dealt with discharges, I immediately realized that this is a quiet discharge. I was looking just in the direction of the refrigerator. A thin, transparent discharge channel was visible. It was transparent, but was clearly visible due to the difference in the refractive index. It was, as it were, a little darker than the surrounding air. It walked in a semicircle from the outlet to the ground about in the middle of the room. I wanted to shout to my wife, but I did not have time. A bright red object the size of a hazelnut crawled out of the socket at a speed of about 1 m/s. A thin red thread of the same color was trailing behind him. After passing through the discharge channel for about a meter, it disappeared with the sound of a New Year's firecracker, scaring my wife. The discharge channel also disappeared.

Comment. The discharge channel went from the outlet to the ground. A bright red object crawled out of the socket.

14. (Likhosherstnykh 1994) U. S. Chaloyan female, Moscow.

"I was sitting at my desk by the balcony door. A cloud came, it got dark, and I decided to turn on the table lamp. I got up, bent over the lamp, plugged the plug into the socket and then, instead of the lamp lighting up, an unusually bright blinding and radiant light flashed around it. My head was inside this glow. I was scared. I thought, I was electrocuted. At the same time, I could neither straighten up, nor recoil, nor scream, some kind of force held me back. After a few moments, I realized that I was alive, that it did not hurt me, that the glow was cold, and I did not feel any pain, only I could not pull my head and shoulder out of this terrible glow. Then I gathered all my willpower and reached out with my right hand, which was out of the glow, to the outlet and turned off the lamp.

After that, the glow did not disappear completely, but weakened, and I was able to move aside and pull my head out of the glowing halo. Then I was even more surprised. Right in front of the face around the top of the lamp was a milky ball with size about forty centimeters, surrounded by rays. The most amazing thing was that this ball consisted, as it were, of separate closed fragments in the form of balls of 15–20 cm, in which clusters of glow pulsed. The movement took place in different directions and seemed to flicker, which made the whole ball seem to move. Then everything disappeared."

Comment. The appearance of a bright glow from an incandescent lamp. The appearance of a glowing ball consisting of separate closed fragments in the form of balls.

15. (Bychkov et al. 2003) E. N. Shevtsova, female, employee of the Central Research Institute of TMI. Recorded from the words of an eyewitness Salnikov V. N. 1986. Abbreviated description.

On Saturday, 6.12.1986 and Sunday, 7.12.1986, it was raining and snowing in Tomsk, there was a positive temperature (in °C). By the night of 7.12.1986, it got colder, a strong wind rose. I was on K. Ilmer Street near house No. 8 from 19.00 to 19.30 local time. When I was walking down the street, I noticed light flashes illuminating low-passing clouds on the opposite side of the street and over houses No. 8 and No. 6. I went to the first entrance of house No. 6. A minute later, I saw a glow over Mir Ave. A trolleybus was going uphill. Above it hung a blue-green ball (like a police patrol car warning light). Sparks flew off from the ball and illuminated the clouds with highlights. The size of the ball is about 40 cm. Its center was located exactly at the point of contact of the trolleybus "whiskers." Throughout the entire path of the trolleybus, within sight, the ball followed him.

I ran out to the bus stop and waited for an oncoming trolleybus. When it turned from 79th Guards Division Street, it had a ball, but as soon as it stopped at the traffic light, the ball disappeared, and reappeared when the trolleybus resumed movement. Turning onto Mir Ave., the trolleybus was again carrying a ball, but of a smaller diameter, about the size of a football. The wind was contrary.

In the morning, I told at work about what I saw, and it turned out that our employee, O. Garbuzova (female) saw the same ball on the trolleybus 3 route near the Sports Palace. The trolleybus drivers could not help noticing the balls, because they drove the cars quietly and carefully.

Comment. The appearance of a ball, the center of which was at the point of contact of the "whiskers" of the trolleybus.

16. (Bychkov and Nikitin 2005) I. S. Arefyeva, female, sent in 2004.

"In November 2002, I (the first observer) and my husband (the second observer) observed BL in my own apartment in Moscow. At 18.00 on a non-working electric stove, (the power supply is turned off) a stainless steel saucepan with salted water was supplied with a glass-ceramic coating. Two standard boilers were lowered into the pot. The first observer performed these actions. At 19.00, the water was brought to an almost complete boil, and at that moment a second observer entered the kitchen. At maximum outburst of emotions of the second observer, the first observer was called into the kitchen and immediately disconnected the boilers from the network. At that moment, both observers saw an unusual phenomenon. An orange ball formed at the bottom of the pan. Its diameter reached 5–7 cm. It had a bright-yellow-blue glow and a halo of pale blue color. The ball shimmered with shades of orange (from the inside) and blue (from above, like wool) colors. It slowly floated out of the pan, floated through the air about a meter for 1 min, and suddenly fell on the table near the metal frame of the table. The moment of the fall was accompanied by a flash of bright color and a crack.

At the place of the fall, two spot traces were found and burned through: the paneling, the countertop, the drawer, and the lower shelves.

The descriptions of the observed phenomenon completely coincided with both observers, except for the emotional assessment. In the second observer, BL caused fear and an increase in anger, and in the first—a desire to calm, stroke, convince to leave.

With the exception of spot marks in the countertop, the table, stove and pan and the boilers remained intact. The potato tasted like a distinct metallic taste, but taking it as food did not cause any digestive disorders."

Comment. The appearance of a ball with a short circuit in the pan. Thermal effects on objects.

17. (Bychkov 2005) V. A. Ivanov, male, Physicist, Ph.D., born in 1948, observation in 1962, survey in 2005. Moscow region, August, pre-storm situation, dense clouds, the veranda of a country house, evening.

"I closed the windows, pulled the antenna out of the TV and put it along the sofa. Suddenly I heard an electric crackling and hissing. A ball 2–3 cm in diameter separated from the antenna. It was bright, blinding white. The ball was round, its borders were blurred, and the border was changing. It began to move along the floor and flew out through the gap between the floor and the door. The width of the gap is 1–1.5 cm. The ball was moving at a speed of 1–2 m/s. The shortest distance to the ball was 50 cm. The duration of observation was 1–1.5 s. The ball passed 2.5 m."

Comment. The appearance of the ball from the antenna. Flew out through the gap between the floor and the door.

18. (Olkhovatov 2008) England at the end of the 1960s.

In the summer, an eyewitness (female) sat in the kitchen and cleaned shoes that had metal pads on them. Suddenly, she heard a loud sound like a pistol shot. The sound repeated two more times, and each time a small flame shot out of the heel. Then the flame jumped on the stove and repeated the same thing on it. After that, the flame formed a fireball the size of a small soccer ball, which then flew away. After a short time, a thunderstorm began.

Comment. The appearance of a ball before a thunderstorm from metal pads.

19. E. N. Kudryavtseva, female: presented and commented by A. I. Nikitin.

"In the summer of 1947. I was on tour with the ensemble in the city of Chernovtsi, Moldavian SSR. After the rehearsal, at 2–3 o'clock in the afternoon, I went to a colleague in the house where she (my friend) temporarily lived. The house is one-storey. We were sitting at the window opposite each other and knitting. The window was open. A thunderstorm was gathering outside, lightning was flashing somewhere far away, but there was still a lull here. Suddenly I looked at my friend and saw that her hair had risen up (on end). She looked at me with her eyes widened with fright, since the same thing happened to me, as it turned out. We told each other this later, and each of us did not feel anything unusual about ourselves. Then we saw that a ball with a diameter of 8–14 cm floated over us. It was matte, like a matte light bulb, glowing with the brightness of a full moon. It moved very slowly. The room was small: a stove stood out against the window, a light bulb (40 W) hung on a long cord in the center. When the ball flew past the lamp, the incandescent hair lit up with a reddish color, although the switch was turned off (we checked). When the ball flew away from the lamp, the filament went out. Then the ball flew around the stove, carefully bypassing all corners and not touching it, turned along the wall and began to approach us again. We sat there, not moving, not talking. When the ball flew away from us for the first time, our hair fell off. Now, at this approach, it stood on end again. The ball floated over our heads and flew out of the window. Soon after, a thunderstorm broke out, the rain was pouring down like a bucket, but it was over quickly." If we take the distance between the floor and the ceiling of the room to be 2.3 m and assume that the ball lightning moved horizontally at the height of the window (about 2 m), then it passed at a distance of 1 m from the girls' heads.

Comment. At popular lectures on the action of electricity, a spectacular experiment is often demonstrated. A girl with loose hair is placed on an insulating stand and offered to touch the pole of an electrostatic generator with a conductive rod. A wire lying on the floor is attached to the other pole. When the voltage rises to 300 kV, the girl's hair rises in the form of a fan (Fig. 2.28). Evaluation, the values of the electric field strength near the girl's head, at which a force arises that can lift her hair turns out to be at least 1 kV/cm. Let us pay attention to another detail of the observation of ball lightning in 1947 in Chernovtsi, when it caused the glow of an incandescent lamp turned off from the mains. This echoes the description of her observation in 1981 in Odessa, when a ball lightning, passing by the lamps of the lighting network, caused them to flare up.

The reason for holding a large charge of the BL may be the presence of a good insulator in the area between the charge carrier and its surface. However, when shorting this insulator with some conductor, the discharge current of the BL can cause melting of the conductor or arc discharge. If the discharge current passes through the human body, it can lead to injury and even death. On 26.07.1753, a tragic event occurred in St. Petersburg, which significantly affected subsequent studies of atmospheric electricity. On this day ball lightning killed Academician of the Russian

Fig. 2.28 Demonstration of the effect of high voltage on a person (Ford 2002)

Imperial Academy of Sciences G. V. Richman. "A pale blue fireball hit the professor right in the forehead, who immediately fell on his back without making a sound." During the examination of the body "a red spot was found on the forehead, from which several drops of blood flowed out, the skin nearby was not damaged. The shoe on his left foot was torn. When the stocking was removed, a blue mark was found in this place. Based on this, we concluded that the electrical force of the thunderstorm, having hit the head, came out through the foot." This description, as well as the descriptions of numerous cases that followed, clearly indicate that G. Richman was struck by an electric current that passed through his body.

20. (Bychkov et al. 2003) I. E. Gortunov, male, Software Engineer, survey1997. Observation at the age of 14 in 1952 at noon in July in the village of Yenishevo, Smolensk region.

The boys and I were fishing. A black cloud moved in and a thunderstorm began. The wooden structure of the mill on a columnar base served as a place of shelter for us. During the torrential rain, there was a very bright flash and suddenly there was a broken trajectory of linear lightning. It hit the body of the earthen dam. At the same time, I felt a strong electric shock on my bare feet. I wanted to jump on a stump, but the paralytic state because of the impact lasted about 5 s. Despite this, I saw a dazzling ball, the size of an orange, rise into the air from the place of the lightning strike, 20 m away from me. It was red, like the color of a ruby laser, shone brighter than 200-W lamp. Its clearly defined surface was clearly visible. The ball hissed and crackled. The ball rose to a height of 3–4 m and over arc, flying about 10 m at a speed of about 1 m per second, it sank to the ground and exploded. In total, it lived for about 15 s. The sound was like a cannon shot.

After the thunderstorm, the boys and I examined the place of the linear lightning strike to the ground. We found a funnel with a diameter of about 7 cm and a corkscrew-shaped hole. The stick lowered down did not reach the end due to the complexity of the channel geometry. The places of the ball lightning explosion could not be found.

Comment. The observer felt a strong electric shock on his bare feet.

21. I. Yu. Numerovskaya, female, 1961 y.b. Observation at the age of about 8 y.o. (together with a neighbor boy). Accountant, survey 2016. Alma-Ata, USSR. In the house on the 4th floor. There was no thunderstorm.

At the moment when an electrical appliance designed for a voltage below the local (127 V) was inserted into the electrical outlet, a bright-white ball with a diameter of 8–10 cm flew out of it. It flew to a distance of 40–50 cm from the wall. Hung in the air at the level of the outlet 5–7 s and went back into the wall. It left a mark in the form of a crack (0.5 mm thin) and about 20 cm long in a plastered brick wall.

General comment on the section

A large number of observations of BL are associated with their occurrence in electrical devices, inlet and outlet from them. This raises the problem of the charge leaving the conductors and creating luminous formations by them.

2.5 Interaction of BL with Various Objects

1. Kirichenko S. I., Assistant Director observed the BL at 14.00 local time on 15.08.1980 in a place 90 km from Krasnodar towards Sochi during the filming of the movie. Russia. The story of 1998.

Twenty minutes after the end of the storm, the sun was shining brightly, there was no wind and it was quite hot, about 30°C. During the filming, I and other employees suddenly noticed a glowing circle about twenty meters away from us, three meters above the ground, like a sunbeam from a mirror. "Look, what it is?" we whispered. It was moving towards us, and we looked around: "Who is directing a bunny to the set? And what if it gets into the actor's eyes and prevents him from working?" And the ball kept moving towards us, and at first we could not understand what it was, because as it approached, it increased in size to the size of a rugby ball and somewhat resembled a comet. It moved like a ball up and down, while oscillating like an air ball not in one straight line, light in appearance, but heavier than the air ball. At the same time, a ball made of jelly was swaying like jelly. Its color was yellowish, bluish, and brownish with orange and rainbow tints, quite bright, like a 100-W incandescent lamp. Here is such a colored jelly. It also resembled a soap bubble, but not transparent, but made of a thick mass.

At first, the ball moved along the top, but then it began to descend towards the ground and entered directly into the dig—the base of a large lighting device. It moved quite fast and walked about twenty meters in about thirty seconds.

The dig was very hot and covered with rows of holes with a diameter of 3–5 cm. BL hit like a pancake, and then seeped through round holes inside the dig. There was a wild roar, like a grenade explosion, and the hundred-kilogram structure fell to the ground. Metal pieces flew off from it, as if a shell had hit it. The metal pieces were twisted. People shied away in different directions. One illuminator got a hot piece in his shoe.

Comment. The BL hit like a pancake, and then seeped through round holes into the dig—the base of a large lighting device. Then there was a wild roar, like a grenade explosion.

2. V. N. Nezamaykin, an engineering student went hunting on the banks of the Dnieper River near the village of Kopos (story of 1992) Russia at the end of June 1972. The sky was clear; there were no thunderstorms and no precipitation. Coming out into the open, on the high bank of the Dnieper, he suddenly saw a glowing object 7 m away from him, in the reeds. It was a ball 10 cm in diameter, bright blue, shining like a 100 or 200-W bulb. Its substance, dense in appearance, fluctuated unsteadily.

The BL hung motionless over the sedge, at a height of 60–70 cm above the ground. Its substance, dense in appearance, fluctuated unsteadily, like a soap bubble.

Without thinking twice, the student shot it by a shotgun with buckshot. The size of the ball increased by one and half times, after which it disintegrated with a pop.

Its pieces of ragged shape, asymmetrical, went out at the surface of the water. All this went on for a minute or a minute and a half.

Comment. The following single case is given in (Stakhanov 1996). "V. A. Bobrin, an Associate Professor from Khabarovsk, met a ball lightning in July 1954 in the evening before a thunderstorm, when he went hunting. It was about 20–25 cm in diameter, white, moved horizontally, repeating the terrain. When it passed at a distance of 15 m, he shot it with a shotgun. There was no effect; the ball only swung. After about 40 s, the BL left the field of view, having passed about 40 m." Based on this observation, Stakhanov makes a fundamental conclusion that the BL is a gaseous medium and its density is approximately equal to the density of air. But from the observation of Nezamaykin V. N., it does not follow that the BL has the weight of air, on the contrary, it is heavier than air if its individual pieces fell, in addition, the extinction of the material is described.

In the article (Likhosherstnykh 1983), it is reported that I. Emelyanov shot at the BL. Lost consciousness. The barrel of the gun was torn.

Tarakovsky A. (Likhosherstnykh 1983) threw a knife at the BL. Lost consciousness. The knife turned out to be melted.

Comment. It follows from the descriptions that the interaction of BL with metal objects is a complex process with the release and transportation of energy.

3. (Bychkov et al. 2003) E. R. Ilyin, Mechanical Engineer Ph.D., born in 1938, survey of 1993. Observation: 1958, August 15–30. Local time 17–18.00 local time, Moscow, in the house on the 4th floor. 3 observers. The reaction is fright. Dry, sunny, warm, no thunderstorms.

A ball with a diameter of 6–8 cm flew into the open window, covered with gauze. The border of the ball was blurry. The intensity of the glow can be compared with an incandescent lamp with a power of 10–15 W. The observation time was 1–2 s. The color is white-yellow, matte. The maximum distance traveled is 6–7 m. The trajectory is horizontal. The BL flew smoothly over the floor at a height of 2–2.5 m. The smallest distance to the observer was 0.5 m. The BL went into the corridor from where the crackling was heard. The observer found nothing there, it smelled of sulfur or burnt.

The shortest distance of the observer to the ball is 1.5–2 m. The path traversed by the ball is 300 m. Observation time: inside the house 30–40 s, outside ~3 min. In total, 5 people observed the object.

Comment. A ball flew into the open window, covered with gauze.

4. (Bychkov and Bychkov 2008) I. M. Naboko female, Physicist, Ph.D., leading researcher, born in 1931, Observation of 1948, July 15, time 14–16.00. Moscow, Tushino, open place. Russia. Watched by 4 people. There was no wind. Survey of 1991.

"It was a hot summer day. A thunderstorm with lightning, thunder and short rain gathered. I was hiding in the doorway of the store. After the rain, I first stood, then walked, and suddenly saw this phenomenon. The BL at an altitude of 6–7 m flew smoothly, straight to the metal truss of the high-voltage transmission line. It flew up to the farm, touched it and discharged with an explosion, there was a bright flash. It left a stain of oxidized metal on the surface of the farm and drops as from welding. We examined these tracks in the evening."

It was a 15–20 cm ball, pulsating and iridescent. The color of the ball was reddish, purplish-blue, and dim. Its glow corresponded to 10–20 W of incandescent light. It changed the shape of a ball to an ellipsoidal one. The longitudinal and transverse dimensions were correlated as 1.0–1.1. The border was shimmering, breathable, and blurred. The observation time was about 60 s. The BL has passed 10 m. The smallest distance from the BL to the observer was 20 m.

Comment. The BL left a stain of oxidized metal on the surface of the metal truss and drops as from welding.

5. (Bychkov and Bychkov 2008) S. V. Maksimenko, male, born in 1958, Physicist, Ph.D. in Physics, 1993 survey. Observation May–June 1984, 14.30–17.30 local time. Nikolaev. Russia. Thunderstorm, rain. BL outdoors. 2 observers.

"There was a thunderstorm outside. There were several lightning bolts, I heard their sound with an interval of 1–3 s, but sitting in the room, by the window, I saw only one of them.

The BL came into my field of vision at the window of the 3rd floor of the building, approximately at a height of 10–12 m above the ground. It was hanging motionless directly outside the window at a distance of 50 cm from the windowpane. It was a sphere slightly larger than a ping-pong ball with a diameter of 4–5 cm. The BL glowed like an incandescent lamp of 20–30 W. Its color was blue-white, dull (as if phosphorescent). The smallest distance from me to the BL was 1.5 m. The BL exploded. The telephone and wiring were located at a distance of 0.6–0.7 m from the object. During the explosion, the insulation on the telephone wire melted, the phone failed.

Comment. The BL exploded. At a distance of 0.6–0.7 m during the explosion, the insulation on the telephone wire melted, the phone failed.

6. (Bychkov and Bychkov 2008) M. T. Dmitriev Chemist. DS in chemistry. Stavropol Territory. Russia.

During a severe thunderstorm, a fireball the size of a football bounced down the street. When in contact with the ground, it knocked out pits half a meter deep and one and a half in diameter. Thus, the ball riddled the entire street for two blocks. Then it burst with a noise and went into the sky like a fiery jet.... The street was paved just before that.

Comment. The ball, when in contact with the ground, knocked out pits half a meter deep and one and a half in diameter.

7. Johnson J. <Brick_96 Hotmail Com>. Live oak, Florida, USA-Monday, 02/27/2006 14:53:42 (PST). Database "Strange Science" about IV.

This happened to me when I was about 10 years old (I'm 66 now). We lived at Conyers Farm, USA. There was a severe thunderstorm at that time. I was on the front porch and it got so unpleasant outside that I turned around to go inside. Shortly before I got to the door, a ball of orange fire the size of a basketball flashed through the screen door. I tried to tell my family, but they wouldn't believe me.

After a few weeks, the screen fell apart and fell out. In addition, they believed me about the observation of BL. The hole was the same size as a basketball hoop.

Comment. The passage of the BL through the screen door mesh.

8. http://www.paralljelnyj-mir.com/products/tri-vstrechi-s-sharovymi-molniyami-/ Summer, 1967. Petersburg. This old house in St. Petersburg still stands at the intersection of Tchaikovsky and Chernyshevsky streets. Its roof is decorated with a small elegant turret with a metal spire.

A short summer thunderstorm has just thundered. However, the rain had not stopped yet, and I had to wait it out on the porch of a neighboring house. Suddenly, lightning flashed, thunder thundered, and at the same moment a blue ball, somewhat smaller than a football, rushed out of nowhere from the sky to the spire. As soon as it came into contact with the iron, there was a bright-white flash, and an explosion thundered. Half of the spire, rather thick, was cut off as if with a knife. When I was climbing the familiar stairs, I noticed that thick smoke was pouring out of the doors of one of the apartments. As it turned out, the TV turned on after the BL hit the spitz caught fire, but the hostess managed to extinguish the flame with a tub of water.

Comment. As soon as the BL came into contact with the iron, there was a bright-white flash, and an explosion thundered. Half of the spire, quite thick, was cut off as if with a knife.

My third meeting with the BL took place in the early summer of 1972 near the village of Krasnitsy, Leningrad region. Russia.

I was returning with my wife by a country road after a thunderstorm to my dacha. It was not far to the village when the movement of a rather large and brightly glowing white ball with a noticeable blue sheen became noticeable to our left. By eye, the diameter of the ball was about half a meter. I watched the object not without fear. Meanwhile, the ball was moving rapidly at a height of 3–4 m from the ground between the road we were walking on and the young forest. The trajectory of the ball was directed downwards towards some dark object lying on the ground near the road. After reaching the object, the ball exploded and disappeared. A frightened girl was running from the forest to the object. Approaching it, I saw a bicycle in front of me, which had no steering wheel, and the tire of the front wheel was smoldering, exuding the smell of burnt rubber. The rim of the wheel was melted.

Comment. The BL came into contact with the bike, after that it did not have a steering wheel, and the tire of the front wheel was smoldering. The rim of the wheel was melted.

9. (Bychkov and Vaulin 2013) Sheichenko Stas. 2001, Alchevsk.

Summer, a gloomy day without a thunderstorm. A brilliant fireball with a diameter of 10 cm flew through the open window. It calmly "floated" around the room and stopped for a moment near the painting, then, crackling, passed through it and disappeared. The painting fell, because along with the lightning, which disappeared, the nail on which the painting hung disappeared.

Comment. The ball "floated" across the room and stopped for a moment near the painting, then, crackling, passed through it and disappeared.

10. (Bychkov and Vaulin 2013) E. N. Vaulina, female, born in 1930. Biologist, head of a Laboratory. The age at the time of observation is approximately 64. Observation of 1991; the second half of August. Until midnight: 20–22.LT. Moscow region, Klinsky District, Suburban areas in the village of Pokrovka. Russia.

The object was in the room. First floor, prefabricated wooden house. The wall is wooden. Bedroom. There was an outlet on the wall near the window; a wire ran along the window. On the street, a metal hook, quite thick, was driven into the wall at the place where the wire passed. A ball appeared from the wire above the outlet. The ball, blue, glowed brightly with blue light and moved very slowly into the room. It was the size of a medium-sized apple, about 8 cm in diameter. It danced a little on the spot. The diameter of the object was 7–10 cm. The longitudinal and transverse diameter of the object practically did not differ. The boundaries of the object: slightly vague. The borders were sharp, but light was coming from the borders, and therefore it seemed that there was no sharp border. The object did not have a core. The object had no halo.

The glow of the BL: very bright, like linear lightning (it was not dark yet). BL was near the window. The smallest distance to the BL was 1.5–2 m. The BL passed 0.3–0.4 m, while it was flowing. The observer jumped out of the room, and when she returned, BL was no longer there. Very slow, light flow. The observation time was about 5 s. The BL was located from the ground at a distance of about 1.5 m. The color of the object is blue. The BL had no structure, a uniform, round dense object.

There was no heat from the BL, and there was no smell. During the observation, the size of the object did not change. During the observation, the color of the object did not change.

The moment of formation of the BL. The object appeared on the wall, where the wire was. Immediately appeared and did not change. I did not see how BL disappeared. I left the room.

Near the BL was an electrical outlet below the place of BL origin by 60 cm, and the wire. There was the impression that the BL originated on the wire.

Whether the appearance of the object affected the operation of radio devices was not recorded. Everything was turned off. Weather conditions at the time of observation: thunderstorm, rain. The wind is not set.

The BL floated horizontally, parallel to the ground. Like a piece of fluff. The trajectory of movement is straight parallel to the floor. After the observer returned to the room, there were no traces left.

Country house. A nail was driven into the wall from the street side. When a linear lightning bolt struck (into a nail), a spherical object with a diameter of 8 cm appeared inside the room on the opposite side of the wall to the nail. The color of the object was bluish; the object was opaque, dense. The lifetime was 10 s. The object was moving smoothly. While I was turning off the electric meter in the next room, the ball disappeared.

Comment. A ball appeared from the wire above the outlet.

11. (Bychkov and Vaulin 2014) Ksyunichka, female. 9.06.2011, 16:09 L.T. Russia.

We were about 14 y.o. We were standing in the courtyard of our four-storey building (a four-storey house), me and my two friends. A thunderstorm was starting, lightning was flashing, but not a drop had fallen yet. Then in front of us the ball descends. It was yellow-red, but with blurred edges. It goes into the asphalt. How we "screamed," probably the whole house heard, in 30 s we overcame all 4 floors. There was a crack in the place where the ball went.

Comment. At the place where the ball went into the asphalt, there was a crack.

12. (Bychkov and Vaulin 2014) Byte. 9.06.2011, 19:34 L.T. Russia.

I have only seen orange BL twice.

One with a loud pop entered the room through the gauze on the window—without damaging it, stood for a couple of seconds and left through the outlet. Everything that was plugged into the outlet—the kettle and the TV—burned down.

Comment. BL entered the room with a loud bang through the gauze on the window—without damaging it.

13. (Bychkov and Vaulin 2014) RIA Novosti Agency Archive 18:19 L.T. 06.24.2002. Russia. BL caused an emergency at the Azot Chemical Plant in the Kemerovo region, on June 23.

As a result of a direct hit of BL into the pipe of the compressor department of the ammonia-1 workshop, the oil block was depressurized, several tens of kilograms of hot oil were released to the surface and ignited.

Comment. Consequences of BL getting into the compressor compartment pipe.

14. (Bychkov and Vaulin 2014) Female. 07.1987, Russia between Moscow and St. Petersburg. Dry thunderstorm (thunder and lightning without rain).

At a height of about 5–6 m, a fireball hangs. Small, the size of a child's ball. It was about 20 m from me to it. I did not see how it appeared. I turned around at the crack, and it was already hanging. The crackling came from it, like the sound of a dry branch breaking. After a couple of seconds, it exploded. Deafening. The volume is comparable to an anti-tank mine, if you heard it, but without a shock wave.

Comment. The crackling came from the BL, like the sound of a dry branch breaking. It exploded. Deafening. The volume is comparable to an anti-tank mine.

15. (Bychkov and Vaulin 2014) 19.20 LT 29.6.2006. Radio station Echo of Moscow, Altai Krai, village of Petukhov Log. Russia.

The BL hit the television antenna on the roof of the building. It went down the antenna, melting it, and entered the house. As a result, the TV burned down. It was turned off at that moment. An electric meter, a fan, and a wall clock also melted. Windows were broken on the veranda of the house. The residents of the house were not injured.

Comment. The BL went down the antenna, melting it, and entered the house. As a result, the TV burned down. It was turned off at that moment.

16. (Bychkov et al. 2015a, b) P. Averchenko, male. 27.04.2015 Section: 1984–1996. "The Soviet period." About the impact of the BL in the building of the "hotel" under construction, in the middle of a gloomy October day in 1991.

"We are standing with the men on the street. The building of the 'hotel' is about four hundred meters away from us…Suddenly a dull explosion is heard from the construction site. I turn around and see a cloud of gray-brown smoke slowly rising over the right side of this building. What is it? I go up to the department, call the supervisor (supervisor's room)—the phone does not answer. I am going to the construction site. In the supervisor's room, the assembled builders discuss what happened, and the crane operator narrates: 'The time is the beginning of lunch. I am going down. Suddenly, out of the corner of my eye I see an incomprehensible glow. I look, and a glowing ball is flying along the high-voltage line. At some point, the ball slowed down, then began to move away from the power line and flew to the crane. I am sitting. I am afraid to move. The ball slowly flew up and ended up between the tower and the cables. It stopped for a few seconds, swayed from side to side and abruptly broke off—hit the wall. Explosion!"

I ask the supervisor: "Why didn't you pick up the phone?"—"Let's go outside." Shows:—"Here was a 'ball' (an electric wire running through the air). And now, look–it's all on the ground."

I look and see a smooth "dotted line" consisting of tubes of light "braid," no longer than ten centimeters each. Inside—not a drop (!) metal. Not in a single piece!.

We inspect the building. The place of the "blow" is clearly highlighted on the facade. You can see a dark, sooty spot and traces down the facade, as if molten metal dripped (or flowed) over the bricks.

We go up to the top floor. Lightning struck the area of the overlap disk of the seventh floor. Reinforced concrete slabs are all in caverns, i.e. in the places of reinforcement passage, in some areas; an unknown force tears out the protective layer of concrete. Therefore, it is all over the floor. On the landing of the stairs of the upper floor, there was a reinforcing mesh. At that time, concreting had not yet been carried out. Rebar, where it was not boiled—burned, i.e. where the rods simply touched each other, or there was a slight gap—simply burned out "on a pencil." On the roof, above the place of the lightning strike, the parapet slabs are broken into pieces. Fault sites are sooty with traces of thermal exposure (the grains of sand are shiny, "glazed," and small spherical droplets of metal).

What else did lightning do: in the opposite wing of the building, on the third floor, in the corridor, the heating pipe burst. (Lightning also struck at the ceiling level of the 7th floor.) In some apartments, in places where heating pipes pass through the ceiling, metal balls are found on the floors—as after welding. This was especially noticeable on the freshly planed floor boards, on which there were dark spots from the hot metal. The windows of several apartments located under the impact site have broken windows.

The supervisor also said that the excited Manager of the kindergarten, whose building was located about a hundred meters from the construction site, called to him and said that she was sitting at the table and was very scared when the phone suddenly, for no reason, jumped almost a meter, then fell to the floor and began to smoke. (The manager decided that something had happened at the construction site, and hurried to check …). In addition, another thing: at the moment of the lightning strike, a team of finishers was coming down the stairs for lunch. How lucky they were that no one was holding on to the stair railing, which at some point in time suddenly sparkled and "splashed" with hot metal. None of the people at the construction site was injured.

Comment. Destruction caused by BL.

17. 17.04.2017 at 15:44, Belorussia, the correspondent of the Minsk-News agency Daria Orel interviewed a Minsk resident Andrei Zotov, a former firefighter, an event until 2010.

There was a fire at the transformer substation "Integral." We decided to wait for the arrival of the electrician, remaining behind the bars separating us from the fire. After a while, BL popped out of there. It walked through the metal fence, melting it, and then struck the top of a colleague's head and burned his hair. Thank God I managed to duck…

Comment. The BL passed through the metal fence, melting it, and then struck the top of the colleague's head and burned his hair.

18. (Bychkov et al. 2003) M. V. Lozovsky, male, 1994. About the observation of BL in mid-July 1983–198; near the village of Voronovo near Moscow, Russia, age 8 y.o.

It was in the afternoon at 17–17.30 LT A thunderstorm broke out with heavy rain and lightning. There was a strong wind, the air temperature was 22–23 °C. My grandmother and I were returning from the forest. Suddenly, in a round green clearing, we saw at a distance of 15 m from us how a ball 20–22 cm in diameter was rolling along a spiral trajectory. The ball was spinning: it will touch the ground and move 5 cm. It resembled a ball of woollen threads. The base is a tangle of red color; it is covered with blue threads. The intensity of the glow is like a lamp of 120 W. We got scared and left quickly. The next day we saw that the surface of the clearing was scorched, its color turned brown-black.

Comment. BL resembled a ball of woollen threads. Burned the surface of the clearing.

19. (Bychkov and Bychkov 2008) I. P. Zavershinskyi, male, Physicist, Ph.D. 20 y.o. at the time of observation, 40 y.o. at the time of the survey in 2003, Samar,

Russia. Observation of 1983 from the garage. Summer-July, the middle of the day, up to 15.00 L.T. It was a very hot day. Then there was a strong thunderstorm with lightning. After the storm, the whole sky was overcast. Observation was limited to 7–8 m of view.

"The ball came into view, moved horizontally, in the street, at a distance of 6–7 m from the observer. Diameter—10 cm. The color is reddish-bluish. The ball glowed irregularly. Lifetime—less than 1 min.

The ball floated along an irregular wavy trajectory, then bumped into an iron gate, hissed and disappeared. There was a blackish, light trace of melting on the gate."

Comment. The ball floated along an irregular wavy trajectory, then bumped into an iron gate, hissed and disappeared.

20. V. B. Gildenburg, male, Physicist, Ph.D., survey 1994, 9 y.o. during observation 06.1946 12–14.00 LT Observation in the basement (school buffet), Nizhny Novgorod. Russia.

"There was a thunderstorm with lightning and rain outside. With every clap of thunder, it was getting closer. Simultaneously, there was a flash and a thunderclap. There was an impression that a red ball of a regular shape with clear boundaries with a diameter of 15 cm hit the window, the glass broke. The object went out. The observation took place for a fraction of a second. It was 5–10 m to the window."

Comment. A red ball of regular shape with clear borders with a diameter of 15 cm hit the window, the glass broke.

21. E. V. Gnezditskaya, female, DS in Biology, born in 1939, V. V. Gnezditsky, male, DS in Biology, born in 1941. Observation on 07.25.1969, 16–18.00 LT Tver region near the village Pestovo, Russia. Survey of 1991. The observation was conducted from an open place on the high bank of the Mologa River. The wind is moderate, partly cloudy with rain. The rain stopped, the sun came out for a while. Before the storm. Cold, T = 12–15 °C. High humidity ~90%.

"Visibility was excellent. A yellow ball with a diameter of 40–60 cm came into view when it was (hovered) over the river at a distance of 1–1.5 km from us. It attracted attention by a complex trajectory. It descended, hung, and fluctuated right-left, up-down. It moved above the earth's surface at first at a height of 15–20 m, then 10–15 m. The ball was moving towards the village. When it reached the village, it moved, touching the poles with electrical wiring. It floated to the house in the center of the village. It sat on the chimney, and then rolled into it. Immediately a column of smoke poured out of the chimney. Before that, it was clear that the stove was not heated in the house.

The horizontal and vertical dimensions of the ball did not differ. The outline was clear, but the borders were blurry. The radiation intensity was 80–100 W of an incandescent lamp. The ball did not rotate. The observation time was 5–7 min. The distance traveled by the ball was 100–200 m."

Comment. The ball landed on the chimney, and then rolled into it. Immediately a column of smoke poured out of the chimney.

22. "Socialist Industry" newspaper, 08.22.1981. Yurt station. Russia.

A freight train passed through the Yurt Station at night. The driver of the electric locomotive V. Arkhipov (male) noticed a white ball with a fiery plume flying towards the train. "What is it?" the driver was surprised. "Comet? He did not have time to finish—the electric locomotive was shaken by a powerful blow; there was a deafening explosion. The driver lost consciousness. A bright flash blinded the assistant. About 250 m the train went without control. The BL pierced the searchlight, melted glass into metal, and blinded railway workers.

Comment. The BL blow into a freight train.

23. (Olkhovatov 2008) 06.16.1819 Kutch, India.

A fireball descended from the hill to the ground. It burned the vegetation.

Comment. The ball burned the vegetation.

24. (Stenhoff 1999) K. Lloyd, personal communication. 1977.

On Monday. March 28, 1977 at 9 p.m. a woman at Newgate Street Village, Hertford, England saw a white light motionless just above a flower bed in the garden of her house, no more than 20 ft (6 m) away from her as she stood in her kitchen The light was between the size of a bicycle lamp and a car headlamp, perhaps about 4 in. (10 cm) in diameter. There was a wire perimeter fence just beyond the flower bed, and this passed close to a substation transformer adjacent to the property. The light disappeared after 3–5 s. There was nothing in the kitchen that could have caused a reflection. The following morning there was a heavy frost.

She discovered a round patch of burnt, blackened grass, partly reduced to ash exactly where the light had appeared. It was about the same size as the light and the grass in front of it was somewhat yellow-brown. The black ash smelled of fresh burning. A weather report (K. Grayling, personal communication 1977) confirmed the presence of cumulonimbus cloud and elevated atmospheric pressure, although there were no reports of lightning or thunder.

Comment. The ball burned the earth.

25. (Stenhoff 1999) R. L. Hooper, personal communication. 1977. An event involving damage to road surface took place in the early summer of 1931. A very severe thunderstorm began at about 8 p.m. at Portrewydd, Gwent, Wales. At midnight or later, two boys sat in a window watching the storm. About 300 yards (270 m) from their house was a paddock surrounded by a brick wall. Suddenly they saw a ball about the size of a football strike the wall and rebound onto the road. There was a shower of sparks and it disappeared. The ball had been in sight for more than 5 s. The next morning they found a hole in the wall and a small crater in the road.

Comment. BL influence on a wall material and asphalt.

26. (Stenhoff 1999) Stenhoff 1981. Around 1 p.m. on a dull and overcast day in February 1981 a woman was cooking lunch on an electric stove in the kitchen of her ground-floor flat, near Warminster, Wiltshire, England.

Suddenly, out of the corner of her eye, she saw in the adjacent room a small blue ball of light about the size of a large marble (from a children's game). Its color was similar to that of a match that just been struck, but it was much brighter. The windows in the room were probably open, but covered with net curtains. The ball traveled swiftly at about 1 ms^{-1} towards her, about 1.2 m above the ground, passing in front of her face, then it veered off towards one of the electric rings at the rear of the stove as if it was attracted to it. She quickly removed a saucepan that was on the ring. Both this ring and the ring in front of it were glowing red-hot. The ball attached itself to the ring, cracking loudly and ejecting sparks. She was very alarmed and turned off the main switch to the stove, whereupon the ball burned itself out. The entire event lasted only a few seconds, and the ball spent most of its lifetime in contact with the ring. At the same time as its appearance, she noticed loud crackling on the radio in the adjacent room, although the radio had intermittent fault, which produced a similar noise. The electric ring no longer worked, but other parts of the stove circuit were not damaged. The ring consisted of an element spiral surrounded by insulating powder, encased in a heat-resistant metal sheath. There was a furrow melted in the sheath of the electric ring 0.14 in width near the surface and 0.045 cm in width at its deepest point was melted in the shell of the burner, the groove was surrounded by a yellow deposit. Some of the melted metal formed two roughly hemispherical globules, solidified at one end of the furrow. The sheath was constructed of a nickel-chromium-iron alloy with a melting point of 1628–1658 K. About 1.46×10^{-7} m^3 of metal was melted, requiring energy somewhat in excess of 600 J.

Comment. The release of BL energy on metal. If we consider BL size comparable with marble (from a children's game) then one has to suppose that its diameter was about 1 cm. In this case, the energy density of BL was about $1.1 \cdot 10^3$ J/cm^3, that is, by about 40 times greater than energy density of average BL.

27. (Stenhoff 1999) (Maidenhead Advertiser, February 18, 1983, E.P. Holland, Personal communication 1983) On February 12, 1983 at 4:30 p.m. in Maidenhead, Berkshire, England, started to snow very lightly. There was no thunderstorm. A woman saw a bright-blue flash go past her glass front door. A second or so later there was a loud bang that shook the house. It sounded like a loud bomb exploding and was heard for several miles around. There were television and citizens' band antennae on the road. In her house there was extensive damage—all electrical devices were destroyed! All the light bulbs smashed, plugs were blown from wall sockets and broken, plaster was ejected from the walls near any wiring, and the telephone and answering machine were destroyed. Some components of a videocassette recorder disappeared altogether. There was a serious fire in the upper part of the house that destroyed most of the contents and the roof. Firemen who attended the scene were adamant that a fireball which had been seen to pass over Maidenhead as the cause.

There was a very unusual hole in the brickwork between the back bedrooms. It appeared as if someone had drilled a hole 2 in. (5 cm) in diameter, and the outer bricks seemed to have been blasted away. There were no burn marks, however, but burn marks were seen near the citizens' band antenna, so it was assumed that this was the point of the strike. Several large holes were blown through the house walls. Paving stones in the garden had been lifted up and smashed in half. Many people in the area suffered damage to televisions and telephones. A woman using a telephone nearby suffered minor facial burns.

Comment. Complex influence of BL.

28. (Stenhoff 1999) Mr. J. Durward reported an event that occurred at Warley, Smethwick, West Midlands, England on August 8, 1975. A severe thunderstorm began about 6 p.m. that evening. Many buildings were damaged. At about7:45 p.m., a lady was in her ground-floor kitchen. Both a louvred window and a door leading to a porch were open. Outside the porch was a conifer tree about 10 ft high.

The woman had just filled a kettle with water and turned towards the gas stove. She saw a ball of light appear between and above the stove and a refrigerator, both of which were next to the open door. Above the stove was a flue leading to an open chimney, sealed with a metal plate inside the kitchen, and above the refrigerator was an air vent.

The ball was round, about the size of the wooden balls used on bowling greens (diameter about 4 in.), with a bright blue to purple core, surrounded by a flame-colored halo. The woman felt burning heat that made her "glow all over" and heard a sound "something like a rattle."

The ball moved towards her at a height of about 39–40 in. Instinctively, she brushed the ball away from her with her left hand. It may have moved downwards about 2 or 3 in. as she struck it. She said, "I can't really say whether it exploded as soon as I touched it, it all happened so fast. I saw it heading for me. As I went to put the kettle on, it seemed as if [the ball] exploded or went off with a bang. I went to knock it away and screamed all at the same time." She estimated that the lifetime of the ball was about 1 s; during this time there was no change in its appearance. The explosion was so loud that her next-door neighbor heard it. She smelled singeing.

After the event, she found a hole melted in her dress and tights. Her legs were not burned, but were numbed and reddened. There was a redness and swelling on her left hand and it seemed as if her gold wedding ring was burning into her finger. The inflammation seemed to appear more quickly than would a scald, and the discomfort from her wedding ring was such that she had to force it off under running cold water. The skin was unbroken. She was only able to wear the ring again after 48 h.

Comment. After the event, an observer found a hole melted in her dress and tights reddened. There was a redness and swelling on her left hand and it seemed as if her gold wedding ring was burning into her finger. The inflammation seemed to appear more quickly than would a scald, and the discomfort from her wedding ring was such that she had to force it off under running cold water.

There was some confusion in her memory of the exact sequence of events. She was not sure whether the ball appeared after or before a clap of thunder, but she thought it more probable that thunder preceded the ball. Her early statement, written 18 days after the event, was as indicated above. However, in correspondence some

6 months later, she suggested that the ball appeared about 5 min after she had placed the kettle on the gas ring of the stove. Given the unreliability of memory, it seems more probable that the earlier statement is correct.

2.5.1 Interaction of the BL with the Tree

1. (Bychkov and Bychkov 2008) L. Maximov, male, Engineer Economist. Observation 1952, July, 12–15, 13–13.30 L.T. Russia, Survey 1990.

"The weather was sunny; a squall wind started 15 min later. A cloud appeared on a gray background. I and 25 other people hid from the storm in the gazebo. A blow was heard, and lightning hit a poplar (the poplar is old, the diameter of the base is 50–60 cm, the height is 30–35 m). Upon impact, the poplar splits to half. The BL fell out of the crevice and walked along the trunk. A burn appeared on the surface of the tree up to its half. The BL separated from the tree. A stallion grazed nearby. The BL passes between the stallion's legs, a sheaf of sparks was visible; after 4 m the BL disappeared."

The diameter of the BL was 10–12 cm. Lifetime 15–20 s. The color was scarlet (fire flame). The distance traveled was 15–20 m.

Comment. BL appeared when linear lightning stroke a tree.

2. (Bychkov and Bychkov 2008) N. E. Afonina, female, Mathematic-Mechanic, Ph.D., Senior Researcher, born in 1943. Observation Abramtsevo, village Radonezh, 55 km from Moscow. Russia. Summer, August. 14–15.00 L.T. survey 2004.

"A weak thunderstorm, there were several thunderclaps. Light. A neighbor said that before that she saw a BL jumping around the site. My daughter and I saw that the BL suddenly appeared in the middle of the spruce. It moved down from the middle of the spruce, from a height of about 12 m. A wire was wound on the Christmas tree. BL went through the wire into the house."

There were traces in the house showing that the BL from the wire got on the wooden wall, fluffed the tree on an area of 50–50 cm. The wooden wall was made of clapboard. It was split, torn, and stuck out in separate pieces. Electrical appliances burned down where there were no fuses, and TV and a charger.

Observation time 1–2 s. The color of the ball was bright white against the background of a dark spruce trunk. The size of the ball was 20 cm.

Comment. BL fluffed the tree over an area of 50 × 50 cm.

3. V. L. Bychkov, male, born in 1951, Physicist, DS. Observation 12.07.2006, Saltykovka Station 15 km from Moscow, Russia, around midnight.

"It was sunny all day, and at 23 p.m. the weather turned bad, and at 23.30 p.m. a thunderstorm with a downpour began. I was sitting 2 m from the window of a country house on the second floor and looked at the flashes of lightning that were

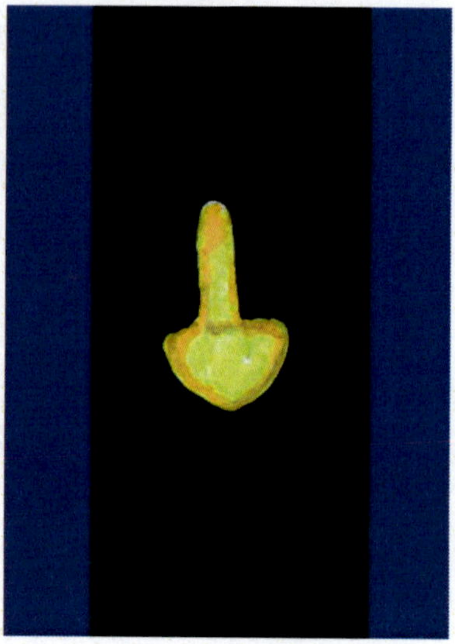

Fig. 2.29 Diagram of the glowing area

reflected on the black trunks of pine trees surrounding the house. Suddenly, after a lightning flash, a glowing area appeared on the trunk of a pine tree near the window (at a height of about 5 m above the ground), located at a distance of 80–90 cm from the house. As if a yellow-green liquid strip appeared, which flowed down the trunk. Therefore, it stretched to a length of about 15 cm with a width of about 5 cm and a thickness of no more than 0.5 cm. At the bottom of the jet (layer), a semicircular area with a diameter of about 6 cm, no more than 1 cm thick, was formed. The next moment, both the layer and the half-ball disappeared. The observation time is 1–2 s. In the morning, I could not find this place on the bark of a pine tree." (Fig. 2.29).

Comment. When the linear lightning struck the tree, a luminous hemisphere appeared.

4. (Bychkov and Vaulin 2014) 04.28.2011, 14.33 L.T. erytryec, Igo. Female. Russia.

"I saw BL. Once as a child. It appeared before the storm. I cannot describe the size—it was far away, 200 m. Most likely, more tennis ball and less volleyball. Orange color. BL seemed to have fallen from the sky in an uneven spiral into the crown of a large tree. The tree burned down. In addition, from the inside. The outer part of the trunk was untouched, and inside there was only ashes. The trunk broke into two parts.

Again 2 years ago (2009), when I went to the peak of Cho Oyu. We were caught in a thunderstorm at an altitude of more than 5000 m, and there were many of these lightnings. Mostly bluish-white and of different sizes. They just appeared out of nowhere; and lightnings from them struck the mountainside, opposite. I can say for sure that I have seen several times how the effect of heated air was observed around ordinary lightning, which manifested from a ball, as from a bonfire, when a distorted image rises and is visible through it. That is, the temperature, most likely, was very big."

Comment. BL seemed to have fallen from the sky in an uneven spiral into the crown of a large tree. The tree burned down. And from the inside. The outer part of the trunk was untouched, and inside there was only ashes. The trunk broke into two parts.

5. Dembitsky NL, male, Ph.D. in technic, Radio Engineer, 45 y.o. at the time of the survey of 1993. August, 1985, 16.00 LT Moscow. Russia. Observation in the open place; green spaces of 50 m—Christmas trees and deciduous trees. Thunderstorm with lightning and downpour.

"One of the lightning struck the top of the Christmas tree. Fragments flew off from it. One of them, which went to the left, was a ball with a diameter of 5 cm. Its color was light white-yellow. The shortest distance to the object was 40 m. The lifetime was less than 1 s. The BL exploded, a dry clap was heard."

Comment. Lightning struck the top of the Christmas tree. Fragments flew off from it. One of them, which went to the left, was a ball.

6. (Pudovkin AK 1999) Male, observation 1996. 18. 00 L.T. Observation time 8–9 s. The distance covered is 62 m. Size: initial 20 cm, final 10 cm. 2 h after the drizzle. Thunderstorm aside.

"There were no lightning bolts. The ball appeared at a height of 12 m, in the middle part of a pine tree. From the moment of the BL appearance on the pine tree, its middle part, about 12 m above the ground and its disappearance in a puddle on a concrete road, no more than 8–9 s passed. The entire visible cycle from the detection to the end of the existence of BL included its direct contact in the zone of the middle part of the pine, vertical descent along the pine, a 90° change in trajectory near the ground. The change in the trajectory led to the partial destruction of the BL. As a result, its volume decreased by about two times—from the original diameter of 20 cm to 10 cm. Then the remaining part of the BL moved along the earth's surface: 12 m over grassy surface, 10 m of pedestrian earthen path and about 40 m of concrete road. Finally, at the last moment of its disappearance on a concrete road, in a pothole with water.

The pulse decay of the BL occurred near the earth, when its trajectory changed by 90° (pine-earth), which was accompanied by a powerful explosion with a very bright short-term flash of light.

About the trajectory of the BL. Its obvious trace of movement has been preserved in the form of a groove of removed bark (Fig. 2.30a) from the middle part of the pine down from a height of 12 m (the roof level of a 4-storey building). The trail ran vertically along the pine tree, expanding to the ground. At the foot of the pine tree, there was a change in the trajectory of movement by 90°, a1 (Fig. 2.30). At this point, the lightning began to collapse with the release of light radiation and explosive acoustic disturbance. Then the ball of a smaller size proceeded over the grass surface (at a distance of 10–12 m) to a pedestrian earthen path, where it again changed its trajectory to an angle of 120°. Further, the movement continued along the trail, and then along a concrete platform and a road parallel to the residential building, with a total distance of 62 m. When moving from the site to the road, the trajectory changed again to 120°. The last section of the path remained straight along the middle part of the road to the pothole with water on the road (in the interval between the I and II entrances of the observation house). This trajectory of the BL movement has interesting features. Its movement along the surface of the pine had an approximately vertical direction to the ground, which was characterized by a trace on the bark. After changing the trajectory near the ground, the option of the minimum distance to the pedestrian path (above the grass) was selected. However, here on this site there was a slope of about 30 cm and this despite the fact that there was a much larger slope from the pine towards the concrete platform, but more heavily overgrown with grass and shrubs.

Once on the hiking trail, the BL began to move along it towards a concrete road of a lower level (by 20 m—a decrease of 1.5 m). Going out on the road with a minimum slope of 40 cm by 40 m, the BL used this slope.

The BL, during its movement, left a significant trace only on the pine in the form of destroyed bark and top dressing to white wood. Starting from the place where the lightning landed on the pine tree, a noticeable groove with a width of 5 mm is visible, expanding to the base of the pine tree up to 50 mm. There were no contaminants or thermal effects on the groove elements (walls, hairs, etc.). It was a clean surface. The geometric expansion of the groove to the base of the pine is most likely due to the corresponding thickening of the pine trunk. The latter fact was confirmed by the fact that there were noticeable trunk thickenings in the form of several ribs on the site. Therefore, starting from the line d (Fig. 2.30a), the groove expands proportionally (the distance is 60 cm).

In the upper part of the pine tree, on the path of the BL movement, there was a branch, which with its one side turned out to be combined with the middle of the groove (zone B). At the same time, the BL bypassed the branch on the left side, while the width of the groove did not change. Below the branch, the vertical alignment of the groove has been preserved.

On the side walls of the groove, there was a step 3' almost along the entire length of the strobe. The section of the groove on a pine tree is shown in Fig. 2.30b, where 1 is the wood of the trunk, 2 is the top dressing, 3 is the bark, 4 is the groove zone.

Eyewitnesses observed that the passage of the BL along the surface of the pine was continuously accompanied by the release of sawdust and dust from the groove at a distance of several meters, similar to a mechanical milling cutter.

When moving along the road, the BL had the appearance of an elongated rounded cylinder with a pointed protrusion in front. Its disappearance occurred in a small depression with water and was accompanied by a faint pop. At the same time, the amount of water was no more than 3–4 l.

There was a bough in the path of the BL (rectilinear vertical trajectory). BL, having made a loop, bypassed the latter and continued to move coaxially to the strobe below the bough. So, if you draw the axis along the entire groove, then it passes approximately through the middle of the bough. The process of "making"

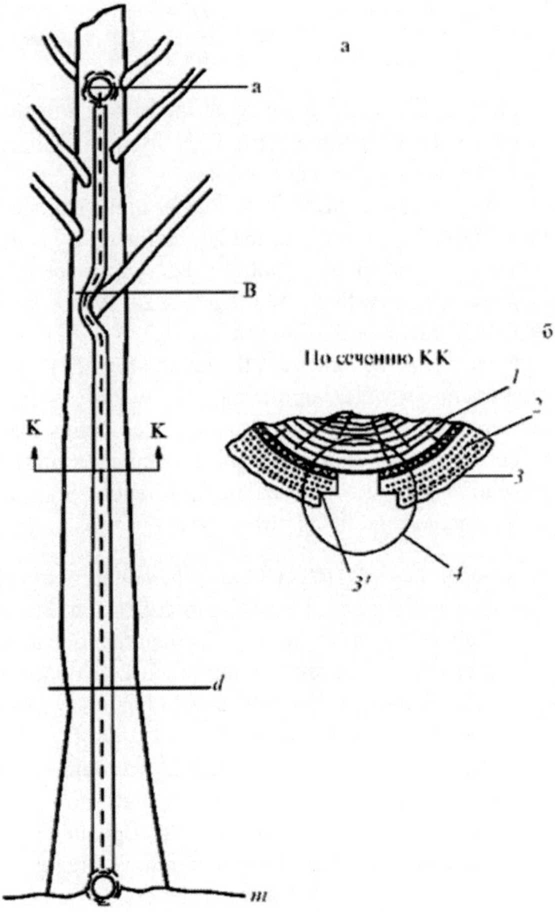

Fig. 2.30 A scheme of the pine-tree impacted by the Ball lightning

the groove was accompanied by the release of fine particles of the crust by several meters.

The last stage of the BL destruction is interesting—it seemed to dive into a puddle, and this was accompanied by a slightly audible pop at a distance of 12 m from it.

When looking at the thinnest hairs of the bark, no burnt places were found. Everything was clean and without changes in the natural colors of the bark, the top dressing and wood."

Comment. A detailed description of the interaction of the BL with the tree.

Let us consider the mechanism of destruction of the tree bark by vapors, which can be realized by heating of water under the bark in the void—channel by the electric current realized by the electric field of BL. At that we consider an energy balance equation of such a heating:

$$U \cdot I \cdot \Delta t = \frac{U^2}{R_w} \cdot \Delta t > C_p \cdot m_w \cdot (T_v - T_0).$$

Here U is the voltage drop at the distance of the order of the BL diameter, or $E \cdot D \sim U$, E is the electric field strength realized in the channel, D is the BL diameter, we assume it about 0.2 m.

I is the electric current in the water in the channel, $\Delta t \sim \frac{D}{v_{bl}}$—time of heating of the distance of the order of the BL diameter. v_{bl} is the velocity of BL, we assume it 5 m/s. Cp is the specific heat of water, which according to biological investigations is close to the distilled water, $Cp \approx 4.2 \cdot 10^3$ kJ/(kg K). R_w is specific resistance of distilled water $R_w \sim 3.5$ Ohm·m. m_w $\rho_w \cdot D \cdot w \cdot d$, ρ_w—water density, w—width of the channel (about 10 mm), d is the depth of water layer (about 2 mm). $T_v = 100$ °C—temperature of water vaporization, $T_0 = 20$ °C—ambient air temperature. Inserting all necessary data into the equation one obtains that the electric field strength has to be $E > 3.6 \cdot 10^5$ V/m. From the text, it is understandable that this field is the projection of BL electric field on the direction of the channel. From the point of view of BL charges and fields discussed above, it is rather modest value.

7. (Stenhoff 1999) (C. G. Palmer, personal communication 1982). Three men were sitting at the bar of a golf club at Selsdon, Surrey, England, on April 14, 1980 at about 7 p.m. There was a violent thunderstorm with much lightning, and the atmosphere was very humid, so the patio doors were open. One of the witnesses wrote: I noticed about 40 yards [37 m] away a pale, orange, nebulous ball which floated into the top of a Lebanon Cedar tree (reputed to have been planted by Elizabeth-I). The size of the ball was approximately … the apparent size of a full moon… After about two seconds, no more, there was an extremely loud clap of thunder, which apparently originated from this ball of light, the result of which was that the glass containing a large gin and tonic which I was holding disintegrated. At that stage, my friends retreated from the open patio doors. The tree, which is a listed tree, had to have concrete poured into its bole and some of the upper branches removed for safety reasons.

Comment. Influencing the tree, which is a listed tree, had to have concrete poured into its bole.

Comment to the section

The interaction of BL with various objects and objects is complex, accompanied by the release of energy. The released energy often goes to the destruction or transformation of the object.

2.6 Interaction of BL with Glass

1. V. V. Demchenko, male, Ph.D., Mathematician, survey 1990. Moscow, Russia.

In August 1972, in the evening, 15 min after a severe thunderstorm, a mathematics student entered the kitchen of his apartment on the third floor of a house in Moscow. Upon entering, he felt the presence of an electric field in the room, such that the hair on his head got up from electrification. He turned on the light. At this moment, 2 m away from him, on the ceiling, in an incandescent lamp, a sphere 25–30 cm in diameter, bright white in color, was formed. It shone like a 100–200 W lamp. It had a core 15–20 cm in diameter. The sphere slowly sank to the linoleum floor and crumbled into small glowing dots. All this lasted 10–15 s.

Patches of burnt linoleum appeared in place of the dots.

The student examined the light bulb. The glass of the bulb did not collapse, the filament evaporated, and its metal condensed on the inner surface of the bulb.

Comment. A sphere formed in the incandescent lamp. The disintegration particles of the sphere left traces of inclusions of burnt linoleum. Impact on glass.

2. (Bychkov and Bychkov 2008) V. A. Kurakin, male, Physicist, Ph.D., 15 y.o. at the time of observation, 48 y.o. at the time of the survey in 1997. 4–5 more people observed. Moscow region, Dmitrovsky District, the village of Bely Rassvet. Russia. The house is wooden, rustic.

"BL appeared in the house in the form of a thread. The thread comes to the glass and disappears. Leaves as if through a glass."

Comment. BL appeared in the form of a thread, went through the glass.

3. N. Yu. Babaeva, female, born in 1957, Physicist, Ph.D. Iowa USA. Observation 3.08.2007.

"On Friday evening, August 3, I'm sitting, listening to music. Suddenly—an unusual lightning outside the window, and at the bottom of the window, the color is like a sodium lamp, but more intense, more yellow-orange. Strange lightning, I thought. I took off my headphones, expecting a terrifying thunder. However, there was no thunder. Moreover, I forgot about it…

In the morning, I found a strange hole in the glass. The windows here are glazed with double panes; the glasses are 1 cm apart from each other. Therefore, at first I thought that the hole was from a large stone. Then I thought, if it is a stone, then why is the second glass intact, not a scratch, nothing? Can a stone, having made such a hole, stop at a distance of 1 cm? As a physicist, I understand that no, it cannot. Inertia. Then I began to examine the edges of the hole. I was horrified. They are smooth, mostly smooth, as if cut with a special tool. They do not feel sharp at all. A little melted at the bottom, see the photo in Figs. 2.31 and 2.32.

I think the BL exploded. I did not see the lightning itself, but I remember a very strange, very intense, orange light at the bottom of the window.

In Fig. 2.33, I present a photo of glass fragments after exposure to BL on glass.

Fig. 2.31 General view from the yard

Fig. 2.32 View from inside the house

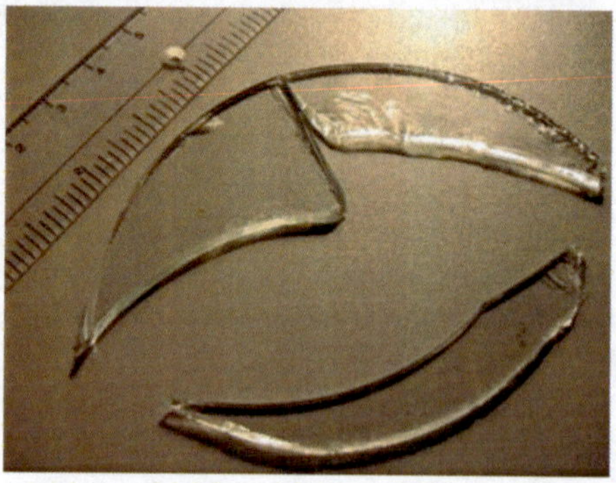

Fig. 2.33 Photo of the glasses

Unfortunately, most of the fragments are between the frames, so I collected only three characteristic outsiders. The size of the calculator—background in width is 7.5 cm. The size of the hole in the glass is 15–16 cm. Please note that the second glass is located at a distance of 11 mm, and is intact, without any scratches. The fragment in the center has a rounded scratch, concentric with an outer rounding. All the curves in contact with the glass in the picture are facing down. By the way, the rounding on the middle fragment is smooth, as if turned on a lathe, but has prominent edges in two places.

About the thunderstorm that day on August 3 at about 9 pm. The storm was well, the most ordinary, a little sparkled and rumbled. There was practically no rain. There must have been a sound from a lightning explosion, but I was listening to music and had headphones on. Again, I was surprised by the color of the flash—bright yellow and orange, and very intense. I didn't see the ball itself."

Comment. Description of the effect of BL on glass.

4. M.G Kasimov, male. End of June 2008. About 15 km from Glazov, Udmurt Republic, Russia. The terrain is elevated. Gardening partnership "Bezvil-1."

In the afternoon, during a thunderstorm, I watched Vityaz TV. Suddenly, the sound disappeared from the TV, but the image was there. Out of the corner of my eye, I noticed something unusual in the upper right corner of the room. Turning my head, I saw a bright ball of bluish-purple light with a diameter of a volleyball ball. The ball slowly descended along a curved line to the window and passed through two windowpanes without damaging them. During the passage of the ball, a quiet buzzing was heard and there was a feeling of cold.

At about the same time, a similar object was observed by the neighbors of M. G. Kasimov in the house opposite the window where the BL flew out. Their electronic clocks were out of order.

It should be noted that about 20 m from the house of M. G. Kasimov there is a transformer booth, and a four-wire 220–380 V power line runs near the house from the side of the mentioned window.

Figure 2.34 shows a diagram of the phenomenon. The number 1 indicates the TV. The trajectory of the BL is shown by a dashed line.

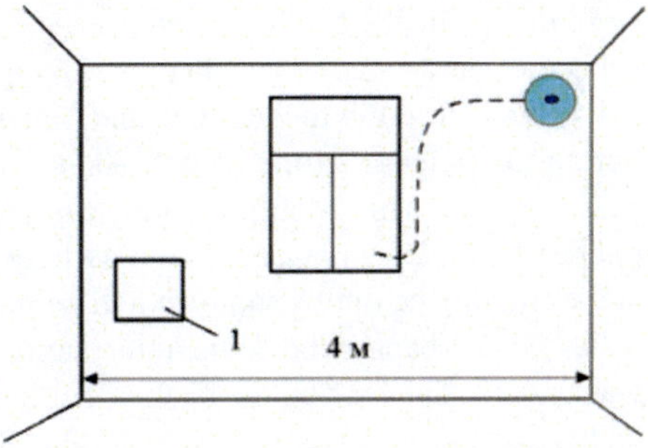

Fig. 2.34 A diagram of the phenomenon

Comment. Description of the passage of the BL through two windowpanes without damaging them.

5. "Strange science." BL database. Kettering, England—Wednesday, 26.07.2006, 10:34:43 (PDT). (translation from Russian.)

In the fall of 1986. I was a company manager in Kettering, England. I was holding a meeting with three executives when the BL passed right through the window glass, immediately to us. It was as if the glass had blown out like a bubble. I got up and touched the glass, which looked perfectly normal as soon as the ball passed through it. I touched the window expecting it to be hot, but it was cold and there was no visible trace of anything in the glass being there.

The ball was about 6 in. in diameter, and after passing through the glass, it fell to the floor and slowly rolled along the carpet for about 20 feet, leaving no traces on the carpet when it disappeared. I cannot remember the smell or the noise when it passed by, but all the computers and the ACD phone system went down.

Comment. Description of the passage of the BL through the glass without damaging it.

6. (Bychkov and Vaulin 2014) Stanislav Filatov, male, Physicist.

During a thunderstorm in the summer of 1937, with the window closed, a BL entered his parents' apartment. BL passed through the window pane unhindered.

Comment. BL passing through a glass without damaging it.

7. www.rusdocs.com/sharovaya-molniya. 27.03.2012.

We gathered at the apartment with the girls, we were 10 y.o. We were sitting on the carpet, collecting some kind of game. A hurricane started, the balcony door was closed. In an instant, like a bang, I jumped on the chair out of surprise. I look, the ball flies in (so bright, right radiant) and into the very center of the carpet. The girls scattered, my eyes widened, the ball circled in the center and flew away… When we came to our senses, we looked at the glass from the balcony door. It was broken through and the carpet was burned in the center.

Comment. BL broke through the window glass.

8. Site: Strange and Unexplained—Ball Lightning.

In the summer of 1960, when Louise Matthews from South Philadelphia, Pennsylvania, was lying on the sofa in her living room, she looked up and saw a huge red ball passing through the window and Venetian blinds, both closed and not damaged during the passage of the object. As the ball, which was making a hissing sound, passed by her, Mrs. Matthews felt a tingling sensation in the back of her neck. She put her hand back in place, but felt nothing. The ball passed through the living room and went into the dining room, exited without damage through a closed window. She called her husband, who came home from work and found that the back of her hand was burned. The hair on the back of her head fell out; the skin became smooth, like on her face.

Comment. The BL came out without damage through a closed window.

9. http://www.sxnarod.com/sharovaya-molniya-t.html 10.06.2005 time: 23.58. Olga Vasilyevna, female. Winter.

The weather that day was clear, frosty. "I was in the garden, repairing greenhouses, when it appeared in the air and hovered. It began to sway, vibrate in the air. The ball spun around the greenhouses, and then rushed into the house." I ran after it. "I heard the glass ringing and ran into the bedroom. I saw that the mirror was broken, there were photos of my husband and children on the floor; the curtain on the window was on fire. I tore it off and started trampling on it, and at that moment I saw a hole in the window pane, as if made by a drill. Through this hole it left."

Comment. A hole appeared in the window pane, as if made by a drill.

10. Amur. info. Blagoveshchensk, No. 52. 9.05.2017, 13:55 LT, Teatralnaya str. The weather on May 23 was cool—about +16° in the afternoon. Cloudy, light short-term rain, there was no thunderstorm. The wind was moderate from the north.

In an apartment on the third floor at about 16 o'clock, a woman heard a bang, but looking out, did not notice anything. The owner examined the balcony more closely and saw a round hole in the glass. In his opinion, there was no reason to believe that the hole was made by any projectile—the edges of the glass were smooth, as if melted. There were small fragments lying around. Another round hole, no more than one centimeter in diameter, was discovered later. The owners of the apartment decided that a BL visited them (Fig. 2.35).

Comment. A hole appeared in the windowpane.

11. UNIAN UGSCHS. Observation of BL in Cherkasy region on 12.05.2016.

Fig. 2.35 A hole in the glass and pieces of the dropped fragment

In the village of Chernobai, Cherkasy region, on the evening of May 12, a BL flew into a residential building during a thunderstorm and exploded. The hostess of the house said that, being in the kitchen during a thunderstorm, she heard a crackling and an incomprehensible hiss behind the wall similar to an electric discharge. The woman followed the sound, and at that moment there was a powerful explosion in the corridor, after which the woman fell to the floor and lost consciousness for a few seconds. Because of the BL explosion, the roof of the house was significantly damaged, as well as ceilings, windows were knocked out by the shock wave. Not only all the electrical appliances burned down, but also all the electrical wiring in the house melted. The pulse from the explosion of the BL for a while disabled electrical appliances in neighboring houses.

As reported by the hostess of the house. "…A fireball the size of a football appeared from the side of the entrance to the attic, where the electrical wiring is located, it passed by me, touching my left forearm. But I did not feel the heat, only later I discovered a small burn, which in a few days almost came off…I look, but there is no plastic door leading to the roof—it was torn from the canopies and badly damaged. There was no roof over my head either. I rushed to the window to open the tulle. I look; people are running to me from the store located opposite. Then I noticed a burnt hole on the tulle, with a radius equal to a soccer ball. In addition, in the lower right corner of the window—through damage. Through it, the "ball" broke out into the street and disappeared…"

Comment. In the lower right corner of the window, the BL left a thorough damage.

12. A. R. Bikmukhametova, female, Physicist, 22 y.o., survey in the village Kolyubakino, Moscow region, Russia.

On the afternoon of July 13 and on the night of July 14, 2016, a strong tornado passed through the village of Kolyubakino in the Ruzsky District of the Moscow region. The path length of this tornado was 49 km, and the maximum width of the destruction lane was up to 570 m.

Many windowpanes, balconies, roofs, trees, and cars were damaged. However, damage to some windows is of particular interest—such as a smooth round hole on the third floor of a residential building. The owners of the apartment were not at home at the time of the hole formation, but neighbors reported that a strong wind rose around 11 p.m., which "tore out the window frame and the balcony door," so one of the neighbors was next to the balcony and tried to fix the door. Suddenly he felt a strong blow, and "the wooden door shattered into splinters." It's difficult to say about the presence of a bright flash, because during the hurricane "lightning struck one after another, and there were flashes everywhere, and we didn't have time to look at them—we moved away from the windows, collected all the documents, covered ourselves with a blanket and prayed that the whole house together with the roof wouldn't be demolished…"—says a resident of a neighboring house. We visited the village of Kolyubakino a few days after the hurricane, and the residents provided us with frames with glass damage of interest to us. As a result of the interaction of the BL with the window glass on the 3rd floor, a smooth round disk with a diameter of 26.2 cm with a thin chip along the edge fell into the space between the outer and inner panes of the frame. The thickness of the glass is 4 mm. The thickness of the glass edge (chip) is 0.15 mm, see Figs. 2.36, 2.37, 2.38, 2.39, and 2.40. The glasses are stored in the laboratory of the V. L. Talrose IEPCF of the Russian Academy of Sciences (Fig. 2.41).

Fig. 2.36 General appearance of the window

Fig. 2.37 Crack-disk in the glass

Fig. 2.38 The disc fallen from the glass

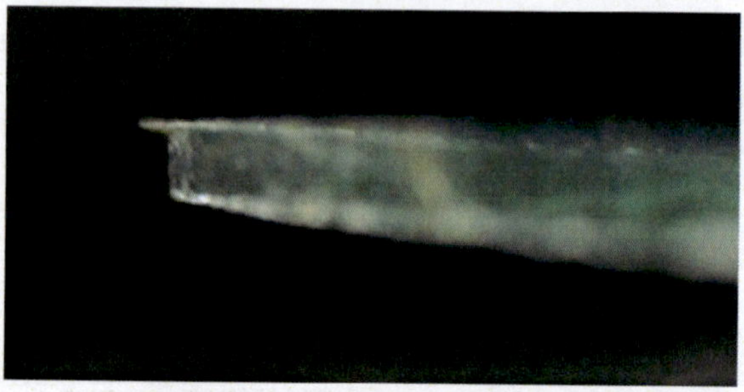

Fig. 2.39 The sharp edge of the disc, side view

Fig. 2.40 Melted edge of the disc

a

b

Fig. 2.41 Shchelkovo disc. **a** On the right there is a hole in the glass, on the left there is a disk that has fallen out of the hole. The dimensions of the axes of the disks (holes) are 8.2 and 7.6 cm. **b** Side view of the disk (pay attention to the sharp edge)

Comment. Because of exposure BL to window glass during a tornado, large holes were knocked out in the window glass.

13. (Bychkov and Vaulin 2014) P. A. Zhilin prof. mathematics-mechanic and his wife I. A. Zhilina. Observation 1972, survey 2003. The place of the event is Lithuania, the village of Laftyushniki, Pobratsky District. Summer, July, evening 20–20.30 LT. The weather is stormy, but there is no rain. Observation in a village house.

"A glowing object (ball), 8–12 cm in diameter, suddenly appeared in an enclosed space. The ball moved first along the wall; then along the ceiling, along the perimeter of the room. It passed ~16 m. It suddenly hit a light bulb on the ceiling, there was a slight crack, and the object flew out in the form of blue splashes (blue flashes). The observation time was about 1 s.

The light bulb turned into a strainer, evenly covered with holes with a diameter of 0.1–0.2 mm.

A neighbor from the next room claimed that at that time in his room, a white "flame" (glow) appeared in the corner, where the electric meter was located."

Comment. Interaction of the BL with an incandescent lamp.

14. (Abrahamson et al. 2002) Indira Murthy, Bangalore, India.

"A Professor Ramasheshan (Physicist), formerly of Raman Research Institute, Bangalore, told me that a BL he witnessed had this unusual feature associated with it. It passed through a glass window and after that, the window turned opaque.

I met Prof. S Ramaseshan and he has given these explanations for your queries on the BL he saw. He saw it some 50 years back as a boy of around ten along with two others, who are now no more. There was a severe thunderstorm on that day. The place was South Calcutta, and the time around 5 in the evening (17.00 L.T.). It was May–June, most probably. He was in a room 12 ft by 12 ft, on the third floor (it was a three storey house). There was a large number of pieces of wooden furniture in the room. A luminous ball 300–400 mm across, luminosity about 50–100 W came right through the glass of a closed window. The opaque part was around 200–300 mm, after its passage. The color was bluish-white. It came in contact with some wooden things and they were charred. It came in contact with the arm of a person in the room, and that part charred and became black. But no death occurred.

Somebody opened a window and it moved out, he remembers. Since he was quite young, and it happened so many years ago, he said "don't give too much credence to the numbers given. The other details you wanted, he simply cannot recall—whether the window glass was weak, etc."

Comment. A luminous ball with a diameter of 300–400 mm, with a luminosity of about 50–100 W, penetrated directly through the glass of a closed window.

15. (Bychkov and Bychkov 2008) NLO No. 8 (55). Yuri Grigorov male. June 1998. "Boiling ball lightning" 1948–1950. The observer at that time was a student of the 2nd–3rd grade of secondary school, Anthracite, Voroshilovgrad region, Ukraine. Summer, July. It was a hot day. Indoors. 2 observers.

"It was a very hot day. A strong thunderstorm broke out, it began to rain. The house went dark. The mother turned off the lights and the radio "so that lightning wouldn't hit." My mother was lying in a big room on the bed dressed. I climbed on the bed in my panties and lay down against the wall. There were several thunderclaps, and there were flashes of lightning. The rain was lashing. There was an impression that lightning flashed very close in the yard, thunder rumbled.

A few seconds later, in the silence that followed, suddenly a sparkling fireball, the color of molten gold, flew into the closed window, absolutely noiselessly, and slowly floated towards us ... the mother also saw it. The ball continued to approach, swaying slightly from side to side. We did not move.

The ball passed my mother and floated over my body. The distance between my chest and it was 5–6 cm, the diameter of the ball was about 10 cm...The ball "came up" to the wall and disappeared (the bed was against the wall, the distance between them was about 5 cm)... After a dozen seconds, it appeared from under the bed and went to the window from where it appeared. When it reached the window, it disappeared. A few seconds later, there was a flash in the yard, and there was a not very loud bang (more like a pop).

...I examined the place where the lightning went under the bed, but there were no traces there. Then I examined the frame and glass in the place of the appearance and disappearance of the ball. The glass was slightly chipped at the edge, there was no winter coating there either.

The hair on my chest was whitish, very small, I did not notice them. However, when the ball was almost above them at a very low height, they all stood...vertically. I even felt some faint tickling on my chest with my skin. My breath almost stopped at that moment, but I felt the freshness of the air.

Visually, it seemed that the ball consisted of a sphere of yellow-golden substance with a fiery-bright rim around. On the one hand, at the first glance at the ball, it seemed extremely bright, so that it was impossible to look at it. However, when it was in front of my face, I looked at it calmly, without any unpleasant sensations for the eyes.

There was some kind of stirring on the surface of the ball, resembling the boiling of particles of mineral water and gas when it was opened (only less intense). Microscopic particles tried to break away from the surface of the ball, but returned back. There was a faint rustling-crackling."

Comment. Suddenly, a sparkling fireball, the color of molten gold, flew into the closed window, absolutely noiselessly.

16. http://www.bufora.org.uk/content/index.php?option=com_content&view=article&id=124:a-case-for-ball-lightning&catid=63:2012-articles-&Itemid=95. Leigh, Lancashire

The woman reported an event that happened in her house on July 1995, when she was sitting on the couch with her boyfriend. They both noticed a glowing ball outside the window and were surprised and very alarmed when this ball passed through the closed windowpane. It was like a blue-green spinning fire the size of a football.

Moved down the trajectory at a speed past the sofa they were sitting on, and then shot through the patio window glass at the other end of their living room. This event happened within a few seconds.

Comment. The ball passed through a closed windowpane.

17. (Bychkov and Vaulin 2014) (f2065) 2010-07-02 13:30:00 LT Moscow. Russia.

The time is most likely either late spring or late autumn, there was a thunderstorm shortly before that. Everyone was sitting in the kitchen. Suddenly, in the window (from behind the edge of the opening, from below), a bright glowing ball with a diameter of 10–20 cm slowly floated out, with a hissing sound. Sailed over the outer ebb along the entire window towards the side window (it was closed), at the window (although it is vertical narrow) rose up and about 2/3 of the height hit the glass and exploded. From this place, the glass split into 3 long cracks, but did not fall out.

Comment. The ball hit the glass and exploded. From this place, the glass has split into 3 long cracks.

18. (Bychkov and Vaulin 2014) yaba 18.05.2010 17:44:16 L.T. Moscow. Russia.

There was a thunderstorm in Moscow yesterday, and I managed to see a fireball in my room. It was like this: in the midst of this thunderstorm, there was a heavy downpour, and several times there was a noticeable rumble, the windows were shaking, bright flashes. I went into the room and for some reason began to pass to the window. At that moment there was a blow, and I saw a huge pillar arc almost the size of a window. Literally a couple of seconds passed.

I moved away and then it appears. It oozes right through the glass! It looks bright, the light is cold. The shape is somewhat like a fish, it cracks like a stitched string transformer of CRT televisions. And it stood, stood, and began to go down to the battery. I stand in the middle of the room and don't move. Then it, having already descended close to the battery, let something like "ropes" to the battery itself and a nearby plug from the amplifier. Then a pop and it disappeared. At that moment, the computer was working, the oscilloscope, and so on, every little thing plugged into the socket. All the equipment is alive.

Comment. BL oozes through the glass.

19. (Turner 1997) Round holes. There are many reports of the appearance of round holes left by BL in the glasses. The work (Turner 1997) can be considered as one of the first on this topic; it describes the case of round holes that appeared in the glasses after the passage of a tornado in Saint-Claude (France), during which numerous BLs were observed (Faye 1890).

A detailed analysis of three glass samples damaged as a result of interaction with BL is described in (Turner 1997). The first sample, which was named "Edinburgh Glass," is a fragment of glass 2.5 mm thick in the shape of a diamond with a side of 15 cm. On one edge of the fragment (about 150°) there is a part of a round hole with a diameter of 4.8 cm. The mirror-reflecting edges of the hole have a conical shape;

the base of the cone is directed into the room. Arc-shaped cracks surround the hole; under the microscope they resemble scratches. No traces of melting were found.

The second sample from (Turner 1997) ("Stockholm Glass") is an intact round disk with a diameter of ~16 cm. Most likely, it fell out of the hole in the outer windowpane into the area between the panes. A hole of exactly the same shape and size remained in the lower corner of the outer windowpane. The dimensions of the original glass were approximately 50 × 110 cm, the center of the hole is located at a distance of about 14 cm from the nearest edges of the glass. Two cracks are visible in the photo of the glass. One of them runs almost normally from the hole to the nearest edge of the glass; the second one starts from the point opposite to the initial point of the crack, and goes tangentially from the hole to the same edge of the glass. This disk has a conical shape with a conical base pointing into the room, as in the case of "Edinburgh Glass."

The third sample described in (Turner 1997) ("Uppsala Glass") represented a round hole made in the outer glass of a double window; at the same time, the inner glass remained intact. The diameter of the hole was about 4 cm, fragments of glass that had fallen out of the hole were found between the window frames. They included a large fragment (half the circumference), three sectors of glass, and two crescent-shaped fragments adjacent to the outer part of the disk. There were no radial cracks in the window glass.

Comment. Description of round holes in glass made by BL.

20. (Shelkunov 2001) Shchelkunov G. P., Ph.D. in technic. 13.04.1994 Shchelkovo City near Moscow, Russia.

In the city, there was sunny and windless weather. At about 5 p.m., a woman in the apartment saw a glimmer of a bright flash in the window, and 1.5–2 s after that she heard the sound of falling glass. She found that a round hole with a size of about 8.2 × 7.6 cm appeared in the outer glass with a thickness of 3 mm, and under it there was a round disk, between the inner and outer glasses, which fell out of the hole (see Fig. 2.1a). The center of the hole was located in the lower part of the window measuring 974,949 cm at a distance of 42 cm from its lower part and 17 cm from its right side. The inner glass remained intact.

The hole in the window glass on the inside (from the side of the room) is an oval, the major axis of which is inclined 24° to the horizon. The hole along this axis is 82.0 mm, and its size in the perpendicular direction is 76.7 mm. On the opposite edges of the hole, there are two trapezoidal chips with a length of 10 mm and a depth of about 200 microns. The base of the trapezoid is 10 mm, and the length of the upper base is 6 mm. Along the edges of the chips (near the bases of the trapezoids), concentric arcs are visible with the center located approximately at the edge of the hole. The left side of the hole is bounded by a crescent-shaped strip with a maximum width of 6 mm in the center; this strip is a trace of a thin glass film cut off by a fallen disk. The edge of the hole on the inside of the glass is rounded, and the wall of the hole has a smooth mirror surface along its entire length. This indicates that this side of the glass has been heated to a softening temperature (about 600 °C). On the contrary, the edge of the hole on the outside of the glass (from the street side) remained sharp. In addition to two large chips on the inner surface of the glass, there are also 4–5 small chips measuring 3 × 3 mm adjacent to the edge of the hole. There are no such chips on the outside of the glass. Here, the dimensions of the hole along the long and short axes of the oval are 80.5 mm and 74.7 mm, respectively. This means that the hole has the shape of a truncated cone with the base turned inside the room. The angle at the base of the cone on the major axis is 76°, and on the perpendicular axis—72°.

Disk. The disk fits the hole perfectly. It is a truncated cone with an extension on the inner surface of the glass. The dimensions of the axes of the disk base are 82.0 and 76.7 mm, and the dimensions of the axis on the outer surface are 80.5 and 75.4 mm, respectively. The angle at the base of the cone along the major axis, as at the hole, is 76°, but on the smaller axis it is increased to 78°. This angle may be greater than the angle of the conical hole. The reason may be that the size of the inner side of the hole has been increased due to the effect of surface tension on the molten glass. The edge of the disk has a smooth mirror surface. The sharp wedge-shaped edge, torn off from the inner glass surface, adjoins the perimeter of the inner surface of the disk, its thickness is about 50 microns, and the height is from 2 to 6 mm. Round potholes are located on the outside of the disk, their depth is about 100 microns, and they are located at a distance of 0.5–1 mm from the edge of the disk. As noted above, there are no such chips on the outer surface of the glass with a hole. At the place of the greatest height of the sharp edge there is a large pothole. It is approximately round with a diameter of about 20 mm, the mirror surface on it is bounded by an arc with a radius of 10 mm, and is bounded by wavy arcs with radii from 12 to 24 mm. The arcs have a common center located on the edge of the disk. The pothole is located on the large axis of the oval, inclined to the horizon at an angle of 24°. It is adjacent to the previously described pothole on the inner surface of the glass with a hole. The center of the pothole arcs on the disk surface coincides with the center of the pothole arcs on the glass with a hole. The circle also passes through this center, on this circle there are two non-spreading cracks about 15 mm long.

Details of physics and chemical investigation one can find in (Bychkov and Nikitin 2014). The analysis of glass samples in contact with BL confirms the main conclusions based on studies of other glasses and also allows to identify new features.

1. The creation of an annular crack was caused by rapid (within a few seconds) local heating of the glass to the softening temperature, followed by rapid cooling of the glass from the side opposite to where the heat source was located.
2. The value of the energy required to heat the surface layer of glass to the softening temperature is from 1 to 7 kJ, and the power of the heat source should not be lower than 0.4–2.4 kW.
3. The glass was heated on the side opposite to the one where the BL was present. This means that the transfer of energy from the BL to the heat source occurred practically without dissipation. Energy can be transmitted in the form of electromagnetic radiation with wavelengths in the spectral windows of glass (400–600 nm, 1–13 cm).

4. The most likely source of heating of the glass surface was plasma generated by an electric discharge in the air. Direct current or high-frequency corona discharge radiation could not serve as a generator supporting the discharge work (because of their low power). The energy source was most likely located inside the BL, and its power could be significantly increased at the time of its death.
5. The absence of traces of radioactivity in the studied samples indicates that in the case of BL there could not be radioactive elements with a half-life of more than 10 years.
6. The presence of only elements characteristic of the composition of the glass on the edge of the disk means that elements such as carbon, copper, zinc, and iron cannot be part of the BL substance.
7. Changes in the percentage of silicon and oxygen in the surface layer of the disk chip may be associated with the reaction of glass with water, which may be part of the BL substance.

21. Mirror event. (Bychkov et al. 2014). Information was received from Mr. Eriks Eisaks about an event that took place after 1974. He texted: "Last weekend I was at my grandmother's. I was talking to Grandma about that day, and… she told me she hadn't seen the BL herself. She was in another room when she heard a loud noise, and this damage in the mirror was the only damage caused by the lightning strike." Since she had already met with BL in her house a few years ago, she came to the conclusion that the mark on her mirror was left by BL. Since she herself did not see the BL, we call the object that left a trace on the mirror a possible BL. Mr. E. Aisaks sent photos of the mirror in his grandmother's room. Then he sent a piece of mirror that was broken.

Figure 2.42 shows a photo of the segment from the mirror. Figure 2.43. shows a photo of the mirror part in front. Figure 2.49 shows a photo of a part of the mirror taken from the reverse side (Fig. 2.44).

From the description of the event, it follows that the mirror was far from the window, so it was the BL, and not the linear lightning, that could leave a mark on the mirror. There is no obvious effect on the front glass side of the mirror when the BL comes into contact with it, see Fig. 2.42.

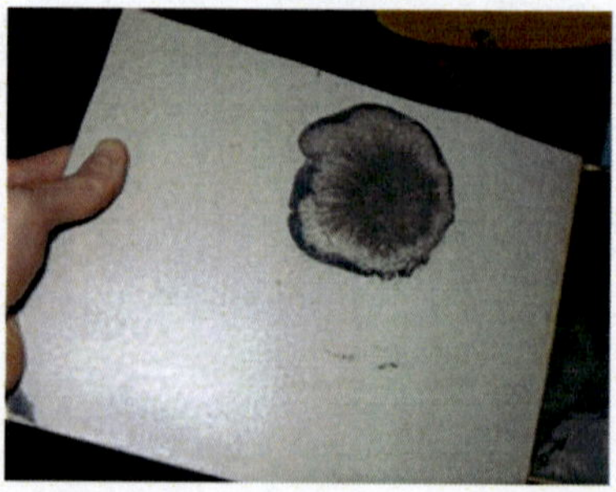

Fig. 2.42 The reverse side of the mirror

Fig. 2.43 The trace of the BL from the outside of the mirror

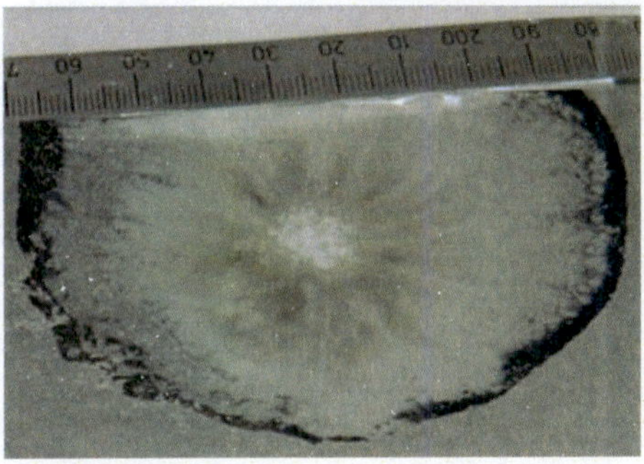

Fig. 2.44 The trace of the BL on the back side of the mirror

Analysis of the back side of the mirror, see Fig. 2.45, shows that part of the metal layer covering the glass was melted. The metal layer was covered with a dielectric film. The film was destroyed, radial traces remained from it.

Figure 2.46 presents a component analysis of the mirror material. The main components are Si, Al, O, Na, and Ca. Si and O are the usual components of glass,

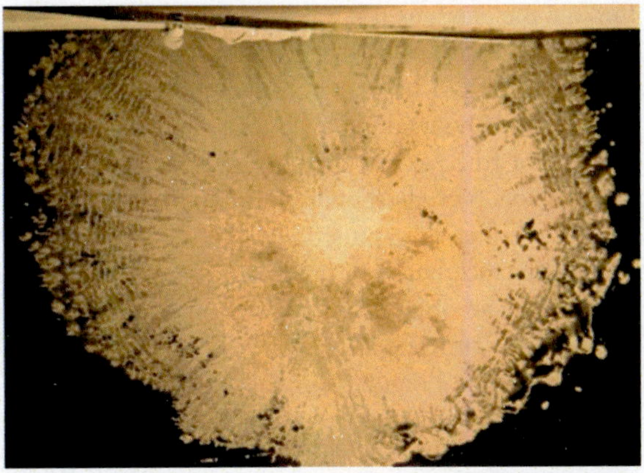

Fig. 2.45 Enlarged image of traces of metal and varnish on the back of the mirror

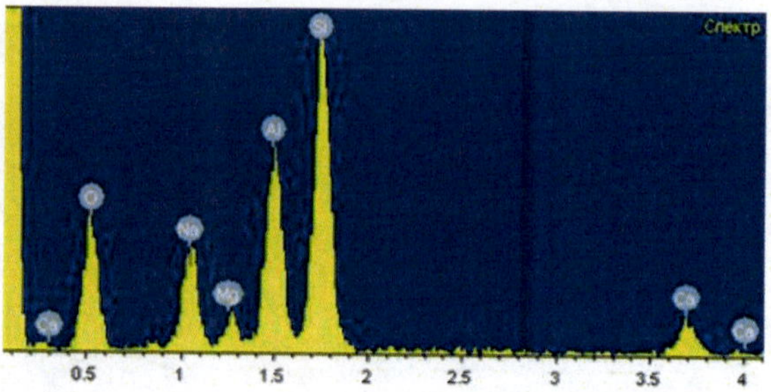

Fig. 2.46 Relative intensity of the mirror components. The horizontal axis shows the energy in keV. Vertical axis in conventional units

Al is the usual component of deposition to create a mirror image. Ca is an element in the composition of silver lacquer for coating Al. Na is a widespread element.

Details of the electron microscopic examination of the mirror surface can be found in (Bychkov and Nikitin 2014). One can say that during the interaction there was heating, destruction (the appearance of cracks on the varnish), melting, and even boiling (bubble marks on the metal) of the metal material and the varnish on the surface.

The presented data show a complex picture of this impact. Most likely, the BL touched the mirror glass from the front side, and on the backside of the mirror realized a strong thermal effect with evaporation and melting of aluminum and varnish—components of the reflective film.

22. (Bychkov et al. 2016a, b) A BL passed through a 2-mm-thick windowpane in front of an eyewitness. The glass had parameters typical for a rural house window of 530 × 350 × 2 mm. Observer I. A. Nosikov (male) in 2004 at the age of 12. The place of the event is Russia, Pskov region, the city of Pustoshka, Pionersky Lane, 5. The observation time is the first half of July, 16:00-18:00 LT. There were no lights on in the house.

"There was a thunderstorm in the evening that day. A spherical object appeared outside a window with a wooden frame. I had a feeling that a glowing object formed outside the window instantly and immediately took the shape of a ball. He 'leaked' through the glass, as if not meeting obstacles, without changing his shape and color. At that moment I was sitting on the sofa, on the opposite side from the window, at a distance of 4–5 m from it. The appearance of the object caused me fright, and I watched it without taking my eyes off and without moving. For a moment, he seemed to hang, or froze at the window, and after that, immediately began to rush between the two windows, as shown in the figure (see Fig. 2.47), gradually approaching the electrical outlet, then pulled into it.

The distance between the windows was 1.5 m. The movement of the ball lasted up to 3 s. The object's speed was 1.5 ± 0.5 m/s. The diameter of the object was about 15–25 cm. The asymmetry of the object and any of its heterogeneity was not noticeable. The boundary of the object/environment section was quite clear. The bright reddish orange color of the ball stood out sharply against the background of the wall of the room darkened because of the storm. The center of the ball was almost

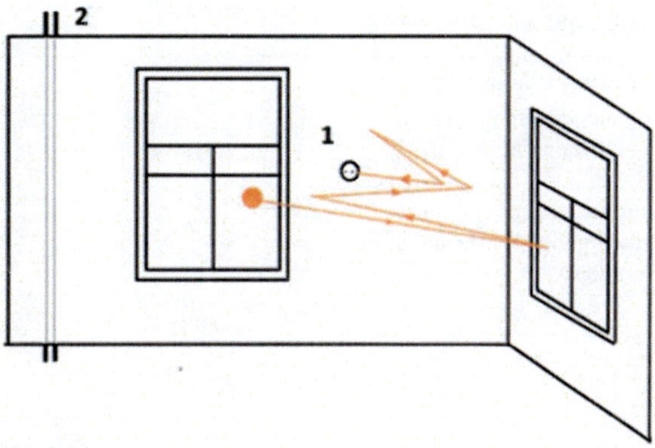

Fig. 2.47 Schematic representation of the room into which the BL penetrated. (1) socket, (2) antenna. The arrows show the movement of the object

white. Its brightness was comparable to that of a 60–90 W desk lamp. I have not seen a noticeable core, halo, or layered structure. The height of its movement above the floor level was 1 m. There was no heat from the object.

In addition to the outlet, there was a TV in the room, and a TV antenna was located outside it. There were no changes in their operation or visible damage.

22.1 ***Results of the glass study.*** Examination of the glass surface in the area where, according to the description of the eyewitness, the BL was in contact with it, allowed to detect a conical depression with several ledges on the outer (street-facing) side of the glass. No other glass defects were found on its entire surface.

The hole was not through. This was determined by the absence of air passage when pumping out the area around the hole using a vacuum pump. Its size on the surface was about 380 microns and the size of the "bottom" was about 160 microns, see Fig. 2.48.

Figure 2.48 shows a section of the outer surface of the glass. In contrast, the glass surface on the opposite side of the window (on the side of the room) has changed noticeably (Fig. 2.49). Colored spots with a characteristic size of 20–40 microns

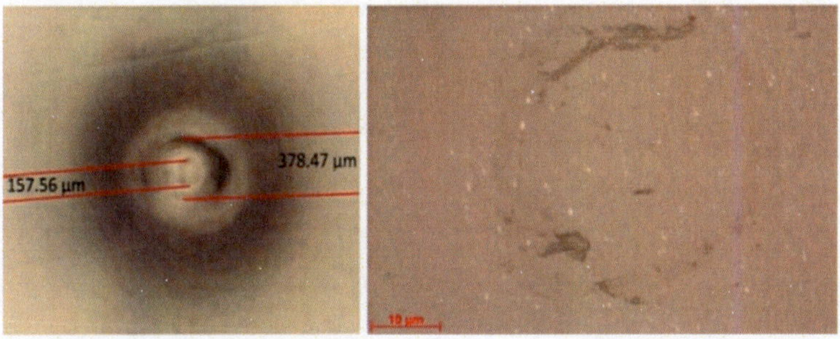

Fig. 2.48 **a** Magnification of 25x, an image of the entrance hole obtained from the Axio Imager Zeiss optical microscope. **b** A snapshot of the area of the outer surface of the glass, magnification 500x, obtained from an Axio Imager Zeiss optical microscope

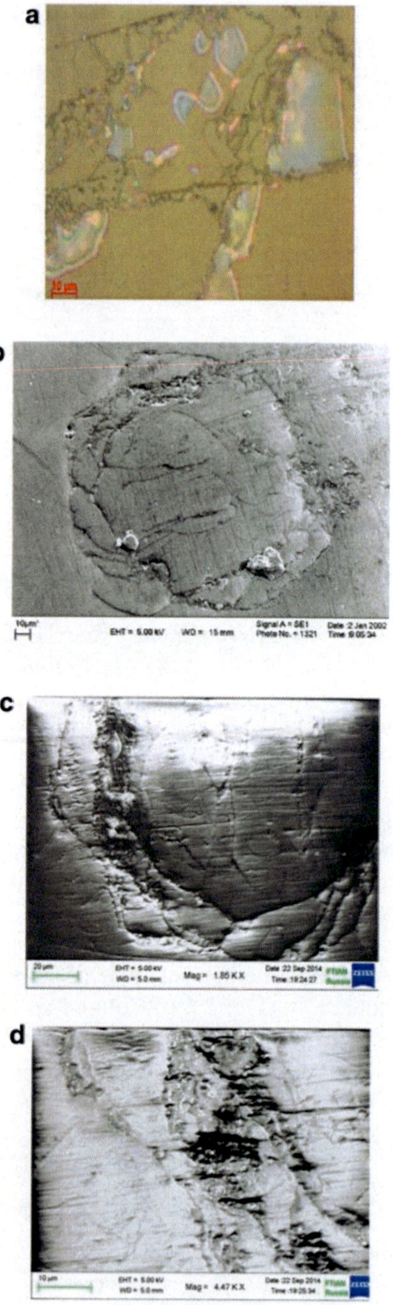

Fig. 2.49 **a** snapshot of the area of the inner surface of the glass magnification 25x, obtained from an optical microscope Axio Imager Zeiss. **b–d**. Images of the surface area on the inner (facing the room) side of the window

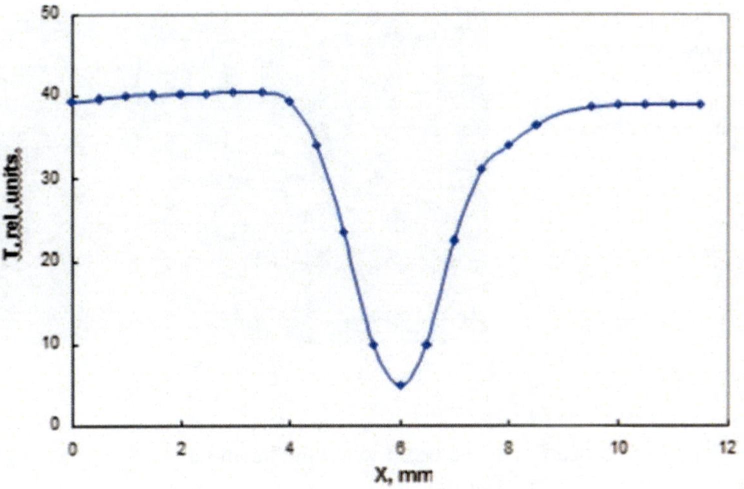

Fig. 2.50 Change in light transmission by glass when scanning glass in a straight line passing through the center of the recess

appeared on the surface, similar to a cracked film. The assumption of the presence of a film is confirmed by the fact that when the focal length was shifted from the surface deep into the glass, the structure of the type shown in Fig. 2.49a disappeared. Figure 2.49b shows a section of glass around the axis of the cavity on the side of the glass facing into the room, taken with a Joel Leo120 scanning electron microscope. In the image, the light areas correspond to the areas protruding above the surface of the glass, and the dark areas are located below the original surface. Gray clusters are dust particles. Figures 2.49c and 2.49d show details of the same section of glass taken at a higher resolution. Defects similar to closed cracks are visible on the nick.

It can be noticed that against the background of the almost unchanged glass surface, the area of 170 × 206 microns is dotted with cracks and slightly raised above the plane of the glass.

Using a Joel Leo 120 scanning electron microscope equipped with an Oxford prefix for microchemical analysis, the elemental composition of the regions in the center and side of the cavity on the outside of the glass was studied. It was found that the ratio of the percentage of the main components of glass in the cavity and outside it turned out to be almost the same. ([O]/[Si] = 2.3 outside the cavern and [O]/[Si] = 2.35 inside the cavern).

A study of changes in the optical characteristics of glass was carried out. In the area around the cavity, a polarized beam of a helium-neon laser with a diameter of 0.5 mm was passed through the glass perpendicular to its plane. The glass could be moved horizontally in two mutually perpendicular directions. When the laser beam hit the area near the cavity, this part of the glass acted as a collecting lens with a focal length of 5 cm. After tightening, the beam diameter increased, so to collect the energy of the expanded beam, a collecting lens with a focal length of 80 mm and an

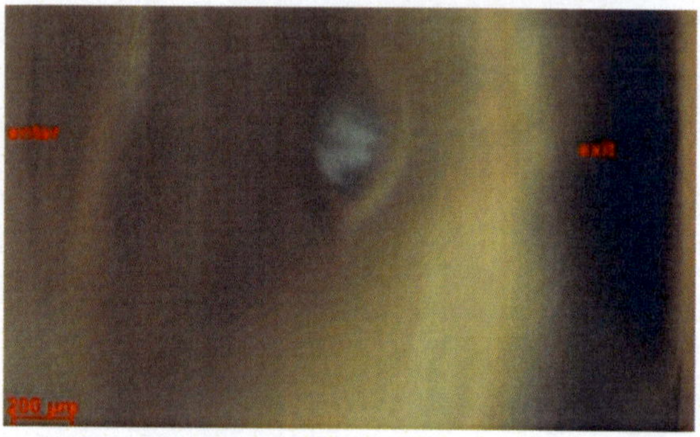

Fig. 2.51 Side view of the area near the cavern. On the left—the place of entry of the ball lightning, on the right—the place of its exit

aperture of 65 mm was installed in front of the DVR. To register a possible change in the plane of polarization of the beam when the beam passes through the glass, a polarizer was located in front of the DVR.

Figure 2.50 shows how the intensity of light passing through the glass changes. It can be seen that the transmission of light in the area near the center of the cavity is 9.4 times less than the transmission of light by unchanged glass, and the size of the area where there is a decrease in transmission (about 5 × 5 mm) noticeably exceeds the diameter of the cavity (0.17 mm). Within the accuracy of our measurements (0.10), no change in the rotation of the polarization plane of the beam in the area near the recess was detected.

Figure 2.51 visually demonstrates the change in the degree of transmission of glass near the cavity. To obtain it, the glass was cut so that the cut plane was located on the side at a distance of 1 mm from the axis of the cavity. The figure shows a view of the slice surface.

22.2 ***Discussion.*** The first conclusion is that the ball lightning caused the glass to heat up before melting. There is nothing original in this conclusion: (Kolosovsky 1981) describes a case when a BL melted a hole in the glass in front of schoolchildren and their teacher.

The peculiarity of our case is that the heating area of the glass was localized in sizes 1–5 mm. This allows us to conclude that the main wavelength of the radiation acting on the glass also lay in this range (1–5 mm).

Measurements showed a sharp decrease in the optical transmission of glass near the heating zone. This can be explained by the rise in the temperature of the glass to the boiling point (1700–1800 °C), and the formation of bubbles filled with gaseous SiO in it with a size of 0.1–0.5 microns. However, this heating did not lead to a change in the elemental composition of the glass. The reason for this, apparently, was that the molten region was isolated from atmospheric oxygen. In the presence of such contact, changes in the elemental composition may occur, see above.

If one assumes that the BL is a material body, then it can pass through the glass if its viscosity decreases sharply due to heating. Thus, it is known that the viscosity of window glass decreases by 10^9 times when it is heated from a temperature of 600 °C to a temperature of 1700 °C (Tareev 1973).

The fact that the area of BL interaction with the glass turned out to be small (0.2–2 mm) allows us to assume that the shell or the elements that make up the CMM have a small size. If we assume that these elements are unipolar charged, we can assume that the forces "pulling" them through molten glass are the electric field created by thunderclouds, or the image forces induced by the BL charge on metal objects in the room. It is known (Mashkovich 1969) that the electrical conductivity of glass increases sharply when it is heated (it will increase 100 times when the temperature rises by 100 K). This contributes to the additional heating of the glass and the passage of charged objects through it. For example, after passing through the glass, a unipolar charged BL is able, due to the action of pondermotive forces on the molten glass, to "tighten" the hole that may have existed when its elements passed through it.

General comment to the section

BL can pass through the glass in three ways: through a defect or a small hole in the glass; through the glass, leaving no holes visible to the eye; knocking out the spherical element of the glass.

2.7 Injuries and Death of Observers and Animals When Exposed to BL

In this section, we present observations of the BL effects on humans and animals that led to the injury or death of the latter. This section is important due to the prevailing opinion that BL is a little dangerous object.

1. (Bychkov et al. 2003) V. A. Rantsev-Kartinov, male, Ph.D. in physics, born in 1939, Russia, near Novosibirsk, survey 2000.

1944, autumn September, evening 20.00 LT Kupino Village near Novosibirsk. The rain stopped. There was no thunderstorm. Quiet, windless. Stuffy, hot. Observation from 30 m from the object in the open air.

A woman went out to fan the iron. At 2–3 m from the woman, a glowing ball appeared (came into the observer's field of view), similar to a cotton ball the size of a fist (7–9 cm). When she brought the iron behind her back, a ball flew into it. Before that, the ball flew for some time after the iron, as if chasing it. The woman fell. This was observed by 8 people. She was picked up, buried in the ground, after 0.5 h her face turned blue, and she died. Observation time 1–2 s.

Comment. Death of a woman after being hit by a BL.

2. (Bychkov and Bychkov 2008) I. I. Aryamova, female, Physical Chemist, 51 y.o. at the time of observation, filling out a questionnaire for the journal Nauka I Zhizn ("Science and Life") in 1993. Observation of 1971, July (second half),

about 14.00 LT Moscow region, Nikolina Gora. Russia. 2 observers. Rain, thunderstorm, strong wind. Indoor surveillance.

"There was a strong thunderstorm. I was sitting in the room. There was a radio receiver on the table next to it with the antenna raised. I turned in that direction and stretched my hand up to remove this rod (antenna), and saw sitting at its end a large, 30–50 cm in diameter, white-yellow ball with a crimson border. The radiation power in comparison with an incandescent lamp was 70–100 W. At that moment, I lost consciousness. Then, when consciousness returned, I felt that I had lost sight and hearing. Vision began to return quicker, and then hearing returned. Interestingly, the right ear, which was closer to the lightning, began to hear faster than the left.

When I came out of the state of shock, my son told me what he had seen. He was about three steps away from me and saw the BL sitting on the end of the rod when I raised my hand up. Then the ball broke up into three parts. Each of them also disintegrated, and the process of crushing and complete disappearance of the lightning passed very quickly.

When I came to myself, I felt that the room smelled of burning. It turned out that the phone burned down. Moreover, its body was completely unchanged. When the fitter opened the lid, everything inside was melted: both the metal and the insulation fused into one lump. Upon further examination, it turned out that there were four damages on the telephone line leading to our house. It should be added that at the moment when we saw the BL, there was a very strong thunderclap. Later we discovered that a large birch tree had fallen about 30 m from the house. Its roots were completely burned."

The nearest distance to the BL was 50 cm, its lifetime was 1–3 s. The BL was at a distance of 2–2.5 m from the ground.

Comment. Loss of consciousness of a woman when interacting with a BL.

3. (Bychkov and Bychkov 2008) Mikhail Kilin, male. NLO No. 6 (53) June 1998. "Strange burn."

1942, the observer at that time was a student of the 9th grade of secondary school, the station Pesyanka Tomsk Railway, Russia. Summer, July. After lunch. It was a hot day. Outdoors, at the railway station.

…Cumulus clouds began to gather in the sky. … Suddenly, a large rain splashed. The sky rumbled louder and more often. It got dark. Suddenly, after one of the most sensitive thunderclaps, something rustled in the birch leaves. Then a light pink-blue ball the size of a small watermelon appeared to my left, slowly approaching the ground. From its depths, myriads of glowing fiery threads splashed out and were drawn inside! There were a lot of them, and they shone so brightly that I had to close my eyes for a while. However, when I opened them, I saw that the ball slowly passed by my head in a copper helmet and, almost without moving, hovered over my shin. I moved my knee a little, and the ball quickly slid over my shin. Then there was a huge explosion! I was thrown forward and hurled along the trail of the BL onto a pile of small chips. As I was falling, I managed to notice how a small ball bounced off my foot, hit a power line pole and rolled along the wires to the factory. There, at that moment, a fire broke out.

I woke up from the cold, my body was shaking as if in a fever, I was sick, green-yellow-red circles loomed in my eyes. With difficulty, I managed to sit on a pile of wet wood chips, and then crawl to the striker under the birch and call for help. At this time, I felt a strong pain in my left leg. I did not have legs, it seemed to me. My leather shoe, torn to shreds, was lying nearby. It smelled of burnt rubber…

…Help has come. My left leg was buried up to the groin in cold clay. The heat from the leg began to come out. In the hospital, the burnt instep of the left leg was liberally lubricated with potassium permanganate…

For three weeks, the swelling in my left groin and the pain in my entire leg did not give me rest either night or day. There was no end in sight to the treatment… However, a friend of my father advised: melted butter, onion, honey, wax—all in equal parts should be boiled well in a Russian oven in earthenware and lubricate the burn with this composition. The medicine helped. After two or three weeks, after seven or six times the skin renewal on the leg, the burn site began to tighten. The final recovery came after the mother-and-stepmother's (plant) leaves began to be applied to the place of the burn.

Comment. Injury of a man when interacting with a BL.

4. (Amirov and Bychkov 2012) Moscow, Russia. Nikolaus von Twickel (N.1-N.3).

Nina Akopova and her husband, Konstantin Akopov, Physicists, woke up early in the morning on 6.08.1979, their daughter Irina slept between them. The trio occupied a small tent at a summer camp in Babushara near the Black Sea, and Nina, worried that a 4-year-old girl might also wake up, carefully lifted the still sleeping child over her body, snuggling up to her husband. …For some reason K. Akopov felt something unhealthy in the air. "My husband suddenly said that there must be something terrible around," Nina recalls. Lightning struck at 6.20 a.m. "There was a strong thunderclap, and almost simultaneously a fiery thread entered the tent," Nina said.

A tent made of orange fabric stood next to their car. Lightning entered through a window covered with a mosquito net, and left both the net and the entire tent untouched.

Nina Akopova, who is now 56 years old, says that she remembers well what happened next, although the event happened 32 years ago and lasted less than a minute.

The luminous thread began to twist, curling up like a ball of wool. However, it was more like a fireball, blindingly bright, although it seemed darker inside.

Nina immediately realized that a BL had entered the tent. The size of a tennis ball, it began to float around the tent. "It was like looking for a way out," she said. She tried to jump up to open the zipper of the tent in front of it. Nevertheless, I could not. "There was some overwhelming energy around me. I just couldn't move on and get on my knees." The bright ball continued to float in the tent. Moving up and down, it slowly turned into a flat, pancake-like shape. "It was like watching a movie in slow motion," Nina said. Suddenly the ball exploded. Nina panicked. "I thought we were all dead. I had to save the baby." Then she noticed that her husband stopped breathing…

Nina had a terrible headache, but she could walk.

The deadly glow struck both parents. Everyone had a major wound. She has dry, burnt flesh near her left elbow, where Nina still has a scar. It turned out that she was lucky because the energy went down her arm and out of her body through her fingers, which also suffered from burns. Konstantin was not so lucky. The lightning action went up through his left arm to his body, through his heart and down his right leg. The burn on his small toe marked the exit. He was 25 y. o. Nina said his body was covered by dark lines, as if his nervous system had been burned from the inside out.

Her daughter, who was thrown away from her in the chaos, received a burn on her forehead, but was alive. Later, doctors discovered that she was partially blinded, but could not say what exactly. To this day, her left eye sees weaker and has a smaller pupil.

Comment. The death of a man and the injury of a girl when interacting with a BL.

5. (Bychkov and Vaulin 2013) E. Kopaneva, female, "FACTS" (Chernihiv—Kiev) 21.06.2013; observer Oksana Lyubchenko, 31, Boryspil.

Frightened by the hurricane, the young woman immediately closed all the windows in the house.

"The lightning flashed incessantly. A real hurricane started on the street. I came to the living room, where my mother, brother and his wife were resting by the fireplace.

Everything happened very quickly and unexpectedly. At some point in the fireplace… a large light ball appeared. Before I knew what it was, the ball flew straight at me. I tried to hide from it with my hands, but I did not have time and ended up in some kind of sparkling cloud. I was not cold or hot. I started screaming. The cloud I was in did not disappear. There was a feeling as if an object had entered me. It was not cold or warm, but I felt it well. This something passed under the rib, then sank into the stomach, then into the legs. In the part of the body where it appeared, everything seemed to turn to stone. I felt something crawling through my internal organs, but I could not stop it. Suddenly, the object that was walking over my body entered my right leg. At that moment, everything flashed, the light went out and the cloud disappeared. I lost consciousness."

"It seemed to Oksana that everything happened for a very long time," says Oksana's brother Egor–"Although in fact it was seconds. I saw my sister go to the fireplace, as a transparent ball flew out of there, flew straight at her and knocked her to the floor. Oksana screamed, and my mother and I were deafened by the sound of an explosion."

"I didn't think at all for the first few minutes. I was brought to my senses by my mother's scream: "Oksana! Where is she? What's wrong with her?"

Oksana: "When I woke up, I immediately felt the smell of burning. The body was limp and felt like cotton. It was hard to breathe.—Surprisingly, lying on the floor, I did not feel anything. Then I stopped feeling my body at all. I wanted to move my legs, but they did not seem to be there. The smell of burning (and I was sure that my burnt hair stank) was suddenly interrupted by a wet towel. Trying to bring me to my senses, my family put it on my face. The body was as black as soot! The synthetic robe that Oksana was wearing was blown to shreds." "Some of its burnt pieces stuck to his sister's body, the rest to the floor in the living room. There was a hole in the place where the sister was standing."

"It's all lightning," says Oksana. "Flying into the fireplace, it hit me, walked over my body, then came out through the leg, broke through the floor and got into a room on the first floor. There, lightning uprooted all the sockets. Broken light bulbs stuck to the ceiling, and in the living room for a long time, I had to tear off pieces of my robe from the floor.

After 5 min, there was a downpour and a thunderstorm outside.

I finally came to my senses already in the intensive care unit. My hands were as black as coals. The lightning left a trace on my chest in the form of a fir branch. However, three days later it disappeared. There were many burns. However, her face was not burned. After a week, almost all the burns came off. Nevertheless, the hair still had to be cut. Now I have minor burns. Lines under the rib, on the stomach, on the leg. Doctors had to put stitches on the leg. The zipper came out through the foot, leaving a deep hole in the heel. I am glad that I no longer feel pain. The next day after the incident, there was a lot of pain in the lower abdomen. This is not surprising—the internal organs have almost doubled in size. The droppers helped."

The examination showed that only the gallbladder was affected. Other organs have increased in size, but not critically.

The daughter was woken up by a deafening crash. "I thought something had exploded," she later said. "I still, when I close my eyes, immediately see this terrible ball with a grimace."

Comment. Injury of a woman when interacting with a BL.

6. (Bychkov and Vaulin 2013) 10.10.2013 09:37. Anna Romantseva, Journalist, 03.12.2013 18:21 LT Greg Slater, Cornwall, UK, male, observer.

A 43-y.o. man was at home with his family when a heavy downpour began. Greg Slater grabbed an umbrella and, running out into the yard, began to examine the metal structure of the gutter. At that moment, there was a deafening explosion, as if lightning had hit the house. "While examining the pipe, I noticed that a gap suddenly appeared in the sky, which was covered with gray clouds, and then I heard a strong rumble. I froze in my tracks, thinking that lightning had struck the cottage. However, noticing the expression on my wife's face, I realized that the lightning had hit me. She said she was looking out the window and saw a BL approaching the umbrella, and then I was flooded with a blinding white light. Slater believes that he was saved from death by slippers with rubberized soles, which he did not have time to change his shoes. After a while, the man felt the effects of the blow." After the explosion, I felt a strong smell of sulfur, and a little later I felt that my neck and shoulders became incredibly stiff. My arms and legs also cramped. The body tingled and pierced with cold. My body seemed to be buzzing from the inside." No one else was injured during the thunderstorm, except for the burned telephone socket and cable.

Comment. The injury of a man during the interaction of the BL with his umbrella.

7. (Bychkov and Vaulin 2013) http://1tv.kz/print/23918. 26. 06.2012-17:16, Kyzylorda. Abnormal weather for the region—rains and thunderstorms—claimed the life of 39-y-o. Kairat Alimbetov, male. The man at that time was with his wife and children a hundred kilometers from Kyzylorda. The family grazed cattle. Kairat was returning home, and in front of his eyes, the yurt was mowed down by a hurricane wind. The man rushed to save his wife and children. At this time, the mobile phone rang. Before he could answer, BL struck him on the spot. Rsaldi Aliyeva, Kairat's mother, said: "My son had just bathed in the river and was walking home. Everything happened quickly. He did not have time to run up to the yurt, as a strong charge, like a fireball, pierced through the body."

Comment. The death of a man during the interaction of a BL with his mobile phone.

8. (Bychkov and Vaulin 2013) The voice of UA with reference to Ukrainian Novelties. 24.05.2012.

43-y.o. Deputy of Novobroskovetsky Village Council of Storozhynetsky District of Chernovtsi region was hospitalized after the shock of the BL. This was reported by I. Malovanyuk.

According to him, the accident occurred on May 24 during a lunch break when the observer was at home. The BL, which flew out of the socket, hit the victim in the shoulder and exited through the leg. The victim is hospitalized; he is in intensive care in a serious condition.

Comment. Injury of a man when interacting with a BL.

9. (Bychkov and Vaulin 2013) ITAR-TASS publications: 8.08. 2003–18:24. Yekaterinburg, Russia. The temperature in the shade exceeds plus 35° with a weak wind. The storm is short-lived.

During a thunderstorm in Yekaterinburg on Friday, a girl was killed by a blow from a BL. The tragedy happened in the afternoon on the beach of the Upper Isetsky Pond. According to eyewitnesses, the BL passed through the wires of the power line, which hangs over the pond, and then it began to descend to the water.

Lightning injured a teenager who was swimming in a pond, and then the ball flew to the shore, where it burned the legs of one and hit another girl in the back, who died on a place. The victims were hospitalized.

Comment. The death of a girl when interacting with a BL.

10. (Bychkov and Vaulin 2013) USA. Joel K <kllgg2 a cs com> Database of a strange scientific page about BL lightning. USA—Thursday, 02.02.2001 at 02:40:27 (PST) in western lower Michigan.

In the spring of 1984 my younger brother was killed by a "fireball." It happened at the beginning of a thunderstorm. The "ball" did not change course, just moved in a straight line about 3 feet from the ground. It hit him right in the chest and tore off his shirt. The heat from the impact was so strong that it burned his internal organs.

Comment. The death of a teenager when interacting with BL.

11. Chesnokov E. V. born in 1945, tester of electric machines. Military Engineer. July 2011. The city of Balashikha, M/O. Russia; time 24.30–0.30 (night). Interview July 2012.

The weather was good all day. There was no thunderstorm. The sky is overcast. Nothing foreshadowed a thunderstorm. Indoor surveillance.

It was hot, but the fan was not on. The room is a garage, the walls are double masonry, the surface is insulated with laminate. There is a grounding. There is a radio equipment. There are no windows; the garage is locked (sealed). It was hot in the room, and I was wearing only swimming trunks (made of synthetics).

I was lying watching TV in the garage. After the news, I noticed on the RC 50 cable Ø 8 mm (from top to bottom with respect to the equipment) a bright glow. It was a white spherical object. Outside is blue, yellow, and red. I looked at it with my eyes. I was not nervous.

There was no glowing shell. There was semi-darkness, lighting when the TV was working. On the background of a dark wall (dark laminate) I saw a perfectly round ball. The diameter of the object is 30–40 mm while it was on the cable. The cable ends with a plug (dad), and this central wire is soldered into the dad. The object appeared on the central cable. Against the background of the object, there was a dark core, 5–6 mm, the color of asphalt, which rolled. The core of the object has not changed.

The object moved away from the cable and moved horizontally in my direction. The distance of 2 m from the object to the observer (from the sofa to the wall), it passed in 1 s. When it moved, there was a crackling sound. The cats, hid who where.

I took a broom and started to brush off the object. I was standing on a laminate (dielectric) and was not grounded. The object stopped at first, as if considering me. The closest distance from the object to my face was 70–80 mm. There was a heat coming from it, similar to that from a short circuit. The radiation of the object is comparable to that from an incandescent lamp of 40–50 W. The object pulsed with a small frequency, comparable to a heartbeat, 60–70 beats per minute.

I got up; it flew around me and, without touching me, began to burn. I felt the pain from the burn on my back and arm. The object moved from the back along the arm to the elbow. It was at a distance of 10–15 mm from the surface of the body. When I felt the pain, I began to wave harder and ran to the door. The door was locked. While I was opening it, the object passed along the leg from the buttock to the foot itself and flew to the other leg. While I was opening the lock, it was flying nearby and burning.

On the left leg, where the shin meets the calf (bone), it acted and left a burn to the bone. (Later, my bone was removed and plastic was inserted.)

Finally I opened the door (the handle at the door is wooden, I did not touch the metal) and it flew out into the street through it. Behind the road, 5 m from the garage, there is a garden. It flew into an apple tree, cracked there and flew away to the high-voltage line. In the yard, while it was flying, it did not crack, and when it was inside the apple tree, it cracked. When the object crackled, small sparks fell out of it and bounced 15–20 mm. There was a flash, and the object disappeared.

Indoors, the TV continued to work, nothing changed.

I called an ambulance; they gave me an anaesthetic injection and took me to the Central Hospital in Balashikha City. They found that 30% of the skin on the legs was burned.

The trunks have two holes with a diameter of 20–25 mm in the scrotum area. The object melted the plastic of the swimming trunks. It went inside the swimming trunks and came out on the other side. The hair inside the trunks is intact and there were no burns. The swimming trunks were left in the hospital. Healing took place half a year.

There were no traces left in the room. The total time the object was in the room was 20–30 s.

Comment. Injury of a man when interacting with a BL.

12. (Bychkov and Vaulin 2014) http://zabmedia.ru/. 08.08.2011 Chita. Russia. During a thunderstorm on August 6, a BL with a diameter of several millimeters flew into the premises of the kebab cafe "Prince's Yard." A small ball "sat down" on skewers with shish kebabs being prepared for the wedding taking place in the next room.

"I took hold of them, I was thrown up first, and then I fell. Lightning struck there," Namik Boiliev, a kebab shop worker, points to a corner of the room. "The tile flew off from the impact. My hand was completely white. When the ambulance arrived, only the hand was already white."

Comment. Injury of a man when interacting with a BL.

13. (Bychkov and Vaulin 2014) www.newsinfo.ru/news/2005-08-13/item/628157/. 13.08.2005

In the Astrakhan region today, Russia. Five vacationers from Ivanovo City received severe burns from a shock during a thunderstorm. This was reported to ITAR-TASS by the Department of the Ministry of Emergency Situations of the city of Volzhsky, Volgograd region, to whose hospital the victims from the neighboring region were taken.

The incident occurred at about 19:00 Moscow time in the floodplain of the Akhtuba River, near the village of Kapustin-Yar. When heavy rain began to fall, tourists from Ivanovo hid first under a tree, and then in a tent. During a thunderstorm, BL hit a tree first, and then ricocheted into a tent. Five of the six tourists who were there received burns of varying severity. Among the victims is a 7-year-old child. They are hospitalized. Doctors assess their condition as "severe, but stable."

Comment. Injury of 6 tourists when interacting with the BL.

14. (Bychkov and Vaulin 2014) www News.Ykt.Ru.17.06.2014 17:48. Saturday, June 14. Russia. In Tattinsky ulus, a 10th grade student almost died after being hit by a BL while talking on a mobile phone in her own house. The girl was in the room, while charging the phone, she called a friend. At some point, the schoolgirl was hit by a BL. The girl was immediately taken to the district hospital of the village of Ytyk-Kyuel, but local doctors decided to send the victim to the Yakut Republican ophthalmological hospital. Reportedly, after a lightning strike, the girl went blind.

Comment. The injury of the girl by BL while charging the phone.

15. (Bychkov and Vaulin 2014) Alexander Boyko. Moscow. 2014. (15:11) August. In the village of Shchapovo in the Trinity District of the capital, Russia. First a little rain fell, then thunder struck, lightning flashed, and the beds were covered with hail.

In the neighboring village of European Valley-2, the BL killed a Moldovan who was filming natural disasters on a mobile phone. People went outside after the hail and began to shoot the ice on a mobile phone. Then a BL flew out of the ground and hovered in the air. Three people rushed into the hallway of the cottage, and the one with the mobile phone continued to shoot the BL. He did not have time to slam the door: lightning crackled, flew into the hallway, and hit him in the head. He died on the place, and his three fellow countrymen received concussion and eye injuries. When one of them regained his sight, he dialled the number of doctors.

The Ministry of Emergency Situations of Russia in the Moscow region confirmed the death of a 55-year-old man and reported that his 51-year-old countryman was hospitalized at the Sklifosovsky Research Institute with a body burn. Two more migrants aged 38 and 53 were sent for outpatient treatment to the district hospital.

Comment. BL killed a man who was filming natural phenomena on a mobile phone.

16. (Bychkov and Vaulin 2014) Chelyabinsk, AN "Access." Russia. 17.08. 2014, in the vicinity of Chelyabinsk City, near the village Starokamyshinsk near Kopeysk. A 16-y. o. boy and his grandfather were riding a motorcycle across the field. The young man was driving a Ural motorcycle, and his 64-year-old passenger was sitting in a wheelchair. At one point, a glowing ball flew next to the motorcycle, blinding both men with a flash of bright light, while there was a strong rumble, as if a thunderbolt had gone literally over their heads. As the 64 y.o. man later recalled, he closed his eyes for a few seconds, and when he opened them, he did not find his grandson driving a motorcycle. About ten meters the motorcycle was moving without a driver. Having stopped the vehicle, the man got out of the wheelchair and a few meters from the motorcycle, in the bushes on the side of the road, found the lifeless body of his grandson. The person died instantly, probably did not even have time to understand what happened. As medical experts later established, the BL hit the young man exactly in the head, the energy potential from the blow was so powerful that even the cap on the guy's head was torn to shreds, and the current came out through his arm.

Comment. BL killed a teenager on a motorcycle.

17. (Bychkov and Vaulin 2014) Region: In the world Topic: Incidents. 31. 03:39, 01.08.2014. In China, a BL hit a farm, killing 170 pigs and injuring a woman, Yun Li. According to the British newspaper "The Daily Mirror." The BL hit a pig farm in southern China, in Hunan Province, as a result of which the farmer's wife Yun Li, 48, was injured (Fig. 2.52).

She told: "It looked like a big fireball, but it suddenly split into two branches and, heating the railing, set the animals on fire. I saw all the colors in front of my eyes and after that, I couldn't see anything for at least 10 min. My left eye turned black and my leg was bleeding." The BL hit the metal railing, sending a massive charge through the entire building, where 400 pigs were located. Of the surviving pigs, 110 needed treatment for burns, but the damage could amount to about 50 thousand dollars. Eyewitnesses claim that some of the pigs were hit by such a strong electric charge that they burst.

Comment. BL killed 170 pigs and injured a woman.

Fig. 2.52 Pigs killed by BL

a

b

18. (Bychkov and Vaulin 2014) erytryec, Igo:28.03.2011, 14:33 L.T. Russia. I saw BL as a child. It appeared before the storm. I cannot describe the size—it was far away, 200 m. Most likely, greater than tennis ball and less volley ball. Orange color. It was as if it had fallen from the sky in an uneven spiral into the crown of a large tree. The tree burned down. In addition, from the inside. The outer part of the trunk was untouched, and inside there was only ashes. The trunk broke into two parts.

Comment. When exposed to BL, the tree burned down, from the inside.

19. (Bychkov and Vaulin 2014) Annals of Plastic Surgery. May (2003) 50: Issue 5. Burn BL. Selvaggi, Gennaro MD; Monstrey, Stan MD, Ph.D.; von Heimburg, Dennis, MD; Hamdi, MD Mustapha; Van Landuyt, Koen MD; Blondeel, Phillip MD, Ph.D.

During a severe storm, a 28-y.o. man and his 5-y.o. daughter received burn wounds after the BL came out through the chimney. These two patients showed signs of electric shock and fire. The father, who lost consciousness, received bilateral superficial burn wounds of the second degree on the zygomatic area and deep burn wounds of the second degree on the right arm (total body surface area, 4%). His daughter had superficial second-degree burn wounds on the left side of her face and deep second-degree and third-degree burn wounds (total body surface area, 30%) on the left side of her neck, both on her forearms and on her back.

Comment. When exposed to BL, the father, who lost consciousness, received bilateral superficial burn wounds. His daughter had superficial burn wounds.

20. (Bychkov et al. 2015a, b) July 2015. Internet: Ball lightning. Victor.

During World War II the BL killed my great-grandmother, in the house, in front of the children. When BL appeared, everyone hid where they could and watched it. It flew for a long time, but when her great-grandmother tried to move to another place, lightning overtook her.

In the 1980s, it happened in the summer kitchen. My grandmother did not run away from the BL, even when it approached a critically close distance. She even held her breath at that moment. As a result, the BL burned her leg flying nearby, but continued her movement towards the entrance doors, and flew out of the threshold of the room, exploded with an explosion of inconceivable volume, either going into the ground, or right in the air.

Comment. When exposed to BL, she burned her leg while flying nearby.

21. (http://forens.ru/index.php/topic/for-for09/21/200906:28) Moscow. Russia.

A thundercloud is approaching the micro-district of the city. It is not raining. On the street, where there are 5-storey buildings, street lights, a lot of tall trees and quite busy traffic, a man is coming from work. Police officers of the police station are standing 50 m away from him. According to the police and other witnesses, suddenly, a large (1–2 m in diameter) dazzling glowing ball appears near the tree next to which the man was passing. Everyone closed their eyes, at that moment there was a deafening bang (not like an explosion). As a result, the trunk of the tree below the crown is singed, and the man is lying on the lawn 2 m from the sidewalk.

External and internal injuries are practically not visible, especially similar to electric markings (histology is presented below). The clothes are all torn and only on the front surface. Only on the foot, there is a slight singeing of the sock. Outwardly nothing specific. If it were not for the witnesses, it is not a fact that the cause of death would have been correctly established. In Figs. 2.53, 2.54, 2.55, photographs of the body and objects affected by BL are presented. The data of the conclusion of the forensic chemical study No.#: "During the forensic chemical examination of objects from the corpse of P … ethyl alcohol was found in a concentration of 1.5%o in the blood, 2.7%o in the urine." The data of the conclusion of the forensic histological study No.#: "Pronounced parenchymal dystrophy of the liver, kidneys, myocardium. Spasm of small myocardial arteries. Granular exudate in the glomerular cavity of the kidney. Acute venous fullness of the kidneys with the phenomena of intravascular hemolysis. Pronounced perivascular edema of the brain. Acute alveolar emphysema, pulmonary edema, focal drainage hemorrhages in lung tissue. The skin preparation coagulates the epidermis, separating it from the basement membrane. In the papillary layer of the skin, blood coagulation in the vessels. No inflammatory reaction was detected. In skin preparation No. 1, there is an abrasion with hemorrhage without signs of reactive inflammation. Vessels are dilated."

The data of the conclusion of the forensic histological examination No. #: Severe parenchymal dystrophy of the liver, kidneys, myocardium. Spasm of small myocardial arteries. Granular exudate in the glomerular cavity of the kidney. Acute venous fullness of the kidneys with the phenomena of intravascular hemolysis. Pronounced perivascular edema of the brain. Acute alveolar emphysema, pulmonary edema, focal drainage hemorrhages in lung tissue. The skin preparation coagulates the epidermis, separating it from the basement membrane. In the papillary layer of the skin, blood coagulation in the vessels. No inflammatory reaction was detected. In skin preparation No. 1, there is an abrasion with hemorrhage without signs of reactive inflammation. The vessels are dilated.

Forensic medical diagnosis. Main: Damage by atmospheric electricity (lightning), thermal burns of the 1st degree in the neck, chest, abdomen, left knee joint (figures by lightning), hair loss in the head, chest, abdomen, acute emphysema of the lungs, venous fullness of the internal organs and brain with hemorrhages under the pulmonary pleura and in the lungs of a diapedetic nature, liquid state of blood, pulmonary edema, cerebral edema.

Concomitant: The presence of alcohol in the blood is 1.5%o, in the urine-2.7%o.

Conclusion: The death of gr. P..., xxx years old, about 1 day old by the time of the examination of the corpse, occurred from exposure to atmospheric electricity (lightning) in a state of alcoholic intoxication. When examining the corpse of gr. P ... in addition to the figures of lightning in the neck, chest, abdomen, left knee joint, penis in the form of thermal burns of the 1st degree, no other damage was found.

Comment. Death of a man when exposed to BL.

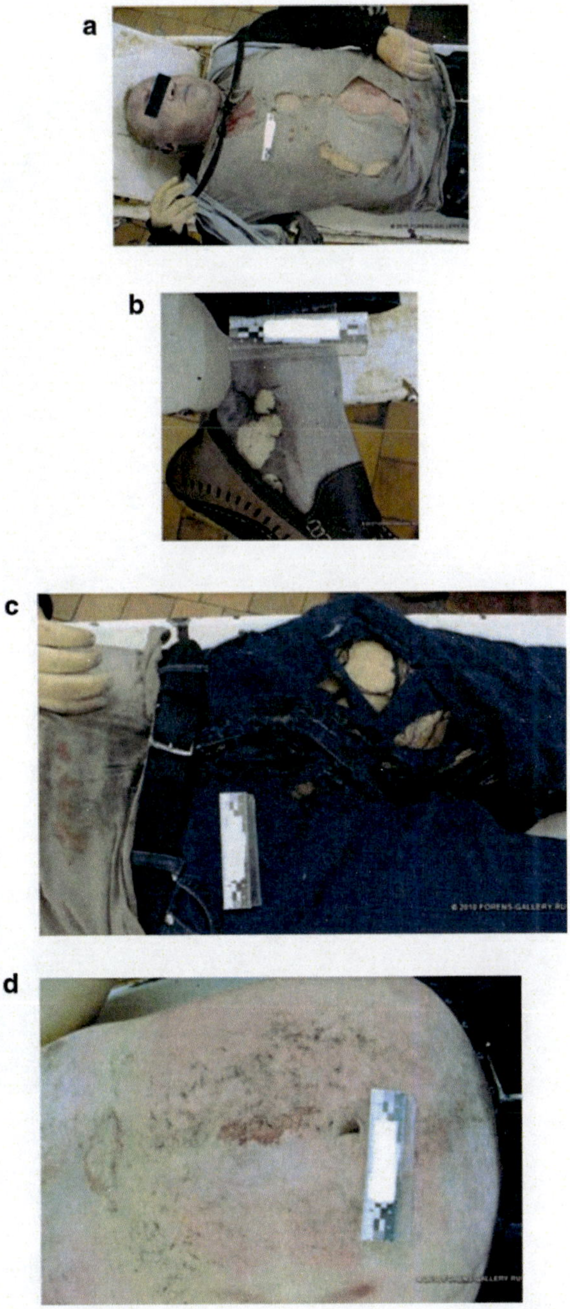

Fig. 2.53 Injuries on a man and traces of BL impact

Fig. 2.54 Yuri Silagadze. The BL burned through his shirt and hit his shoulder

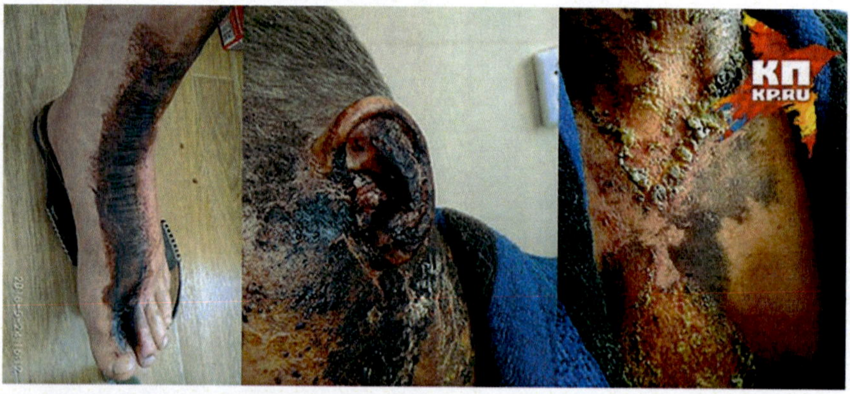

Fig. 2.55 Photo of the observer's injuries

22. (Bychkov and Vaulin 2014) Echo of Moscow. Moscow. Russia 20.06.2015

BL killed a man hiding under a tree during a thunderstorm. It discharged itself into a mobile phone, which rang in the man's hands.

Comment. Death of a man when exposed to BL. It discharged into a mobile phone.

23. (Bychkov and Vaulin 2014) http://myslo.ru/news/arhiv/article-6496. Tula, Russia.

Severe thunderstorm in Tula in the evening of 17.09.2015. More than in other areas of the city, bad weather was rampant on Oblique Mountain (Kosaya gora).

BL was seen in different parts of the village. At about seven in the evening, 44-y.o. welder Andrey Ponomarev died from a blow from a BL. "My husband was returning from work at Oktyabrskaya Street. The accident occurred in front of a woman whom Andrey overtook at the store 'Solnechnyi'. As soon as her husband was a couple of steps ahead of her, a BL appeared in the air. Lightning struck Andrey. He died without regaining consciousness." According to eyewitnesses, Andrei Ponomarev involuntarily covered a woman and a child who was with her.

Comment. Death of a man when exposed to BL.

24. (Bychkov and Vaulin 2014) Yankina Liana (22.06.2014, 23:02). The man was hit five times by a BL 46-y.o., Special Forces soldier Yuri Silagadze, lives in the Krasnodar Town of Ust-Labinsk, Russia.

"For the first time it happened on May 15, 2010,—It was at night. It started raining with a thunderstorm, my mother asked me to run outside to take off my underwear. I grabbed the rope on which the things were hanging, and saw a small glowing ball right in front of me. Everything happened so quickly that I didn't even have time to understand anything, but only felt a sharp blow to the chest from the right side."

Yuri had called an ambulance. However, apart from slightly elevated blood pressure and burns, the patient did not complain about anything, and the next morning the man was discharged from the hospital.

A month later, Silagadze was watching TV when a thunderstorm started and there was interference on the screen. Yuri went to fix the antenna. A bright flash, a blow to the left shoulder… Again, a lightning burn, and again the doctors, except for increased pressure, did not diagnose anything in the patient.

Two years later, heavy rain again caught Yuri in the yard. Next: blow, burn, ambulance.

This time it burned harder—from the chin to the stomach. Nevertheless, Silagadze survived.

And a year later it was raining with a thunderstorm in Ust-Labinsk, and Yuri left home. "I asked him not to go out, but he didn't seem to hear me," says Yuri's mother Asya Ivanovna. "And a few minutes later he became a victim of BL again." "Doctors began to examine. But they never found any deviations." Silagadze is now a healer—treats headaches, sleep disorders. They even take heart patients and epileptics to him.

"In conclusion, they wrote: thermal burn." Yuri Silagadze was bandaged at the burn site and treated with antiseptic. The patient had no other complaints, so hospitalization was not required. The conclusion states "thermal burn" because the doctor cannot write "lightning strike." We find out the origin of certain wounds only if it is an industrial injury.

Comment. The man was hit 5 times by a BL.

25. (Bychkov and Vaulin 2014) 30.06.2015 Newspaper “KUZBASS,” SOCIETY. Russia.

A hurricane with an almost tropical downpour hit Kuzbass on the evening of June 24. In the air, dried out by several days of heat, the BL seemed to have emerged from the void. And it exploded just a meter and a half away from people … “ I can’t believe it happened to me,”—admits 25-y.o. Pavel Aksenov (male). A team of power engineers-installers set up towers near Leninsk-Kuznetsky (city) … on the 24th in the morning they hung on a power transmission pole, fastened with safety belts on shift … “The height of the support is about 30 m … Dima and I are above everyone, Anton is lower, Oleg insured everyone on earth. It’s a long way from us to the high-voltage wires. The heat was so intense, it was clear: there would be a thunderstorm in a few hours.

At 11.20 a fireball appeared in front of the installers. It formed in front of us … then the BL, after hanging for a few seconds, exploded. And with such a powerful bang, a roar that my ears were blocked, ‘bunnies’ flashed in my eyes…One of the arrows of the exploding lightning hit Pasha, who was holding on to a piece of support at that moment. However Dima was not hurt, because he raised a brush with paint and broke away from the support, swaying, floating away …—Lightning struck me here,” Pasha shows a line on his chest, going obliquely—straight to the heart. “But it did not go into the heart, but passed over the top-into the left hand raised in work, burned the edge of the left palm… In addition, it swept down the support, exploding further and further, heading towards the substation, losing power. What saved me? The edge of the cross touches the ‘autograph’, the deadly trajectory of lightning… And the cross was twisted.” He says that he was stunned for a long time, that he began to see well only after a day…We met with him and his comrades a day after the lightning strike and the transfer of the man from the intensive care unit to a regular ward. We wonder how lucky the man is. He did not lose consciousness. He got with complaints of impaired sensitivity of the forearm and hand, periodic rare convulsions of the left leg”—doctors explained at Kuzbass scientific and clinical center for the health of miners in Leninsk-Kuznetsk. “On the palm, at the little finger, a burn of I-II degree, spot wounds”…Doctors say that Pasha was lucky, only pinpoint wounds on the palm remained from the lightning strike.

Comment. The injury of a man by BL.

26. (Bychkov et al. 2015a, b) 17.08. 2016. In the village of Cape Nyaruy-Salya, Yamalo-Nenets Autonomous District, Russia, a tragedy occurred.

A BL flew into the plague and exploded inside. A 7-y.o old boy was killed; and girls 12- and 16-y.o received second- and third-degree burns, they were taken to the hospital.

Comment. The death of a 7-y.o. boy and burns of 12- and 16-y.o girls.

27. (Bychkov et al. 2015a, b) “Notebook Krasnodar.” In Timashevsky District of Kuban 4.08.2016. Russia.

BL hit the company of young people. The incident happened in a local park near the stadium, where three men were hiding from the rain under a tree.

According to eyewitnesses, during a severe thunderstorm, a glowing ball fell from the sky and hit people. One person died on the place, two were taken away by ambulance. It is known that the deceased 34-y.o. man was talking over the phone at the time of the thunderstorm. His companion, a 33-y.o. man, is in a serious condition in intensive care and it is unlikely that he will survive, since more than 50% of his body is burned. The third victim, a 29-y.o. man, was stunned by lightning and is now in hospital.

Comment. The deceased 34-year-old man was talking on the phone at the time of the thunderstorm. Two people were taken to the hospital.

28. (Likhosherstnykh 1994) P. S. Tyurnikov, male, oil well driller, Chardzhou. Russia. Rain, thunderstorm.

"I was having lunch in the cultural booth and looked out the window. There was a thunderstorm. After another lightning bolt, a white ball the size of a football with a brightly colored core surfaced from behind the grass at a distance of 60–70 m. The ball began to approach the telephone pole. When it approached the wire, I turned to the master to draw his attention to this strange phenomenon (I had no idea what it was). As soon as I turned around, I felt as if someone had hit me in the back with a sledgehammer, and something flashed in front of my eyes, and the ebony phone box shattered into pieces, and some fragments hit me in the face. As I was falling, I was surprised to think who hit me and why? I woke up an hour later. I was lying on my right side. My left leg was twisted so much that my head was below the knee. I did not hear anything, I just saw people moving their lips. After I was unbent, I began to hear a little. The examination showed that I had a burn on my back, running in a strip from my left shoulder blade to my armpit, then to my stomach near the navel and then down. Two socks were burned on the left foot, but there were no burns. The watch lying in my breast pocket crumbled, as if it had been specially disassembled.

At the hospital, 26 fragments were removed from me. There was a through hole in the ear membrane, and since then I have not heard with my left ear. Our electricians said that I was 'lucky': my clothes were wet from the rain, and therefore the BL passed over my clothes, and not over my body."

Comment. The injury of a man by BL.

29. (Likhosherstnykh 1994) D. V. Vasiliev, male, S. Stakhovka, female. After a thunderstorm. Russia.

"Without bringing the TV plug to the outlet by seventy centimeters, I suddenly saw that a small ball popped out of the outlet and 'sat' on my hand. Feeling the pain, as if from an electric shock, I lost consciousness. When I came to, I was lying on a pile of plaster knee-high. When I heard my daughter's scream (she was seven y.o.), I began to crawl on the floor, trying to find my daughter, but the room was dark with dust and smoke. Finally, when I crawled under the table, I took a guide to the door, but the door was not in place (later it turned out that it was blown off by an explosion). As it also turned out later, the daughter was thrown by the explosion into another room, where her grandfather picked her up.

Everyone forgot about me, saving the girl. When I got out into the yard, I saw many people around my daughter lying on the ground. When I approached my daughter, my wife showed me her (daughter's) hand, in which a 150 mm rusty nail was sticking out. I bent down, pulled out this nail and threw it under my feet. Then an ambulance arrived. Everything went well in general, except that after that the daughter began to be terrified of thunderstorms.

The damage in the house was as follows. The double ceiling and the roof above it a hole pierced—the hole was about two meters in size. On the walls of the room where the wiring took place, there were real trenches, as if plowed by a plow. There was a funnel where the electric meter was hanging, as if from a shell explosion. There was also a funnel on the opposite wall, but already from the impact of a torn-off electric meter, the fragments of which were lying right there."

Comment. The injury of a man and a girl SM.

30. V. Smelyansky, male. Rabochaya gazeta. 1.12.2006, No. 11(61) Guest from Perun (in abbreviation). The place is 100 km from Ufa. The White River. Wooden shed on the riverbank. Russia.

Fishing was really a success…, but suddenly the sky frowned, and from the leaden-gray clouds, rare still rare raindrops fell. The observers hastily gathered up the gear and ran to the barn. After a couple of minutes, the elements played out in earnest. The rain drummed fiercely on the roof. Thunder rumbled. It quickly got dark. 'The storm must have charged for a long time,' Viktor Leutin (male) remarked, 'we should at least light a lantern.'

He reached for his backpack, but at that moment, two lightning bolts flashed near the barn at once. Then, in the semi-darkness, we saw a glowing ball right above our heads. Slightly crackling, it slowly descended along some kind of spiral trajectory. Numb, we watched him in amazement. I remember that the size of the ball was slightly larger than a football and a strong smell of ozone emanated from it. However, we did not feel any heat. Suddenly, changing the direction, the ball began to move to the corner where gears were lying. It touched a metal box with spare fishing lines and hooks, hissed and jumped up again.

Victor, whose body was apparently numb from the uncomfortable position, moved, sighed, and the lightning, as if attracted by his movements, smoothly glided over his head. The guys screamed, and the fireball burst with a bang and scattered sparks.

Having thrown things in the shed, we grabbed Viktor lying motionless and took him by boat to the hospital of the district center. Fortunately, already in the emergency department, he woke up and spoke to us.

The doctors who examined my friend concluded that his leather cap saved him. Lightning burned through it, but his head remained intact. True, a perfectly round, even circle of gray hair appeared on the top of Victor's head. It did not disappear even after the haircut to zero.

Later Leutin said that he did not lose consciousness—he clearly saw us and heard voices, but it seemed to him that we were moving and talking, as if in slow motion. After the collision with the BL, he began to stutter slightly, but he has no other health disorders.

Returning to the barn for our gears and fish, we found that all the hooks had melted in the box on which the fireball had dived. The plastic boxes in which they were lying, and the coils of fishing line remained intact.

One more thing: among the groceries in Victor's backpack were several raw eggs wrapped in newspaper. They all turned out to be empty—only shell, but without a single crack. Discussion of observation. The effect of BL radiation on metal objects is well known and is shown in the collected observations. However, the effect of BL on the condition of chicken eggs turned out to be a new property.

Discussion. To find out the nature of the action DS in physics V. L. Bychkov with DS in physics A. I. Nikitin did the following experiment. Tightly wrapped a fresh egg with a piece of newspaper and put it in a microwave oven with a capacity of 1.5 kW for 10 s. When the egg was taken out, it turned out to be in the cracks, under which the cooked inside was located. All this showed that the microwave radiation only led to heating of the inside of the egg.

Nevertheless, the reliability of the description made it seem that the observation did not contain fantasy. An analysis of the literature on the properties of chicken eggs led to this conclusion. Thus, the article (Mikhailov 1998) describes the complex structure of a chicken egg. Here are the main facts relevant to this observation. The shell is permeated with numerous narrow tubes—pore, or respiratory, channels, simply pores. There are thousands of pores in the egg, and gas exchange is carried out through them: oxygen enters, carbon dioxide leaves. Water is stored in protein with a "reserve," and evaporation is used for special needs. Due to this partial evaporation of water, a free space is gradually formed on the wide pole of the egg, which is called an air chamber. Decomposition products are released in a dry, poorly soluble form (otherwise they could poison the embryo)—this is not urea and not ammonia, but "dry" uric acid. It is intuitively used by housewives: they pierce a hole in the shell with a needle, after which the egg dries and becomes hollow.

Based on this, it can be assumed that under the influence of BL radiation, the pores in the shell increased, oxygen, carbon dioxide, and water vapor freely left the egg, and the decay products turned into uric acid. Therefore, the egg was dehydrated and turned into a hollow formation.

Comment. The injury of a man by BL.

31. Yu. K. Sedov, male. Description of the event from the words of the Taimyr television journalist R. P. Yaptun, female, who interviewed eyewitnesses. She also reported that there were no other cases of sightings of BL in Taimyr. Summer: July–early August, 1979. Taimyr Peninsula, Ust-Yenisei District of the Taimyr Autonomous Okrug, Russia. The place of action is the summer Nenets camp on the bank of the mouth of the Yenisei River, where it is convenient to engage in fishing and hunting feathered game.

In the afternoon, the Nenets family waited out the storm, settling down in the chum, the entrance to which was open. The rain stopped and the sun came out. The family consisted of six people: a husband, a wife, a grandmother, and quite adult children—two sons and a daughter.

Suddenly one of the sons, sitting at the entrance, noticed a small, glowing ball floating in the air a few dozen meters from the chum. The 15 cm ball seemed to him like a Christmas tree toy, about which he informed the others. The ball, glowing with a faint blue-violet light, meanwhile approached the chum and flew inside. After making one flyby of the chum over the heads of people sitting on the floor, the ball was attracted to the cast-iron stove standing in the center of the chum. There was a bright flash with a loud pop; ash flew from the stove all over the chum. The ball, having decreased in size by about half, flew out into the upper hole of the chum and "exploded" with a loud sound at the neighboring chum. At its entrance there was a dog injured during the "explosion."

There were no eyewitnesses of the "explosion" from which the dog suffered. The first thing that immediately caught my eye was that the shaggy dog seemed to have been cut short. One of its ears became noticeably longer than the other and remained that way. Its coordination of movements was disrupted; the dog began to move clumsily.

Upon careful examination of the first chum, it turned out that all the aluminum dishes that were next to the stove melted and froze in the form of ingots. At the same time, one hundred pieces of cardboard (punk) cartridges with shot in factory packaging were externally preserved, but when the wads were removed from them, it turned out that the shot also froze into a single lead ingot. Such cartridges were no longer suitable for hunting birds.

But most of all people suffered. Everyone's eyesight deteriorated greatly: one man went completely blind, the others partially. Their eyesight was restored only by winter.

Within 2 years, all the men died, the women (except grandmother) are alive and continue to live in the tundra.

According to the old Nenets beliefs, this event was understood as a heavenly punishment for the guilty family. Summer camps have not been arranged at this place since then.

Comment. All those who were exposed to BL had their eyesight severely deteriorated: one man went completely blind, the others partially. Their eyesight was restored only by winter.

32. (Olkhovatov 2008) A note by the United Press Agency, in the Decatur Evening Herald newspaper. The end of December 1928. Greendale, New York, USA.

It is claimed that a meteor fell on a house and killed a woman and a child. Farmers living in the vicinity of Greendale, New York, are convinced that a meteor fell from the sky on Friday evening and set fire to a farmer's house, resulting in the burning of a woman with a 1-year-old child and six other people were injured. Mr. J.Hicks said on Saturday that he was standing in front of his shop and saw a fireball "shot" from the sky, which fell on the roof of Mr. V. Pitor's house. Mrs. Peter and her 1-year-old nephew were killed. Others who were in the house, aged from 5- to 28-y.o., received severe burns.

Comment. The death of a woman with a child in a fire caused by BL. Burns of many observers.

33. (1982) Attention: Ball lightning! Technic of Youth (Technique of Youth). No. 3, pp. 38–41. E. A. Asonov, male, radio equipment adjuster. 1978, August. Karelian Isthmus, Sosnovo District. Russia.

"We lived in tents on the shore of the lake. On one of the warm, clear days of the end of the month, a dark cloud suddenly ran up, it turned cold sharply, and large hail began to fall. Suddenly there was thunder overhead.

A lightning rod mounted on a tall pine tree was struck by lightning. Immediately from the place where the grounding of the lightning rod began, a bright flame flared towards the side, which took the form of a ball with a diameter of 12–15 cm. First a reddish, then a white-yellow ball swelled almost twice and became a dazzling white, brighter than a 20 kilowatt xenon lamp. Breaking away from the lightning rod, the ball flew at a height of about 1.3 m. I realized that it was a BL.

When the dazzling ball approached me by five meters, the hair on my head rose. I felt that even the hairs on my arms and chest stretched towards the lightning, as it happens when setting up high-voltage equipment. Moreover, 'threads' of electrical discharges stretched from the chest towards the BL. Fortunately, the ball passed me in 3–4 m. I clearly saw that its brightly glowing central part was surrounded by a very thin (less than cm) dark zone, around which a radiant halo 3–5 cm wide shone.

The ball passed 30–50 cm from the back cover of the Zenit camera standing on a tripod and headed down to the lake at a walking speed, exactly following the will of the winding path. Then it went dark in my eyes. Vision was restored only after 2–3 weeks, but the eyes had to be treated for another two months, and the general malaise lasted six months."

Zenit was standing on a tripod, and the shining ball passed 30–50 cm behind the device. Everything is intact outside in the device. Only the plate on the frame translation lever and on the tripod parts is swollen. Some parts inside the case are damaged. The plastic cassette and the receiving coil are melted and severely deformed; the film is warped. A part of the polymer-frosted glass of the viewfinder was melted. And the curtain and other parts are not damaged. This type of damage interested the employee of the forensic examination laboratory D. Naumov. His experiments showed that such damage could occur at 150–300 C. Nevertheless, how such a heating occurred inside the case in a short time of passing the BL and why half of the glass is melted and half is intact is unclear.

Comment. Injury of the observer.

34. (Bychkov et al. 2003) M. E. Salkova, female. Moscow. Russian.

"The incident occurred in a nursery that went to the summer cottage. There was a thunderstorm. I was standing in the kitchen, leaning against the door jamb (the kitchen was an extension to the house).

Suddenly I felt a pain in my hand, as if someone had hit it, and immediately a very bright ball the size of two fists flashed before my eyes near the hand. The next moment, the ball moved away from the hand and flew in the direction of another door, passing next to another employee and not hurting her. Flying out through the door, the ball hit a birch tree, which fell on the titan: both immediately shattered.

My hand began to turn black right in front of my eyes and there was a terrible pain with tingling. The staff buried my arm up to my shoulder in the ground and began to pour cold water. The blackness began to fade. I have been lying in bed with this hand for two weeks.

The gold wedding ring that was on the finger later fell apart."

Comment. Injury of the observer. The gold wedding ring fell apart.

35. 2011–07-27 3:44:00 RIA "Siberia Kyzyl. 24.07. Tes-Khem District of Tuva. Russia.

During a severe thunderstorm, right above the camp of the family of Sayyn Sandykmaa, the shepherd of the State Unitary Enterprise "Choduraa" was formed a BL. At this time, the sheep rushed away from the yurt to take shelter from the rain under the pine trees near the shepherd's camp. BL was carried away after the animals huddled under the trees. The glowing ball collided with one of the pines and a powerful explosion was heard. Its strength was such that one of the sheep weighing under 30 kg was thrown 10 m. The trunk of a 15-m pine tree was split like a splinter by a knife, from the top to the top. When the storm subsided, it turned out that 84 animals died from the lightning discharge.

Comment. BL killed 84 animals.

36. (Tutt 2004), 2004 V. de Fonville. 09/10/1945, the village of Salagnac, France.

A fireball flew into the kitchen of one house. One of the cooks shouted to the other to "throw this thing out of the kitchen," but the second cook just let the BL fly out of the kitchen and, as it turned out, did the right thing—the lightning went to the pigsty, where one of the pigs tried to smell it, after which the fireball exploded. The pig died, and the pigsty was significantly damaged.

Comment. BL killed a pig.

37. (Stenhoff 1999) A terrible thunderstorm raged over the city of Uralsk (now Ural, Western Kazakhstan) on 05/22/1901.

It was a holiday, and the streets were filled with people. Closer to 5 o'clock in the afternoon, several young guys and girls, total 21 people, took cover in the lobby of the house, and a 17-y.o. girl K. sat on the threshold with her back to the street.

Suddenly there was a strong clap of thunder, and a dazzling sparkling fireball appeared in front of the door, gradually descending to the place where they were. Touching the head of the girl K., who immediately bent down, the ball fell to the ground in the middle of a group of people, walked around everyone in a circle, then continued its way into the room of the owner of the house, touched his shoes and singed them. Brought the room into complete disorder, broke through the wall into the stove in the next room, hit the iron chimney and tore it out with such force that it crashed into the opposite wall, and finally flew out through the broken window. That is what they found after the first fright passed. The young girl was still sitting with her head bowed, it seemed that she was asleep. A few people were walking in the courtyard and did not see or hear anything, the rest were lying in the lobby in a deep faint.

The girl K. was dead. The fireball hit her in the back of her neck, went further down her back and left thigh, leaving a black mark in its path. There was a wound on one arm with a small amount of blood, one of the shoes was torn off the leg, and there was a small hole in one of the stockings. All the victims went deaf.

Comment. Death of the girl and injuries of people. Impact to objects.

38. (Stenhoff 1999) G. C. Dijkhuis, male, physicist Ph.D. 1993. Amsterdam 3.06. 1992.

The ball hit the man in the forehead. At 11:30 p.m., he was standing in a room on the second floor not far from the open exit to the balcony, covered with a metal mesh to protect against mosquitoes. The weather was wet, it was raining heavily, and lightning flashed nearby. Immediately after that, he saw a bright green ball the size of a football above a tree growing near a nearby business building. This ball, which was visible for only a fraction of a second, hit him on the forehead, causing him to fall backwards to the floor. He was stunned, but he did not lose consciousness. When his wife, hearing the noise, entered the room, they both smelled ozone. Nothing was damaged in the room, but a red spot appeared on the victim's neck. After a medical consultation, he received a referral for examination. A brain scan using the magnetic resonance method showed the presence of damage in one of the parts of the anterior lobe.

Comment. Effects on the human brain.

39. KP-Irkutsk, date: 24.05.2018-22:31 Angarsk. Russia. 21. 05. 2018.

"The lightning was like a football, so bright and dazzling. I instinctively covered my head by my hands, and then I heard the car alarm go off," 69-y.o. Siberian Antonina Kuravleva (female) recalls the chronology of that terrible day. She and her son were at the dacha. On TV the weather is good, so observers decided that the son, who had a day off, would help his mother with the garden. Gardening is located nearby—just a half-hour drive from the house in Angarsk.

"I was in the garden, and Andrey was doing the formwork," says Antonina Ivanovna. "The weather suddenly turned bad. Clouds, hail. The son went out the gate, admired the hail all the time… At this moment, and covered! The ball lightning came from the forest." The Kuravlevs' area is just closest to it. The woman went to the "screaming" car and saw her son lying on the ground. He was unconscious. Sneakers, turned into rags, lay nearby. All the clothes were burned.

"He was wearing a warm jacket that day because he was ill the day before," the Siberian woman says "There's not a trace of it left. If it were not for the cross…, then the son would not be alive. Yes, yes, the lightning hit right into the cross and chain! The charge was such that the metal melted! Then lightning struck the left leg."

The son was lying with all black skin unconscious, I tried to bring him to his senses, the pensioner continues. "Literally for a second he opened his eyes and said that he could not feel his legs."

Neighbors ran to the woman's cry and called an ambulance. The man spent a couple of days in intensive care and today the Siberian, who works as an electrician, was transferred to the ward. Doctors assess his condition as of moderate severity. Now there will be a long treatment. There are serious burns on the man's body. Nevertheless, he does not remember the moment of the lightning strike and does not want to talk to journalists yet.

General comment to the section

We showed that BL strikes lead in a number of cases to the death of people, and in many cases to the serious injuries of observers. Recently, fatal BL strikes by observers often occur during thunderstorms in mobile phones and other devices. So one can conclude that BL is a very dangerous phenomenon.

2.8 Moving Objects When Exposed to BL

In this section, we present observational data on the BL movement of objects.

1. (Likhosherstnykh 1904) V. A. Lysenko, male, Gorlovka. Russia. July, no rain. The observer was fishing in a remote place on the bank of the Samara River.

"Suddenly, a bright-yellow ball appeared on the opposite shore at a distance of seventy meters. I knew that in such cases it is impossible to move, and therefore froze. At this time, the fish pecked. I looked and saw 5–6 m away from me, half a meter above the water, the same ball. Now it was possible to see that it was flattened from the poles and some kind of boiling was happening inside it. I remember very little further. Suddenly I felt that I was falling somewhere.

I came to myself a hundred meters from the place where I was fishing. There was a tree nearby and the ball was hanging about a meter above the ground. Its shell in its cellular structure resembled a section of a honeycomb. I lost consciousness again and woke up fifty meters from the tree. The ball flew away from me."

Comment. Moving the observer. The shell of the ball in its cellular structure resembled a section of a honeycomb.

2. (Nikolaev 1999) Tomsk group of anomalous phenomena. Galtsov phenomenon. The village of Galtsovka, Altai Krai, Russia.

The BL, the size of a football, flew over the village at an altitude of two to three dozen meters. The first shed with reinforced concrete pillars that came in its way was crushed and toppled.

Flying over a residential house covered with slate, the BL tore off the slate from the roof of the house together with nails (from the surface of about several hundred square meters) and lifted all this slate into the air, dragging it with it, scattering it all over the village.

Flying over the tractor station, the BL crushed one frame welded from the corners and covered with a tarpaulin, and the other frame, when approaching it, the BL first dragged along the ground, and when the BL overtook the frame, it was lifted by the BL and moved 300 m. The frame itself remained intact. The total weight of the frame was at least 100 kg. (We do not provide photos because of their low quality.)

Comment. Moving of 100 kg metal frame.

3. (Egely 1987 case 321). Ms. Nemese reported the following sighting near their goose farm in Fabiansebestyen on June 27.

"The storm passed, and the weather was already calm, cloudy, around 20:00. Suddenly, I noticed a floating, grayish spherical object, above the stream. It was approximately 2 m in diameter. It was about 30–40 m away from us and was moving towards our building at the speed of a running person (presumably 7 m/s auth.).

I told my husband to come and look at it, but 2–3 s later, by the time it reached the window, we heard a loud crash above our heads. In addition, 35 × 15 m the roof of the building was carried away (raised towards the auth.) and fell down on the area of the neighbor's oat field. The brick chimney and its 1.5-m-long iron extension fell, and the back wall of the building (one brick 12 cm thick) collapsed. No traces of fire or heat were found."

According to (Egely 1987 case 321), a girl on a nearby farm (approximately 100 m away) saw a reddish ball moving away from the destroyed house immediately after its loud crash. It swam only above the oat field and moved horizontally. The owners noted a strange thing: a steel reinforced concrete pillar broke at ground level and was directed towards the building after the events. The average weight of the roof (shriveled slate, thermal insulation, wooden supports) is estimated to be approximately 20 kg/m^2, and (the roof) was raised at least 1 m according to witnesses. Therefore, the total work was approximately 10,000 J, but the energy density cannot be estimated because the BL has floated away.

Comment. Simplified analysis of observation data (Bychkov et al. 2014). From the description, it follows that the BL flew onto the roof of the building. The diameter of the BL was 2 m. The BL knocked down a pillar. The mass of the roof was about 10,500 kg. At the same time, thunderstorm conditions suggest the presence of an external electric field of magnitude $E = 3 \cdot 10^3$–$3 \cdot 10^5$ V/m (Bychkov et al. 2010). There were no explosions or other effects of plasma or chemical type, since no traces of soot and melting were found. According to the description, the ball floating over the stream had a diameter of about 2 m and was at a distance of 30 m from the house, moving at a speed of about 7 m/s. After 2–3 s, the BL was above the roof of the house. This description suggests that the eyewitness may be mistaken in determining the time of the process involving BL (factor ~ 3). Figure 2.56 shows the event plan (all dimensions are given in m.)

Assuming that the farmhouse has a standard height of 8 m (1st, 2nd floor + attic), the lightning was initially at the eye level of the observer and it is possible to make

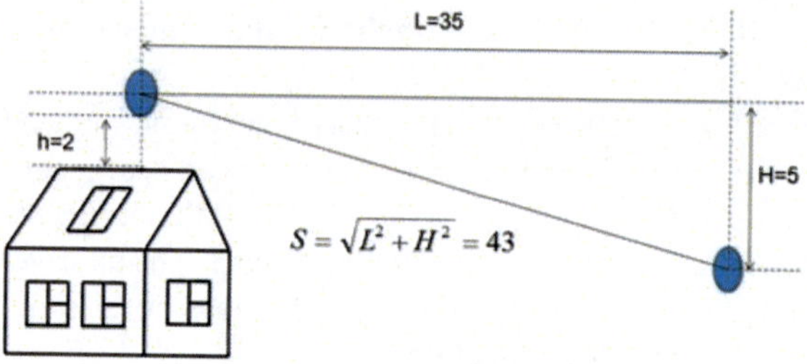

Fig. 2.56 The scheme of interaction of the BL with the house. Here L is the distance from the house to the position of the BL at the initial moment of observation, H is the difference in heights at which the BL was at the initial and final moments of observation, h is the height to which the BL raised the roof, and S is the distance traveled by the BL during the event. All distances are given in m

an estimate of the speed v_{av} of the BL. To do this, let's assume that the BL traveled the path S in time $t = 2$ s, and in time $t = 1$ s, colliding with the roof, lifted it to a height of 1 m, $v_{av} = S/t \approx 22$ m/s.

At the same time, the roof weighing 10,500 kg was raised by 1 m by the BL, after which it was dropped down to the neighbor's plot. It can be assumed that the BL produced an explosion or released a jet, which caused the explosion, but these assumptions immediately entail the need to include a BL model, which is not part of the task of this study. In this case, within the framework of the simplest model, we will assume that the BL transmitted an impulse to the roof when it was in a state of indifferent equilibrium, which took place during the electrical action of the BL on the roof.

Let's write down the law of conservation of momentum for the BL before the interaction with the roof and after the interaction:

$$M \cdot v_{av} = (M + m_r) \cdot v_r, \tag{2.8.1}$$

where M, m_r are the masses of the BL and the roof, respectively, v_r is the speed of the roof + BL system after interaction. The following expression for the mass of a ball lightning comes

$$M = m_r/(v_{av}/v_r - 1). \tag{2.8.2}$$

From which follows $M \approx m_r v_r/v_{av}$ at $v_{av}/v_r >> 1$, i.e. $m_r >> M$. It follows from this expression that when estimating the mass of the BL, the most important parameter is the speed at which the BL-roof system moved. Despite the complexity of its definition from the description, we assume that the BL moves under the influence of air, i.e. wind. It is these values that correspond to the observed velocities of the BL (Bychkov et al. 2010). When the BL touches the roof, in a state of indifferent equilibrium, it can move the roof (if it is not specially reinforced) by a length equal to half the length of the roof, after which it falls. Since the description does not specify

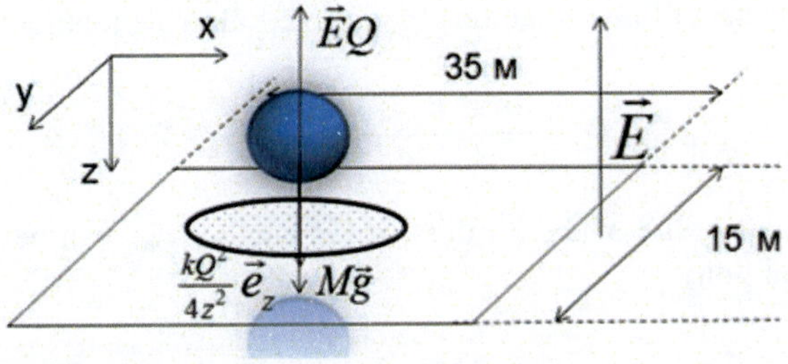

Fig. 2.57 The scheme of interaction of the BL with the roof. Here $\vec{E}$ is the electric field strength of the Earth's field, M is the mass of the BL, Q is its charge, and z is the distance from the BL to the roof

the side of the roof in the direction of which its fall occurred, we take a value of ~7 m (at the time of the roof shift along the smaller side) for evaluation. The time of movement of the BL-roof system is assumed to be equal to the time after which the observer approached the window, or about 3 and 9 s (taking into account the possible error in determining the time by the observer). From these values, we obtain the speed of the system turns out to be about 2.3 m/s and 0.8 m/s. Substituting the values m_r v_{av}, v_r in (2.8.2), we get $M \approx 1200$ kg in the first case and $M \approx 400$ kg in the second.

Having determined the estimated weight of the BL, we estimate the thunderstorm conditions and the charge of the BL, under which the levitation of the BL would be possible.

Considering the roof as a plane and using the image method, we estimate the charge of the ball by which we model the BL, with a radius of 1 m, and also estimate the electric field strength of the Earth's field near the roof.

Figure 2.57 shows the scheme of interaction of the BL with the roof of the house. For the estimates, we write an equation of force projections on the Oz-axis describing the interaction of the BL charge:

$$\frac{kQ^2}{4\varepsilon z^2} + Mg_z = E_z Q, \tag{2.8.3}$$

where $k = 9 \cdot 10^9 \frac{N \cdot M^2}{C^2}$ is a constant in Coulomb's law, g_z is the acceleration of gravity, E_z is the electric field strength of the Earth, Q is the charge of the BL, M is its mass, $\varepsilon_r = 2$ is the dielectric constant of the roof material (plastic and organic matter), and the effective dielectric constant of this material is $\varepsilon = (\varepsilon_r + 1)/(\varepsilon_r - 1_r)) = 3$. Equation (2.8.3) is a quadratic equation with respect to Q, it allows to determine the minimum value of the field strength required to maintain an object in the air, as well as the corresponding charge of the BL. To do this, we rewrite Eq. (2.8.3) in the form:

$$Q^2 - \frac{4\varepsilon E_z z^2}{k} Q + \frac{4\varepsilon z^2 M g_z}{k} = 0. \tag{2.8.4}$$

The discriminant of this equation satisfies the following relation:

$$D = \frac{4\varepsilon^2 z^4 E_z^2}{k^2} - \frac{4\varepsilon z^2 M g_z}{k} \geq 0. \quad (2.8.5)$$

The minimum value of the Earth's field strength E_{zmin} can be found from the following condition:

$$D = 0 \text{ or } E_{zmin} = \sqrt{\frac{M g_z k}{\varepsilon z^2}}.$$

Substituting the value of $M = 400$ kg (the most favorable option), we get $E_{z\ \min} \approx 1.7 \cdot 10^6$ V/m. It is known that the strength of the breakdown field near the Earth's surface reaches a value of $E_{z\ \min} \approx 3.0 \cdot 10^6$ V/m, i.e. the obtained value indicates the possible conditions of the electric field strength during observation.

Determine the corresponding value of the BL charge:

$$Q_{min} = \frac{2z^2 E_{zmin}}{k} \cong 1.3 \cdot 10^{-3} C.$$

In this case, the charge is the intensity of the electric field created by the BL on its surface:

$$\vec{E} = \frac{Q}{4\pi \varepsilon_0 R^3} \vec{R},$$

from where, substituting the values of the charge Q and the radius R of the BL, one gets the value $E = 11.7 \cdot 10^6$ V/m. The resulting value significantly exceeds the breakdown electric field in the atmosphere. The question of a mechanism that ensures a long lifetime of such an object without the implementation of breakdown conditions will be discussed below.

4. (Nikolaev 2001) 1966, settlement (village) Voronovo, Omskaya oblast. Russia. Winter.

BL moving at high altitude along the road stopped over the two Chirikov brothers and began to decline. The brothers felt an electric shock on their arms and legs (their muscles cramped), and some unknown force began to press on them from above. It was difficult to lift arm, some kind of weight was pressing on shoulders, as if a person was sitting on them. The ball was decreasing, and the thermal effect of this object was increasing. It became stuffy; the brothers unbuttoned their sheepskin coats and took off their mittens. When the angular size of the object increased to 0.17 radians (with this angle value, an object measuring 1 m is at a distance of 6 m), it stopped decreasing. The force pressure on the brothers stopped, but the thermal effect and the brightness of the object became difficult to bear. After that, the object began to be deleted. The brothers were electrocuted again, and some unknown force pulled them up. They grabbed each other's hands, but this force lifted them up, significantly easing their pressure on the ground. As the object was removed, the force decreased. Having risen to a great height, the object flew away. Within a radius of 70 m from the place of the object's descent, the snow melted and became loose.

Comment. They grabbed each other's hands, but this force lifted them up, significantly easing their pressure on the ground.

5. (Nikolaev 2001) February 1985. Near Petrozavodsk, Russia.

The drivers of the locomotive leading the empty freight cars saw a glowing ball with a diameter of 2 m. The ball was flying next to the locomotive. When the train began to slow down on a steep climb, the ball suddenly moved across the train and appeared in front of the locomotive at a distance of several meters from it. The train began to gain speed involuntarily. The driver began to brake and turned off the fuel supply to the engine, but the locomotive did not slow down. The ascent was replaced by a descent. The driver turned on the braking system, but it did not help. The ball continued to drag the train at a speed of about 50 km/h. The driver reported by radio to the station that due to the intervention of an incomprehensible object, he could not slow down. Since, fortunately, there were no obstacles to movement; the station attendant gave the train a "green light." Coming out on the platform, she saw an outlandish sight: a bright-white ball joined the locomotive, in the center of which a fiery red disk stood out slightly to the right of the axis of movement. Just before the station, the ball flew away for a few seconds. Nevertheless, when the train came out on a blind stretch, it again fell in front of the locomotive. A sharp braking followed, the speed of the train dropped from 50 to 20 km/h, and then the train began to accelerate again. The driver's attempts to slow down the train did not lead to anything. The ball dragged the train for several tens of kilometers, all this time the controller in the cab of the car was at zero. That is, there was no diesel locomotive's own traction. As a result, the strange object allowed saving 300 L of fuel. At a distance of 50 km from the station, the ball broke away from the locomotive and flew away.

Comment. Well-known and very popular observation. However, there is not enough information in the description to conclude the reliability of the observation.

Comment to the section

We presented observations that can be explained by taking into account the anomalous parameters of the BL, which can be realized in the anomalous electric fields of the thunderstorm atmosphere.

2.9 BL Meeting with Airplanes

Before creation of aeronautic devices in the end of nineteenth–the beginning of twentieth century, the BL were observed near a surface of the Earth. With development of aircraft, people could transfer an area of an observation directly to a zone of BLs formation—storm clouds. It has led to detection of new BL properties. Besides a pure scientific interest, researches of BL interaction with planes have the important practical value as it is closely connected with a safety of a crew and passengers. Knowledge of BL behavior features in planes and accumulation of data about similar cases when the meeting with the BL have not brought to ruination of an aircraft, which are the main result of these researches.

The basic results of researches of influence of a BL on planes are summed to the following: (1) The BL is capable of getting inside the all-metal frame of an aircraft ("Faraday cell") through the existing apertures or made by it. (2) As a rule, it gets into a nose part of an aircraft, moves along an axis of a fuselage, and leaves outside through an aperture in a tail part of the aircraft. (3) Sometimes it passes through an internal space of an aircraft without an explosion, and sometimes it blows up. (4) After an explosion of a BL in the aircraft for some time, an internal telecommunication vanishes and the navigating devices fail. (5) In some cases before penetrating inside of an aircraft, the BL was observed being on external elements of an aircraft (on wings, on a fuselage). At that, it moved in an external stream of an incident air with a speed of a flying aircraft. (6) BLs in storm clouds were observed at heights from 500 m up to 10,000 m. (7) Sometimes from planes were observed shining objects in the size up to 100 m. These objects emitted narrow rays of light.

To meetings of BL with planes and to the analysis of a danger degree of such meetings, the seventh chapter of the book of Stenhoff (1999) is devoted. Below we result descriptions of BL observation of other cases in a statement of skilled pilots. Majority of these messages have been published in Russian and, hence, were inaccessible to English-speaking readers. As required we have supplied each description by a brief comment.

Part of the data in this section was obtained at the Main Military Meteorological Center of the USSR, the other was collected from a survey of eyewitnesses, literature, and the Internet.

In 1991, Professor B. M. Smirnov wrote a letter to the commander of the Main Military Meteorological Center of the USSR V. I. Litvinov about permission for me (DS V. L. Bychkov) to visit the Center and view data on aircraft collisions with the BL. The Commander of the Center allowed visiting the center. For 2 or 3 days, I analyzed the reports in the Journals that existed in the Center. Their sizes were about 100 × 50 × 10 cm. I do not remember, but I think I looked through about seven or eight journals. They collected information about all possible events related to the meteorological activity of the atmosphere and the flight of the crew. The data was on both sides of the sheets. On each page there were six to seven cases written in different ways: sometimes in details, sometimes without them. When there was no information about the plane, it meant a military fighter plane.

As for the "static electricity," which includes all the data about the BL in the Center, it was necessary to view all events with between cloud or other types of discharges. The task was to find all cases resembling BL written in some kind of military slang, but it was also difficult to separate BL events and discharges between clouds. They are often described in the same way—a pop and a flash. (Why in this order I do not know, maybe this is due to the fact that the sound of the explosion is transmitted through headphones to the crew earlier than people see it (the visible reaction is slower.) Sometimes only the season could help distinguish these events: BLs usually appear in the period from May to August, and between cloud discharges are usually observed from September to April, when pilots have the main flight hours and more frequent flights, and, consequently, there is a high probability of observing luminous formations. However, in the period from May to August, there is a problem of their separation. Thus, of the 49 cases found in the records, in 22 cases, BL was reliably observed—and in 13 cases—presumably, and in 14 cases we could not distinguish the phenomena.

The commander also said that pilots usually prefer not to inform military meteorologists about meetings with BL and other static electricity events. Pilots are strictly forbidden to enter areas of thunderstorm activity, and they have to fly around rain sources or cumulus clouds. The pilots had rules for flying in a stormy atmosphere, if they violated them, they were punished in various ways. For example, they were not provided with apartments for their families, and they were forced to live in dormitories.

So, generally speaking, the events that we have are those when some traces remained on the surface of the aircraft or above its surface. In five cases, the destruction of aircraft occurred, in two of them it was due to the impact of the BL. We should note that the crews had time to leave the plane, because electrical discharges and lightning strikes usually destroy navigation systems, and the plane is still moving, and people have time to save.

1. (Akkuratov 1991) February 1946. Vologda oblast, Russia.

This story took place in February 1946 above a large forest of the Vologda oblast. In the four-motor airplane "Petlyakov-8," the cabin of navigators is settled down in the nose part. The navigator of the 1-st class Zubov was sitting in the right armchair, the author—the navigator Akkuratov occupied the left lateral seat in three meters behind him. The commander of the ship Zadkov and the co-pilot Samokhin piloted for several years. Under their cabin the cabin of the radio operator, where the fifth member of the crew Kuksin was sitting, has been located. They were in clouds at height of 1200 m, the temperature behind the board was −14 °C, the airplane was not swinging … Suddenly near a head of Zubov has flashed dazzling a white ball, and has hung, pulsing, and rocking. In a brain of Akkuratov, the author of the report, a guess has flashed: "The BL! But whence it has appeared from? It is winter, no a thunderstorm. And how the ball could appear in a tight cabin of the airplane?" … The pulsing clot of fire, having hung for a while nearby Zubov, smoothly, as if unwillingly, has moved along the left side of the cabin aside the author. Screwing up his eyes from a glaring light, he has instinctively pressed to a wall, having squeezed in his hand a celluloid navigator calculator ruler. The devil ball, all pulsing and rocking, came nearer to his armchair. Up to it there was 30–40 cm. The observer has not felt a heat, but for him was clear the twingeing in the top part of a head under a helmet with earphones. In the cabin smelled sharply by ozone … Muscles of the author have strained, along a backbone has run an unpleasant cold. Everything, having grown dumb, tensely watched for strange and dangerous "a heavenly guest." It is difficult to tell how much time has passed; it seemed very long, as suddenly the ball, changing color to green-golden, began to depart from the head of Akkuratov slowly.

Not moving by eyes he watched its movement. Descending, the ball was smoothly floating to an entrance into the flight radio operator cabin.... The ball rolled under his sitting and has blown up with a terrible roar. Dazzling sparks have hidden the radio operator. The black caustic smoke has filled both cabins; the telecommunication has broken ... "What has happened? Where is a fire?"—coughing from a smoke, Zadkov has grasped greedily air from the open lateral window. "The BL! The heat insulation of the radio operator cabin!" ... People have gone downwards where the navigator and the radio operator were struggling with fire. They three together have soon brought down a flame from a burning covering ...Legs of the radio operator's seat have been melted off...

The author attentively has examined the navigator's cabin. All windows and hatches have been closely densely down, there was no slit. Even a small aperture where a flare pistol is usually inserted has been densely closed. He has called to pilots: "Have you seen anything before the BL penetration into the airplane?" The co-pilot has informed, that he saw, how on the right wing near the running green fire there was a bright-white ball. He has thought that there was a short circuit of an electric lamp, but the flash has not disappeared, as usually happens. The ball has slowly creeped over a frontal edge of a wing and has disappeared under the nose part of the craft. He had no time to say anything by the phone when the loud crack has sounded, a black smoke has moved down into the pilot cockpit, and the telecommunication has broken.

After landing in Moscow a special commission surveyed the airplane. It has confirmed a fact of BL penetration into the airplane and those destructions which it has made. The weather bureau has checked up carefully the prognostic and actual weather maps. In the European part of the Soviet Union, thunderstorms have not been revealed anywhere.

Comment. Ball lightning was observed in an absence of a thunderstorm in the airplane flying at height of 1200 m. At opening of windows there was not depressurization of cabins, therefore, quite probably, that the BL has entered in the airplane through an unidentified aperture. The BL has appeared on a wing and moved together with the airplane, overcoming resistance of an air stream. The BL did not cause sensation of heat, and the "twingeing of the head" could be caused by an electric field of its charge. Before the explosion of the BL in the cabin the smell of ozone was felt. At explosion, apparently, because of the discharge current impact there was an ignition of organic materials. Actions of the airplane's crew in the described situation were correct.

2. (Akkuratov 1991) Flight Moscow-Arkhangelsk, Russia.

Again the BL has penetrated into the modern airplane, absolutely hermetically sealed, on the line Moscow-Arkhangelsk under similar weather conditions, but with a temperature of air of zero degree. The airplane IL-62 was at height of 2000 m in the continuous overcast. There was a plentiful snowfall and the icing of the airplane has begun. Suddenly from under the control panel of pilots the bright-blue flame has been pulled out. Simultaneously a loud, similar to a pistol shot, dry click has sounded... The flame in a form of a ball has departed by a head of the flight engineer to the passenger's saloon ... The bright clot of a flame has flown by through all the passenger's saloon and has disappeared in the tail part of the airplane... All devices worked regularly, but some shrill howling sound was torn from behind the control panel. At survey of the airplane in the nose has shown a fragmentary hole through which the adult person could get through, and in the tail—a small aperture burnt through in the metal covering. The locator aerials fairing has been scorched also.

Comment. The unusual fact in this case is formation of an aperture of about 50 cm in size in the covering of the airplane made by the BL. At that it has not been destroyed and has existed for some time. A movement of the ball along the saloon to the tail is typical for BLs. Its exit outside through the aperture of a small size made by the BL is typical too. The danger of the described case is the opportunity of the depressurization of the airplane at flight at high altitudes.

3. (Stakhanov 1996, p. 36) Russia.

During a stormy weather behind a window of the passenger airplane IL-14 to the right of the board there appeared a white with a pink shade shining ball as a football in size. It has passed along the board of the airplane. After landing of the airplane on its fuselage several apertures in diameter about a thimble were revealed.

Comment. The BL moved slowly enough with respect to the airplane. During the contact it has had time to evaporate some part of metal airplane covering.

4. (Stakhanov 1996, p. 95). Riga. Autumn of 1967. Russia.

The pilot Kuznetsov has met a BL in the autumn of 1967 in the airplane at height of 7000 m near Riga. Overcast was not a stormy—5–6 points. In the cockpit has appeared a BL of a bright-white color with a small aura of a tennis ball in size. During 20–30 s it has passed from an input of a radio antenna up to a back of the pilot seat the distance of 1.5–2 m ... After that it has blown up, it has burnt out a part of the armor (about 3 cm^3) ... Protective circuits of the airplane's radio stations have failed. After landing it was revealed that almost on all acting aerials were burnt out black strips.

Comment. The BL was observed at a high altitude. The speed of its movement inside the airplane was no greater than 0.1 m/s. At its explosion energy no smaller than 180 kJ was released, it has been transferred to the wall, apparently, in the form of a pulse of electromagnetic radiation. Safety devices of the radio stations also have fused under the action of this pulse.

5. (Stakhanov 1996, p. 104) Russia.

In dense overcast at height of 1600 m, the airplane IL-12 has collided with a BL. The shining ball of 10–15 cm in diameter has struck in the nose part of the airplane, therefore inside the airplane between the covering and the aerial, the spark has slipped and a radio transmitter has been fused.

Comment. The BL was lost outside of the airplane. Apparently, during a moment of destruction, it has emitted a pulse of the high-frequency radio emission which has led to occurrence of the spark in the cabin.

6. (Stakhanov 1996, p. 122) August of 1975. Russia.

The navigator Osipov was flying in the airplane TU-104. At height of 900 m, there was an overcast of 10 points, the temperature behind a board was 180 °C. When the airplane has entered into a cloud at height of 1200 m, on a glazing of the cockpit have appeared electric discharges of the bluish color of 15–20 cm in length; they were separating and flew back. In 3–4 s after that has appeared a green-blue ball of 10–12 cm in diameter near the edge of the glazing. After this a deafening explosion has sounded, and the glass was enveloped by a flame. The radio station and a compass have been burned down. After landing, a small reddish burn on a duralumin crosspiece of the glazing has been found. The glass has not been damaged.

Comment. The author of the report, apparently, observed the act of a BL birth. This act was preceded by an occurrence of the St. Elm's fire discharge type on the glass. The flash accepted for a flame was a result of BL fragments scattering. During the moment of the BL explosion has happened the generation of a pulse of a high-frequency radiation.

7. (Grigoriev 2006, p.9) Route Moscow-Berlin.

The pilot Levin, a member of the flight crew, was executing a flight on a route Moscow-Berlin. In the area Smolensk–Minsk, he has noticed on a floor of a fuselage a red ball (as though heated) of 3–4 cm in diameter. By brightness it was similar to the sun ascending in clouds or in a fog. It was not so clear, if it was rolling or flying: on the unpainted floor of a cabin of the airplane has not been found any trace. Moving over the floor of the airplane, the ball has found to itself a way out in a hole in the floor and a deaf clap has fallen to the mechanism of the outlet aerial. As a result, the wire of the aerial has burned down, and the ball either has flown away or has fallen to the ground. A speed of the ball during the observation varied: at first the ball almost stood, and then it has quickly disappeared in the hatch. Everything has happened for few seconds.

Comment. In the report, there are no data about weather conditions and a story of a BL appearance in the airplane. Its behavior does not differ from a behavior on the ground. The BL can leave an airplane through natural apertures.

8. (Grigoriev 2006, p. 30) Tiksi. Russia.

The radio operator Zhukov served in the polar aviation. In the beginning of May 1950 at 14–16 h of local time he was flying at the altitude of 1800 m in the airplane LI-2 from Tiksi to the settlement Chersky. Approximately, the half of the way the airplane went in clouds (overcast of 8–10 points, temperature 0 °C); to the south in 10—15 km usual storm activity was observed. In headphones more often lightning discharges were heard. The radio operator has shown to the navigator a display of the thunderstorm on the aerial: he has disconnected the aerial from the radio-complex

ARK-5 and, bringing it to the terminal and then removing it, he draw a spark out in length of about 5 millimeters. As there was bad passage of radio waves, the work was conducted with the outlet aerial on short and middle waves. The outlet aerial represents a bimetallic cable with a cross section about of 3 mm, length about of 60 m, a winch, a pipe, a metal axis with the directing roller located in the cabin of the radio operator at height of 30–35 cm from the floor.

Suddenly in front of the airplane a huge fiery sphere has appeared. It was similar to a dense spot of white-yellow color of 1–1.5 m in diameter. It was bright as a projector of average power. At the same moment, the explosion which has muffled a noise of two motors has sounded, something has struck the radio operator in a leg. Pilots have dumped the gas and began to make their way to the ground. Having passed the clouds have revealed that all devices worked normally, the inflate was in norm. The airplane has returned to the former altitude. The survey of the equipment has revealed that the switching relay has been scattered in sparks, the entrance contour of the receiver US-9 has been burned down, the outlet aerial has disappeared, and the axis with the directing roller which has struck the radio operator in the leg has been welded, as with a help of a gas welder. This piece of metal was still hot within 5–7 min. After all that has happened in the cabin, there left a light smell of the burned down isolation.

Comment. The explosion of a BL has taken place outside of the airplane, presumably in the location of the outlet aerial. The pulse of the high-frequency radiation had broken inductive elements of the relay and the radio receiver, and the cylindrical axis with a roller has appeared heated up to a fusion temperature. Energy which is necessary for fusion of a steel roller of weight 120 g is about 90 kJ. The cable of the outlet aerial has been burnt. Elements of the radio equipment and automatics removed from a place of explosion have not suffered.

9. (Grigoriev 2006, p. 43) Moldavia. Russia.

In summer of 1947, the major Tcherepanov at 18–19 o'clock of the local time flew on the double seat small airplane PO-2 from Odessa to the city of Bolgrad Moldavian SSR. Weather in Odessa was clear. The airplane flew along the Black Sea coast through estuaries. Approximately, on half of the way, there appeared the stormy clouds lightened by flashes of lightnings. The airplane flew in clouds at the altitude of 500–600 m. To the right under angle about 30° to the airplane's route, the multi-colored balls of different sizes—about a football, greater or smaller began to appear. These balls were of very bright coloring: inside bright-white, of the blinding light; on edges-multi-colored, as a rainbow. They arose and disappeared instantly in one or in another place, absolutely beside, or at some distance. The impression was such, that one of the balls just will hit the airplane, but this has not happened.

Comment. The case of mass formation of BLs in the cloud at the altitude of 500–600 m is described. Spheres had the bright core surrounded by the color cover, presumably due to a luminescence of a corona discharge on the surface of a BL. Balls avoided collisions with the airplane that is possible if they had an electric charge the sign of which coincided with the main sign of charges on the airplane.

10. (Grigoriev 2006, p. 43) Aktyubinsk. Russia.

On August 6, 1976, the pilot Tregubenko flied on the airplane AN-12 from Aktyubinsk City to Chita City. The flight took place at the altitude of 6000 m, at the speed of 600 km/h. In the area of Novosibirsk City, the airplane has got into the storm cloud and began to gain height to leave from the thunderstorm. But the cloud was powerful, and at the altitude of 8000 m the airplane still could not leave it. The airplane was thrown there and back. Outside except for engine's noise the sounds reminding a roar of a close cannonade became audible. In front of a window stood and flew by fiery ball of bright-yellow colors of 30–40 cm in diameter. Balls with a terrible roar burst scattering sheaves of sparks and lighting the whole side of the airplane. At that the airplane was shivering so that it seemed that it was about to collapse. All this lasted 20 min, and then the thunderstorm remained behind.

Comment. The case of BLs mass formation in the area of a storm cloud in extent about of 200 km at the altitude of 6000–8000 m is described. At the explosion, elements of BLs' internal structure (sparks) flew out into air, they possessed a mechanical impulse. This impulse was transferred to the body of the airplane.

11. (Grigoriev 2006, p. 43) Spassk-Dalnyi. Russia.

In the autumn of 1960, the pilot Lukonin had to make an emergency landing to the airdrome near the city Spassk-Dalnyi near the Chinese border. In the area of the emergency landing, there was an overcast of 10 points, a storm rainfall and lightnings sparkled. The top edge of clouds was at height of 10500–11000 m. The continuous cumulus—rainy clouds reminding great waves with smooth humpbacked downfalls in the ocean—were there. For the first time, the pilot has entered into a crest of a cloud, but the airplane, having begun to vibrate, has begun to shake, and by the intensive turbulent movement of air it has been thrown upwards, as a chip. Then the pilot has directed the airplane downwards between two waves. The airplane vibrated, it was crumpled. Linear lightnings "tore" the clouds. The airplane went downwards almost without a fuel. And at that very moment the pilot has seen something an unusual: inside of a storm cloud was floating a set of BLs. They went aboard the plane, as shells of a large apple in size; with hissing they were touching a cap of the airplane. It happened at the altitude from 4500 to 6000 m. It seemed that it was lighter inside the cloud, than outside it.

Comment. The case of BLs mass formation inside a powerful stormy cloud at height of 4500–6000 m is described. There are bases to consider, that to BLs formation favors an intensive turbulent movement of air inside of the stormy cloud.

12. (Leonov 1965) Russia.

The airplane IL-12 of civil air-fleet made flight at the altitude about 4000 m in the absence of an intensive storm activity. Suddenly directly at the route of the airplane there appeared an orange-red ball which promptly having approached the airplane. It has turned away from the pilot's cockpit and has got in the blade of the right motor. An explosion has sounded, the airplane has been strongly shaken, however, the motor worked smoothly. The navigator has found out that a radio compass and other devices on an electromagnetic basis have been failed, and the radio communication was broken. After a while a work of devices and communications was restored by itself, any damages to them were not revealed. In the airport, it has been found that one blade of the propeller of the right motor in a place of the BL impact was melted off.

Comment. Ball Lightning was observed in the absence of a thunderstorm. If to accept a weight of the melted off part of the propeller equal to 100 g the energy, transferred to the propeller at explosion of the BL has been about 80 kJ. The pulse of the electromagnetic radiation has been weakened by an external cover of the airplane's frame and has caused only temporary overload of electromagnetic devices.

13. (Singer 1971, p. 61) Russia.

In 1956 Soviet propeller airplane LI-2 has been struck by a BL when it flew at the altitude of 3300 m in cumulus-rainy clouds. The airplane has just passed through the cold storm front and has been strongly iced over. The height of clouds reached 5500–6000 m, the temperature of air at height of flight was –2°C to –4°C, the airplane was strongly stirred. From the front an orange-red fiery ball of 25–30 cm in diameter was quickly approaching the airplane. At 30–40 cm from its nose, the ball has turned to the left and has flown by the cockpit. During this moment it has touched from above the right propeller and has blown up with a dazzling white flash. The sound of explosion has muffled a noise of the motor. To the right along a fuselage, a fiery stream has moved by, and the airplane has sharply rushed upwards. During the flight through the thundercloud a strong radio noise was marked. When the radio operator after the lightning discharge wanted to switch-off the aerial, he has been struck by a current. The only damage in the airplane was revealed on the blade of the propeller: its edge has been melted off on the site of 40 mm in length and of 5–10 mm in width on a distance of 30 mm from the end of the blade. The melted offsite was surrounded by a layer of soot which was easily erased by a hand.

Comment. The meeting of the airplane with a BL in a cumulus-rainy cloud at the altitude of 3300 m is described. The BL has blown up at impact to the propeller. The air stream after the explosion of the BL has affected the movement of the airplane. The volume of the melted off a blade's part of the propeller is approximately equal to 2 cm^3. Energy necessary for heating before fusion of 2 cm^3 of duralumin is nearly 5 kJ. An occurrence of soot has been caused by the carbon dioxide decomposition in the electric discharge accompanying the explosion of the BL.

14. (Singer 1971, p. 62) Irkutsk City, Russia.

In 1959, in the USSR, a BL in diameter of 10 cm has struck an airplane near Irkutsk City. The airplane flied with a speed of 400 km/h in the dense clouds located between two layers at heights of 400–4000 and 8000–10000 m. At height of 1400 m in rough weather, the ends of wings were not visible, and air was very restless. Wet snow has quickly closed up forward windows and wings. In front of the airplane in the direction of the ground, a usual lightning has struck. During this moment the airplane has experienced a strong impact, but it has not lost flight control. To the right of the airplane, a fiery ball with "rain" of sparks has moved by. The pilot has lifted the airplane from clouds to the altitude of 4000 m. The equipment of the cockpit was strongly magnetized, and the compass has deviated from a correct direction by 100°C. Radio worked, but the radio compass gave incorrect indications. An inspection of the airplane after the flight has shown that on the right board in the forward part, the fuselage rivets were melted off but the covering has not been destroyed. The size of the part has been approximately 150 mm × 1000 mm.

Comment. Magnetization of elements could happen both because of a linear lightning current, and because of a BL discharge current which has passed touching the right board of the airplane. Fusion of rivets could be caused by the vortical electric currents generated by the BL explosion.

15. (Singer 1971, p. 62) 1960. USA.

In 1960 a BL has appeared in the airplane-refueller KS-97 of the US Air Forces, which flew with a full store of a fuel. Flight happened in clouds at the altitude of 5400 m at light deposits at temperature above zero. On a fringe of forward windows there appeared St. Elms fires. Through a windshield the yellow-white ball about of 45 cm in diameter has penetrated. The ball has moved between the first and the second pilots with a speed of "running man." The ball has flown by over a pass by the navigator and the flight mechanic. Approximately, in 3 s, the attendant has reported from the tail compartment that the fiery ball has rolled through the back cargo compartment and over the right wing has left into clouds. The ball did not make any sound.

Comment. The case of a BL penetration into a cabin of the airplane and its exit from the tail compartment is described. The BL during its "quiet life," apparently, had a low temperature and was unable to ignite subjects. In case of BL explosion inside of the airplane-refueller of a consequence would be tragic.

16. (Singer 1971, p. 63) Route New York to Washington. 1963. USA.

In 1963 a regular airplane of Eastern Airlines company flew from New York to Washington, passing through a thunderstorm, which suddenly from different directions has been surrounded by a bright flash, accompanied by a loud noise. After several seconds from the pilot's cockpit, a bluish-white sphere of 22 cm in diameter of has flown off and has flown along a pass at height of 75 cm from a floor. It moved rectilinearly with a speed of 1.5 m/s and has passed at the distance of 50 cm from the observer. The BL represented a regular sphere with hardly darker edges. A light force from it corresponded to a bulb of 5–10 W power. The sphere did not radiate heat and did not rotate.

Comment. The case of BL observation inside the interior is described. As well as in other similar cases, the BL moved from a nose to a tail of the airplane, being at some distance from subjects. Properties of a BL in an airplane do not differ from its properties near a terrestrial surface.

17. (Singer 1971, p. 64) 1965, Kolsky Peninsula, Russia.

In 1965, at flight above Kolsky Peninsula in dense layer-rain clouds, Soviet airplane LI-2 has collided with a BL of 60–80 cm in diameter. The airplane passed a northern part of a cyclone. The ends of wings were not visible. The temperature behind a board was –5 °C. At impact of the ball in a nose part of the airplane, the sound reminding a rifle-shot has sounded. On a covering, red strips of discharges have slipped. After the collision of the airplane with the ball, the radio compass began to turn, and an arrow of a magnetic compass within 3–5 min was randomly moving forward and back. When it, finally, has stopped, its indications have appeared to be incorrect. All radio equipment also has failed. At inspection of the airplane after landing, the damaged rivets in a forward part of a fuselage and two holes of 1.5–2 cm in diameter melted through in the back edge of the elevator were revealed.

Comment. At impact of BL in an external covering of the airplane, the currents over the body which have caused an occurrence of magnetic fields have passed. Apparently, the charge of the BL has been transferred to the airplane. It also flew down from the body within several minutes generating at that fading magnetic fields. The pulse of the high-frequency electromagnetic field has caused a fusion of rivets and formation of hole in the frame. If to accept a volume of melted through metal equal to 2.5 cm^3, then energy necessary for its fusion is nearly 6 kJ.

18. (Imianitov and Tikhiy 1980, p. 41) Russia.

The airplane LI-2 flew in layered clouds. There was no thunderstorm; there was not even a weak bumpiness. Unexpectedly, from under a control panel, between the first and the second pilots, has flown up a white fiery ball of 30 cm diameter. The pilot has ordered to the mechanic to remove the ball. He has taken a piece of a canvas and began to drive the ball to a tail of the airplane. All this has taken a minute or two. Then on the ground in the tail were revealed two holes of 3 cm in size near an alarm lamp. To hide the incident from the heads which considered that the crew has violated the instruction and made the flight in the stormy cloud, the holes have been secretly riveted.

Comment. The BL was observed in the absence of a thunderstorm. As the airplane safely continued to fly with two holes in the tail, it is not excluded that the BL has entered the cabin also through an unidentified aperture. The mechanic risked his life, banishing the BL, but, fortunately, all has finished well. The BL has melted through holes in the covering, having remained the whole. It has spent energy for it nearly 6 kJ.

19. (Imianitov and Tikhiy 1980, p. 43) Yakutia, Russia.

The airplane LI-2 flew above the mountain area of Yakutia in cumulus clouds at altitude of 3600 m. There was a weak icing, average, not a stormy bumpiness. And suddenly something has sparkled ahead, and from under a control panel there has appeared a light fiery ball about a volley ball in size. It has bumped for some seconds in front of the board and has disappeared in the passenger saloon. There was nobody there. Later, already on the ground, an aperture of 8 mm in diameter has been found in a nose of the craft, and a lot of fine holes around it. It was impossible to find the exit outlet.

Comment. The BL was formed in the absence of a thunderstorm, but, however, as a result of some flash. It has passed into the cabin through apertures the size of which is close to the size of its structure elements. The BL could quietly go out in the saloon of the airplane.

20. (Imianitov and Tikhiy 1980, p. 43) Russia.

A BL has got inside the tight airplane IL-18. It has appeared from under a control panel, has rolled in the first, and then in the second saloon and there it has disappeared. A plastic fairing of a radar aerial proved to be all in small holes, as a sieve for a flour.

Comment. In case the BL consists of elements of the small size, it could enter the airplane through the apertures done by it in the fairing of the aerial.

21. (Imianitov and Tikhiy 1980, p. 44) Summer of 1965, Russia.

In summer of 1965, the AN-12 airplane flew in the east direction at altitude of 6000 m. On a line there was a thunderstorm, and pilots began to bypass it. "Flares"—images of deposit zones on the radar screen remained from the right side. The airplane was at a distance of 70–100 km from them. From the right side, there was an overcast (the front passed from the north to the east), pilots have entered into it. The storm hearths have been removed from the airplane up to 100 km. In the layer-rain clouds, there were drizzling precipitations. There was no bumpiness, the flight was quiet. Strong electrification of the airplane was observed (a crash in headphones with characteristic squeak). St. Elms fires were observed on glasses. The pilot was looking at the screen of the radar. Suddenly, the screen has brightly flashed and in 1.5–2 s after that the loud sound of explosion has been heard. The glass of the navigator's cabin has been captured by a ring of a flame. The flame was of red-yellow color. It, most likely, was a BL which has blown up at a contact with a nose of the navigator's cabin. The sound of explosion was in the beginning sharp, and then rolling, as an impact by a sledge hammer on a tun. In four minutes, the airplane has left from the clouds. The radar was in full serviceability. After landing, places of the covering melting in points where there were rivets have been found. Besides behind the rivets there was stretched the dark thin coat of 1.5–3 cm in length.

Comment. The meeting of the airplane with the BL during a drizzling rain is described. However, the condition of weather (electrification, St. Elms fires) reminds stormy condition. Selective melting of the covering near rivets, apparently, is obliged to generating of vortical currents at explosion of the BL.

22. (Imianitov and Tikhiy 1980, p. 46) August 1970. Russia.

In August 1970, in the airplane TU-104 at the altitude of 3300 m at climbing in a cabin of a crew, there appeared a fiery ball of a small diameter moving to the right board of the airplane. Having floated above a table of the radio operator, the ball "has discharged" (has disappeared) in a connection place of the open aerial and the transmitter. After landing, a cable's breakage of the aerial was revealed at external survey. Other damages were not revealed. The equipment and devices of the airplane worked normally. The airplane flew in not stormy clouds.

Comment. After landing a cable's breakage of the aerial was revealed.

23. (Imianitov and Tikhiy 1980, p. 46) April 1970. Russia.

In April 1970, at climbing at flight in not stormy clouds, the crew of the airplane AN-24 has noticed an orange ball which moved from the nose part of a fuselage over the left board. In several seconds, the crew has heard an indistinct explosion then one of the radio stations has failed. After landing a damage of the top fairing keel was revealed; in the nose fairing of the radar, there was a hole of 2–3 mm in diameter; to the left on the board of a fuselage some rivets have been melted off and two through apertures of 2 mm in diameter were formed.

Comment. After landing, a damage of the top fairing keel was revealed.

24. (Imianitov and Tikhiy 1980, p. 47) Russia.

In front of the airplane IL-18, piloted by the commander of the ship Matyushin, approximately in two minutes after rising from the airport of Riga at the altitude about 600 m has suddenly appeared an orange ball of 50 cm in diameter. After this the crew of the airplane was blinded by a bright-blue flash. The thunder similar to the gun shot has sounded; it muffled a sound of motors. The eyewitness who was at this time on the ground near the place of incident saw, how in 15–20 s after the IL-18, has disappeared in low dark clouds, at height of 50–100 m was flying a bright-white ball. Behind the ball there was a long foggy trace. In 2–3 s after the ball has flown by, the eyewitness has heard a thunderclap over the city center. The BL which was described by the ground observer, according to other people observed it, has a diameter about one meter. It had colors of a rainbow with prevalence of yellow color and has been surrounded by a white foggy layer. According to the meteorological analysis, this day in the area of Riga there was not a thunderstorm and it could not be. Crews of other planes being in clouds marked a strong electrification of planes.

Comment. The unique case of simultaneous observation of the BL from a board of the airplane and from a surface of the ground is described. It is possible to see that estimations of its size have practically coincided; however, a distance up to it the terrestrial observer has determined with the big error. The BL was formed in the absence of a thunderstorm, but in conditions of the atmosphere's electrification increased level.

25. (Abrahamson et al. 2002) Alaska Airlines, USA.

Jim Mills, commercial airline pilot of Alaska Airlines, USA, during a daytime trip in a Fairchild Metroliner III observed, along with a number of passengers on board, a ball lightning phenomenon. The aircraft was flying at 16,000 ft(4877 m) near McCall, Idaho, USA, when it was struck by a bolt of lightning near the nose area. After the strike, a bright glowing object slowly drifted down the aisle of the aircraft, floated to the rear and disappeared with an audible "pop." Investigation showed that the aircraft was struck on the right front nose baggage access door, as well on the stall avoidance system (SAS) vane, also on the right side of the plane. The strike welded the SAS vane in place, rendering the system inoperative. It also damaged the internal components of the system. The strike may have indeed vaporized some small amount of aluminum, as two or three pinholes were present on the nose baggage door, but this does not explain how the ball lightning could enter the pressure vessel of the aircraft. This aircraft is pressurized and the ball lightning would have to pass through a sealed pressure bulkhead to enter the cabin area. No loss of pressure was experienced.

Comment. Apparently, the birth of a fireball has occurred as a result of impact in the airplane of a linear lightning. Not clearly, by what type of a lightning: by a linear of BL small apertures were burnt through in the body. In any case through them, there could pass a BL consisting of elements of a small size. As the BL can fly against a wind, it was capable to move in a direction against an air stream which went through the aperture outside. By the description of the BL destruction process, its energy was small.

26. (Bychkov and Amirov 2010) June 1984 near Vorkuta, Russia.

In June 1984 near Vorkuta airplane AN-12 flew in the rare part of the cyclone with crossing of secondary cold fronts. At crossing of layer cloudiness of 150 m thickness at height of 2900 ma collision of the plane with ball lighting took place. The crew felt a strong blow and saw a flesh of orange color with a bluish rim. The navigator was blinded by the flash. After impact of ball lightning the radio station ARK-11 began to work unstably and a self-inflammation of the radio-locator took place.

Comment. After impact of ball lightning the radio station ARK-11 began to work unstably and a self-inflammation of the radio-locator took place.

27. (Bychkov and Amirov 2010) June 1984, Germany.

In June 1984, in Germany, pilot Gromyko flew in the airplane SU-17. At height 900 m, cloudiness was 10 units, a visibility was 3–5 km. During descending the plane appeared in the zone of intensive electrification (over on earth radio station). A sphere was formed in the nose part of the fuselage. It had come into the cabin and strongly hit a radio station. It was broken. The pilot was blinded and got multiple electric burns of the face and other parts of his body. The pilot lost capability to control the airplane and has baled out.

Comment. Probably, that the increased level of air ionization above the radio station promoted a formation of the BL on a nose of the airplane (without a help of a linear lightning impact). The explosion of the BL has led to a temporary worsening of health of the pilot that has forced him to leave the airplane.

28. (Bychkov and Amirov 2010) December 1983. Russia.

In December 1983, pilot Koval flew in Baltics in airplane MIG-23 through cloudiness 2–3 points. Height of upper cloud level was 350 m, visibility—10 km, temperature –1 °C to –4 °C, wind—western 6–9 m/s. At descending at landing course at height 620 mat velocity of 440 km/h at going out of clouds, he has seen a bright green flash in the nose part of the airplane and heard an indistinct blow in the lower part of the fuselage. A lighting of the cockpit has gone out and then recovered again. Engine speed has dropped by 20%, the generator has gone out. The pilot has baled out.

Comment. Apparently, the airplane has collided with the BL which has put out of action the engine.

29. (Bychkov and Amirov 2010) March 1985, Chirchik, Uzbekistan.

In March 1985 near Chirchik, Uzbekistan, the pilot Korobov flew in the airplane SU-17. Cloudiness was 10 points, visibility—6–10 km. At the altitude of 7300 mat, the airplane's velocity of 900 km/h, the pilot discovered a yellow luminescence over the lip of the air intake. It had an ellipsoid form. At its disappearance, there were a strong blow under the cabin and a strong drag. The engine has stopped. The pilot has baled out.

Comment. The engine has stopped. The pilot has baled out.

30. (Bychkov and Amirov 2010) April 1986. Latvia.

In April 1986, during a fly of airplane AN-12 near Chernyakhovsk, Latvia in cloudiness 10 units at a height 2400 m an electrification of the plane has started. The discharge took place. The crew in the moment of the discharge has seen the arch-type flash from the left console of the wing to the nose of the airplane and heard a flap. After flight the investigation has revealed melting of rivets in the nose and on the wing has revealed melting of rivets in the nose and on the wing.

31. (Bychkov and Amirov 2010) July 1987, Latvia.

In July 1987 during a fly of airplane AN-12 near Riga, Latvia in cloudiness 8–10 units at the altitude 2400 m the crew has heard a flap and has seen a flash in the nose part on the left nose part of the fuselage. After the flight, an investigation has revealed point-type burns in the nose part on the length about 2 m and on the end of the stabilizer to the left.

Comment. After flight, the investigation has revealed point-type burns in the nose part on the length about 2 m and on the end of the stabilizer to the left.

32. (Bychkov and Amirov 2010) April 1988 near Chimkent, Uzbekistan.

In April 1988 near Chimkent, Uzbekistan, the pilot Savchenko flew in airplane SU-17. Cloudiness was 8–10 points at the altitude 1200 m, temperature + 14 °C to + 16 °C, there was a weak rain. Weak cumuli cloudiness was below 600 m. At height of 200–300 m over the earth, a glowing ball of 6 cm in diameter has appeared on the receiver of the air pressure. Then it has separated and moved to the tail part of the airplane. After flight, the investigation has revealed damages of the antenna fairing and radio-locator.

Comment. After flight, the investigation has revealed damages of the antenna fairing and radio-locator.

33. (Bychkov and Amirov 2010) August 1988 near Ksiva, Poland.

In August 1988 near Ksiva, Poland, the pilot Fyodorov flew in airplane SU-24. Cloudiness was 6–10 units at height 1200 m, temperature +14 °C, it was a weak rain. The pilot was making his flight at height of 1500 m in clouds. In the left motionless part of the cabin light, he has seen a luminescent spot with a reddish tint, and simultaneously he has heard a flap. Gray smoke has appeared in a region of the panel board. The navigator in this moment has seen a bright flash at the screen of the radio-locator. Then the screen has gone out, and a smoke has appeared in the region of the radio-locator control. After flight, the investigation has revealed braking of covering of the fin end, a damage of the electronic block, and a damage of the blister. The fin and the stabilizer were burnt by electrical discharges.

Comment. After flight, the investigation has revealed braking of covering of the fin end, a damage of the electronic block, and a damage of the blister. The fin and the stabilizer were burnt by electrical discharges.

34. (Bychkov and Amirov 2010) June 1990 near Bzheg, Poland.

In June 1990, the pilot Ignatov flew in the airplane SU-24 near Bzheg, Poland. Visibility was 2–3 km, temperature 0 °C, there was haze, and raining in some places. Plane flew at the altitude of 3050 m with a velocity of 550 km/h. The crew has noticed a shining ball near intake of the air pressure receiver. The ball has immediately moved to the fore edge of the cockpit light and has exploded. After flight, the investigation has revealed insufficient melting of the antenna end on the discharger and an absence of the static discharger on the left wing.

Comment. After flight, the investigation has revealed insufficient melting of the antenna end on the discharger and an absence of the static discharger on the left wing.

35. (Bychkov and Amirov 2010) April 1978 in Russia.

The event took place in April 1978 in Russia in the airplane TU-22. At the altitude 1000–1500 m, the cloudiness was 5–6 points, visibility 4 km. In a daytime, the crew has seen a BL of 20 cm diameter at filling bar. The ball existed during 20 s and decayed. In this moment, the crew heard in earphones a hard cracking and has felt a blow. After flight, the investigation has revealed a damage of discharging antenna of 10×14 cm, a diameter of the melted hole was about 3 cm.

Comment. After flight, the investigation has revealed a damage of discharging antenna of 10×14 cm, a diameter of the melted hole was about 3 cm.

36. (Bychkov and Amirov 2010) April 1979, Russia.

In April 1979 at flight of the airplane AN-2 in clouds, a cracking has appeared in the earphones. The navigator has seen a bright-white ball of 10–15 cm in diameter on the glass of the cabin. The ball began to move upwards to glassing of the pilot cockpit. There its diameter became 40 cm. Coming to the antenna, the ball separated into two parts with a flap. The crew has felt a smell of burn. After flight the investigation has revealed melting of 32 rivets.

Comment. The ball separated into two parts with a flap.

37. (Bychkov and Amirov 2010) April 1980, Russia.

In April 1980, a plane was flying at the cloudiness of 10 points at an altitude of 500–800 m. There was haze, raining, temperature +2 °C to +3 °C. At height of 400 m, the crew has observed a flash during 1 min 45 s in front to the left, at the distance of 10–12 m. All devices worked normally. After flight, the investigation has revealed a through hole of 1–2 mm in diameter in the end of the blade, melting of rivets on the junction and on the end of the upper covering. Melting of metal on the edges was 1.5–2 mm wide and 50–70 mm long.

Comment. After flight, the investigation has revealed a through hole of 1–2 mm in diameter in the end of the blade, melting of rivets on the junction and on the end of the upper covering. Melting of metal on the edges was 1.5–2 mm wide and 50–70 mm long.

38. (Bychkov and Amirov 2010) August 1982 near a settlement Chkalovsk in Moscow region, Russia.

In August 1982 near a settlement Chkalovsk in Moscow region, Russia, the airplane TU-134 came into the cloudiness area at the altitude of 2700 m. The crew has seen a bright red ball and after that a discharge. After flight, the investigation has revealed burn of static electricity fairing on the plane.

Comment. After flight, the investigation has revealed burn of static electricity fairing on the plane.

39. (Bychkov and Amirov 2010) June 1982, Germany.

The event took place in June 1982 in Germany on a board of the airplane TU-16. There was cloudiness of 8–10 points at the altitude of 500 m, visibility 6–10 km, weak cumulus raining conditions, haze, and raining. At the altitude of 1500 m and velocity 550 km, in thunder clouds, the crew has felt an increasing noise and a crackling in earphones and has seen a flash and felt a dull blow to the right in the board of the plane. The decease of engine velocity took place. After flight the investigation has revealed breaking of 2/3 part of the antenna in the upper part of the airplane's fin.

Comment. After flight the investigation has revealed breaking of 2/3 part of the antenna in the upper part of the airplane's fin.

40. (Bychkov and Amirov 2010) October 1981 Brand, Germany.

An event took place in October 1981 in the airplane SU-7 in Brand, Germany. Cloudiness was 10 points from the altitude 600 m to 2400 m, visibility 8 km. At the altitude of 880 m and velocity 500 km/h, the pilot has seen a light flash in the fore part of an air intake. It has a form of a ball. Then sparks have appeared under the panel board and a smoke has appeared in the cockpit. Then the surging of the engine took place with a rise of temperature of going out gases up to 900 K. The pilot has baled out.

Comment. The pilot has baled out.

41. (Bychkov and Amirov 2010) Sochi, Russia.

The pilot Petrovich told about an event which took place between 1990 and 2000 near Sochi, Russia in the plane TU-134: "I was working as a co-pilot. In Sochi at taking off (at the altitude of 1500–2000 m) we got a ball lightning. I was piloting. I took my eyes from devices, and have seen a red-yellow ball flying into my face. Instinctively I have hidden my head behind a panel board. The ball went through the whole airplane. It entered near the place where I was sitting then it made an explosion in the saloon and vent out into a region of the right engine. It melted some rivets, cut static brushes, and left small black points at places of entering and exit. Pale senior stewardess said that something has exploded as a garnet in the saloon."

Comment. The description contains typical features of a BL influence fireball on the airplanes. In the Sochi airport (Adler), a taking off of the airplane takes place aside the sea where often there is a storm activity (sometimes even at clear weather on the coast). On an entrance and on an exit from the airplane, the BL has left small apertures which have appeared enough large for its pass, but did not cause depressurization of the airplane. A movement of the BL has appeared typical too—from a nose to a tail of the airplane self-summer that it is possible to consider as some law.

42. (Chernobrov 2003, p. 88) 1971, Moscow, Russia.

In 1971 the airplane IL-14 flew to Moscow. At the altitude of 3600 m, a bright red ball in the size by 4 times greater than the football has begun to move. Then the ball has flashed, its sizes have increased by several times. There have appeared the pouring "boiling" jets of light which have been starting from the bowels of the ball. The ball has taken the form of a disk, has caught up the airplane, has hung, and has "sat" on the right plane of the wing.

Comment. The ball has taken the form of a disk, has caught up the airplane, has hung, and has "sat" on the right plane of the wing.

43. (Chernobrov 2003, p. 130) September 1977, Khabarovsk City, Russia.

In September 1977 Zynaida Shashkova (fem) flew on the airplane IL-62 from Khabarovsk City to Moscow. At 7 o'clock in the morning by local time she has seen through a window as in the direction opposite to the airplane's flight, on a small distance from it has flown by an unknown round subject of black color. On it there were many projecting needles of bright red color which were shining as the heated iron. Behind from this subject was going a long fiery path. This subject has not rendered any influence on passengers; everybody was in a good state of health.

Comment. Behind from this subject was going a long fiery path.

44. (Chernobrov 2003, p. 161) June 27, 1978, Route Leningrad through Aktyubinsk to Moscow.

On June 27, 1978 the airplane-laboratory AN-30 with a glass floor was coming back from Leningrad through Aktyubinsk to Moscow. A speed of the flight was 380 km/h, altitude—6200 m. At 21.52 aboard-engineer, sitting between the pilots, has noticed an ideal white ball moving directly to the nose of the plane. It was at the distance about 700 m. The pilot has increased a speed of the plane, changed direction to the right and up, then to the right and down. The ball continued to move directly to the plane. The plane descended to 5000 m, the ball approached to a distance of 50 m. Then the ball slowly approached the right wing, without changing of luminosity. It moved under the wing and glued under the back edge of the wing in a region of the engine. It was ideal circle, white, opaque ball. Its diameter in this position was estimated as 3 m. During the next four minutes it slowly moved to the edge of the wing at that it was increasing in sizes and as if loosing density. At the end of the fourth minute, it increased in diameter up to 10 m, stars became visible through it. Then the ball has disappeared. During the observation no blows in regions of the plane engines were detected. After 15 min, the autopilot was finished in the airplane. The plane began to sharply fall aside. This occurs when an error is gradually stored in the autopilot, and ailerons take it away at once. The commander and the board mechanic shut down the autopilot and with difficulty manually lead the plane into the normal mode.

Comment. Behind from this subject was going a long fiery path.

45. According to data of Korezhin. October 1981. Near Kaliningrad.

In October 1981, Korotkov piloted the airplane-missile carrier near Kaliningrad (the Baltic Sea). At the altitude of 1800 m and velocity 520 km/h, he began to descend with a velocity of 15 m/s. It was a light cloudiness. At the flight altitude of 1300 m, he noticed by the side vision a ball of matches-fire color. At landing start, the pilot saw in front of him a fireball of 5 m in diameter. The ball was light-red color with a core of a brown ring of 1 min diameter and thickness of 15 cm. It was also fire, but darker. The observation time was one second or more. For some time, the ball was moving ahead of the plane at 5 m from the observer, without touching the plane. Then it has disappeared from the field of vision, and a dull explosion sounded in the tail side of the airplane. There was a smell of burn. The airplane's engine has immediately stopped, and the radio compass and the speed counter stopped working. The velocity dropped down and the airplane went down. Korotkov managed to start the engine at a height of 300 m from earth. The airplane's investigation has shown destruction of force elements of the covering and of the fin top back part fairing. A preliminary analysis of damages was as follows: impact of load created by the high-velocity air head in conditions of overheating.

Comment. A preliminary analysis of damages was as follows: impact of load created by the high-velocity air head in conditions of overheating.

46. According to data of Korezhin. April 1984. Russia.

In April 1984, the crew of the airplane MIG-24 was executing a training flight. They were informed that at a screen of radio location station there were traces of an unknown object near a zone of the flight. After some time the airplane as if encountered something. The crew felt a strong push. The engine has stopped, and the airplane began to fall down. The crew has baled out.

Comment. The engine has stopped, and the airplane began to fall down. The crew has baled out.

47. (Dmitriev et al. 1985) Route Erevan-Kiev, 1976.

Observer Vereshyagina (female) in summer of 1976 was the stewardess in the airplane TU-134 flight on the course. The weather at takeoff from Erevan was between clear and cloudy. At ascending over the Sevan Lake (it is located at the height of 1850 m over the sea level) approximately at 16 o'clock of Moscow time, she occasionally has seen a bright flash in the window. In spite of hermetic cabin, she heard a strong sound, as if something hit the fuselage and cut its part. After the bright flash, the airplane was pushed up-down, and then it was shaking in sides. The pilots informed that a BL has hit the airplane. At landing in Borispol Airport (Kiev), the airplane has been repaired.

Comment. After the bright flash, the airplane was pushed up-down, and then it was shaking in sides.

48. (Dmitriev et al. 1985) Minsk 1949.

Melnikov, a Military Serviceman, as a member of a crew, consisting of 5 men with passengers on the board, took off in the middle of summer 1949 from the airport Bykovo (near Moscow) at 9.00 am. In the district of city Minsk, they have met a cloudiness at height 3000 m, which was spreading like a sharply cut wall from the altitude above 3000 m to earth. The plane went into the cloudiness, it was calm, and was not shaking. After 20–30 min of the flight in the clouds, an orange-red circle has flashed in front of the cabin, it was unexpected and incomprehensible. Everything was still calm without shaking. Suddenly something has flashed behind them. There was an air nozzle behind crew chairs above in the fuselage. The BL has entered through it, moved as the red flame (red tape) through the cabin where the radioman and the board technician were sitting, then it moved as a red tongue through the passenger's saloon, and went out through the ventilation nozzle in the upper part of the fuselage (in the saloon). At that it did not make any harm inside the airplane. The radio connection went out of order after the BL went out of the plane. It proved to be that the wire antenna went out of order. It was going over the fuselage from the nose to the tail of the airplane. A part of it was melted (it was discovered after the plane landing in Minsk). Later it was found that one duralumin strut at the tail unit near the diving rudder was melted (only its ends left) in the place where the radio antenna was going. It was revealed after the percale covering was put off the tail unit, but the percale covering was not damaged.

Comment. It proved to be that the wire antenna went out of order. It was going over the fuselage from the nose to the tail of the airplane. A part of it was melted.

49. (Dmitriev et al. 1985) Russia.

The airplane IL-2 had a flight at low altitude. A yellow sphere of a football diameter has appeared far from the airplane. It was not clear whether the ball or airplane was approaching. After several minutes, the ball has behaved as if it was attached to the nose part of the airplane. The airplane was shaken a little. Then the ball has separated and in some seconds it entered the half opened small window elongating into the pilot cabin. The door from the cabin to the airplane was opened, and the ball went into it moving over the middle part. Then the ball has turned to the door and elongating went out of the slot. The airplane landed successfully. The melting was discovered on the nose.

Comment. The melting was discovered on the nose.

50. (Dmitriev et al. 1985) Summer of 1967 Caucasus, Russia.

In summer of 1967, there was an intensive impact of BL to an airplane near Elbrus (Caucasus). The control of the airplane was disturbed. At the after flight investigation, these were discovered through holes round the light of the cabin. Their diameter was about 1 cm. Edges of the holes were smooth as if they were accurately drilled without traces of burn through. They were different from the bullet traces.

Comment. Formation of round apertures in sheets of metal, apparently, occurs with smooth edges because of a local evaporation of a material under the action of a powerful pulse of the electromagnetic radiation arising at explosion of a BL. In the literature, many cases of evaporation of ring metal subjects by a BL are described.

51. (Dmitriev et al. 1985) Palanga, Lietuva.

The airplane was flying in the region Palanga (near the Baltic Sea) in daytime behind the clouds in conditions of the strong high-altitude turbulence. The plane was shaking. Altitude was several thousand meters. Members of the crew suddenly noticed a bright ring that as if surrounded the plane. The ring was far as it seemed to the crew. Its borders were fuzzy; the ring was located in the horizontal plane. Then the ring began to become narrower and it moved to the plane. This incomprehensible phenomenon caused the fear of the crew. At the moment of touching the airplane by this ring, the commander has noticed as the fireball suddenly rolled over the plane and dived down.

Comment. Complex appearance of BL.

52. (Vostrukhin 1985) The flight № 8352 Tbilisi-Rostov-Tallin.

The flight № 8352 Tbilisi-Rostov-Tallin of the airplane "TU-134A" was performed by the crew of the Estonian management of the Ministry of Civil Aviation of the USSR. The commander—Igor Alekseyevich Tcherkashin. The co-pilot—Gennady Ivanovich Lazurin. The navigator—Egor Mihajlovich Ognyov. The flight mechanic—Gennady Mihajlovich Kozlov.

At 4.10 in the morning, the airplane was 120 km to Minsk. The plane as if did not fly—it stood in the center of the Universe. No rustle in ear phones. They happened to be alone in the transparent air, inside a block of a black glass with holes of stars. Glancing around his part of the sky, the co-pilot noticed at the top right side of him an unblinking large star. But not the star—a yellow spot about the size of a five-copeck (3 cm) coin, extended along the edges. "What of it, he has quietly said to himself, the light refraction in the atmosphere or something else." A thin beam of light arose from the spot and fell freely on the earth. Then the pilot nudged the mechanic: "Look, Mihajlovich, at that." The mechanic took a look at it and said: "Commander, we must inform the earth." And the beam of light suddenly expanded, turning into a bright light cone. Since this moment everybody could see an event on the starboard side. A second cone, wider, but paler than the first one has appeared. Then the third cone, wide and absolutely bright, has appeared. "Wait, the commander shrugged, what to report? It is necessary to look, what will be further. And in general what can it be?" Pilots ought to understand that a distance could not be estimated by an eye. And yet all four had the same sense that the unknown object was hanging over the earth at the distance of 40–50 km. The co-pilot quickly started to draw the unusual phenomenon. It was incredible, but on the earth, lit by a cone-shaped beam, everything was clearly seen—houses, roads. What a power must have this "searchlight"?! The "searchlight" beam got up from the ground and stared at the airplane. Now they saw the blinding white point surrounded by concentric color circles. The commander still hesitated: to report about the event or not? But then has happened something that has put an end to the doubts. The white point has flashed, and a green cloud has appeared in its place. "It has started the engines and is getting out,"—the co-pilot said, involuntarily translating the phenomenon into a known and an ordinary life of the pilot. And the commander thought that the object has begun with a huge speed to get nearer to the airplane, cutting across its course at a sharp angle. In short, it rushed across. Tcherkashin has shouted to the navigator: "Call up the earth!". But—a strange coincidence—after first words of Ognyov the object has stopped. It seemed to the commander that it stopped to come nearer. "It stopped to move away,"—the co-pilot decided. The Minsk air traffic controller took note of the crew message and politely told that, unfortunately, he saw nothing—neither on the screen of the surveillance radar, nor in the sky. "Now, said Lazurin, being offended they will tell, that we are mad." And the green cloud suddenly has fallen down, lower the altitude of the airplane. Then it vertically has risen up. It rushed about to the right, to the left, and once again downwards-upwards. And, finally, it has stopped precisely opposite to the airplane. It flew with the airplane, as though it was fastened there—at an altitude of ten thousand meters with the speed of 800 kilometers per hour. "The escort of honor, Tcherkashin murmured, what a privilege for us! "The lights "were dancing" inside the cloud—they flared up and faded away, like the Christmas tree lights. Then the fiery zigzags have crept across. The navigator honestly informed the earth about everything. The excited voice of the air traffic controller answered him: "I'm observing flashes on the horizon. Where do you see your cloud?". The navigator answered. "Coincides"—the air traffic controller said. The cloud continued to vary. A "tail," similar to a tornado, has grown from it—wide above, thin near the ground. "Comma" turned out. Then the tail began to rise "in the horizon," and the cloud has changed from the elliptic to the quadrangular. "Look, the co-pilot said, it imitates us." Really, now they were escorted by the pointed "cloudy plane"—a plane without wings, with a slant tail. It lighted yellow and green. Where the real airplane has a nozzle, there the dense kernel was sensed. … Meanwhile one more real airplane has entered the control zone of the Minsk air traffic controller. "TU-134" flew from Leningrad to meet the Tallinn crew. The distance between two liners was about 100 km. From such a distance, the huge cloudy airplane should be noticed. However, the commander of the other "TU" answered Tcherkashin's question that he observes nothing. The air traffic controller of Minsk, who now well saw the cloudy plane, gave to Leningrad's crew the coordinate and the direction in which they could find the unusual phenomenon. But they seemed dazed. And only 15 km before the meeting they have begun to see clearly and in accuracy have described the cloudy airplane. Much later Tcherkashin's crew, trying to explain what they had seen, assumed that the light from the object was polarized, and it was propagating not in all directions. Together with the cloudy airplane they have passed Riga, Vilnius—air traffic controllers of these cities consecutively detected a strange tandem. Flying over the Chudskoe and the Pskovskoe Lakes, Tcherkashin's crew could estimate the sizes of the cloudy plane. These two oblong-shaped lakes are divided by a small bridge land. "TU-134" moved 120 kilometers to their left, and the cloudy airplane to the right, close to Tartu. From the place, where the dense kernel was sensed, a beam again has appeared. The light spot fell upon clouds, crawled along the ground. The object involuntarily gave its coordinates. Now it was possible to estimate that its length was equal to the length of the Pskovskoe Lake. They flew together all the way to Tallinn. And after the landing, their own Tallinn air traffic controller told them some more curious details. On the screen of the surveillance radar station of the Tallinn Airport, the "TU-134A" was observed not alone. Behind its light label on the screen two more labels crept over the screen though there were no other airplanes in air. In addition, these two labels were seen constantly, as they had to be. But the light spot of "TU" was disappearing, and appearing again. "I could have understood, the traffic controller said, if you 'were blinking' on the screen of a landing locator. But on the surveillance radar we have no such things, this cannot be."

Comment. The crew of passenger airplane TU-134A for a long time observed a shining object of the large size, moving near the airplane. The object emitted a ray of light in a form of several cones enclosed each other. The center of the beam was white around which concentric color rings were observed. Such a form of a polychrome beam could arise, if the reason of the light generation was the synchrotron radiation of electrons, moving with different speeds. In this case, the external ring of the beam should be red, and the center of the beam where there is a summation of all colors of a rainbow—white. Quite probably that the object was a BL of large size.

53. (Bychkov and Vaulin 2014) 11/21/2014, 21:41IN "contact" from the article "Ball lightning or UFO: the causes of the crash of the AN-24 aircraft in Perm."

Yuri Sytnik remembers how once a BL flew into the cockpit of his plane. At an altitude of 3.9 thousand meters, the lightning first came out, as if it appeared and hovered in front of the glass. After a few seconds, it flew into the cabin, although everything seems to be sealed. Here is such a glowing yellow-red ball. The only thing I said to the crew: "Softly, softly, softly, softly, softly," and the lightning floated smoothly into the cabin. It felt like the plane was just flying, then stopped, all the instruments were spinning like that, there was just nothing, and then it flew on. There was a smell of burning. A flight attendant ran in, shouted: "We're on fire!" I say, "Engineer, quickly!" The engineer looked back and forth. The chickens that were warming up for the passengers to feed them (because only 15–20 min had passed there after takeoff) were all charred." He was lucky: the plasma ball left the plane in the same way as it entered—through the wall.

Comment. The plasma ball left the plane in the same way as it entered—through the wall.

54. 30.03.2019 17:05/"Society. Tsargrad." Former aircraft commander, flight safety specialist Alexander Romanov.

"The meeting of an airplane with lightning is not too frequent, but it is quite a familiar phenomenon for any pilot. If we consider the problem from the point of view of safety, modern aircraft are equipped with electrostatic dischargers that divert static electricity into the air. In addition, the onboard systems are well shielded, the devices are also protected. Very often, passengers do not notice at all that lightning has hit the liner. There remains only a small burn on the skin at the point of impact." Romanov worked for AIR INDIA for several years. Due to weather conditions in South Asia, thunderstorms are especially frequent. Sometimes discharges constantly run across the glass, and this is a truly fascinating picture. "One day Mikhail Gorbachev was flying with us, and a BL got into the plane. The ball flew through the cabin and came out in the tail section. When we landed, we explored the ship for a long time, looking for where the blow fell. The spark arresters were melted a little, but nothing more. The passengers—and Gorbachev among them—may have been scared at first, but after landing, they were very animated. It's not every day you encounter such an adventure in the air."

Comment. A BL got into the plane. The ball flew through the cabin and came out in the tail section.

55. Zvezda TV Channel 2019. Honored Pilot of Russia, former flight director of Vnukovo Airlines Yuri Sytnik: "Planes are made so that they withstand lightning strikes. Every year, up to a hundred cases when lightning hits an airplane. At an altitude of 3600 m in Domodedovo on the Tu-154, my BL flew in and exploded in the kitchen area. The plane is intact, I flew to my destination and sat down."

Comment. A BL hit the plane and exploded in the kitchen area.

Discussion

Ball lightning meeting with aircraft (observation processing)

In addition to purely scientific interest, studies of the interaction of BL with aircraft are of great practical importance, since they are closely related to the safety of the crew and passengers. Knowledge of the peculiarities of BL behavior in airplanes and the accumulation of data on such cases when a meeting with BL did not lead to the death of the aircraft are the main result of these studies.

The term BL is usually understood as autonomous objects that appear in the air without an obvious source of energy. During the analysis, it is necessary to conduct a special analysis of a luminous spherical object observed by witnesses in order to distinguish BL from flashes of different nature: mirages (Menzel 1962), St. Elmo's lights, etc. For all such objects, we further use the term fireball (FB).

The main results of studies of the effect of BL on aircraft are as follows: (1) BL are able to penetrate into the all-metal frame of an aircraft ("Faraday cell") through existing or holes made by them. (2) As a rule, they penetrate into the nose of the aircraft, move along the axis of the fuselage, and exit through the hole in the tail of the aircraft. (3) Sometimes they pass through the interior of the aircraft without an explosion, rarely explode. (4) After the explosion of the BL in the plane, the internal communication disappears for a while, and the navigation devices fail. (5) In some cases, before penetration into the aircraft, BL hits were observed on the external elements of the aircraft (on the wings, on the fuselage). At the same time, the BL moved in the external flow of incoming air at the speed of a flying aircraft. (6) BL in thunderclouds were observed at altitudes from 500 to 104 m. (7) Sometimes luminous objects up to 100 m and bigger in size were observed from airplanes. These objects emitted narrow beams of light.

The seventh chapter of the book (Stenhoff 1999) is devoted to the meetings of BL with airplanes and the analysis of the degree of danger of such meetings, here we partially reproduce the materials from this chapter. Finding out the features of the behavior of BL in airplanes and the accumulation of data on such cases when a meeting with BL did not lead to the death of the aircraft are the main result of these studies.

For the analysis, we selected data from flight observations, which are usually consistent with our previous analysis of the BL in terms of the dependence of the lifetime on the diameter (Amirov and Bychkov 1991, 1994a, b). Therefore, we excluded data on FB with dimensions of more than 5 m, a lifetime of more than 500 s and an increase in speed to values exceeding supersonic. We agree with Menzel (1962) that large spheres with an extremely long lifetime (about 45 min) can represent special type of mirages formed as a result of reflection of solar or lunar rays on flat ice crystals in the air. We also shared the obvious observations of the BL: all the BL inside the planes had the usual description for the BL, they appeared unexpectedly, disappeared with an explosion or extinction. In two cases, the BL entered the plane from the outside. According to the pilots, in some cases, the BL appeared "out of nowhere" inside the aircraft, they also observed how the BL penetrated through the holes into the cockpit. BLs appeared in thunderstorms and in good weather, as well as when the plane was flying over the areas of radio transmission station operation. In all these cases, the witnesses could not estimate the duration of observation, but according to the descriptions, it was in the range of 5–20 s, which correlates with the usual statistics for the lifetime of the BL (Amirov and Bychkov 1991, 1994a, b). The results of these observations are presented in Tables 2.1 and 2.2.

To clarify the role of BL and between cloud discharges, we conducted a correlation analysis of the frequencies of FB observation from aircraft in the DB SCB Stakhanov-Keul-Bychkov Data Bank (DB SCB) (Amirov and Bychkov 1994 a,b), between cloud discharges (Brylev et al. 1989) and BL observations (Amirov and Bychkov 1991) (without observations from aircraft). The corresponding frequency distributions by month are shown in Figs. 2.58, 2.59, and 2.60.

Table 2.1 BL form

Inside/outside aircraft		
Type	Size (cm)	Number
Sphere	5–20	12/1
Tape-ribbon	–	1
Tape-ribbon Sphere/complex	–	1/1

Table 2.2 BL parameters outside aircrafts

Type	Number	Size (cm)	Lifetime (s)
	52	10–40	0.01–3
Oval	3	–	100–120
Flash	1	500	1–3
Sphere with blurred relief	1	100	60

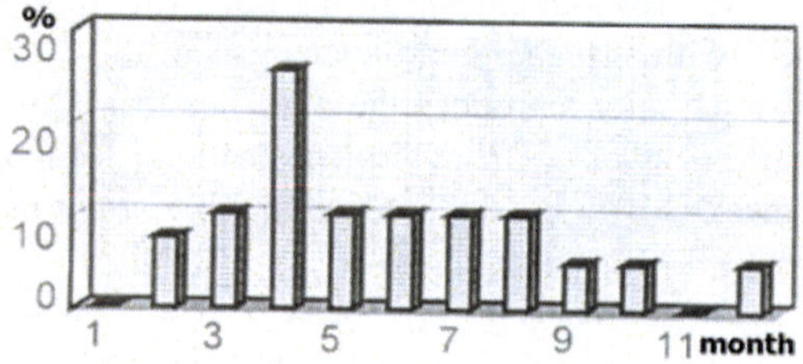

Fig. 2.58 The frequency of observation of FB outside the aircraft during the year (SCB DB)

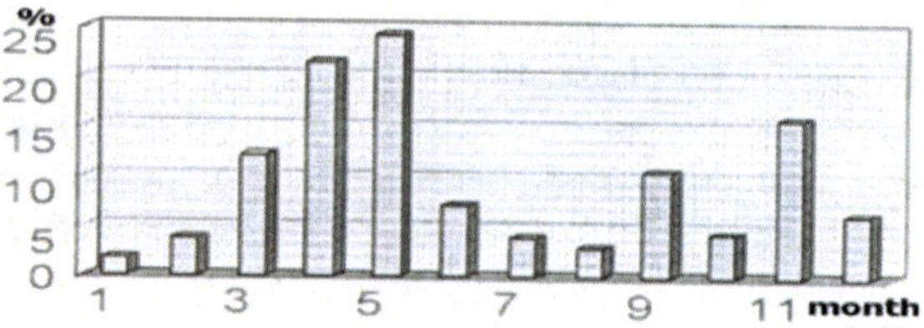

Fig. 2.59 Frequency of (Clowd spheres-OS) OS observation in stratified clouds during the year (Brylev et al. 1989)

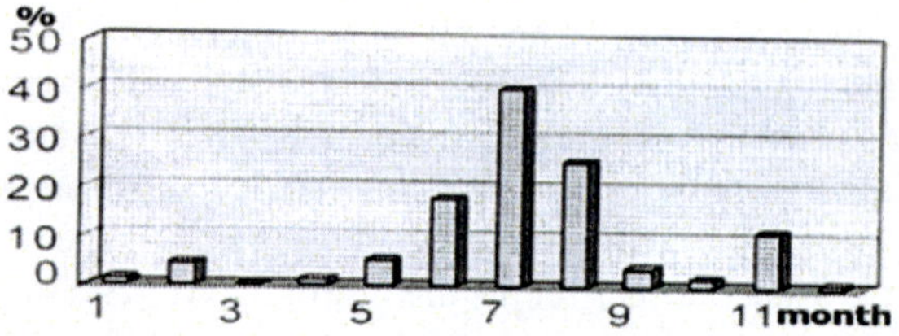

Fig. 2.60 Frequency of BL observation during the year (Bychkov and Amirov 1991)

1. The analysis of the FB observed outside the aircraft was more complex. Usually FB appear in difficult meteorological conditions. The lifetime is very short. The descriptions of these events are similar to each other. Pilots usually saw a bright flash and then heard a muffled bang, so it was almost impossible to distinguish the BL from the description of discharge flashes in the inter-cloud region in layered clouds (Brylev et al.1989). Only in one case (Kogan-Beletskii 1957) [7] it was obvious that the FB was a BL that collided with the propeller of the aircraft.

To clarify the role of BL and between cloud discharges, we conducted a correlation analysis of the frequencies of FB observation from aircraft in the DB SCB (Amirov and Bychkov 1994 a, b), between cloud discharges (Brylev et al.1989) and BL observations (Amirov and Bychkov 1991) (without observations from aircraft).

The corresponding frequency distributions by month are shown in Figs. 2.58, 2.59, and 2.60.

In 30 cases, the post-flight inspection revealed small-destroyed areas on the surface of aircraft parts and the destruction of electrical circuit elements. In 6 cases, collisions lead to the death of aircraft. In 5 of them, the electronic circuit of the aircraft was destroyed, 4 events occurred from October to March, and 1 event occurred in July. Only in one case it was clear from the description that the BL suddenly appeared on the control panel of the SU-17 aircraft, and then hit the pilot, the event took place in February. The pilot was injured; the marks on the body were typical of an electric shock. The pilot left the plane. Typical effects of the interaction of OS = FB with aircraft are presented in Table 2.3. Estimates made on the basis of data on heating and evaporation of aluminum for typical hole sizes give an energy supply in the range $1\text{–}2 \cdot 10^3$ J. The destruction of the fairings and the effect of static discharge can be estimated approximately as $(1\text{–}2) \cdot 10^5$ J. These values are consistent with typical values for BL, but the descriptions of destruction also resemble the action between cloud discharges. Of the 73 analyzed BL events observed outside/inside aircraft, only in 18 cases it is possible to speak unambiguously about BL.

Analysis of the presented data

BL data in the cabin of the aircraft. The analysis shows that BL can appear inside aircraft indoors, usually in the cockpit, where the largest number of electrical flight control devices are located. During the operation of this device, an electrical breakdown may occur with the destruction of insulating materials. Therefore, such BLs can be formed when a short circuit in electrical devices destroys various materials.

External conditions of the aircraft. The above explanation can be applied to explain the origin of BL in the summer, when linear lightning streamers destroy objects made of various materials in nature and on the surface of airplanes. In this case, the BL

Table 2.3 Effects of FB interaction with aircrafts

Effect		Number of cases	Type of destruction
1	Fire	3	Small
2	*Spraying, melting*	15	*Small*
3	Destruction of fairings and drains	8	*Small*
4	Welding marks and burnt holes	6	*Small*
5	Rupture of the plane's surface	1	*Small*
6	Destruction of the radio communication system	11	*Small*
7	Generator failure	1	*Plane crash*
8	*Engine failure*	3	*Plane crash*
9	*Failure of aerobatics devices*	1	*Plane crash*
10	Failure of the radio station, injury of the pilot by the BL	1	*Plane crash*

is the result of the formation and charging of a complex frame in the conditions of atmospheric electricity. In winter, the most likely occurrence is between cloud discharges. According to (Brylev et al. 1989), between clouds, the electric field can reach $E \sim 10^5$ V/m, and under these conditions, aircraft charging can reach breakdown fields near sharp metal parts of the aircraft. This can cause the appearance of corona discharges near aircraft and provoke inter-cloud discharges to their surface. Pilots can consider these phenomena as flashes, FB and BL.

All the results obtained show that near airplanes, in inter-cloud conditions with strong electric fields, both BL and inter-cloud discharges can appear. The first usually appears in summer, the second in winter. Both phenomena have a large amount of electrical energy and can cause serious destruction of aircraft. BL can also appear inside aircraft as a result of destruction-breakdown-charging processes that occur when an electrical circuit is destroyed. Their high energy has an electrostatic or electrodynamic nature.

International information on BL risks (Doe et al.2009)

Scientific publications and mass media focused on individual BL incidents and did not analyze general patterns as a risk assessment for flight safety. Due to the lack of established incident reporting channels, most BL observations are still transmitted as "aviation professional knowledge."

To overcome this unsatisfactory condition, the authors (Doe et al.2009) collected and analyzed an international data bank of 87 stories about airplanes and BL for the period from 1938 to 2007. Of the entire sample, 36 (41%) cases occurred in Russia/RF/USSR, 24 (28%) in the USA/Canada, 23 (26%) in Europe, and 4 (5%) in the Asia-Pacific region. Meetings took place with various types of military (USA: C-54/141, B-52, KR-97/135 Stratotankers, C-130, P-3 Orion, RF/USSR: PO-2, IL, SU, TU, MIG; Nimrod, Saab-105) and civilian aircraft (USA: DC-3/6, Metroliner, B-727/737/757/777), (RF/USSR AN, TU); VC-10, Fokker F-28, CRJ-200, as well as general aviation aircraft (C-172, Falcon-20).

Reports on BL show an annual maximum from April to August. At the impact of the BL, 15 aircraft rose, 7 descended; most were at cruising altitude. 42 (48%) reported BL outside, 37 (43%) inside, and 7 (8%) both inside and outside the aircraft. In 34 (39%) cases, no damage was reported, 39 objects (45%) caused minor damage, 11 major damage (13%), and 3 resulted in the loss of military aircraft. 3 objects caused minor, 1 serious injury to the crew. 23 cases of damage were related to BL inside the fuselage; all 4 cases of crew injury were there. The average size of the BL is described as 25 cm, sometimes more than 1 m. The color of the BL is 30% in yellow-red, 10% in the blue-green region of the spectrum, and 8% in white. The average duration of BL observation is about 10 s, sometimes more than 1 min. 33 (38%) incidents ended with the explosion of the object. Thunderstorm conditions were reported by 25 (29%) observers, 9 (10%) reported that there was no thunderstorm. Approximately 50% of the impacts are on the airframe, some of them cause minor damage, potentially threatening the crew and passengers.

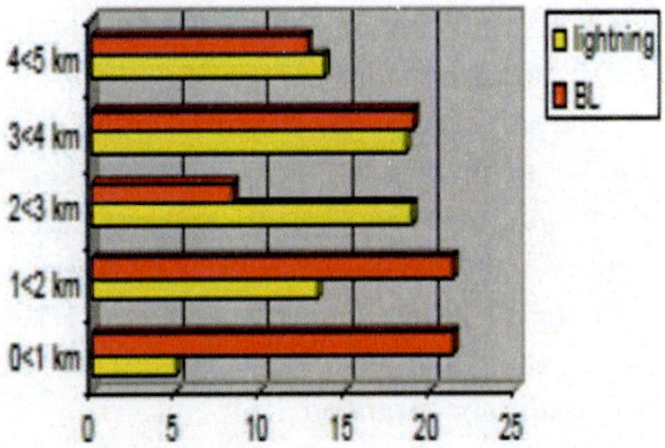

Fig. 2.61 Comparison of the heights of the appearance of lightning discharges (yellow) and BL (red)

Structural damage covered by the media is extremely rare. Although the BL does not pose a significant risk to air traffic, the authors (Doe et al.2009) propose routine reports on BL incidents/accidents and investigation of BL damage/injuries.

International results on the interaction of lightning with aircraft (Rakov and Uman 2003) were compared with reports on BL aircraft (46 with flight altitude data). When linear lightning interacts with an airplane, the main one is from 2 to 4 km, while incidents with BL are most frequent in the range from 0 to 2 km, see Fig. 2.61.

Comparing the results of the interaction of linear lightning with aircraft (Rakov and Uman 2003) with reports on BL, general schemes (for example, flight level—BL maxima at 1000 and 3000 m) indicate BL as an accompanying effect—a secondary electrical thunderstorm phenomenon. Its danger is much less than the danger of linear lightning.

2.9.1 BL and the Sea

In this section, we present observations of BL at sea near ships and the impact of BL on them. Here are the cases known to us that can be identified with BL. We also give descriptions of possible analogues of BL that appear in complex hydrodynamic conditions.

Note that many Russian events were taken from Russian translations.

1. (Skryagin 2000) Robert Boyle describes. The British warship Albemarle, 1681, 300 miles west of Cape Cod.

"During a violent storm, after a violent thunderclap, a flaming mass fell from the clouds onto the ship." The sailors testified that because of this, the mainmast of the ship was split, and the longboat standing on the deck was broken into pieces. "The mass burned for several minutes, emitting a smell similar to the smell of gunpowder…" The fire that started on the ship was extinguished.

Comment. The flaming mass burned for several minutes.

2. (Skryagin 2000) 1749. Merchant ship "Dover," three miles west of Cape Lizard.

The sailors noticed a glowing blue ball hovering over the water right ahead. Its size corresponded to the average millstone. 15 m from the ship, it soared up and exploded. At the same time, the sound resembled a volley of several guns. The mainmast was blown to pieces, the mainmast cracked at the spur, the deck superstructure was destroyed, the deck itself, and the side of the ship were severely damaged. The sailors standing on the deck were knocked down, they temporarily lost their sight, and one of them received severe burns. For several minutes, the crew of the "Dover" felt a pungent smell, reminiscent of the smell of gunpowder. The ship leaked heavily and had to be stranded.

Comment. A glowing blue ball. Its size corresponded to the average millstone.

3. (Skryagin 2000) On 24.07.1798, the frigate "Resistance" was anchored near Bank Island in the Indian Ocean. The appearance of the BL led to the fact that the Frigate was blown to pieces. Of the 294 crew members, four survived.

Comment. The explosion of the BL.

4. (Skryagin 2000) 1840. English packet boat "Poland."

During the passage across the North Atlantic before the storm, a luminous ball with a diameter (according to eyewitnesses) of about half a meter descended from a thundercloud onto the ship. It grazed the chain top sheet and exploded. Due to the fire that started on the deck, cotton caught fire in the holds of the ship. The packet could not be saved, the crew was picked up from the boats by a passing vessel a few days later.

Comment. The explosion of a BL with a diameter of about half a meter.

5. (Skryagin 2000) 1840. English customs cutter "Chichester."

A BL from the sky penetrated through the crack of the door into the cabin of the ship, where the officers were having breakfast at that time. It hung over the table for a few seconds. All the glass objects of the table setting turned out to be melted, but people were not injured. Then the ball went through the table, deck, and bottom into the water. The team had to work on pumps. Later it turned out that under the hole in the bottom of the vessel, the copper sheet of the skin turned out to be burnt.

Comment. The BL hung over the table for a few seconds. All the glass objects of the table setting turned out to be melted.

6. (Skryagin 2000) 29.08. 1862. The ship of the British Royal Navy is the Nelson. 43° south latitude and 160° east longitude.

Captain Meiklejohn. The ship was caught in a storm. For several minutes, the fires of St. Elm shone on the clothes of its masts. At this time, seven meters from the leeward side, a glowing ball descended from the sky into the water. A bright orange light illuminated the deck of the Nelson, but there was no explosion when the ball touched the water. The place where this object fell glowed for five minutes.

Comment. The place where the glowing ball fell glowed for five minutes.

7. (Skryagin 2000) 6.05.1890 English cargo steamer "Keikoura," 36° south latitude and 50° west longitude. Captain of the steamer Crutchley.

A luminous ball the size of a fist descended onto the open running bridge of the ship. A few seconds later, it exploded between the officer of the watch and the helmsman, without harming them. The sound of the explosion could be compared to the sound of a pistol shot. Some items of the navigation bridge glowed blue for about five seconds.

Comment. A glowing ball the size of a fist exploded.

8. (Skryagin 2000) September 1896. Night. The German steamer Helgoland.

The sailors saw a giant glowing ball descending from the sky onto the ship. The sailors standing on the deck felt its warmth. The ball touched the water a few meters from the side of the steamer and exploded with a strong roar. Waves were observed at the explosion site for several minutes.

Comment. The ball exploded. Waves were observed at the explosion site for several minutes.

9. (Skryagin 2000) 10.01.1899 The English passenger steamer Sardinia. The North Atlantic.

The BL exploded one and a half meters from the upper deck. The mate of the watch, who was nearby, fell to the deck in a state of shock and lost his sight for several hours.

Comment. After the explosion of the BL, the mate of the watch fell to the deck in a state of shock and lost his sight for several hours.

10. (Skryagin 2000) Magazine "Around the world" No. 28 for 1908.

"The sailing ship Eclipse with a capacity of 1,469 tons, sailing from New York to San Francisco, crashed under the following circumstances. After a three-month safe voyage, on the 85th day of the journey, in calm and clear weather, lightning suddenly flashed in the air, and thunder rumbled over the heads of the sailors, resembling shots from heavy artillery guns. A few seconds later, an aerolite (BL) hit the top of the mainmast, shattering it to pieces. Then it plunged swiftly down, breaking through the deck, the forecastle and the keel, and sank into the sea with an ominous hiss. Judging by the size of the hole punched in the deck, the meteor was slightly larger than a human head. The fire that started on the ship was soon extinguished. But the sailors failed to plug the leak. After working four days with pumps, without sleep and rest, they moved into the boats and left the ship, which almost immediately went down."

Comment. An aerolite (BL) hit the top of the mainmast, shattering it to pieces.

11. (Olkhovatov 2008) Report of the British [Scientific] Association in 1861. Brig "Victoria," Mediterranean Sea. 1861. Nice weather.

The sails were suddenly thrown over, as if under the influence of an unexpected squall, and for the next 2 h they were affected by this strongest wind, which then just as unexpectedly stopped. The phenomenon was accompanied by an unbearable smell of sulfur and unthinkable heat.

At the same time, 3 luminous bodies emerged from the sea surface at a distance of about half a mile from the ship and remained visible for about 10 min.

There was an eyewitness who at the same time saw from Mount Lebanon (near Beirut) 2 luminous bodies that were 5 times larger than the moon. Each had an "appendage," and it is possible that the bodies were connected to each other by some radiant formations. The brightness of the glow was such that it was painful to look at the bodies. The duration of the phenomenon was about 1 h.

At the same time, a similar glow was observed by another eyewitness from Antalya (Turkey) for 20–30 min. There it seemed that the glowing object "broke apart" some time after its appearance.

Comment. Three luminous bodies emerged from the sea surface at a distance of about half a mile from the ship and remained visible for about 10 min.

12. (Olkhovatov 2008) Monthly Weather Review magazine for March 1887. The Atlantic Ocean. Captain of the Dutch ship S. Svort 19.03.1887

During a strong storm, I saw an unusual object that consisted of two balls, one of which was glowing, and the other was "very black." A glowing ball, which seemed oblong, hovered over the ship. It became dark, as night, but the sea seemed to be "burning." The object fell into the water with a roar, raising huge waves that rolled over the ship.

The atmosphere was so suffocating that everyone on board was suffocating. Immediately after the incident, dense blocks of ice fell on the deck, and the ship's rigging also turned out to be covered with ice, although the thermometer showed 19° C. The readings of the barometer could not be removed, as they changed all the time. The survey of the vessel did not reveal any damage, but on the side of the vessel, in which the object fell, the surface of the vessel was blackened and shells formed on the copper coatings. After the meteor (BL) fell, the wind strength increased to hurricane strength.

Comment. An object that consisted of two balls, one of which was glowing, and the other was "very black."

13. (Olkhovatov 2008) Hydrographic Service of the USA. 10.03.1885 The bark Innervich under the command of Captain Waters at midnight on 02/24/1885, 37°, S. S. and 170°. 15 min V. D.

The captain was awakened by the mate and went on deck, where he saw that the sky was turning a fiery color. Quite suddenly, a large fiery body appeared over the ship, completely blinding the observers. When this body fell about 45 m downwind, it hissed so much that this sound even covered the sound of the explosion. The ship shook all over. After that, a mass of white foam was rapidly approaching the ship. The noise from this approaching volume of water was described by eyewitnesses as deafening. Once again, the bark received a powerful blow. The brief respite before that was used to fill the sails, and this roaring mass of water swept ahead. Another "wide streak of flame" swept along the mizzen mast and spilled into myriads of sparks. The strange reddening of the sky lasted for 20 min.

Comment. Explosion of a large fiery body.

14. (Olkhovatov 2008) Journal "Nature" for 1891. Bark "Killarney." Storm.

At 9 p.m. on 10/15/1890 during a severe thunderstorm, the wind suddenly changed. Accompanied by thunderous sounds, a large "mass" fell about 180 m from the vessel. The roar from its fall was very noticeably different from the sounds of lightning discharge. The fountain of water from the fall reached about 12 m.

Comment. The fall of a large mass during a thunderstorm.

15. (Olkhovatov 2008) 7.04.1889 The British ship Tropic. A certain object fell right at the side of the ship.

At 5 a.m. during the storm, an unusually bright lightning occurred, followed by deafening thunder. The sea turned blood-red. A hissing sound came from above. The sound became stronger and stronger, and with it the glow intensified. It soon became so bright that none of the crew members could see any details, except perhaps for "fiery sparks." Above them was a large sphere of fire "the size of a ball." This sphere sloped downwards towards the ship. It fell into the sea just a few meters from the ship, filling the air with flakes of fire. Millions of sparks falling from the footprint of the sphere illuminated the ship for several minutes.

A few minutes later, the object sank into the sea. The faces and eyes of the sailors "squinted" from pain, and the side of the steamer closest to the sphere was warm.

Comment. A large fiery sphere "the size of a ball" sloped down towards the ship. It fell into the sea just a few meters from the ship, filling the air with flakes of fire.

16. (Olkhovatov 2008) On 03/29/1895, the steamer "Nexmore" USA was located near the southern tip of the Newfoundland Bank.

The weather was good in the morning, but it worsened in the afternoon of the first day. Suddenly, a dense black fog formed, which completely enveloped the steamer. At that moment, without any warning, a terrifying roar was heard from the bow mast. The brightest flash of white color accompanied it, and small particles resembling white ash began to fall on the deck.

The crew was shocked by what had happened, and the ship's captain, Richardson, who was on the bridge, stopped the ship. He was so stunned by what had happened that at first he thought that someone had fired a projectile at his ship. However, initially no visible damage was found, and the steamer continued its journey to the American continent. A subsequent, more detailed inspection revealed that the tree on the front mast was split in all directions. The paint was burned the entire length of the mast.

Immediately after the explosion, torrential rain began, lasting 20 min. When it stopped, the fog lifted and the sun shone brightly, but half an hour later the steamer again fell into this fog. There was no lightning before or after this mysterious explosion.

Comment. An explosion during a dense black fog.

17. (Olkhovatov 2008) In the first half of July 1908. 200 miles east of Philadelphia. Dutch steamer "Ocean."

The captain reported that the "big meteor" almost hit the ship, a large wave threw the ship, a cloud of "poisonous gas" forced the crew to seek refuge in the hold, "specific brown dust" fell on the ship. The "meteor shower" lasted for several minutes, and the sea shone to the horizon.

Comment. The fall of the "Meteor Shower" lasted for several minutes, and the sea shone to the horizon.

18. (Olkhovatov 2008) The newspaper "Indianapolis Star." February 1910. The steamer Trafalgar was sailing from Port Talbot to Bastia with a cargo of coal. 10 miles from the Wolf Rock (approximately 50 deg. s. s. and 6 deg. s. d.).

According to the captain, "We were hit by a comet or lightning. the ship shook violently and the sound of an explosion was heard. The foremast turned into a flaming mass of flame and the whole ship lit up. At the same moment, the crew saw a large fiery body with a tail 9–12 m long, which fell into the sea about 6 m from the ship. His appearance was accompanied by a strong hiss, and at the moment of its disappearance, a column of water rose into the air. Immediately after that, crew members began to run out of the forecastle shouting that there was a fire there. The entire interior of the ship shone brightly with a bright light. Everything in the engine compartment glowed with a pale blue light, millions of sparks were falling. The whole crew jumped on deck. It so happened that the second mate of the captain was measuring the depth with a lot at that time and received a strong blow from a steel rod; which he was holding in his hands. The phenomenon was not long-lasting."

When the crew recovered from the experience, it turned out that all the compasses were demagnetized and "lying." In such circumstances, the captain decided to head to the nearest port of Falmouth.

When the compasses were moved to the shore, they began to show correctly and very accurately. There was a suspicion that the whole ship turned out to be strongly magnetized.

Comment. A large fiery body with a tail 9–12 m long, which fell into the sea about 6 m from the ship. The second mate of the captain was measuring the depth with a lot at that time and received a strong blow from a steel rod.

19. (Olkhovatov 2008) The Papazov couple from Bulgaria during a circumnavigation of the world on a small yacht. 11/28/1979, 8.5 deg. s.l. and 120–130 deg. w.l.

About three o'clock in the morning, when the wind began to die down and they were removing the sails, some glowing body plopped into the water a hundred meters from the yacht. A water column shot up to the sky like a geyser, and the light disappeared into the depths of the ocean in 1–2 s. The resulting wave doused the Papazovs and rocked the yacht. Everything happened so quickly that they "didn't have time to utter a word."

Comment. The glowing body plopped into the water a hundred meters from the yacht.

20. (Azhazha 2013) 08.19.1845 The Baltic Sea.

From the brig "Agamemnon" near the island of Bornholm. A "strong flame" was observed, with an area of 10 square meters, bursting out of a meter-long object on the water. It was not extinguished by the incoming waves.

Comment. A "strong flame" was observed, with an area of 10 square meters, bursting out of a meter-long object on the water.

21. (Azhazha 2013) 13 08.1845 The Baltic Sea.

From the ship Gangut, Vice Admiral Lazarev II observed the meteor flight at the Dageror lighthouse.

Comment. "Meteor flight" is most likely a horizontal flight of the BL.

22. (Azhazha 2013) 04.04.1901 The Persian Gulf. (S).

The captain of the ship "Kilve" Housizon, at a report to the Royal Meteorological Society, announced the observation of luminous circles at a low altitude above the sea surface for 15 min in the morning. They appeared in front of the ship in flashes of light, alternating with each other at a speed of 60 miles per hour.

Comment. Observation of luminous circles at a low altitude above the sea surface.

23. (Azhazha 2013) 02.11.1922 Atlantic Ocean, Brazil.

Officer Fr. S. Blessing told the New York Times about observing the passage of a strange meteor for three minutes and 20 s. At 21.30 L.T., he noticed it 10° above the horizon. It flew at high speed, made a large arc in front of the ship, standing off the coast of Brazil in Hoboken, and went over the horizon to the east. It was the size of the moon.

The ball had a dazzling bright red color and was surrounded by a shining halo. A long trail of fire traced behind him. The ship and everything around for 350 m were flooded with light. It turned out that this phenomenon was seen both in Buenos Aires and in Montevideo.

Comment. The ball had a dazzling bright red color and was surrounded by a shining halo.

24. (Azhazha 2013) 06.18.1845 The Mediterranean Sea.

The crew of the Brig Victoria of the British Navy at a point with coordinates 36°40′56’ s.l., 13°44′36’ v.l. observed half a mile from the ship (about 900 m) the exit from the sea of three luminous disks connected to each other by light rays. The object was five times larger than the moon and glowed with a golden light. The duration of observation is at least 10 min.

Professor Baden-Powell collected the testimony of the crews of other ships within a radius of 15,000 km and published it in the British Association magazine in 1860, then in 1886. Observers from Syria and Malta saw a double object. The total observation time is about 1 h. F. Hewlett, a member of the Royal Astronomical Society, also saw this object from Adalia (Asia Minor).

Comment. The exit from the sea of three luminous disks connected to each other by light rays.

25. (Azhazha 2013) 07.16.1864 South China Sea, Gulf of Thailand (S).

Watchman Timofeev from the clipper "Vestnik" at night at the entrance to the bay noticed two bright pulsating light spots ahead on the horizon. He steered the ship between the light formations. These were glowing and rotating in different direction wheels in the form of 10–12 curved rays 20 m thick, half coming out of the water. The rotation speed is about 10 revolutions per minute. The rimless wheels were about 460 m in diameter. When they were left astern, another glowing rotating wheel was seen ahead and to the left, but only of a smaller size. The total observation time is about 20 min.

Comment. Glowing and rotating in different direction wheels in the form of 10–12 curved beams 20 m thick.

26. (Azhazha 2013) 05.15.1879 Indian Ocean, Persian Gulf. The captain of the British warship "Valcher" D. E. Pringle reported the observation at 9.40 pm at a point with coordinates: 26° 26' s.l. and 53°11' e.l. luminous wheels on both sides.

They rotated in different directions at a speed of 20 revolutions per minute. The spokes of the wheels were illuminated, had a width of about 8 m, and were 30 m from the side of the ship. The diameter of the wheels was about 40 m. The wheels slowly sank and disappeared into the depths after 35 min. In cables (185 m), this picture was also observed by the crew of the steamer "Hawk."

Comment. Glowing wheels on both sides of the ship.

27. (Azhazha 2013) 1840–1899. The Persian Gulf.

During this period, at least six observations of silently rotating luminous disks were recorded on the surface of the sea.

Comment. Silently rotating luminous disks.

28. (Azhazha 2013) 1906. Indian Ocean, Gulf of Oman.

An officer of the British ship D. Carnegie wrote to the journal "Royal Meteorological Society" about the observation of a stationary luminous underwater object, from which rays moving at a distance of 6–7 m began to emanate at close approach. They were not delayed by the hull of the ship and passed to the other side. The flashes of light moved at a speed of 200 km/h. Analysis of water samples under a microscope showed nothing abnormal.

Comment. A glowing underwater object.

29. (Azhazha 2013) 11.03.1907 Indian Ocean, Strait of Malacca.

The crew of the Delta ship observed at 2 p.m. rays of light rotating on the surface of the water around the center, which moved about 7 miles (13 km) next to the ship. After half an hour, the phenomenon stopped. The length of the beams was 300 m.

Comment. Rays of light rotating on the surface of the water around the center.

30. (Azhazha 2013) 08.23.1908. The Sea of Okhotsk. 2 cases.

Military doctor F. Derbek from the ship "Okhotsk" gave a description of a series of luminous spots (circles) that began at a point with coordinates 57° 03' s.l. and 155° 50' e.d. at 23.00 o'clock. A bright-white-green spot first appeared under the stern of the ship. For some time, it increased in size and moved along with the ship. Then the spot quickly moved to the side and forward, reaching the horizon in three minutes. There was a moment when four spots were tracked simultaneously. The speed of the spots was calculated—180 km/h.

Mid-September 1908—similar phenomena were observed in the same area.

Comment. Glowing spots (circles).

31. (Azhazha 2013) 1908. South China Sea, Gulf of Thailand.

The navigator of the Conseller, D. Turner, made an entry in the logbook about the ship passing by a rotating radiation 30 m wide.

Comment. Rotating radiation with a width of 30 m.

32. (Azhazha 2013) 1900–1909. The Persian Gulf.

Luminous rotating (spots, circles) with a diameter of up to 40 m were observed at least five times.

Comment. Rotating "wheels."

33. (Azhazha 2013) 1925. The northern part of the Indian Ocean.

The commander of the French ship "Prussin" testified: "Around the ship, everything glowed from a set of glowing balls floating from the depths, with a diameter of 20–25 cm. They rose at a speed of 0.5 m per second with an interval of 20–30 s. At the water's edge, they swelled sharply by half, and on the surface turned into disks with a diameter of up to 120 m. Bright silver-green circles after a while went out in the dark."

Comment. A lot of glowing balls floating from the depths, with a diameter of 20–25 cm.

34. (Azhazha 2013) 1928. The Sea of Japan.

The crew of the English ship "King George" witnessed a meeting with an underwater glowing object. "The 'perfect circle' was going to the intersection of the course. Upon reaching the ship, the circle took the form of an oval, 300 m long. When we stopped moving, the object took the shape of a circle again, and then continued moving in the same direction."

Comment. The circle took the form of an oval, 300 m long.

Fig. 2.62 LLO near sea ship "Kaiba Sikharo," Brasil, 1962

35. (Klimov2008) A. I. Klimov DS in physics basing on materials of Popovich M. 2003 UFO over the planet Earth, SPbGPU Publisher, St Petersburg; Hill R. P. 1995 Unconventional flying objects, HR, Canada and Pravdivtsev V. 2007 Underwater phantoms; UFO's diving/Video. The first Russian TV program made the report.

According to many observations large-scale long-lived plasmoids can be located at altitudes of 10–20 km from 6 min to 10 h, and can travel at cruise speeds up to 4000 km/s. The same objects exist in seawater at depths up to 10 km, where they can move at cruise speeds up to 300 km/h. Their parameters: diameter 3–10 m; color—red, orange, silver, rainbow. Sea water depth 0–10 km, lifetime 0.1–10 h (Fig. 2.62).

Comment. (Together with DS A. I. Nikitin) According to Kapitsa's model, a plasmoid appears in the antinode of a standing electromagnetic wave between a cloud (or ionosphere) and the earth. If this distance is $H = 1$–10 km, the wavelength is also 1–10 km, and the frequency $f = c/H = 30$–300 kHz (here c is the speed of light). However, it is known that the typical diameter of a ball lightning is $D =$ 20–30 cm. According to the author, this is due to the resonance of a standing ion-acoustic wave that moves in the air at a speed of $Ca = 2 \cdot 10^3$ m/s. At a frequency f = 30–300 kHz, the length of the ion-acoustic wave $\lambda a = Ca/f = 7$–70 cm. The size of the BL is close to λa. Large-sized plasmoids exist in the ionosphere at an altitude of $H = 100$ km. The frequency of a standing wave with a length of 100 km is equal to $f = c/H = 3$ kHz. The length of the acoustic-optical wave (and, therefore, the size of a large plasmoid) $Ca/f = 0.7$ m. The speed of movement of such plasmoids is comparable to the speed of movement of an acoustic-optical wave of 2 km/s. Apart from the difficulties of explaining the movement of large-sized plasmoids at low altitudes and in water, the author's model can serve as a basis for explaining the nature of these objects. But to do this, it is necessary to suppose the presence of strong electromagnetic fields in the atmosphere that can cause an electrical breakdown of the air. However, the fields in the thunderstorm atmosphere below clouds do not rise greater than 10^4 V/m, in clear weather they are below 1–10 V/m. Therefore, the question about nature of such plasmoids is still open.

2.10 BL and Geotectonic Events

Various kinds of light phenomena are observed in connection with earthquakes. Below are some of the most interesting examples from the author's point of view, collected in (Olkhovatov 2008) from various sources related to tectonic activity, translation from data in Russian.

2.10.1 Luminous Formations Generated by Geotectonic Activity

1. (Olkhovatov 2008) 1663, Quebec, Canada.

Before the earthquake, we saw lights, torches, fireballs, which sometimes fell to the ground, and sometimes dissolved in the air. Sometimes they were seen 6–7 times a night.

Comment. Multiple appearance of fireballs.

2. (Olkhovatov 2008) 1672, winter, Edo (now Tokyo), Japan.

The local earthquake was accompanied by the appearance of a fireball resembling a paper flashlight. It was flying through the sky in an easterly direction.

Comment. The appearance of a fireball.

3. (Olkhovatov 2008) 9.10.1677 Region. Hititi, Iwaki and Owari, Japan.

That night, a huge tsunami hit the coast. Three fireballs were noted, which jumped out of the sea and flew away to the northwest.

Comment. The appearance of fireballs from the sea.

4. (Olkhovatov 2008) 1698, M. Tosa, Japan.

In connection with the earthquake, fiery objects resembling wheels were observed, which flew in different directions.

Comment. The appearance of fireballs in the form of wheels.

5. (Olkhovatov 2008) 4.11.1704, Zurich, Switzerland.

The earthquake was preceded by the passage of a bright meteor.
Comment. The meteor flight is most likely a flight of the BL.

6. (Olkhovatov 2008) 1707. The center and south of Japan.

During the tsunami generated by one of the strongest earthquakes in the history of Japan, mysterious round glowing objects were visible over the approaching waves.
Comment. Round glowing objects were visible over the waves.

7. (Olkhovatov 2008) 15.04.1752, Stavanger, Norway.

The octagonal glow in the sky was emitting fireballs from its corners.
Comment. Octagonal glow.

8. (Olkhovatov 2008) 13.11.1761, Geneva, Switzerland.

During the earthquake, a huge ball appeared, which then turned into a glowing plume and disappeared with an explosion.
Comment. The ball turned into a glowing plume.

9. (Olkhovatov 2008) 18.11.1795, England.

When there was a push in the Derby, a fireball flew by. There were flashes in the sky.
Comment. A fireball flew by. There were flashes in the sky.

10. (Olkhovatov 2008) 13.08.1816, Dunkeld, Scotland.

During the push, a "meteor" flew from east to west, towards the direction of the push.
Comment. Most likely, the BL flew horizontally.

11. (Olkhovatov 2008) 02.20.1818 England.

In the town of Kerton, a "meteor" the size of a "cannonball" with a tail flew at high speed through the sky during an earthquake.
Comment. A ball with a tail.

12. (Olkhovatov 2008) 20.11.1822, Chile.

Meteors were observed flying in the same direction as the tremors.
Comment. Most likely, the bumblebees were flying horizontally.

13. (Olkhovatov 2008) 20.08.1823, Ragusa (now Dubrovnik in Croatia).

The sky suddenly darkened, and then returned to its normal state. A fiery "meteor" appeared over the city and fell into the sea, followed by an earthquake that destroyed many houses. The sea moved away from the shore for 1 mile, then returned.
Comment. Most likely, the BL flew horizontally.

14. (Olkhovatov 2008) 4 and 5.11. 1854, the center and south of Japan.

A powerful earthquake was followed by a tsunami. The sky from the side where the tsunami came from was glowing. It was also reported about a fireball hanging motionless over the sea for 2 nights, about a moving fire column.

Comment. A long-lived fireball.

15. (Olkhovatov 2008) 11.11.1855, M. Ansei, Japan.

During the earthquake, the glow resembled lightning, which soon disappeared. Before noon, something like a vertical rainbow was visible to the southeast. The fisherman 3 or 4 times observed flashes from the sea, similar to lightning in the direction of Tokyo. A group of 19 people who were at sea before the earthquake suddenly saw a glow in the northeast, which was so bright that colored patterns could easily be distinguished on their clothes. Soon, a terrible roar sounded from the depths of the sea, as if gravel jets were hitting the bottom of the boat. At the same time, a fiery mass, accompanied by sounds, flew across the sky. Soon after, they felt an earthquake.

The observation of some dark object flying across the sky was reported by several other eyewitnesses.

Comment. A fiery mass and a dark object.

16. (Olkhovatov 2008) October 8.10.1857, USA. An earthquake was felt in Illinois.

Over St. Louis at 4.28 am, something exploded with a terrible roar and shook the earth. Eyewitnesses reported that a "large dazzling meteor" swept across the sky. In addition, several small tremors occurred prior to this major tremor. According to the catalogs, the earthquake had a magnitude of 4.9.

Comment. A "large dazzling meteor" swept horizontally across the sky.

17. (Olkhovatov 2008) 10.12.1868, Kvishet, Caucasus.

At 1.00. 03 L.T., an underground shock was heard in Kvishet, and at 5.00. 32 L.T. in the morning there was a strong earthquake with a rumble. It went from southwest to northeast. At midnight from December 8 to December 9, a fireball "the size of the Moon" flew through the city in the same direction and illuminated the city for 1–2 s "as in the daytime." After its passage, a fiery streak was visible in the sky for some time. Following this phenomenon, an underground rumble was heard.

Comment. A fireball flew horizontally across the sky.

18. (Olkhovatov 2008) 10.12.1868. Kussary tract, Caucasus.

At 5:48 a.m. an earthquake occurred, on December 8 at 23 p.m., a luminous object was observed—a bright-white fireball with a long and wide tail at the end. It made a rather long arc from south to north on the western part of the sky. A few seconds later, from the side where it disappeared over the horizon, a crackling was heard and a glow was visible.

Comment. A white ball with a long and wide tail at the end was observed.

19. (Olkhovatov 2008) 02.19.1873 (old. style), Kola (68 deg. 53 min n.l. and 33 deg. 1 min e.l.), Russia.

At 4 o'clock in the morning it was light. The weather was calm. Suddenly it became dark, and then a huge ball of dark purple color appeared on the eastern side of the sky and disappeared in the west. At that moment, an underground shock rang out and an earthquake began, lasting 5 min. It was so strong that the houses staggered and all the utensils fell.

Comment. A huge ball of dark purple color was observed moving horizontally.

20. (Olkhovatov 2008) 12.12.1880, near Seattle, USA.

Immediately before the earthquake, two very large fireballs were visible in the air, located at a short distance from each other. After a moment, they rushed down and towards each other and, colliding, disappeared. The next moment there was a push. In addition, there are reports of a meteor falling into a mountain range.

Comment. Collision of two fireballs.

21. (Olkhovatov 2008) 4.01.1886, Chembar, Penza Province.

In the first hour of the night, when a strong wind suddenly rose, a "meteor" flew "very low" over the city, which with a crash resembling a strong thunderclap exploded outside the city on the road. This explosion killed a horse, and the surviving driver was so scared that the only thing he could say was that a fiery serpent flew by and killed the horse. Ten minutes later, there was a sound like an explosion. After a short time—an even stronger sound, accompanied by a vibration of the soil. The roof of one of the houses collapsed, and pipes collapsed at another house, fluctuations of lamps were noticeable in the rooms. On the lake, the ice was broken and the ice floes piled one on top of the other. Despite the small area of the city, in some places, the earthquake was not noticeable at all, but in others it was difficult to stand on your feet. This earthquake was also felt in a place 20 km away from the town of Chembar on Spitsiny farms.

Comment. BL killed a horse.

22. (Olkhovatov 2008) 31.08.1886, Charleston, South Carolina, USA.

Before sunset, the sun was covered with a black mass of clouds. "Meteors" and glowing clouds appeared. The lighting effects began to intensify. Then there was an earthquake. In turn, a few minutes after the aftershock.

On September 4, a bright "meteor" swept over Charleston, leaving a trail of fire behind it. After that, "meteors" often continued to appear over Charleston. So during a strong aftershock on October 22, about 50 "meteors" were observed in the sky. Reports of "meteors" were also received in the evening of October 24, 28, and 29.

Comment. Plural birth of BLs.

23. (Olkhovatov 2008) 31.08.1886, Columbia, South Carolina. Two bright meteors were observed over the city.

Comment. Observation of two BL.

24. (Olkhovatov 2008) 1886. Atlanta pcs. Georgia, midnight from October 23 to 24. The meteor exploded, illuminating the ground above the city.

Comment. The explosion of the BL.

25. (Olkhovatov 2008) 18.08.1892, England. An earthquake in Pembroke with a magnitude of about 5 points. During the earthquake, reports of a "meteor" were received from several places.

Comment. Plural birth of BL.

26. (Olkhovatov 2008) 17.12.1896, Worcester, England.

During the earthquake (at 5.30 am), a bright glow was visible in the sky, which, like a "meteor," crossed a significant part of the area of concussions. It was so bright that you could find a pin on the ground. Weak concussions were noted several hours before and after the main shock.

Comment. The bright glow of the BL as a "meteor" crossed a significant part of the area of concussions.

27. (Olkhovatov 2008) 8.09.1905, Calabria, Italy.

At first, "meteors" began to fall, and 3/4 h later a large "meteor" fell, which exploded and an earthquake occurred.

Comment. The explosion of the BL.

28. (Olkhovatov 2008) 04.18.1906, San Francisco, USA.

Flames and a flickering glow were observed. On one of the streets, a dark funnel-like formation was seen hanging in the air, which was illuminated by flashing lights.

Comment. Observation of a dark funnel-like formation.

29. (Olkhovatov 2008) 11.16.1911, Germany.

Powerful light phenomena were observed in connection with the earthquake. In a largely cloudless sky, such phenomena were observed as a streak of light; fiery rays; a moving fiery wave; sparkling; a cloud of light; a gray-yellow light cloud of an elongated oval shape with vague outlines; like the morning dawn or the aurora borealis; BL, sometimes seemed to come out of the ground; the lights of St. Elm. At about 10 p.m., a bright glow was observed over Lake Bodensee, moving at high speed from northwest to southeast at an altitude of 0.5–1 m.

Comment. BL coming out of the ground.

30. (Olkhovatov 2008) 1.09.1923, Tokyo, Japan.

Fishermen saw several columns of fire in the sea near the entrance to Tokyo Bay. An employee of the Central Meteorological Observatory saw a stationary fireball in the sky over Tokyo. Nikagawa in the morning before the earthquake saw flashes of light on the sea.

Comment. Flashes of glow on the sea.

31. (Olkhovatov 2008) 05.23.1925, Tazima, Japan.

During one of the strongest night aftershocks, a glowing ball was observed, which quickly flew across the sky like a "meteor."

Comment. During one of the aftershocks, a glowing ball was observed.

32. (Olkhovatov 2008) 14.01.1929, Inner Mongolia, China.

A red fireball resembling a ball of lightning rose up and instantly disappeared. Then a strong rumbling was heard, which did not stop until the earthquake itself.

Comment. Observation of a fireball resembling a BL.

33. (Olkhovatov 2008) 11.26.1930, Izu District, Japan.

A large number of reports have been received about various light phenomena in connection with this earthquake. The first observations of light phenomena took place about 12 h before the earthquake: a fireball flying at high speed across the sky. There were reports of glowing beams, columns. Fireballs and funnel-shaped glow (including "curved") were observed, as a rule, in the vicinity of the epicenter. The duration of some forms of glow reached several minutes. The description was compiled by a fisherman who happened to see the process of the appearance of glow in its development. The glow first appeared over the mountain (the caldera of the old volcano) Hakone and began to spread to the south, towards Mount Amagi. Then it turned back to Mount Hakone. This glow had a white color and was visible only near the surface of the earth. The fisherman did not have time to be amazed at the strange phenomenon, as he felt an earthquake. Another fisherman saw a large fireball that appeared near Mount Wasidu and flew to the south. And another one reported a fireball that jumped out of the water. In the caldera of the volcano (mount) Hakone, very bright fireballs were observed, which were "lined up" in a row. A large number of lights accompanied landslides. The estimate of the power of some types of glow produced by the Japanese scientist Torahika Terad has values up to 3 MW. It was reported that after the disappearance of the fireball, the sky began to glow a little pink. During the same earthquake, streaks of light and glowing balls were visible in the sky.

Comment. Observations of funnel-shaped glow and the appearance of a fireball that jumped out of the water.

34. (Olkhovatov 2008) On 12.20.1930, a local earthquake in Hiroshima Prefecture, Japan.

The glow was observed by several people. It was reported about the red-brown glow of the sky, a fireball that moved upwards, then disappeared, after which a pinkish glow appeared from all sides. A message was also received about a vertical beam of light illuminating the clouds.

Comment. Observations of a fireball and other types of glow.

35. (Olkhovatov 2008) 06.17.1931, Tokyo neighborhood, Japan.

There were many light sources and different types. Among them were: "a lamp that fell from heaven," a funnel-shaped glow and a fireball rising up and disappearing with the sound of an explosion, after which the lower part of the sky was lit up pink-red for a while.

Comment. Observation of a funnel-shaped glow.

36. (Olkhovatov 2008) 3.03.1933, Sanriku District, Japan.

A large number of reports of glow both before the earthquake and in connection with the tsunami that followed 1 h later. Two days before the earthquake, they saw "light" and "light rays." At the time of the earthquake, a glow resembling lightning was seen near the epicenter. We also saw a whitish-green "arch." Remarkably bright lights were observed in the vicinity of the coastal town of Hachinohe. It is located on relatively young rocks. Some types of glow were visible directly on the ground and water surface. During the shock, large numbers of different glowing bodies appeared on the surface of the sea for a couple of seconds and disappeared. Their number grew with the intensity of the push. Bluish-white flames have also been reported emanating from the water. A large number of lights an hour later accompanied the tsunami. A strange fireball appeared above the water at a height of 7–10 m; the surface of the sea emitted a bright light resembling lightning or a searchlight beam. Some of these phenomena were accompanied by the sound of an explosion. There is also a report of a bluish-white glowing ball that jumped out of the sea in connection with the tsunami.

Comment. Observation of a large number of different luminous bodies on the sea surface. Some of these phenomena were accompanied by the sound of an explosion. A message about a bluish-white glowing ball that appeared from the sea.

37. (Olkhovatov 2008) 1952, September–October, France.

The epicenter of the earthquake on September 29 was located about 50 km north of Strasbourg, in a place known for its thermal springs. Immediately before the earthquake in the city of Wissembourg (near the epicenter), a glowing ball was seen flying to the east. The ball was first bright white, then turned red, and finally black. It disappeared with a bang.

Comment. Observation of the color change of a glowing ball.

38. (Olkhovatov 2008) On 8.10.1952, a stronger earthquake occurred in the city of Wissembourg, felt in the city of Strasbourg.

Just before the earthquake, a glowing ball was seen flying. It flew very quickly from west to east, producing a sound like a cannon shot.

Comment. Observation of a glowing ball before an earthquake.

39. (Olkhovatov 2008) 1965–1968, M. Matsushiro, Japan.

During this swarm of small and moderate earthquakes, many cases of luminescence were observed. The duration reached tens of seconds. The following features of the glow are noted: (a) a luminous hemisphere with a diameter of 20–200 m in contact with the earth. The sphere is white, but reflections from clouds can be colored. (b) The glow usually follows an earthquake and lasts from 10 s to 2 min. (c) The glow occurs exclusively in certain areas that do not coincide with the epicenter. More often they occur on the tops of mountains in quartz-diorite fractured rock. (d) Usually the glow is followed by spheres, the most intense in the range of 10–20 kHz. (e) Meteorological conditions affect the probability of the appearance of the glow. (f) The magnetometers of the local observatory do not react. There is evidence that the glow was associated with the flow of magmatic water.

Comment. Observation of luminous spheres.

40. (Olkhovatov 2008) 1.10.1969, Santa Rosa, California, USA.

In connection with the earthquake, various lights were observed, which resembled: lightning, electric sparks, St. Elmo's lights, fireballs, "meteors." There were sounds of explosions.

Comment. Observation of the glowing lights of St. Elmo, fireballs and "meteors."

41. (Olkhovatov 2008) 05.05.1970, prov. Yunnan, China.

Before the earthquake, lightning of a bluish hue began to sparkle, red and green bands of glow appeared, fireballs, and other light phenomena of various kinds (in total, more than 30 types of glow).

Comment. Observation of 30 types of luminous objects.

42. (Olkhovatov 2008) 11.14.1970, village Vardan near Sochi, USSR.

The earthquake had a magnitude of about 5.2 with a hypocenter depth of about 9 km. It happened at 5 a.m. local time, almost in complete darkness, in the village. A bright flash, similar to a very large flash during electric welding, was observed in the village immediately before the shock. It was repeated several times with less force, and after each flash a push was felt. After the tremors subsided, a large glow remained for some time in the northwest of the village. A glow was also observed, similar in intensity to sunlight and spreading from south to north. In addition, there was a phosphorescent glow of the sea and many glowing dots on the beach. The phenomenon was accompanied by the "rustle of pebbles." Local residents who jumped out into the street at the moment of the push noticed glowing formations on dry ground, as well as on an asphalt highway, about 10 by 2 cm in size, consisting as if of a dust-like dry luminous mass, which deformed when rubbed with its sole, and then took its former shape. According to other reports, after the shock, the earth's surface was covered by a thin layer of ash or drizzle, which soon disappeared. The air smelled of sulfur. The intensity of the earthquake in the village has about 7 points (minor damage).

Comment. The appearance of luminous formations on the asphalt highway, consisting as it were of a dust-like dry luminous mass, which deformed when rubbed with the sole, and then assumed its former shape.

43. (Olkhovatov 2008) 04.28.1971, prov. Yunnan, China.

Before the earthquake, they saw a fiery light and a red fireball rising up, shaped like a pumpkin.

Comment. The appearance of a red fireball shaped like a pumpkin.

44. (Olkhovatov 2008) 02/14/1975, prov. Liaoning, China.

There were "fire pillars" rising into the sky, "electric welding fire," "flares," fireballs, glowing spots, etc. Noteworthy is the appearance before the earthquake of a huge flame rapidly rising to the sky.

Especially characteristic was the appearance of fireballs, which often resembled flares (they were occasionally observed 2 years before the earthquake). Therefore, over the floodplain of the Liaohe River and its fairway in the mountainous terrain, fireballs appeared, resembling flares coming out of wells gushing with sand and bubbling with water. Regularly such glowing balls began to appear from February 3, and the maximum number of them was registered in the time of the earthquake. In one place, about 10 min before the earthquake, a huge red fireball began to rise rapidly from the ground. It was surrounded by a sort of luminous cloud. Nearby saw fiery red balls that rolled over the surface of the earth. Some of them got off the ground at the same time. The balls were the size of a basketball and when they met any obstacle, they exploded. After that, the sky turned red and an earthquake started after 4–5 s.

During this earthquake, cases of the impact of such luminous formations on people were noted. So, when a 12-y.o. girl stumbled and fell, she was illuminated by a white "fallen lightning" at that time. On the second day, both cheeks of the girl began to swell and acquired a purple-burgundy color, blisters appeared. Her mother, who ran after the girl, also had purple-maroon spots on her cheeks after 2–3 days. A similar story happened to some other eyewitnesses—there were burns of varying degrees on the face, the whole face acquired a dark purple color, and a few days later cracks appeared, and the skin peeled off. Gradually, however, everything passed.

Comment. The effect of luminous formations on people.

45. (Olkhovatov 2008) 6.04.1976, Inner Mongolia, China.

For 1–2 min before the earthquake, light phenomena similar to the flashing of lightning or the light emitted during electric welding were visible. These phenomena were in the form of ribbons, fireballs, in the form of an arc, stripes of white, red, earthy gray, and dirty yellow.

Comment. Observation of luminous formations in the form of ribbons.

46. (Olkhovatov 2008) 05/29/1976, prov. Yunnan, China.

More than 60 different types of light phenomena have been reported. The observations were made during earthquake aftershocks by seismologist Lee Yun Shan. After the main shock, he lived in a tent in the epicenter zone for a couple of nights. One night, when it was raining and low clouds hung over, he left the tent at midnight. To his surprise, he saw a fireball about 50 m in diameter two hundred meters away from him. Unable to believe his eyes, he woke up a colleague, and both watched this ball for almost half an hour, rechecking their observations from each other. The following morning, an aftershock of magnitude five occurred. In the place where the fireball was observed, a large amount of remnants of "boiling sand" was found in the middle of the rice field.

Comment. Observation of a fireball about 50 m in diameter. A large amount of remnants of "boiling sand" was found in the middle of the rice field.

47. (Olkhovatov 2008) 06/28/1976, prov. Hebei, China.

Light phenomena accompanying the Tangshan earthquake were observed at a distance of up to 200 km from the epicenter. Half an hour before the earthquake, a flying fireball was observed, which, having flared up, in an instant turned into a strip emitting flames.

Comment. Turning a fireball into a streak.

48. (Olkhovatov 2008) 08.16.1976, prov. Sichuan, China.

Light phenomena on a large scale began to appear about a month before the earthquake. A characteristic feature of this earthquake was the appearance before it and simultaneously with it of a large number of light phenomena in the form of stripes and pillars. Especially Chinese seismologists distinguish a large number of fireballs both before and during the earthquake.

Comment. The appearance of a large number of fireballs both before and during the earthquake.

49. (Olkhovatov 2008) 22.07.1976, prov. Sichuan, China.

On July 22, at 21:30, employees of the department dealing with seismological problems observed a fireball that, having broken away from the surface of the earth about 100 m away from them, soared into the air. Its original diameter was about 1 m. Having come off the surface of the earth, it took the form of a large wash basin, and a plume stretched behind it, like a comet. Having risen to a height of about 10–15 m, the fireball began to descend down an arc-shaped trajectory, similar to a parabolic one, and its size was reduced to a chicken egg. The glow faded, then became stronger again. Small wisps of white smoke swirled around the glow and a soft crackling sound was heard. There was also a smell resembling the smell of garlic or sulfur. The colors of the ball were blue and white, but red dominated. After the ball rose into the sky, it became light around, and all this lasted about 4 s.

No effect of the ball on radio receivers, radio reception quality, local electric lighting, geomagnetic measurement system, radio compass, and telluric currents was observed. The next day, when investigating the area where the fireball was observed, no traces of radioactivity or other anomalies were found, but a small funnel-shaped hole was found in the ground.

Comment. Interesting dynamics of changing the shape of a glowing ball.

50. (Olkhovatov 2008) 07.23.1976, prov. Sichuan, China.

A number of cases have been recorded when fireballs scorched vegetation, brought people into a state of shock. Fireballs of both "round" and "flattened" shapes were observed; some of these balls charred growing peas. In another case, a fireball appeared near a house, rose up along a tree and burned a hole in the roof of the house. In total, about a thousand fireballs were observed before the earthquake, including 50 during one evening alone. During the day, small balls consisting of "smoke" were observed. According to Chinese seismologists, it is possible that this is the same phenomenon as observed at night. The most frequent fireballs were observed along the intersections of riverbeds and faults.

Comment. Before the earthquake, about a thousand fireballs were observed, fireballs scorched vegetation.

51. (Olkhovatov 2008) 01.20.1988 Australia.

At 4 o'clock in the morning on the Perth-Adelaide Highway passing along the coast, south of the Nallarbor Plain. The road was quiet; there were only a couple of trucks and a car in which the Knowles family (mother and three adult sons) were traveling from Perth to Adelaide on a stretch of several kilometers long. Suddenly, in front of them, they saw something more than a glowing funnel, which, as they approached, began to resemble an egg lying in a cone-shaped egg glass. This formation swayed from side to side, glowed and disappeared from time to time! In the end, they felt, although they did not see anything, that this thing was just on the roof of their car. It seemed to them that the car was literally "sucked" off the road, and then thrown down again. There were numerous strange effects, such as a strange smell resembling Bakelite and a black powder that penetrated the car. Two days later, a strong intraplate earthquake with a magnitude of tremors up to 6.7 units occurred 1400 km north of the scene.

Comment. Interesting dynamics of changing the shape of the luminous formation.

52. (Olkhovatov 2008) 1988 and 1989, Quebec, Canada.

From November 23, 1988 to the end of January 1989, a swarm of earthquakes occurred in the Canadian province of Quebec. The strongest occurred on November 25 at 23. 46 L.T. at a depth of 28 km and had a magnitude of 6.5. It was preceded by a foreshock on November 23 at 9.12 L.T. at the same depth with a magnitude of 4.8. 4 h after the main shock, an aftershock of magnitude 4.1 occurred and during the next 4.5 h—several more aftershocks with a magnitude of 2.5–2.9. Further aftershocks were occasionally observed until the end of January 1989. A total of 54 seismic events were registered. Unusual luminescence has been observed since 1.10. 1988. by 21.01. 1989. There were 3 types of glow: silent sparks, diffuse glow, reminiscent of dawn when it covered a significant part of the sky, vertical and horizontal luminous stripes, similar to the Northern lights.

Fireballs with a diameter of several meters were also observed, which periodically popped out of the ground at distances up to several meters from eyewitnesses. Others moved or remained stationary at altitudes up to several hundred meters. Some eyewitnesses reported "dripping luminous droplets" that quickly disappeared a few meters below under the stationary fireballs. There were only 2 reports of the observation of flames on the ground without any signs of cracks—one in the snow, the other on the pavement (parking place). The most common colors were orange, yellow, white, and green. The duration of the glow reached 12 min. Most of the glow was observed in the vicinity of the time of the two main shocks, the rest of the observed glow correlated with the dates of aftershocks.

Comment. Fireballs with a diameter of several meters were observed, and "dripping" of luminous droplets.

53. (Olkhovatov 2008) 2005. Atmospheric Research.

Two eyewitnesses on the Rock-ko Mountain (northeastern part of Kobe) saw a moving luminous object a few seconds before the main seismic shock. Regardless of them, fishermen in Osaka Bay (opposite side of Mount Rocco) also noticed a moving glowing object. Scientists estimated the diameter of the object at about 100 m. The same or, perhaps, slightly higher was the altitude of the flight. According to the fishermen, at the end of the flight, the object fell on Mount Rocco, and before the fall itself, lightning struck the mountain from it. In addition, there were several other eyewitnesses who probably saw the same object. The direction of flight may have coincided with the direction of the fault.

Comment. A luminous object with a diameter of 100 m was observed.

54. (Olkhovatov 2008) 17.08.1999, Turkey.

Before a strong (magnitude 7.4) earthquake, a number of phenomena were registered. So, 3 weeks before the main shock, glowing formations were observed several times in the vicinity of the future epicenter. In one of the cases, the duration of the glow was 20 min, the height was about 540 m, and the volume of the luminous area was about 35 cubic meters. The stones that were on the ground under the glowing object were burned, and the surface of some flint-containing stones was melted. A luminous emitting object with a diameter of 3.3 m was observed.

Comment. A luminous emitting object with a diameter of 3.3 m was observed.

55. (Olkhovatov 2008) F. Montgomery, Meteorologist. Inside the tornado "Black Well," Oklahoma, 05/25. 1955.

Observation of objects 120 m wide, located 250 m above the ground. They were as dazzling as a blowtorch. They were spinning dizzily.

Comment. Observation of objects 120 m wide.

56. (Olkhovatov 2008) R. Hall, Meteorologist got inside the Texas tornado of 1948.

He saw a column of light that was separated from the transparent dark walls. This column was formed by rings, which, when lowered, turned into a BL.

Comment. The column was formed by rings, which, when lowered, turned into a BL.

2.10.2 BL in Japan

The Ball Lightning Information Center (Japan) collected 2,060 observations from all Japanese islands by 1988 (Ohtsuki and Ofuruton 1989). From a detailed analysis of the data for the year, it was concluded that Japanese BL differ from continental ones, which were reported in Europe (1923) and the United States (McNalley).

Almost all BLs appear in clear or cloudy weather, but very few in thunderstorm conditions. However, in the presence of thunderstorms, the appearance of Japanese BL is strongly associated with thunderstorms in space and time. The distributions of thunderstorms and BL by area and months closely coincide. There are not many differences between Japanese and continental BLs in size, color, shape lifetime, and sound. It is not yet known why BLs in Japan are created in clear or cloudy weather, but very rarely in thunderstorms. Rather, their appearance is associated with seismic conditions. The following are typical descriptions of BL observations in Japan from Ohtsuki and Ofuruton 1989) without comments (note translation from Russian).

1. Mr. S. S. is 32 years old. Tokyo. 1980, June, 20:30.

It was a clear summer evening. I was waiting for the train at Kichijoji Station (Tokyo). I saw an orange-colored lightning ball rising at some distance. It was lighter than the helicopter lights. After 5 or 6 s, it changed course and accelerated. It flew across the sky for 1 or 2 s and disappeared without a sound.

2. Mr. S. S. Tokyo. 21.07.1987. The weather was cloudy.

When I was driving home and turned the corner, I saw a lightning bolt floating at a height of 1 m above the road. I approached it, but it left me. I ran after it for 1 min, but could not catch up. It collapsed and disappeared away from the road, as if a balloon had exploded. The shape of the lightning was a sphere with a small tail. The color of its center was orange and red, and the tail was blue. The diameter was 25–30 cm.

3. Mr. Yu. Sano. Kanagawa. July of the late 1970s.

It was a good day and the temperature was 27–28 °C. Around 7.00 pm.

I was walking down the slope of Den'engfu (Tokyo) with my friend, and we saw a red ball moving slowly and horizontally among the trees. Its diameter was 20–30 cm. The whole phenomenon lasted about 10 s, then the ball disappeared without a sound.

4. Mr. Yamamuro, 74 years old. Tokyo. June 1927. 19:30. I was 13 years old.

It was a clear evening. I was standing on the road near the railway bridge of Ebisu Station (Tokyo). I saw a lightning ball with a long tail floating and moving towards the railway bridge.

The color of its center was orange, its surroundings were light yellow, and its tail was blue. Diameter 30 cm, tail length 3–4 m.

I ran after it until it disappeared. It was moving over the railway bridge and disappeared into the trees near the railway bridge.

5. Mr. A. Hayashi, 74 years old. Tokyo. 30.01.1987.

It was a clear winter night. When I was walking to Todoroki (Tokyo) on my way home, I saw the BL moving slowly and horizontally. Next to the BL there were electrical wires, the height of which was about 5 m. The BL moved away from the wires and along them.

The color of its center was red, and its surroundings were yellow. Its shape was a sphere with several small tails. The whole phenomenon lasted about 2 s. I didn't hear a sound.

6. Dr. Ofuruton H., Physicist, 29 years old, Tokyo. Summer of the late 1960s in Tokyo. The weather was good.

I saw a little red BL floating and coming out of the window. I could see it for about 2 s and it disappeared from my field of vision. Its diameter was about 3 cm. I did not hear a sound.

7. Mr. K. Wada is 43 years old. Maebashi. 30.08.19874.20–5.00 am. The weather was good.

My wife and I went to Mount Kirigamine (Nagano Prefect) to photograph the sunrise. We walked along the path from the hut for about 20 min.

When we turned around to see the hut, suddenly an orange color of light began to shine in the sky to the west. The light became brighter and brighter and reached the greatest brightness. After 10 s, the light became fainter, turned white, and disappeared.

A minute later, it appeared in white. It became brighter and brighter, turned orange, and reached its greatest brightness. It lasted 10–20 s. The light grew fainter and fainter and disappeared.

My wife told me there was another light on the right side and I saw a light there. She told me that this phenomenon appeared many times, but I saw it once. Moreover, she told me that it also changed color, but I did not see it.

8. Mr. M. Shibata and Ms. M. Shibata, Nagoya. 1980. I do not remember the season. The weather was normal, not rainy, and not windy. 8 p.m.

We saw an orange BL floating and moving right over the roof of a neighbor's house.

Its shape was a sphere, bigger than a ping-pong ball, and smaller than a baseball. It moved horizontally for 3 s. We did not hear a sound.

9. Mrs. H. Sato (Saitama). Summer around 1982. I do not remember the weather. Our family was walking along the banks of the Arakawa River in Tokyo. About 8 pm.

A fireball appeared in the sky. The color of its center was green or blue, and its surroundings were white. It moved horizontally, then descended, and disappeared behind the building. I did not hear a sound.

10. Ms. M. Takada, Nagoya. December 4, 1983. The weather was snowy in the morning and good in the afternoon.

My husband was driving the car, and I was sitting next to him. The car was traveling on an expressway in a mountainous area between Kyoto and Nagoya.

The red BL crossed the front window from right to left. Its diameter was 30 cm. The whole phenomenon lasted 2–3 s.

11. Ms. M. T., Kanagawa. 7.07.1977. At about 13:30, the weather was stormy.

I was sleeping in my bedroom. Suddenly I heard a loud noise, and lightning flashed and fell to me from the ceiling, I felt a strong blow on my arm. Noise, flash, shock, they happened simultaneously. At that time, the television screen of our house turned white, and I smelled something burning.

The next morning I found out that the TV antenna was broken, and the wire securing the antenna was cut off. In addition, the telephone wire was broken, the fuses burned out and the plug of the TV remote was broken. The TV was beyond repair. The dog outside the house was dead.

2.10.3 Artifacts and Various Traces of BL

1. (Olkhovatov 2008) “Chicago Tribune” (2 issues) April 1879.

Over one of the Chicago neighborhoods during rain and lightning. There was an explosion in the air of a large fireball. This ball was shrouded as if in smoke. The roar of the explosion was comparable to a shot from a cannon and even almost knocked down one passerby. A hail of shrapnel fell on the streets, but no one was injured. Lightning struck 2 nearby rooms, where it caused some destruction, and at the same time into a telephone wire. A fire started.

The picked up fragments of the object resembled slag were dark on the outside and almost white on the inside. They were collected in a row of 70 cubic decimeters, and some of the fragments were still very hot. The largest were in two fists. The most powerful explosion.

The object occurred over the lake.

Shortly after the events described, an article appeared in the American Journal of Science claiming that a chemical study of the fallen substance showed that it did not correspond to a meteorite.

Comment. Description of BL residues.

2. (Olkhovatov 2008) Around 1855. Several eyewitnesses.

They reported that they saw small glass-like formations falling from the sky to the ground. This was accompanied by sound and light phenomena.

In the mid-1960s, the journals "Science" and "Nature" published articles on the analysis of preserved samples: they turned out to be of terrestrial origin.

Comment. Description of the remnants of atmospheric objects.

3. (Olkhovatov 2008). 13.06.1909, Ongar, England. «Nature».

During a severe thunderstorm, the meteor stone with a loud explosion fell into the stable paddock, plunging 8 inches into the gravel. A space of several feet near this place was "perforated" with small fragments. The main mass weighed about half a kilogram. There is no such meteorite in the catalogues of falls.

Comment. Description of the remnants of the "meteor stone."

4. (Olkhovatov 2008) 03/14/1912, St. Albans, England. «Nature».

During a severe thunderstorm, the aerolite fell. It happened within a few meters from the eyewitness, who handed over the "aerolite" to specialists. The stone weighed more than 2 kg. Subsequent analysis revealed that the stone is not a meteorite.

Comment. Description of the remnants of "aerolite."

5. (Olkhovatov 2008) On 03.21.1950, Mongolian border guards observed.

The object broke up into many pieces, which were collected in a total of almost 28 kg. However, they turned out to be not "cosmic matter," but slag with inclusions of fused quartz.

Comment. Description of the remains of the crumbled object.

6. (Olkhovatov 2008) 23.01.1974 North Wales. A letter from Ron Maddison with a message from R. Edwards.

The night was the first when a completely clear sky was visible for a considerable time. At about 8.15 pm, he went to the south side of the house to look at the sky. Suddenly, behind him and above his head, he heard a whistling sound, reminiscent of the sound created by a large flock of crows flying by.

However, the difference was that some reverberation was felt in the air, as from a train passing by the station, if the sound of its engine was not heard.

After R. Edwards raised his eyes up, he saw that a glowing ball was flying over his head. The size of it looked like a football from a distance of about 6 m. For several seconds, he managed to examine the ball through binoculars. It looked, indeed, like a round ball. Its color was pale yellow, but in the front part the ball was almost "incandescent white."

However, after the object changed its course, R. Edwards saw "embers" separating from the main body. They seemed to roll down the stream, like boulders rolling off a cliff. The "embers" trailed behind the ball, forming the thinnest "cord," which eventually broke. Then the ball and its particles completely disappeared. R. Edwards once again stressed that at a height of perhaps 170 m, the glowing ball in its reduced size and its trace cord disappeared simultaneously.

Unexpectedly, high in the western side of the sky, R. Edwards saw what looked like a "shooting star" flying in a gentle arc from west to east. It left behind a yellowish glowing trail. It was high in the sky, perhaps at an altitude of about 6 km.

When he put on his glasses, he saw that the object seemed to be throwing yellow pieces back down the stream, which stretched for a greater distance than in the case of the first object.

Then this second object disappeared completely as suddenly as the first one. In both cases, the phenomena were silent.

Comment. A glowing ball and its trace-a cord.

7. (Olkhovatov 2008) "Popular Mechanics" May 2001. 12.17.1977 the following case occurred in the USA.

A couple of eyewitnesses (there were more eyewitnesses in total) drove into the town of Council Bluff on December. According to them, about half a mile ahead they saw a glowing red ball falling over the tops of the trees. When eyewitnesses drove up to the crash site, they found that the frozen ground was covered with a thick layer of red-orange mass of molten metal.

The glow of the molten metal was also observed 15 min later, when one of the chiefs of the local fire department arrived at the scene. After the metal cooled down, its samples were taken. The total metal turned out to be about 20 kg. The samples were studied in various laboratories. They showed the similarity of carbon steel with slag. Because of further research, both the meteoritic origin of the metal and the fact that it could be a fragment of space debris were excluded.

Comment. BL is a red-orange mass of molten metal and its composition.

8. [1] "New York Times" March 1897. West Virginia, USA.

"A meteor exploded over New Martinsville." The sound of the explosion was similar to the roar from a volley of heavy artillery and was heard at a distance of 30 km. The fireball was "cylindrical in shape," and the smoke gave it the appearance of a burning ball. The hissing sound could be heard for miles.

When the "meteor" exploded, its pieces flew in all directions. The walls were studded with its fragments. One horse's head was blown apart and almost torn off by a "meteor fragment," and the horse in the neighboring stable was completely deaf.

One eyewitness was knocked down by the "force of air" produced by the rapid movement of the body before its explosion. The explosion knocked him unconscious. The appearance of the "meteor" was accompanied by a rumbling noise, followed a moment later by a hissing and the appearance of a fireball that appeared, hissing and smoking. Then there was an explosion. It was not possible to find a corresponding meteorite in meteorite catalogs.

Comment. The BL is cylindrical in shape.

9. (Gromyko 2004a, b) and from the correspondence with Prof. Churilov G. N. The fact of observation of BL occurred on 07.19.2003.

The woman was in the kitchen preparing lunch sitting at the table with her back to the window. Her sister, who entered the kitchen, saw a fireball moving towards the center of the kitchen from the corner between the window and the vent, screamed. Both, amazed by what they saw, looked at the fireball, which seemed to them the size of a football. A few seconds later, there was a deafening explosion and red-hot metal balls scattered on the glass coffee table and the floor. The glowing balls quickly darkened, leaving burn marks on the linoleum floor. Recovering from fright, the women collected most of the cooled metal balls, which were handed over to the author.

A second description of the same event. A fireball floated into the room on the second floor, which seemed to eyewitnesses the size of a football. The place of BL penetration into the kitchen was not recorded. After a few seconds, a deafening explosion was heard and red-hot metal balls of about 15 pieces, average diameter ~ 1 mm, scattered on the floor. For a long time in the ears of eyewitnesses, there was a ringing from the explosion of BL, and the smell of ozone in the room.

Comment. Description of the study of the composition of BL.

Some of the metal balls were handed over to Prof. G. N. Churilov, who performed electrophysical studies of the presented material with a team of employees. The following results of experimental studies have been obtained:

1. BL balls are hollow spheres of pure iron, 0.3–1.2 mm in diameter, with a wall thickness of about 10 microns (see Fig. 2.63a, b).

A comparative analysis of the structure and chemical composition of this object was carried out.

The main difference is that the balls were hollow. This indicates that they were formed in a gas environment at high pressure. This type of spheroidal particles were previously obtained in plasma.

Most of the cooled metal balls were collected and analyzed at the Institute of Physics of the Siberian Branch of the Russian Academy of Sciences in Krasnoyarsk.

2. The type of magnetization curves shows the absence of hysteresis, i.e. indicates the amorphous structure of the material of the balls.

Fig. 2.63 Balls of BL (magnification 100), transferred for research to the IF SB RAS (**a**), and (**b**) (without magnification), remaining for repeated studies

The results of the performed analyses are as follows:

1. The collected BL balls are hollow spheres made of pure iron.
2. The results of comparing the magnetization of the BL and α·Fe balls show that the shape of the curves is identical; there is no hysteresis in the studied materials.
3. As a result of experimentally taken spectrograms for Fe balls by device of chemical investigation, clear resonant lines of graphite and iron were revealed, whereas similar spectral lines were not observed in the sample of Fe balls of BL, which also confirms the amorphous structure of the sample of Fe balls of BL.

The experimentally removed dependence of forced Raman scattering shows that the structure of the molecules of BL Fe balls have characteristic features that cause the nonlinearity of the sample, expressed in a large number of harmonics. For Fe balls, nonlinearity is not observed by device of chemical investigation with a wide range of influence on the sample.

2.10.4 Geophysical Phenomena and BL

This section provides data about objects that most resemble BL. Their appearance, shape, or other properties distinguish them from a set of BL with average parameters typical for most observations. We provide data on the holes that appeared when exposed to BL. We also describe the events associated with the appearance of crop circles and their connection with BL. Note in this section we use events description in translation from Russia.

Abnormal properties of BL

(Olkhovatov 2008) Height 611 m. Researchers: V. V. Dvuzhilny, an employee of the Meteorite Commission of the Far Eastern Branch of the USSR Academy of Sciences and DS V. N. Salnikov, a Geologist from the Tomsk Polytechnic Institute, et al. Dalnegorsk, Russia.

On 29.01.1986, residents of the small Far Eastern town of Dalnegorsk at 19:55 local time noticed a reddish ball flying from the southwest. The ball flew parallel to the earth's surface, silently. The speed calculated according to eyewitnesses turned out to be equal to 15 m/s. It did not have a tail or a train. Flying over the town, the ball fell, "dipping" at an angle of 60–70°, on the slope of Limestone Mountain (height 611 m). There was no sound of an explosion during the fall, and only one eyewitness heard a weak thud (Fig. 2.64).

When the ball fell, it demolished part of the cliff ledge with a volume of 2–3 cubic meters, scattering it into small fragments.

The fall ended with "burning" for an hour, and the brightness of the glow was comparable to a short circuit on the power line. At that time, six times within half an hour, the ball slowly rose and fell, gradually shifting to the north. One of the

Fig. 2.64 A https://vk.com/wall-193776923_12880 03.11.21. The same mountain, from the other side, the photo: N. Bulkin

eyewitnesses claimed that after that the ball rose and flew away to the northeast. The crash site was examined 2 days later.

There was no snow on an area of about 2 × 2 m, and heavily sooty, cracked flint fragments were lying around the place.

On them, particles of silvery metal, as if sprayed in the form of small balls, were noticeable. There was a burnt stump on the edge of the platform.

Three main types of residues were found at the site of the event: lead balls in the form of the above-mentioned plaque, iron balls (about 70 g of metal were found in total) and the so-called net.

The analysis of the remains was carried out in several academic centers and 11 research institutes. According to V. N. Salnikov and V. V. Dvuzhilny, the main research results are as follows: (1) The balls in the form of a metal plaque (60 g) were mainly lead alloys containing various additives, including rare earth elements. The dimensions are from 0.1 to 4 mm across. Along with the balls, "drops" and other "irregular" forms of a similar composition were also found. Almost all the balls had holes (recesses). Approximately, 10% of these formations contained fused fragments of rocks in the form of grains of sand. Isotopic analysis of lead corresponds to the lead of the Kholodnenskoye deposit in the northern Baikal region. (2) Iron balls (15 g.) had 2–4 mm in diameter and were magnetized. They were very solid and could be processed only with a diamond tool. They differed from each other in composition: iron with an admixture of aluminum, iron with an admixture of chromium, with an admixture of tungsten and cobalt. They also contained a significant amount of carbon and oxygen and were classified as white cast iron according to the micro-scale. According to X-ray diffraction analysis, they were classified as alpha iron. (3) So called the "net" is a glass-like black and brittle mass with many holes. Quartz threads of a thickness of 17 microns were found on many samples of the net. A gold thread was found inside one of the threads. The results of chemical analyses of the "net" were characterized by high instability in time. Salnikov classified "net" as a remnant of a degraded rock and a tree stump. There is practically no water in the rock samples from the fall site. The silicon content is slightly reduced when enriched with oxides of various metals. The magnitude of the residual magnetization is increased.

The event took place in an area rich in a wide variety of metal ores, and, in particular, the crash site is located on the territory of the ore field. The direction of the object's flight coincided with the direction of two deep faults.

According to Dvuzhilny, on February 8, at about 20.30 min local time (8 days after the ball fell), two more objects appeared over the town of Izvestnyakova. They did four circles and disappeared.

On 10.28.1987, 32 objects of various shapes appeared over various areas of the eastern coast of Primorye: cigar-shaped, cylindrical, and spherical. Four of them passed over the town of Lime, three hovered over Dalnegorsk, and five "illuminated" the adjacent mountains. They moved silently at altitudes of several hundred meters, skirting the terrain and following the valleys of rivers and ravines.

When flying over houses, they caused failures in the operation of televisions and even the central telegraph. As Dvuzhilny wrote, an analysis of these and other similar cases in Primorye revealed that the objects were observed mainly during the full moon and usually over faults and ore bodies.

As V. V. Dvuzhilny wrote in 1989, a certain "field" has been preserved at the crash site for 3 years. All living creatures avoided it, it also affects humans. The composition of the blood changes, the pulse increases, blood pressure increases, and coordination of movements worsens. Failures in the operation of radio-electronic and photographic equipment are noted. For example, there was no image on the film if the shooting was conducted at the crash site, but the shot near the mountain was of good quality.

Comment. Description of complex geophysical phenomena at the place of the ball falling.

1. (Olkhovatov 2008) Sasovo, Ryazan region. Russia. On the night of 11–12.0.4.1991, an explosion occurred near Sasovo, Ryazan region, which led to the formation of a crater and destruction in the city. After a thorough study of the phenomenon, almost all versions disappeared, leaving as the most likely assumption that it was a geophysical (tectonic) phenomenon.

In the case of the Sasovo explosion, the flight of one of the "bolides" 30 km from Sasovo was accompanied not only by a rumble, but also by a shaking of the earth. In the area of the epicenter of the Sasovsky explosion, before the explosion, the earth was shaking, and, moreover, judging by the testimony of eyewitnesses. At first the glow seemed to fly to the explosion site, then a bright flash occurred in this place (a pillar of fire appeared), and then the explosion itself occurred. There have been reports of a glowing streak in the sky, a sense of fear, and even a sense of the end of the world.

In Sasovo, the manifestation of the phenomenon was spatially heterogeneous—for example, the distribution of damage had a clearly expressed azimuthal (from the epicenter) dependence. Note that at some distance from the epicenter, the formation of holes in the ground occurred.

Approximately at the end of June–beginning of July 1992, 8 km from the site of the first Sasovsky explosion, in a sparsely populated area, a second, similar explosion occurred, of slightly lower power.

Comment. Description of complex geophysical phenomena.

2. (Olkhovatov 2008) 29.03.1990 near Petrozavodsk. Russia. A logging site with burnt tree roots and singed (selectively) foliage in the suburbs of Petrozavodsk was examined by DS V. N. Salnikov.

At 4 o'clock in the morning on 29.0.3.1990, an eyewitness suddenly had a heaviness in his head. He turned to the window and saw a flash of light. There was a pop, and then the formation in the form of a cylinder of "white-straw" color rose above the forest and went into the clouds. Smoke billowed over the forest, then it began to rain. After some time, a forest dump measuring 30 × 25 m was found in a swampy area. According to Salnikov, the trees were turned out in a right spiral with 2 branches. On some trees, the bark was burned in the form of vertical stripes. The roots of the trees are burned with a concentric strip 10–15 cm wide, as if they were cut by an autogen. Some of the roots were charred from above, sometimes under the bark (but the bark is intact). The outer part of the collapse spiral was characterized not only by burnt trees, but also by concentric burnt circles around standing trees. We add that the meteorological situation was seismic.

Comment. Description of complex geophysical phenomena. A cylinder of "white-straw" color rose above the forest and went into the clouds.

3. (Olkhovatov 2008) 14.01.1993. Evening. Poland, Krakow and its surroundings.

Residents watched the flight of a bright fireball. Immediately after that, a strong explosion was heard near the small village of Jerzmanowice, 20 km from Krakow. It turned out that a significant part of the limestone rock located there was destroyed and scattered at distances of up to 200 m or more. Zigzag branching furrows ran from the broken rock in the ground.

A nearby geophysical observatory recorded electromagnetic and seismic disturbances. In addition, electrical wiring was melted in the village houses. No traces of a meteorite fall or thermal changes were found in the collected material at the site of the event.

According to the Czech astronomer Zdenek Tseplecha, subsequent studies revealed that the event was associated not with one, but with two or even three "fireballs" that followed each other at intervals of about 1 min along close trajectories. Therefore, the first trajectory began at an altitude of 2 km and ended at an altitude of 0.5 km. The average speed of the "cars" was several km/s.

Comment. Description of complex geophysical phenomena. With appearance of fireballs.

4. (Tutt 2004) Mauritius. 05.24.1948.

Some force dug a trench through a dirt tennis court. "The trench ran from north to south, its length was sixty feet, width—from one to two and a half feet, and depth—from one to four inches.

The soil extracted from the court was dropped to the ground fifty feet from the end of the trench, and individual pieces of soil weighing at least a pound were transferred by an unknown force to about thirty feet. The surface of the ground turned slightly black, as if the earth was exposed to high temperature, in addition, for about two minutes there was a crackling sound, as if a reed was burning. One person claimed to have seen a fireball about two feet in diameter flying from a football field to a tennis court; this ball passed through a fence made of metal mesh, leaving no traces."

Comment. The appearance of a trench. One person claimed to have seen a fireball about two feet in diameter.

5. (Tutt 2004) Toledo, Ohio, the twentieth century, the 60s of the twentieth century.

"Everyone was thoroughly shaken, after that, holes the size of a baseball appeared in the walls of our and several other trailers. For a while, we watched in fascination the delightful blue glow around the tornado and how orange balls and lightning burst from the tip of the tornado. The "tail" of the tornado reminded me of the trunk of an elephant: it bent down to the ground, as if to take food, and then rose up—as if the elephant was bringing the trunk with food to his mouth. When the trunk was at the top, the tornado was not dangerous, but as soon as it descended, trouble began. My son and I watched orange fireballs flying along Race Way Park, then they flew up to the stable, and the roof came off it."

Comment. The appearance of a BL from a tornado.

Holes in the Ground and observations of the BL

1. (Olkhovatov 2008) "New York Times" July 1886: A meteor stone arrived in Wisconsin and hid under a barley field. USA.

Eyewitnesses of the phenomenon working in the field were greatly surprised by a loud sound that resembled a train. The volume of the sound increased to deafening in a few seconds. It seemed that the sound came from the sky, looking at which, eyewitnesses saw a large "giant ball of smoke" descending rapidly.

The ball fell a couple of dozen meters from the witnesses, plunging deep into the ground. The size of the hole was about 1 m in circumference, the depth could not be measured even with the help of the longest poles. This "meteor stone" does not appear in meteorite catalogs.

Comment. Description of the appearance of the hole when the BL hits the ground.

2. (Olkhovatov 2008) USA, Storrs, October 1909. Local newspaper.

Just after midnight, a powerful thunderclap was heard—although the only one, but so loud that it woke up the entire local population. The witnesses are an agricultural college clerk and a football coach.

They saw a giant fireball descending, followed by a glowing trail. There was a hiss. The ball hit the telegraph pole, splitting and twisting (note the twisting) the latter. Then it flew into Mr. Chadwick's house, soared up to the roof, tore down the skirting board, broke the windows, made holes in the ceilings and left the house above the door. Those in the house were thrown to the floor by the explosion. No traces of fire were found.

In hot pursuit, several holes were found in the ground at a distance of 8–9 m from each other.

Comment. Description of the appearance of holes in the possible impact of BL on the ground.

3. (Olkhovatov 2008) USA, Associated Press, May 1922.

A meteor that flashed across the sky of the American state of Virginia at night and, falling, shook houses in half a dozen cities was found 12 miles from the town of Blackstone. Falling into an oak grove, it formed a 500-square-foot crater, burying several trees in the crater during its fall. The "meteor" probably consisted of some kind of metal. Local residents reported that when the object fell, a band of fire rose up, which was visible for miles, and a powerful explosion broke out. It is assumed that the weight of the object can reach 20 tons. This object is not a meteorite because this "meteorite" did not get into any meteorite catalogs.

Comment. Description of the appearance of a crater with a possible impact of a BL meteor on the earth.

4. (Olkhovatov 2008) USA, Ochard, September 1927.

In the morning, a huge fireball fell from the sky and hit the ground in the middle of the street. There was an explosion, the buildings shook and the surrounding area was covered with smoke. A hole about a meter across and a meter and a half deep was formed at the crash site. It was crusted, as if molten metal had been poured on it. The previous night there was a thunderstorm. There is no such meteorite (Olkhovatov 2008).

Comment. The appearance of a hole when the BL falls to the ground.

5. (Tsarev and Tsareva 1982) 1994. Muromtsevsky District of the Omsk region. Russia.

A pit with perfect smooth walls with a diameter of 5 m and a depth of 4.5 m was found. There is not a single crumb of earth around. One of the residents saw a glowing ball descending to the ground.

Comment. The appearance of a hole when the BL falls to the ground.

6. (Tsarev and Tsareva 1982) Russia, Bashkiria, oilmen (two brigades).

Observers saw several glowing dark-red balls hovering near the ground. No one dared to approach the vibrating objects.

The next day, deep wells with a meter diameter were discovered.

Comment. The appearance of deep wells with a meter diameter by the impact of dark-red balls on the ground.

7. (Tsarev and Tsareva 1982) Kostenki Village, Voronezh region, 1990. April. Russia.

Fig. 2.65 Photo of the formed well, in the Voronezh region in 1990

A resident of Fomina E. I. saw from the window of her house how a red ball hung over her garden at night, intensely glowing and blinding her eyes. In the morning, she found a deep hole in this place. The well was a perfectly round hole with a diameter of 10 cm with well-preserved, as if cut off walls. The hole was 5.4 m deep (Fig. 2.65). There was not a crumb of soil near the hole in the ground. It is estimated that the weight of the rock that occupies the volume of the well is about 120 kg.

The radiometer readings were within the natural background.

A few months later, a zone with a diameter of 8 m with more dense vegetation appeared around the well. The potato tops were 10–15 cm taller and the leaves were juicier compared to potatoes in another part of the garden.

Spectral analysis of potato tubers showed a two- to three-fold decrease in the content of manganese, zinc, and phosphorus. The content of copper and phosphorus in the tops also decreased by 50% compared to the control. However, the content of zirconium and yttrium in tubers increased by 12 and 6 times, respectively.

Comment. The appearance of a hole when the BL influences the ground.

8. Tsarev IV (1995) Who does lightning choose? “Labor,” 28.06.1992. “Thunder and lightning.”

“Thunderstorms in our places are an ordinary thing,” says I. A Stanovoi from the village of Sokhachi in the Koropsky District of Chernihiv region. It’s like something is drawing lightning here. Then they hit the chimney, then the barn is set on fire, then the tree is broken into chips. I happened to see BL several times.

Once a glowing ball floated very close to me—it seemed to reach out and get it. It moved easily and smoothly, but suddenly stopped and hit the ground with a heavy weight. There was a deep hole with scorched edges, about 15 cm in diameter. We tried to get the bottom, but the six-meter pole was not enough…Then I noticed that lightning often chooses this place.

Comment. The appearance of a hole when the BL is exposed to the Ground.

Suppose that a BL with a charge $q = 10^{-4}$ or $q = 10^{-3}$ C has descended to the ground, we estimate the force with which a ball with a radius $r = 0.1$ m will act on the soil with a relative permittivity of $\varepsilon_2 = 3$. To do this, we will use the known ratio for the image strength of a charged ball

$$F_{im} = \frac{q^2(\varepsilon_2 - \varepsilon_0)}{16\pi \cdot r^2 \varepsilon_0(\varepsilon_2 + \varepsilon_0)}$$ where $\varepsilon_0$is the dielectric constant of the vacuum.

Substituting these values into this formula, we get that a BL with a charge $q = 10^{-4}$ C will act with a force of $3.5 \cdot 10$ N, which is equivalent to the action of $3.6 \cdot 10^2$ kg mass, and in the case of $q = q = 10^{-3}$ C will act with a force of 3.5 10^5 N, which is equivalent to the action of 3.6 10^4 kg mass, which shows that the action of such a BL can push deep holes in the soil.

BL, crop circles and on the surface of the Earth

In this section, we consider observations of crop circles in connection with the appearance of BL. There are observations in which the appearance of swirling areas of grass was recorded when exposed to vortices (Tsarev and Tsareva 1982), note that a number of cases are translate from Russian.

1. (Tsarev and Tsareva 1982) Steve Alexander, 1990. Great Britain.

…I saw my first lap on the field in 1981 in Hampshire. Since then I have been regularly visiting places where pictograms appear. Previously, I was sure that these were traces of unidentified flying objects. However, in 1990 my life and perception of witch circles has changed once and for all. This happened when I visited a place called Milk Hill. When I was videotaping a fallen crop of corn, a small but very bright ball of light appeared in the lens, which seemed to be moving purposefully over the field, flashing like a ray of sunlight reflected from a foil. I took it on film…

The mysterious ball was gradually receding. On the way, it flew over a tractor that was working in a field. Then, in an interview with journalists, the driver confirmed that he saw a glowing ball the size of a football. Moreover, he reported that when this object flew over his head, the tractor engine stalled.

The footage of this video of mine was shown on television (later they were included in documentaries more than once).

Subsequently, similar balls of light were photographed and filmed by other people. There is even a video showing two glowing balls the size of a football, circling in a dance over a wheat field and immediately the stems of plants bend to the ground, forming an image.

Comment. When I was shooting a dead crop of corn on a video camera, a small but very bright light ball appeared in the lens.

2. (Tsarev and Tsareva 1982) 06.28.1996 France.

Eyewitnesses reported a red glowing ball; it descended into the district of Montsoreau. The next day, a local farmer Alain Rouillet found fallen ears of corn and a yellow oily liquid in the center of a dried round spot in his field.

Comment. Eyewitnesses reported a red glowing ball.

3. (Tsarev and Tsareva 1982) Delgado P. and Andrews K. Great Britain.

The terrified couple watched for some time on the edge of the field a brightly glowing object resembling a giant Ferris wheel. Then "it" rose up and disappeared into the clouds. The next day, the peasants found a new circle in this field.

Comment. The appearance of a circle after observing a glowing object resembling a giant Ferris wheel above it.

4. (Tsarev and Tsareva 1982) 26.06.1990 Rural life. Russia.

Machine operators of the Limansky collective farm in the Shcherbinovsky District of the Krasnodar Territory saw flashes similar to electric welding flashes in the steppe. In the morning, a circle with a diameter of 30 m was found at this place, formed by wheat rolled to the ground, inside of which there were two circles of smaller diameter. Entering this circle, two eyewitnesses almost fainted from a severe headache.

Comment. The appearance of circles in the place of flashes.

5. (Tsarev and Tsareva 1982) Kingston D., 1976. Great Britain.

"I became interested in crop calipers in 1976. It was then that I saw my first circle and those who created it. It happened in the town of Clay Hill. It was at night.

Three glowing balls (about six feet in diameter) circled around and above us for three hours. From time to time, then connecting into one object, and then separating again. I remember how one of the balloons rapidly descended to the ground, hovered about thirty feet above our heads, and then sped off into the field again…

When dawn came, I saw a patch of bent wheat on the field. We carefully examined the plants—no violations, not a single broken stem! This spot was a perfect circle thirty feet in diameter."

Comment. The appearance of circles in the place of the flight of glowing balls.

6. (Azhazha 2013) The Caspian Sea. August–September 1980. Russia.

From a personal conversation with the pilot, Lieutenant Colonel M., it turned out that an object once hovered quite high in the coastal area of Krasnovodsk (Turkmenistan). He "probed" the territory of two radars on the hill, after which ring traces of melted soil remained on the ground in several places.

Comment. The appearance of melted soil under the place of the object's flight.

7. (Tutt 2004) Gwyneth Tom, farmer, Gloucestershire. Great Britain. June 1988.

On a warm summer evening, Gwynett was driving along the edge of a wheat field when all the electrical appliances of the car suddenly failed. The farmer got out of the car to try to establish the cause of this incident, and heard an "unusual buzzing sound." Glancing at the wheat field, Tom noticed a dimly glowing red ball the size of a football over the field.

The ball did not look like a solid body—it seemed to consist of a swarm of red sparks, and these sparks came in a continuous stream from the surface of a wheat field. The farmer observed this phenomenon for about two minutes, all this time a characteristic buzzing was clearly heard. Then the ball suddenly disappeared, at the same time the electrical equipment of the car started working as if nothing had happened. Gwinnett did not inspect the field, drove on, but when he returned to the place the next morning, he noticed a twenty-foot circle of curved ears under the point where the ball was hanging.

Comment. The appearance under the point where the ball hung, a twenty-foot circle of curved ears.

8. (Abakumov et al. 2017) The Adyghe Republic. Russia. The place of the studied impact of BL with soil is a field near the village of Dondukovskaya in the Giaginsky District of the Adyghe Republic. 2015.

Samples are provided by observers. According to eyewitnesses, a hole appeared in the soil under the site of the appearance of the BL, Fig. 2.65. The parameters of the hole are shown in Figs. 2.66 and 2.67. The volume of the monolith is 21.56 dm^3, the mass of the monolith is 28.59 kg, and the density of the composition is 1.33 g/cm^3.

Discussion. Experimental studies of a natural sample.

The methods used for the study of samples: measurement of electrical conductivity of samples, waxing (determination of porosity of samples), and measurement of magnetization.

As a result of the studies, a change in the electrical conductivity of soil samples exposed to BL was revealed by more than 2.5 times (see Fig. 2.68a). The density of the monolith decreased from a value of 1.9 g/cm^3 to 1.5 g/cm^3 as it moved from the dense wall of the hole in the monolith to the periphery of the piece of soil, to external control (see Fig. 2.68b, c). The value of the maximum density exceeds the value of the external control density by 1.2 times.

Fig. 2.66 The hole and its parameters in the interaction of the BL with the soil

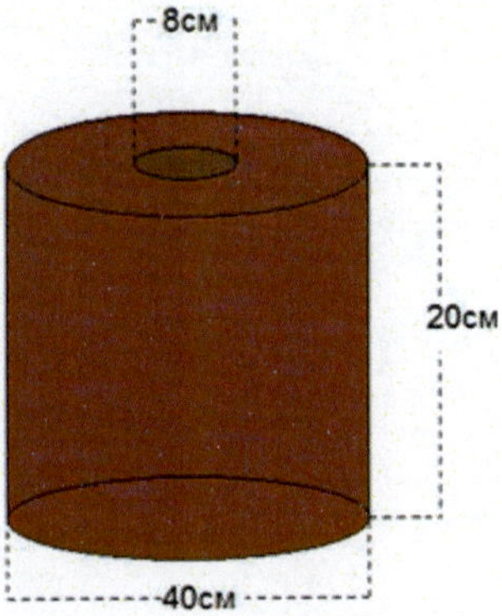

Fig. 2.67 The volume of the monolith

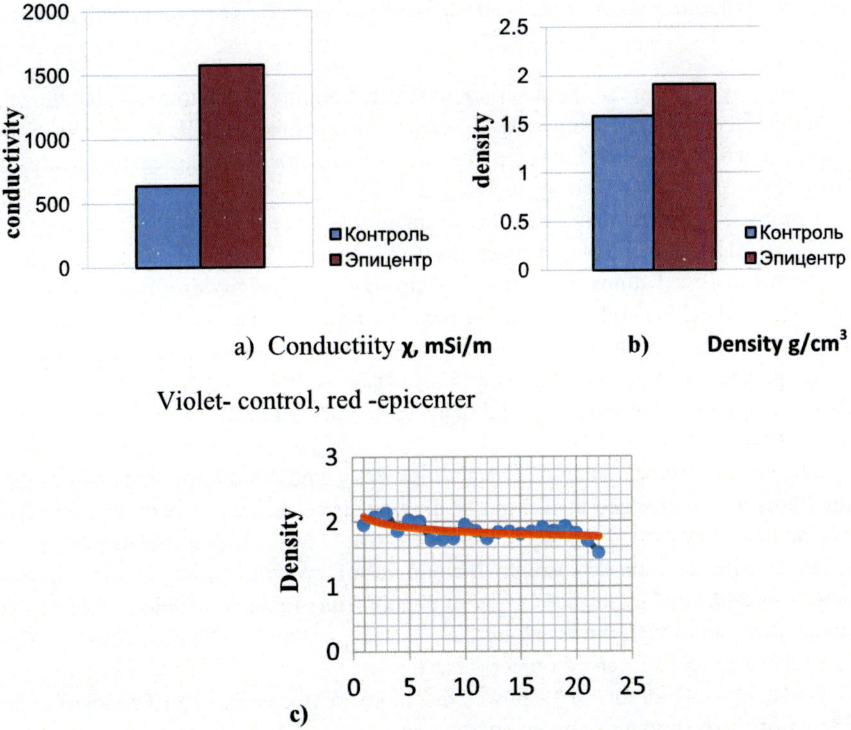

Fig. 2.68 **a** The difference in the values of the average electrical conductivity (χ, mSi/cm) of samples from the epicenter (dense part of the monolith) and external control samples. **b** The difference in the values of the maximum density g/cm^3 of samples from the epicenter (dense part of the monolith) (red) and external control samples (violet). **c** The change in the density of samples as they move away from the wall, cm. Blue line—density, g/cm^3, red line—density approximation line

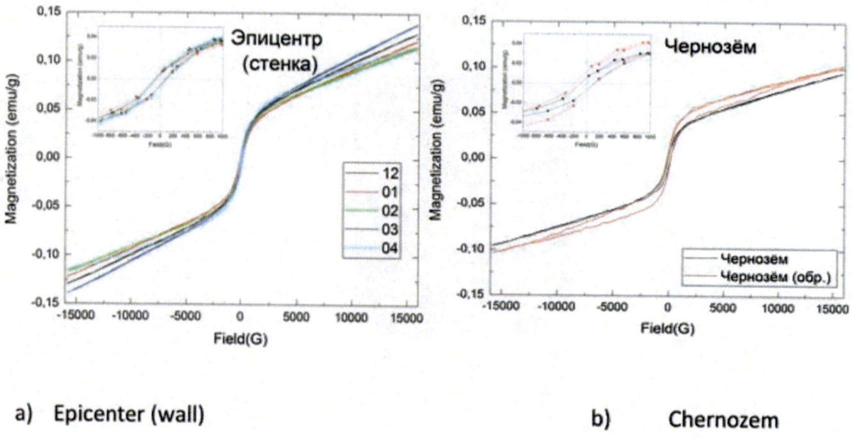

Fig. 2.69 Magnetization curves of samples: **a** monolith walls, **b** chernozem of external control

The study of magnetic characteristics using a vibrating magnetometer also showed a change in the magnetization curve for samples exposed to BL compared to the chernozem of external control (see Fig. 2.69). The porosity of aggregates as a result of compaction was reduced by an average of 17%.

Comment. The following considerations indicate the possibility of studying the effects of BL. Crop circles or pictograms (CP) and holes in the ground and soil cover according to observations (Fierani 2014) appear during the growing season of plants, in the period—May–July, i.e. during the maximum appearance of BL. In Europe and North America, this period corresponds to the greatest thunderstorm activity, accompanied by ion flows to the Earth's surface, which are captured by plants. At the same time, it is obvious that the fields of cereals become huge antennas for ion absorption.

The ears collected inside the CP (Fierani 2014) under a microscope showed that the stems were deformed as if rapid heating occurred inside the stem, as if electromagnetic energy heated the liquid inside the plant, converting it into steam, which when heated caused an explosion in the nodules of the stem and as a result its explosion in this place (Fierani 2014). These phenomena are analogous to explosions of tree bark when the BL moves along their surface, see above. That is, when passing strongly electrically charged objects near trees.

Soil analysis (Fierani 2014) showed that there are two types of dust deposits in the CP region. In the first type, magnetic ironstone is present in 600 higher concentrations than under normal conditions. Another type of sediment is silicon SiO_2, but not in the form of ordinary sand particles, but in the form of small balls, as if silicon crystals would be melted by a powerful flow of energy. In the dark, such a powder emits a yellow-green light, such radiation can be explained by assuming that the silicon crystals were excited by an external source in a short time.

Two more features should be noted. Observers claim (Abakumov et al. 2017) that CP of natural origin have the form of round rings or rings connected by straight lines.

CP were observed in the UK in areas where megalithic structures of ancient people are located, and in the Krasnodar Territory—near mounds in the form of hills.

In this regard, attention should be paid to studies of the soil cover from the sides of aircraft (Doyle 1970), which showed that buried underground remains betray their presence through crops or wild grasses growing above them. It is obvious that subsurface conditions can be favorable or unfavorable for plants. Above the former ditches and other depressions, where the soil is more fertile, there is more moisture and it is easier for the roots to penetrate into the depth, the vegetation is more lush, higher, and thicker. Over stone floors, foundations of buildings or over powerful roads, it is noticeably poorer and weaker. Plant signs are especially well manifested thanks to plants with a branched root system, whose roots go far into all recesses with more fertile (i.e. once disturbed) soil. In this respect, all crops are good. Beetroot reacts well, etc.

Based on their stated facts, it can be assumed that both flows of charged particles to the Earth's surface in a highly charged thunderstorm atmosphere and processes occurring inside the Earth in the soil cover can participate in the appearance of CP inside the soil cover, due to various physics and chemical processes, as well as in places of activity of ancient people, with the construction of various local objects of vital activity and objects of worship using various materials (metals and dielectrics), an electrode system arises that locally concentrates ion flows and accelerates them, directing them to a quasi-electrode formed underground. In this case, the shape of circles and lines becomes unsurprising, since they are inherent in the forms of channels, the place of rituals and roads of ancient people. In more detail, plasma-chemical processes that can lead to intense heating and ignition of organic components will be discussed later.

Globular nodules

Recently, there have been attempts to link BL with other spherical objects, which differ fundamentally in all but form properties. Thus, in (Tarasenko 2008), the hypothesis was considered that underground thunderstorms with the formation of underground BL eventually lead to the appearance of spherical nodules. Questions immediately arise about the nature of underground thunderstorms and the mechanism of BL formation. The answers to which are currently missing.

The hypothesis discussed in (Mikhailov 1998) seems more plausible. 35–40 million years ago, massive eruptions and lava emissions began. Gas-saturated lava flows and ash layers have covered significant areas of the earth's surface in various places in Asia and Latin America.

At a temperature of 600–700 °C, vitreous particles were formed in a thick layer of ash. The round core of volcanic glass, when cooled, released a variety of gases that spread evenly in different directions from the center of crystallization and, as a result, formed spherical voids. Neighboring particles began to crystallize around almost every such core, filling these voids, because of which a sphere consisting of solid particles—future stone balls—gradually formed. The sizes of such spherical bodies depended on a number of factors—the thickness of the ash layer, the temperature in its depth, the characteristics of its chemical composition, the cooling rate, etc.

Spherical nodules with diameters from 50 cm to 5 m have been found in the Caucasus, Kazakhstan, and Costa Rica. Over time, water, rain, and wind eroded the relatively loose, non-crystallized layers of ash covering the crystal balls during their formation.

2.10.5 Luminous Objects of Hessdalen

In this section, we consider the phenomena perceived by observers as BL, which they are not. At the same time, the proof of this fact requires some analysis. Let us start with Hessdalen's "lights" (Strand 1993), Zou Y-S (1995).

Hessdalen's "Lights." Hessdalen is a small valley in the middle part of Norway, located southwest of Trondheim. It attracted the attention of specialists after, starting in December 1981, unknown lights began to appear in it. These lights could be observed in any part of the valley, over the mountains surrounding it, and just in the sky. They had different shapes and colors.

Local residents made many descriptions of the observed lights, and some of them were supported by photographs, see Fig. 2.70a, b

Usually the lights were visible in the evening, at night or early in the morning, that is, at night. Sometimes you could see several lights in one day in different parts of the valley. The largest number of observations occurred in autumn, in spring there were significantly fewer of them, and in summer, there were no reports of lights seen at all.

The regular appearance of lights attracted the attention of specialists, and in 1984–1985 a group of scientists from different countries took part in the Hessdalen project. Equipped with modern equipment (radar, seismograph, night vision device, spectro-analyzer, Geiger counter, magnetograph, laser), specialists conducted a series of observations in the valley in the period from January 21 to 02.26.1984. They received a large number of visual and visual-radar observations. Seismic and magnetic activity was recorded simultaneously.

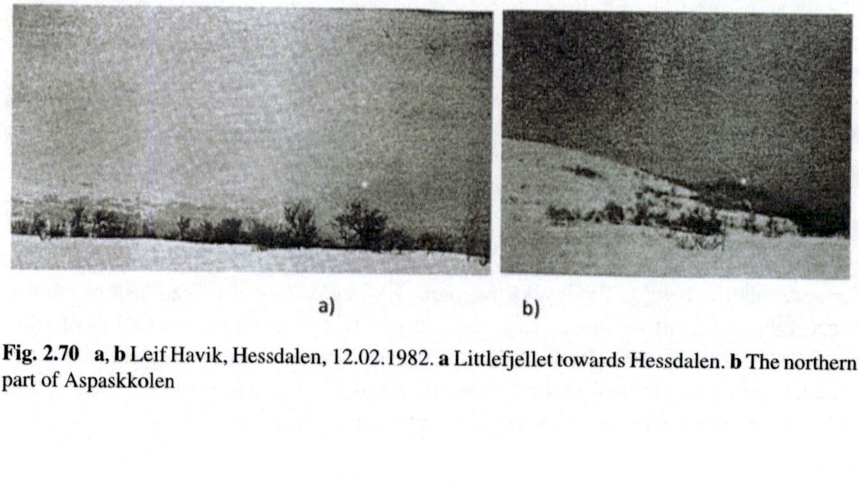

a) b)

Fig. 2.70 a, b Leif Havik, Hessdalen, 12.02.1982. **a** Littlefjellet towards Hessdalen. **b** The northern part of Aspaskkolen

Specialists in BL were invited to the seminar by E.Strand (Strand 1993). It was clear from the discussions that the Hessdalen lights are very different from the usual BL.

Analyzing the expedition reports, the observed lights could be divided into several types. In Bychkov et al. (2006), we divided the objects by their lifetime (usually close to the observation time). In principle, the division can be carried out by color, shape, size, type of movements performed, etc. However, there was no detailed information on these data.

Long-lived lights were assigned to the first class, which were usually observed hanging motionless for 20–40 min low above the horizon. These lights were observed in the evening by the project participants, in the evening and in the morning by local residents earlier. Moreover, these lights could appear for several days in a row.

The second type consists of lights whose observation time did not exceed four minutes. They moved quickly, but also quickly disappeared. These lights could be seen in various parts of the valley.

Knowledge of the properties of BL, as well as the results of research (Bychkov et al. 2006) based on theories of artificial missile clouds (Platov and Rubtsov 1999) and optical illusions in the atmosphere (Novoseltsev 1996) led to the conclusion that these are not BL, but some objects, most likely of artificial origin, such as fuel residues of missiles or aircrafts.

2.10.6 Local Heating of Objects

The influence of thunderstorm and geotectonic activity. It is known that during periods of thunderstorm activity, the local values of electric fields under clouds on the Earth's surface increase to 10^3–10^4 V/m or more (Hrgian 1976). In particular, an electric field of the order of 4.5×10^4 V/m is realized under the thunderstorm clouds of a thunderstorm cell carrying an electric charge of 5 Cl, located at an altitude of 1 km above the Earth's surface, and with abnormal charges of a thunderstorm cell this value can be several times greater, and even accompanied by a glow of the grass surface (Hrgian 1976; Reiter 1992; Enomoto and Hashimoto 1990). With the development of gas discharges in abnormal electric fields, the ignition of natural organic materials—vapors of heavy esters and components of decomposition of organic hydrocarbons—is possible. Strong electric fields can arise because of tectonic processes in the subsurface layer, with corresponding processes of ionization and charging of various objects, including trees and foliage. It is gas discharges in pointed objects, on which high electrical voltage can be realized (Reiter 1992)] that can ignite rising vapors and streams of organic components.

Under normal atmospheric conditions, the average concentration of charged particles (ions) in the air is about 10^3 cm^{-3} at the rate of ionization by fast particles of the atmospheric background Q = 4–10 cm^{-3} s^{-1} (Enomoto and Hashimoto 1990; Ardelyan et al. 2017a). This rate increases with the development of geotectonic activity, and near the moist surfaces of mountains it becomes several orders of magnitude larger, associated with the rapid processes of electron detachment from negative ions (Ardelyan et al. 2017b). The presence of high-frequency electric fields affecting the environment has been observed in nature under conditions of geotectonic activity (Enomoto and Hashimoto 1990; Zou 1995; Pulinets et al. 1998).

Thunderstorm electric fields and geotectonic activity are realized on relatively long (hours) and spatial (meters) scales. Strong ionization, heating, and ignition can also occur with local amplification of the electric field on pointed objects, as we show below.

Studies of anomalous air ionization during periods of tectonic activity are of interest for solving problems of early diagnosis of seismic activity from the ionosphere (Pulinets et al. 1998). Before earthquakes, an increase in the intensity of ionization by fast particles is observed to a level corresponding to the rate of formation of electron-ion pairs $Q = 10^3$ cm^{-3} s^{-1} (Pulinets et al. 1998), and luminescence in the area of geological faults (Zou 1995), (Pulinets et al. 1998), (Grigoriev et al. 1988). Glow can occur due to the appearance of plasma near small particles and pointed natural objects, caused by an increase in the local value of the electric field strength at the surface of the particles to breakdown values (28–30 kV/cm at atmospheric pressure 760 mmHg and ambient temperature 290 K). According to (Enomoto and Hashimoto 1990; Zou 1995), during periods of tectonic activity, the average value of the electric field strength on the Earth's surface increases from 1 V/cm under normal conditions to 10–100 V/cm and above (Pulinets et al. 1998; Grigoriev et al. 1988).

The average concentration of charged particles (ions) in normal atmospheric conditions is about 10^3 cm (Ardelyan et al. 2017a). Under conditions of geotectonic activity, it will increase and create favorable conditions for the formation of cluster ions. In turn, cluster ions are the centers of condensation of water molecules and the formation of droplets, which affects the meteorological characteristics of moist air. Such droplet formation provides remote diagnostics of ionization processes on the Earth's surface (Pulinets et al. 1998).

Plasma chemistry of humid air

To study the chemical kinetics of discharge in moist air, we applied a detailed plasma-chemical program developed in (Ardelyan et al. 2013). It contains air chemistry and electron-molecular reactions in an external electric field in air and water. The plasma of moist air in this simulation includes only water molecules in the gas phase without large aerosol particles. Reactions involving 27 components of air plasma and 23 reagents containing hydrogen are used, including: positive ions O^+, $O_2{}^+$, $O_4{}^+$, $O_6{}^+$, $O_8{}^+$, NO^+, H^+, $H_2{}^+$, OH^+, $HO_2{}^+$, H_2O^+, $O_2{}^+(H_2O)$, H_3O^+, $H_3O^+(H_2O)$, $H_3O^+(OH)$, $H_3O^+(H_2O)_2$, $H^+(H_2O)_2$, $H^+(H_2O)_3$, $H^+(H_2O)_4$; negative ions O^-, $O_2{}^-$, $O_3{}^-$, $O_2{}^-(H_2O)$, $O_2{}^-(H_2O)_2$, H^-, OH^-; atoms O and H; molecules H_2, O_2, H_2O, O_3; and free radicals OH, HO_2, H_2O_2. Excited states of atoms and molecules $O(^1D_1)$, $O(^1S_0)$, $O_2(^1\Delta_g)$, $O_2(b^1\Sigma_g^+)$, $O_2(v)$, $N_2(v)$, $N_2(B^3\Pi_g)$, $N_2(X^1\Sigma_g^+)$. Electrons and molecular particles, including nitrogen. The estimated number of reactions for dry air is 200 and 250 for wet air components containing hydrogen. We also considered the reactions of ion recombination of each positive ion with each negative ion and calculated the rate constants of these reactions, since they have not yet been investigated for all combinations of ions. The model includes reactions of three-particle electron attachment to the O_2 molecule in the presence of O_2 and H_2O as a third body, electron detachment from O^- and $O_2{}^-$, recharge and ion-ion recombination with all positive ions, data on them are presented in (Ardelyan et al. 2013). The cooling of plasma electrons due to elastic collisions with water, oxygen, and nitrogen molecules, as well as the excitation of their vibrational and rotational degrees of freedom, is taken into account. It is assumed that the removal of the excitation of vibrational and rotational levels occurs mainly in collisions with molecules. The corresponding terms for heating and cooling in the electron energy equations were interpolated relative to the electron temperature T_e. Gas heating and cooling in exothermic and endothermic plasma-chemical reactions, as well as Joule heating, were also taken into account. Plasma is considered an ideal gas with an adiabatic index of 1.4. The results were obtained under the condition of constant density, which is true for large scales. In the case of humid air, the proportion of water vapor concentration is 1%. The input power W (the rate of ionization of air by fast electrons) in a gas, which we usually use in units of eV/(cm^3 s), and the rate of excitation of the molecule Q are related by equation (Ardelyan et al. 2013) $Q = W/(Ui)$, where Ui is the cost of ionization, in air $Ui = 31.6$ eV. Ionization and excitation of molecules by fast electrons are taken from (Ardelyan et al. 2013).

Calculation results

Figure 2.71 shows the results of calculations for dry air at electric field strength $E = 1$ V/cm (for the supplied energy $Q = 4$ and 100 cm^{-3}c^{-1} concentrations of negative and positive ions with normal background excitation $Q = 4$ cm^{-3}c^{-1} reach 10^3 cm^{-3}). With an increase in background excitation to $Q = 100$ cm^{-3} s^{-1}, achieved under geotectonic conditions, the concentration of air plasma (negative and positive ions) increases by an order of magnitude to 10^4 cm^{-3}, and the main ions are $O_2{}^-$ and NO^+ ions. The remaining positive air ions make a significant contribution to the charge balance at the initial stage of entering the quasi-stationary background state. The anomalous value of the external electric field ($E = 4$ kV/cm) at the considered excitation rates by fast particles Q reduces by two orders of magnitude the time for the concentration of positive NO^+ ions to reach a quasi-stationary value. NO^+ ions reach a quasi-stationary value of 1.6×10^4 cm^{-3} in 10^2 s (1.67 min), and with the usual background value of the electric field strength, the transition time to a quasi-stationary state exceeds 104 s (165 min). Thus, in the case of optical diagnostics, atmospheric disturbances with an anomalous field will be detected much earlier.

The situation changes significantly when considering humid air with a concentration of 0.1% water molecules. The results of the corresponding calculations are shown in Fig. 2.72 for an ordinary background field $E = 1$ V/cm for two values of the fast particle flux $Q = 4$ and 100 cm^{-3} s^{-1}.

The quasi-stationary values of plasma concentrations practically do not change in comparison with those obtained in calculations for dry air. The main particles in the charge balance are cluster ions, which include water molecules. Cluster ions are

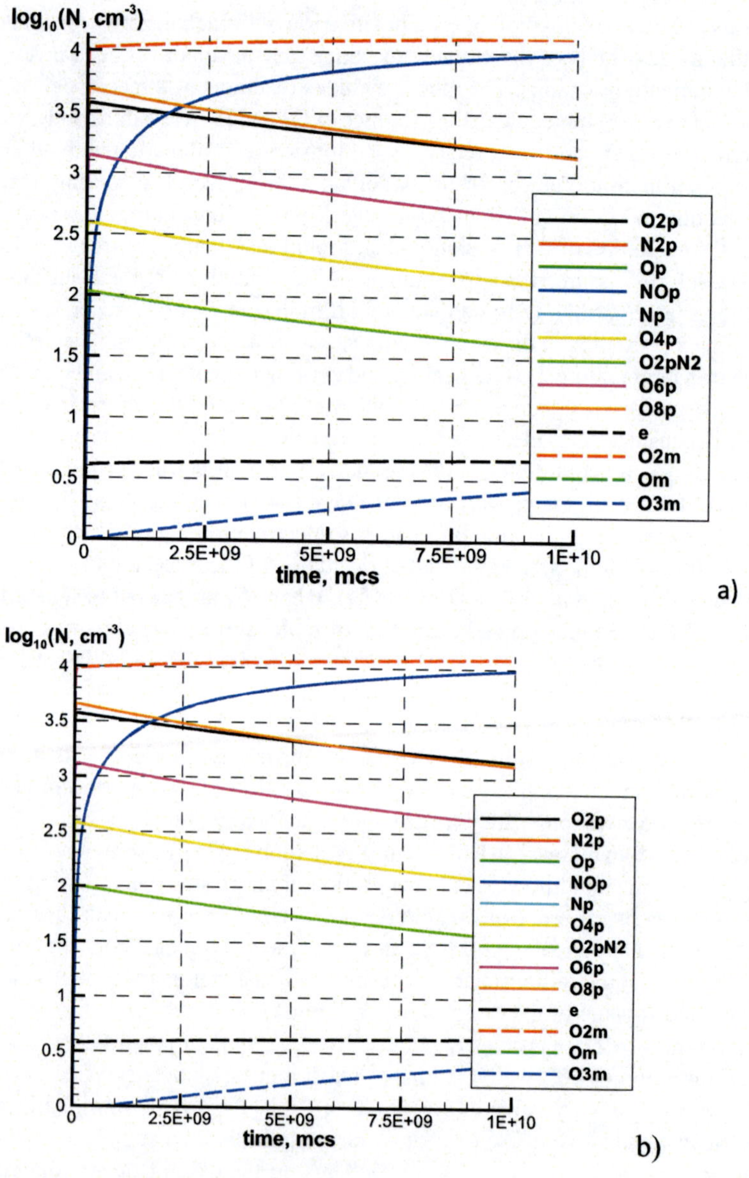

Fig. 2.71 Time dependence of the concentrations of charged particles in dry air at $Q = 100\ \text{cm}^{-3}\ \text{s}^{-1}$ at two values of the electric field strength: **a** $E = 1$ V/cm (above), **b** $E = 4$ kV/cm (below)

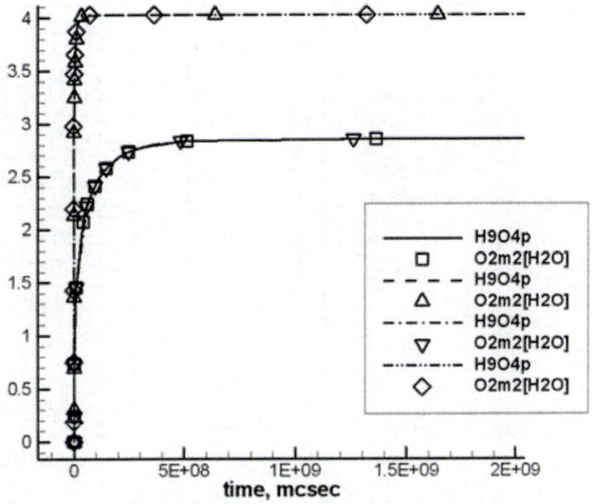

Fig. 2.72 Dependence of charged particles concentration in humid air on time. The concentration of water molecules is 0.1%, the electric field strength is $E = 1$ V/cm. $Q = 4$, 100 cm^{-3} s^{-1}

the centers of condensation of water vapor, which plays an important role in remote diagnostics.

With an abnormal electric field, the charge balance graphs practically coincide with those shown in Fig. 2.71 for a normal background field. The same applies to the balance of charges with an increased concentration of water molecules equal to 1% in the air mixture.

Figure 2.73 shows the results of heating the air under various conditions. Figure 2.73a shows that ionization at long times has several modes: pre-breakdown, intermediate, and rapid ionization mode with gas heating. The pre-breakdown ionization regime provided by fast background electrons takes place at small external electric fields; here the adhesion processes (three-particle and dissociative) reduce the electron concentration. With prolonged heating of the volume occurs, see Fig. 2.73b): at high temperature, the attachment of electrons stops and rapid ionization occurs. At large values of the external electric field, we see three ionization modes.

The pre-breakdown ionization mode occurs at a low gas temperature (see Fig. 2.73a, b); the intermediate ionization mode occurs with a slow temperature increase, where the attachment of electrons to oxygen molecules reduces the electron concentration, and the plasma is electronegative with a predominance of ions at concentrations of 10^8–10^{10} cm^{-3}; then rapid heating occurs and the electron concentration begins to grow rapidly—this means that the attachment processes stop at high temperature.

The horizontal sections of the curves in Fig. 2.73a, b after the completion of the rapid ionization regime with gas heating correspond to the establishment of a quasi-equilibrium state in the thermal balance, accompanied by the balance of electrons during the time of thermal conductivity or cooling due to gas expansion. The latter processes can be taken into account only in a complex gas dynamic model. High gas temperatures ($T \sim 2500$ K) in a short time ($t < 10^3$ s) can be realized at $E > 13$ kV/cm.

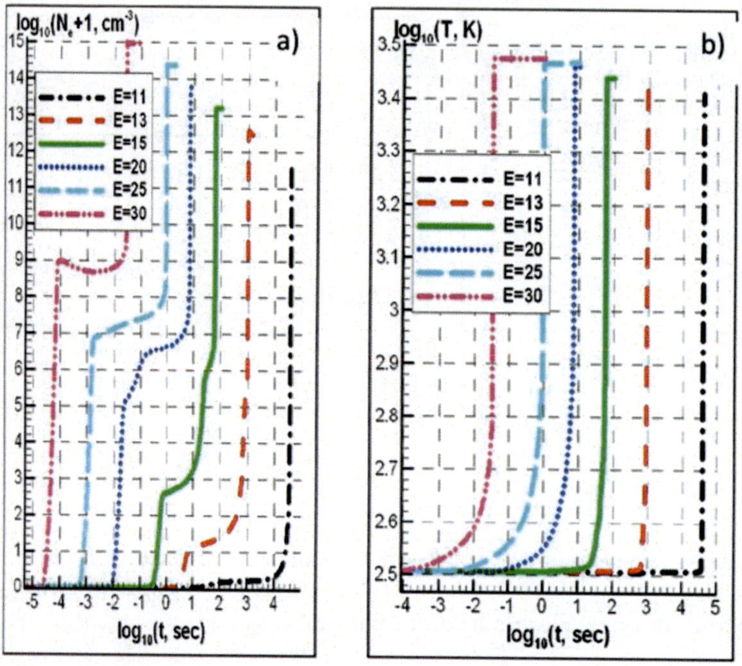

Fig. 2.73 Time dependences of **a** electron concentration—on the left and **b** gas temperature—on the right, at the power set $W = 3660$ eV/(cm^3 s). E, kV/cm. Dry air

Therefore, there appears condition of ignition of flammable substances at contact with such plasma.

The role of pointed objects

It is possible to estimate the parameters of a pointed object, near which the ionization of the electric field in the gas increases. To do this, we will use the results of (Landau and Lifshits 1982) on increasing local electric fields due to polarization. In accordance with it, the formula takes place near the vertices of conducting spheroids in the electric.

$$E_{\mathrm{surf}} = E_{\mathrm{ext}}/k,$$

where E_{surf} is the electric field strength at the top of the spheroid, and the intensity of the external electric field.

In this case, the coefficient k depends on the ratio of the transverse b and longitudinal a dimensions of the spheroid (Landau and Lifshits 1982)

$$k \approx (b/a)^2 \cdot \left(\ln\left(4/(b/a)^2\right)/2 - 1\right).$$

Using this formula, it is possible to find an electric field $E = 15$ kV/cm at the top of the spheroid, which can be realized in an ambient electric field equal to $E =$ 0.45 kV/cm, realized during thunderstorms or geotectonic processes.

This assessment shows that thin wires, needles, pieces of metal fittings, etc. can lead to local rapid ionization with an increase in local temperature, that is, they can be initiators of fires in thunderstorm and geotectonic conditions.

The same conclusions apply to the analysis of the causes of crop circles. Let the length of the object (blade of grass) be 20 cm and the thickness is 2 mm, so $(a/b) \approx$ 100, then $k \approx 2000$. This is enough to realize a strong electric field near the blade of grass, and to provide heating of the area to a temperature at which the stem of the blade of grass will explode.

In the presence of BL, which realize electric fields of $E_0 = 31$–33 kV/cm near the surface of the spikelets, rapid heating of the air can occur in the air. For $t = 10^{-6}$ – 10^{-2} s, the temperature in ideal conditions can rise to 3000 K.

For example, the electric field on the surface of a large BL with a radius of $Rb =$ 1 m, which will be realized by the heating of objects at a distance $Rd = 3$ m, can be determined on the basis of the expression

$$E_x =_0 (R_d/R_b)^2,$$

or $E_x = 2.7 \cdot 10^7$ V/m and charge $q = 2.7 \cdot 10^{-2}$ C. In other words, BLs can serve an indicator of suitable areas for the occurrence of pictograms, if the corn in these areas has differences (size, chemical composition) from the ears surrounding this area.

In this case, it is possible to formulate the conditions for the formation of circles: the presence of a pattern—an area inside the soil filled with fertile material; the presence of plants that stand out by their size, growing on the soil on top of the pattern, the presence of a high electric field that causes plasma-chemical heating of plants, and the presence of a hydrodynamic atmospheric vortex that causes helical movement of plants and leads to possible precipitation of gelemeteors (see below) inside the area.

Glowing balls in the woods

Linear lightning strikes on coniferous trees or evergreen leaves with needle-like leaves often show some specific damage in the form of tiny pieces of bark saturated with oleoresin, and sometimes charred particles that apparently burned during penetration, sometimes get stuck in the needles of surrounding trees (Zacher Nature Notes). Oily wood products are extremely flammable. Therefore, it is quite possible that very small particles of these flammable substances as a result of lightning discharge contribute to the ignition of forest fires. It is possible that the burning particles ejected in this way can either get stuck in the needles (which themselves are filled with flammable oleoresins) and form a kind of burning torch stuck into the needle itself, or small pockets of resin literally explode from the very beginning of the passage of an electric discharge.

Microscopic examination of some of these needle clusters showed (Zacher Nature Notes) the presence of many embedded cortical particles. They have an irregular shape, often triangular, of a strange shape and size, some are charred, some are not. The enormous pressure that causes the explosive effect of the lightning shock wave often penetrates deep into the needles. According to one possible explanation, a kind of exploding fireball is formed from flammable substances divided into these tiny particles, of which there are thousands. Explosions of coal dust in mines and explosions of flour dust in mills can be rough analogues of this gorenje caused by separation into small particles. In the case of coniferous trees, we are dealing additionally with combustible resinous or resinous properties of fuel.

A fireball of small particles ignited by lightning can be one of the means of igniting dry needles and grass on the ground or bark, live needles, and moss on trees. In support of this hypothesis, forest fire observers reported (Zacher Nature Notes) objects described as fireballs at the foot of trees after passing a lightning flash.

Conclusion to the section

The background conditions of dry and moist air under normal and thunderstorm atmospheric conditions and under conditions of geotectonic activity are considered. Numerical simulation was carried out on the basis of the concentrations of the main charged particles under these conditions. The plasma concentration reaches values of the order of 10^3 cm^{-3} in normal air and 10^4 cm^{-3} in geotectonic activity conditions.

Calculations of ionization parameters that can be realized during thunderstorms and geotectonic activity show high levels of electron concentration and gas temperature that can be realized at moderate electric fields. The local gas temperature reaches values of ~2500 K. At this temperature, the ignition of hydrocarbon materials, combustible gases, etc. is possible. The role of pointed objects should be taken into account. In particular, with the large ratio of the longitudinal and transverse dimensions of pointed objects, there will be an increase in the electric field near the tip, local ionization, and heating of the gas. They can lead to the ignition of hydrocarbon materials.

The analysis showed that in humid air conditions near mountains with geotectonic activity, ionization and electron separation in weak electric fields can occur, which can lead to local heating of the gas.

This may be the reason for the appearance of lights (similar to Hessdalen lights). Promote the formation of crop circles and the realization of fireballs in coniferous forests.

2.10.7 Gelly-Like Meteors and Similar Phenomena

A number of descriptions on ball lightning include information about gray, silver, or pearl-colored objects. Objects fall during the rainy season in different places in the Northern Hemisphere. These gelatinous objects can possess sizes up to 1 m or more, and fall to Earth during periods of rain and thunder. Such instances have been well recorded in Russia, Great Britain, Germany, and USA (Corliss 1995; McKenny 1910; Chernobrov 2003; Bychkov 2005a, b, 2006; Buckland 2011; National Geographic Television 2010; Petrisko 1998).

These objects are not just spherical but are complex gelatinous formations. In different places they have different names, such as "tremblers" (since their appearance resembles trembling fungi (Tremellales) (Corliss 1995; Bychkov 2005) "sky," "cloud," "sky jelly," "star jelly," "pwdre ser," etc.

In the scientific literature, they are called gel- or jelly-like meteors, because they are covered with jelly-like mucus. Their descriptions are also found in fiction, as Walter Scott mentioned them in the story Talisman (Corliss 1995).

A number of eyewitnesses affirmed that the fallen jelly-like masses had taste of fresh cold water and in 1–2 h or during a day they turned to water or evaporated. Sometimes observers compared these masses with the light «frog caviar», but without «grains» (Corliss 1995). At the same time (Corliss 1995; Chernobrov 2003; Bychkov 2005a, b), according to eyewitnesses, it was affirmed that these masses sometimes smelled badly. In the Internet, there were photos of such objects (see below) and even of red-orange color (Fig. 2.74).

Recently in the Internet (Buckland 2011; National Geographic Television 2010), there was a discussion of fantastic ideas about extraterrestrial—a meteoric origination of these objects (star gels) observable during the rains, accompanied meteoric downpours, or opportunities of an origin of their slime from «the frog caviar». In the film (National Geographic Television 2010), without any justification, they are associated with red rains in India (Fig. 2.75).

Though both in the Internet and in the scientific literature (Corliss 1995; Bychkov 2005, 2006) continues to be discussed more realistic opportunity of their biological formation, for example, at participation of blue-green algae. Therefore, in (Bychkov 2006), the origin of these objects was connected with the appearance of the blue-green algae bacteria (of Gloeotrichia sort) in clouds due to air streams from Earth, and their survival at the expense of the biological particles brought by airflows.

Clarification of this natural phenomenon is necessary for its separation from the objects connected with ball lightning, UFO, and other atmospheric phenomena. Besides they represent an independent natural-scientific interest connected with propagation of various bacteria, algae, and other large heterogeneous-macroscopic objects in clouds, and their possible impact on human life on Earth.

Observations

Here we represent a set of this phenomenon observations (events) for its better understanding and for facilitation of a following analysis:

1. (McKenny HT 1910) "In 1908 I was with my wife and one of my boys in Ingleborough, where we found the 'pwdre ser' lying on the short grass close to the stream, a little way above Gaping Ghyl Hope. For the first time I felt grateful to the inconsiderate tourist who left broken bottles about, for I was able to pack the jelly in the bottom of one, tie the cover on, and carry it down from the fell. I sent it, with the sod on which it appeared to have grown, to my colleague, Mr. E. A. Newell Arber, with a brief sketch of my story… M. Arber

Fig. 2.74 Gelly-type meteors—Star Jelly typical appearances

that it was not nostoc (a genus of algae (green–blue algae V.B.)), and said that he had sent it to Mr. Brookes, in the Botany School, who reported that it was a mass of bacteria…".

2. After the dense black cloud has passed above the village Schuknavolok, the USSR, 1932 (Bychkov 2006), on the ground remained a jelly-like substance. Peasants gathered in bottles and used as a medicine.
3. In the Kaluga area, Russia, May, the end of day, 1998 (Chernobrov VA 2003) the tractor operator Yudenkov observed descending of a large shining oval body to height about 30 m. The observer has lost consciousness. Having come in consciousness, the eyewitness has found on an arable land on a place of the object flight a strip in width in some tens of meters, consisting of slime mass

Fig. 2.75 Red rain in India—based on materials (National Geographic Television 2010)

(as if a web on the ground). In several minutes the web disappeared, and on its place was left a set of the most thin, gray needles up to several centimeters in the length. It was not possible to analyze the needles.

4. During the summer of 1912 at a location near Sagaiskoe, Karatuzovsky, Russia, the "sky" had "fallen down to earth in a meadow" (Bychkov 2006). The object in question was brought to the settlement and put into a cup. It was up to 200 mm in length and up to 60–100 mm thick, whitish, semi-transparent, and of jelly consistence. The "sky" seemed to consist of separated linked hailstones with sizes greater than 1 cm. The next day this mass had disappeared to surprise of the inhabitants.
5. During the summer of 1920 at Gorodok, Minusinsk District, after a week of heavy rains in the summer evening when from the sky low clouds were moving away, suddenly on the skirts of the village in a side street from above has fallen a "trembler" (Bychkov 2006). Its falling was noticed by the local inhabitant F. A. Yushkova who has taken it in hands. According to Yushkova, the fallen mass was grayish-bluish color, cold, elastic-viscous, and opaque. That evening in the village has fallen some more «tremblers» in the size up to a bucket each. Next morning they still could be seen on their places.
6. Another description of such a gelatinous object was made by N. D. Isidorchik, a resident of Abakan (Bychkov 2006). "In end of June 1941 several kilometers from Karatuz, a group of women noticed falling of a large muddy-white mass. A round jelly-like clot had fallen from almost cloudless sky on a flat area with a light noise. Its diameter was about 600 mm and 200 mm thick. It had slight rainbow color. No inclusions had been noticed in a mobile mass. When women were going back to the village in the evening, this substance was not at the place."
7. Newspaper Komsomolskaya Pravda. 23.10.2013 The trembling substance was picked up in the winter—in February 2013. The substance that was lying on the grass was transferred to the laboratory of the Center for Biological Diversity

(Centre for UK Biodiversity (AMC), UK at the Natural History Museum. The research ended in vain.

8. Newspaper Komsomolskaya Pravda. 23.10.2013 Scientific journal. "Report of the British Association for the Advancement of Science" "On 8.10.1844, near Koblenz, a German gentleman accompanied by another person was walking late in the evening in the dark on a dry plowed field. They saw a glowing body coming down right next to them—no further than 20 m. And they heard it clearly, with a noise, hit the ground. They marked the place, returned there early the next morning and found a jelly-like mass of a grayish color."
9. Newspaper Komsomolskaya Pravda for 10.23.2013. 09.26.1950, four police officers from Philadelphia spotted a glowing object falling on the field. Approaching the crash site, they saw a pulsating mass with a diameter of about two meters. It was then filled with light, then faded. One of the policemen, who dared to touch that, claimed that the mass was soft to the touch, similar to gelatin. He tore off a piece that melted in the palm of his hand like snow. After half an hour, the strange mass disappeared without a trace.
10. BBC News. Dorset online. 26.01.2012 Bournemouth, Dorset. Blue balls 3 cm in diameter fell on a garden in Bournemouth during a downpour on Thursday afternoon. Steve Hornsby found about a dozen balloons in his garden. He said: "[They are] difficult to pick up, I had to take a spoon and throw them into a jam jar… The sky became really dark yellow… For a few seconds there was instant hail, and I thought, "What's that in the grass?"… They have an outer shell with a softer inner layer, but they are odorless, not sticky and do not melt." Walking through the garden, he found that there were many more blue spheres scattered on the grass. The Meteorological Department stated that the jelly-like substance is not "meteorological."
11. Komsomolskaya Pravda. 23.10.2013. The trembling substance was picked up in the winter—in February 2013… The substance that was lying on the grass was transferred to the laboratory of the Center for Biological Diversity (Centre for UK Biodiversity (AMC), UK at the Natural History Museum. The research ended in vain.
12. Komsomolskaya Pravda. 23.10.2013. Scientific journal. "Report of the British Association for the Advancement of Science" "On 8.10.1844, near Koblenz, a German gentleman accompanied by another person was walking late in the evening in the dark on a dry plowed field. They saw a glowing body coming down right next to them - no further than 20 m. And they heard it clearly, with a noise, hit the ground. They marked the place, returned there early the next morning and found a jelly-like mass of a grayish color."
13. A jelly-like object has fallen from the sky in the city of Oakville, Washington, the USA, August 7, 1994 (Buckland 2011; National Geographic Television 2010).

Sunny Barclift has found small clots of transparent gel on a farm of her mother. Unlike the majority of "star gels." Barclift has sent samples to the hospital from where they were promptly forwarded to the Washington State Department of Health for further analysis. The analysis has shown that in the gel there were many harmful bacteria. In the report have been listed Pseudomonas fluorescens, Enterobacter cloacae, and human leukocytes. Within several days after the first downpour Barklift's mother became ill (according to her doctor, she had "a severe inner ear infection"). Besides has died more than a dozen of kittens on the premises. Neither the FAA, nearby McCord Air Force Base, nor the State Health Department had an explanation for the falling gel. Three days later a police officer's patrol car was covered by more falling gel. After that the officer, and other city inhabitant, became so sick that they were hospitalized.

This event has been in detail described in a film (National Geographic Television 2010). In it the accent was emphasized on an opportunity of infection of city dwellers by emissions from AF helicopters, and use of city dwellers as experimental animals. Properties of bacteria features from a point of view of bacteria's participation in inter-cloud processes were not considered and a possibility of the star gels biological nature was not mentioned.

In the film (National Geographic Television 2010), red rains have been also shown, and was mentioned their possible connection with star gels.

Analysis of the phenomenon

The reason of this phenomenon can be connected with filling of clouds by particles of a dust of the organic and inorganic origin, including spores, pollen of plants, and other remains which are risen by air streams and are delivered to clouds artificially, and occupy clouds (Yablokov 2005; Sattler et al. 2001). Being nuclei of heterogeneous ice- and hail-forming, they in vortical cloudy cells form hailstones; flakes; little pieces of ice nearby; or on which can take place physical, chemical, and biological processes.

Colloidal particles

1. The analysis of the literature devoted to colloidal particles shows (Kroit 1955), (Shreider 1986), (Papkov 1974), that a number of organic and inorganic substances can form gels. Organic threads (cellulose) and particles of clay (Kroit 1955) (kaolin and minerals of bentonite type) with the size of 0.1–0.2 mm relate to them. The last particles can be noticed with an open eye; however, from the descriptions of "star jelly" follows that their components are invisible. Therefore, such particles cannot participate in formation of the 'Star jelly" but they can create nuclei of heterogeneous ice-forming and indirectly influence the new structure's forming. As to particles of wood (Papkov 1974) in the cloudy conditions at sufficiently low temperature they, most likely, have not enough time to be dissolved in water drops and to form gels.

Let uniform gel particles without an internal structure are formed. In this case, it can be shown (Byalko 1983) that they can be destroyed due to the turbulent friction forces

$$F_{fric}\, \rho_1\, v^2 R^2, \tag{2.16.1}$$

which become greater than the force of the surface tension

$$F_{surf}\ \sigma_0 R, \tag{2.16.2}$$

where R is a radius of a drop (by which we consider the nucleus of the structure), ρ_1 is air density, σ_0, is the surface tension of a liquid of the drop, and the drop velocity v is determined at equality of the gravitation and the friction forces:

$$v\ \sqrt{g\,\rho_0\,R/\rho_1}, \tag{2.16.3}$$

where g is free fall acceleration and ρ_0 is density of the drop's liquid. So these components cannot form the structure. From (2.16.1) to (2.16.3), it follows that at nuclei radius greater $R > 0.2$–0.3 cm (for example of water or alcohol) they will be disturbed by the turbulence and destroyed by air at falling down at their radiuses by 3–4 times greater than R, i.e. at $R \sim 1$ cm. These particles also cannot be considered as participants in creation of the "Star jelly" structure.

Let us note that particles of fibrous structure which could adsorb water are generated in many of high-temperature processes (Shreider 1986). The elongated fibers of carbon are formed at catalytic decomposition of carbon-containing gases at temperature from 600 up to 1300 K. Such gases are used in many industrial processes, and thread-like particles formed in their result can be sufficiently widely widespread. For example, it is known that they are formed during work of the enterprises of the oil, petrochemical, and nuclear industry (Shreider 1986) where deposition of the carbon fibers negatively affects efficiency of a heat transfer in various installations. Special researches of fibrous particles of the carbon formed at pyrometallurgical refinement in copper-smelting furnaces have shown (Shreider 1986) that fibers from 10 to $2 \cdot 10^3$ nm in diameter and from in length from 10^2 to 10^6 nm are formed. Certainly these particles will be united and consolidated at movement in a cloud, until they drop down to the ground. However, absence of a slime occurrence on their mechanism also does not allow such particles to be considered for participation in formation of large jelly-like air-meteors.

Particles in a cloud are usually strongly different with respect to their motion character. Brownian particles with sizes of several μm move under collisions with air molecules; the greater particles move down. Following (Bychkov 2006) let us consider a falling particle in a cloud. In the case of the spherical particle, its velocity for laminar motion and allowing for the Stokes force (and disregarding its charge and inter-cloud electric field) take the following form (Landau and Lifshitz 1988):

$$u_{sp} = \frac{2 \cdot \rho_p \cdot g \cdot R_{sp}^2}{9 \cdot \eta}, \tag{2.16.4}$$

where R_{sp} is its radius, ρ_p is the density of the particle, η is the physical air viscosity, and g is free fall acceleration.

In the case of the cylindrical particle, its velocity u_c for a length L greater than its radius R_c (the radius of the cylinder), $L \gg R_c$ is (2.16.4):

$$\frac{u_c}{\ln(3.7 \cdot \eta/(\rho_a \cdot R_c \cdot u_c))} = \frac{\rho_p \cdot g \cdot R_c^2}{4 \cdot \eta}, \tag{2.16.5}$$

where ρ_a is air density, other variables are the same as in (2.16.4).

Estimates for particles of 10 μm–0.1 mm show that cylindrical particles have a much smaller velocity as they fall (up to 10^4 times for the micron diameter size particles) than the spherical ones, it is well seen at deposition of dusty particles in light. It means that the suspended particles of elongated or thread form stay in the cloud cells longer, until they conglomerate with other particles turn to water and fall down to the earth in the form of drops. The situation becomes interesting if one supposes that there are coils or nets of micron size threads in a cloud (particles of clay have a form of plates, therefore they do not form a grid). In this case, they can stay in the cloud cell for an extremely long time. In this case, the question remains of increasing their volume to the macroscopic size typical of the gel-like meteors described above.

Biological aerosols

The situation becomes extremely interesting if these micron size threads represent the algae or bacteria that can propagate inside a cloud. For example, the effective size of a bacterium is $r \sim 10$ μm (Fedorov 1979) and it has to reproduce ~10^3 daughter bacteria in order for the bacteria to conglomerate and become equal in mass to the typical inter-cloud water drop, with the radius $r \sim 0.1$ mm.

Work (Sattler et al. 2001) has shown that bacteria can propagate in supercool clouds and they daily increase their mass up to 20% even at very cold conditions at ambient temperature $T = 0$ °C. In favorite conditions at temperatures $T > 4$ °C, the propagation of bacteria can by orders of magnitude be greater [19–20]. In clouds the highest temperature is located in the lower part, its mean value is about $T = 2$ °C (Rogers 1976; MacGorman and Rust 1998). So one has to expect growth of biological formation there. Algae and bacteria can survive in extremely cold conditions (Seckbech 2007). Some bacteria, such as Pseudomonas Syringae (Lindow et al. 1982), can even participate in creation of ice clusters and be nuclei of ice-forming. This shows that these bacteria can be candidates for nuclei of star jelly.

Let us estimate a number density of bacteria, which can be located in a spherical structure of radius $R_{str} = 10$ cm. At that, we consider that the typical sizes of the individual bacterium are length $l_b = 20$ μm and thickness ~ 2–5 μm. The structure consists of water drops and bacteria connecting them. At that, we consider a chain of bacteria between two drops. The structure begins to fall down if the size of drops is so large that they begin to move the net of bacteria with them. Using formula (2.16.4) one can find out that water drop of radius $r = 10^{-3}$ cm falls down with a velocity $1.2 \cdot 10^{-2}$ m/s, that is, very slow, i.e. it leaves a 1 km height cloud during ≈23 h, so we will not consider it. At the same time water drop of radius $r = 10^{-2}$ cm falls down with a velocity 1.2 m/s, so it leaves this cloud during ≈14 min, and this time is comparable with time of thunderstorm events. So we consider drops of water of $r = 10^{-2}$ cm. It is known that water content of a thunderstorm cloud is $M = 10^{-3}$ kg/m^3 (Rogers 1976), then the total number of drops N_d is

$$N_d = \frac{M}{\rho_w} \cdot \frac{R_{srt}^3}{r^3}, \tag{2.16.6}$$

where ρ_w is density of water.

The mean distance between the drops L is

$$L = \left(\frac{4 \cdot \pi}{3} \cdot \frac{1}{N_d}\right)^{1/3} \cdot R_{str}, \tag{2.16.7}$$

the number of bacteria between two drops is

$$N_b = L/l_b, \tag{2.16.8}$$

the total number of bacteria is

$$N_t = N_d \cdot N_b. \tag{2.16.9}$$

Inserting (2.16.6)-(2.16.8) into (2.16.9) and dividing it by the volume of the structure, one finds the following formula for number density of bacteria in the structure:

$$n_d = \left(M/\left(\rho_w \cdot \frac{4 \cdot \pi}{3}\right)\right)^{2/3} /(l_b \cdot r^2). \tag{2.16.10}$$

This formula gives the number density of bacteria in the structure $n_d \approx 2 \cdot 10^3$ cm^{-3} what is of the magnitude order of bacteria experimentally determined in clouds (Sattler et al. 2001).

Green-blue algae (cyanobacteria)

The analysis of possible applicants from bacteria (Fedorov 1977a, b; Seckbech 2007) shows that blue-green algae or its mutants under abnormal conditions can be of interest in the given problem. Blue-green algae (Cyanophyta) is a group of the most ancient autotroph plants and organisms in general. They meet in every possible habitats on all continents and reservoirs of the earth. They are with rare exception, microscopic plants: monocelled, colonial, and multicellular (filamentous). Colors of thalluses can be various (blue-green, olive-green, pink, or violet). For these algae is typical absence in a cell of the shaped core. The protoplasm filling a cell is divided into two layers: external, adjoining to an environment in which pigments are dissolved, and internal, in which the nuclear substance is localized. Cellular environments often become slimy. Sexual process is not present. Duplication of filamentary forms occurs by a disintegration of a filament in separate fragments, of monocelled ones—by division of the cell in two. Further, we consider two kinds which form rather large colonies.

Algae can survive in clouds since they are stable both against radiation and to strong decrease in temperature. At that time of their duplication is several hours (Fedorov 1977b). The typical sizes of one specimen: length ~ 20 μm, thickness ~ 2–5 μm. In principle, presence of the algae can lead to occurrence of gels. However, according to events N.1 and N.13 in samples of a material sent to laboratories, blue-green algae had not been detected. So we can look also for other candidates besides it.

Nostoc

Colonies macroscopical, more often from 1 mm up to 1 cm in a diameter, no often larger. Many kinds have outstretched irregular forms, from the outside gristly, inside soft, jelly-like, or firm. Annual occur in freshwater reservoirs and in soil. See Fig. 2.76.

Gloeotrichia

Jelly-like colonies, regular spherical or shapeless, continuous or hollow, attached or freely floating. Inside of a colony the radial arrangement numerous thread-like multicellular filaments are clearly visible.

Pseudomonas

About bacteria of Pseudomonas sort is known that they are widely spread in the nature (Fedorov 1977a; Rainey 2007; Mosby's Medical Dictionary 2009). They can be met in air, ground, sea, and freshwater reservoirs, sewage and in silt, oil, and on gas deposits. Pseudomonas have been found out on foodstuff, bodies of animals, plants, and also in purulent wounds and excrement of sick mammals (Fedorov 1977a).

Fig. 2.76 Schematic image and photo of the Nostoc colony (Fedorov 1979)

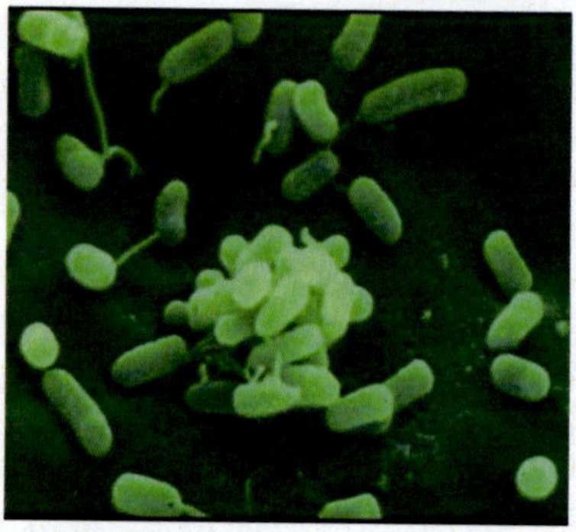

Fig. 2.77 Ps.Fluorescens—Schematic image. Mobile soil bacteria, from the website www.scienceclarified.com (modified)

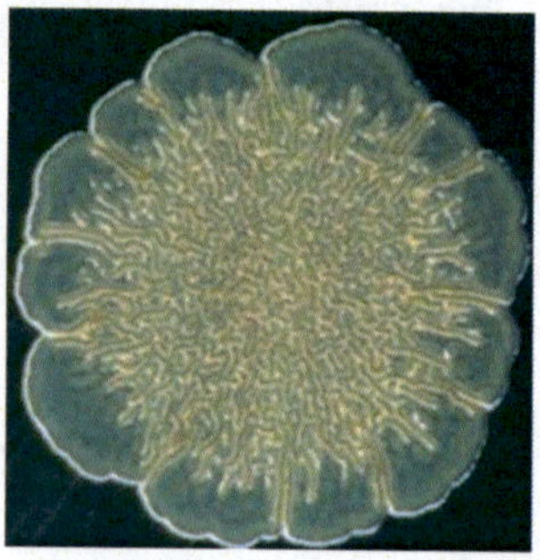

Fig. 2.78 Colony of mutant bacteria Ps. Fluorescens on the surface of the nutrient medium www.eurekalert.org (modified)

Pseudomonas are mobile and have the form of straight or curved sticks and two polarly located flagellums, their size 2–3 × 0.6 μm, see Figs. 2.77 and 2.78.

Cells in cultures are often united in small lumps or the grains surrounded by a thick mucous membrane, sometimes sharply outlined—zoo-gels. Formation of plentiful slime singles out this sort of bacteria. The Pseudomonas "collective" is formed of bacteria-mutants (Rainey 2007), which release the increased quantity of the substances promoting gluing of cells. Such bacteria after division cannot "come unstuck" from each other. At that single cells are in thickness of a broth, and stuck together emerge on a surface where there is much more of oxygen. The film («bacterial mat») on a surface of environment is formed as a result.

Majority of bacteria of sort Pseudomonas possess heterotrophic type of a metabolism, i.e. for construction of a body the ready organic substance is required to them. Biosynthetic processes thus are realized due to an exchange of the oxidizing type. Some representatives of this sort can exist due to anaerobic nitrate breath, others use energy of hydrogen oxidation. Many kinds of Pseudomonas form pigments, various on colors and the chemical nature; some of them synthesize vitamins, antibiotics, and toxins. Ps. Fluorescens sometimes is used for cleaning of soil, water, capacities, and sewage of the industrial enterprises from pollution by oil products in northern regions since they can be developed at sufficiently low temperatures 3–15 °C.

Let us note also Ps. Aurantiaca (Fedorov 1977a). Cells of this kind of bacteria are stick-like, their size is also 2–3 × 0.6 μm, they are mobile, have on the end 2–5 flagellums. Colonies on nutrient mediums have orange or red-yellow colors, smooth, glitter, flat, or convex. The pigment diffuses into a substratum. Cultures well grow on usual nutrient mediums. Optimum temperature of growth is nearly 25 °C. The structure of an orange pigment includes some substances of the various chemical nature.

We also note the kind Ps. Syringae. These bacteria possess abilities to form ice clusters. Present in jelly-like substance in Events N 3 and N 4 of the ice clusters and hailstones is explainable at presence of the ice-forming nuclei in the form of these bacteria.

The presented descriptions of properties of bacteria Ps. Fluorescens, Ps. Aurantiaca, and Syringae allow to select them as applicants for a role of the bacteria participating in formation of star gels of various colors and structure.

Hypothesis of the gelatinous meteors physical nature

Taking all these facts into account, one can formulate the following scenario of the gelatinous meteor appearance (Bychkov 2012, 2014). The specimen of bacteria uplifted from the ground on a level of clouds by streams of air (or delivered there artificially) is grasped by cloudy cells where they propagate, eating organic particles or the plants also grasped by streams of air. There they create the filaments covered by slime or coils of the bacteria interconnected with each other, i.e. they create a skeleton for complex structure.

This takes place during the time when this coil or complex could be regarded as a filtered one, i.e. when the mean free path of molecules λ_{air} in air

$$\lambda_{air}\ (N_0\,\sigma_m)^{-1}$$

is smaller than the mean free path λ_{str} with respect to the structure particles, $\lambda_{str}\ (N_{str}\,\sigma_{str})^{-1}$,

where N_0, N_{str} are, respectively, a concentration of particles in air and in the structure, and σ_m, σ_{str} are, respectively, elastic cross section of air molecules and the cross section of the structure molecules, or

$$\lambda_{air} > \lambda_{str}.$$

If then, the structure becomes impermeable, it begins to fall with a sail-type structure with rarefied density. It means that it lasts until filamentary element in the cloud can be considered an element independent of others. However, everything changes, when the slime (a product of bacteria duplication) densely covers apertures in a grid.

For typical conditions in the atmosphere at the altitude of cumulus clouds existence, 2–5 km, the mean free path of air molecules is $\lambda_{air} \sim 10^{-5}$ cm. Hence, at $\lambda_{str} > 10$ μm coils or the grid elements covered by slime become impenetrable for air and fall, but as objects with density considerably smaller, than at drops of water. Until this time they can easily be besieged on particles of the ice or the hailstones, creating large, interconnected, glued together structures. These structures can grasp drops of water in a cloud and turn to a complex structure with a skeleton of the bacteria, slime, water and the ice particles, or the hailstones. At that the skeleton of bacteria will be invisible. Bacterial filaments and films can cover and stick together hailstones, forming greater macroscopical areas.

Interpretation

Using this hypothesis one can explain observation N.1 and N.5 in which it is said that the masses, which dropped down on the ground, included bacteria. It is possible to explain also a smell of the object: at destruction of bacteria the putrefactive smell of products of biological disintegration can be formed. Bluish, whitish, and even orange colors (Ps. Aurantiaca) also are typical for the bacteria Ps. Fluorescens.

The evaporation of water from the conglomerate occurs more quickly than of the waters of the same volume, due to presence in a conglomerate of a plenty of cavities unfilled by water. At that, bacteria aspire to leave into ground, which for them is a habitat of existence that will promote fast disappearance of the object.

Case of gathering of gel from the observation N.2 one can explain, namely, by the ability of these bacteria to clean water and soil.

In the description of the object from the observation N.3 long filaments of muculent mass (as if a web on the ground) and the most thin, gray thin needles which have quickly disappeared by many features reminds the description of long ice particles arising in clouds (Rogers 1976) and quickly thawing on the ground. Large particles in the description N.4 also correspond to hailstones, formed in clouds (Rogers 1976). Here it is possible to assume a mixture of several kinds of bacteria Ps. One of them formed crystals (Ps. Syringae), others—mucous aggregates.

In case of occurrence of red or red-orange gels, or bloody rains the bacteria Ps. Aurantiaca are applicants for a role of a color component.

Let us address again to the event N.13. In particular to the bacteria Enterobacter mentioned in it. Enterobacter is common variety of the stick-like bacteria of family Enterobacteriaceae of (Mosby's Medical Dictionary 2009) found in excrement of people and animals, dairy products, sewage, ground, and water. E. Cloacae and E. Aerogenes are the inter-hospital pathogenic microorganisms responsible for many infections such as bacteremia, infections of the lower respiratory ways, infections of uric ways, and a septic arthritis. They grow on environments for enterobacteria at temperature 37 °C, pH 7, forming muddy smooth colonies. Their appearance is represented in Fig. 2.79. The description of these colonies mismatches the description of gelatinous meteors mucous surface.

Therefore, it is possible to assume that " Enterobacter cloacae" and human leukocytes from the description N.13, most likely, have been brought into clouds together with Ps. Fluorescens from the Earth, and then survived and existed in the slime

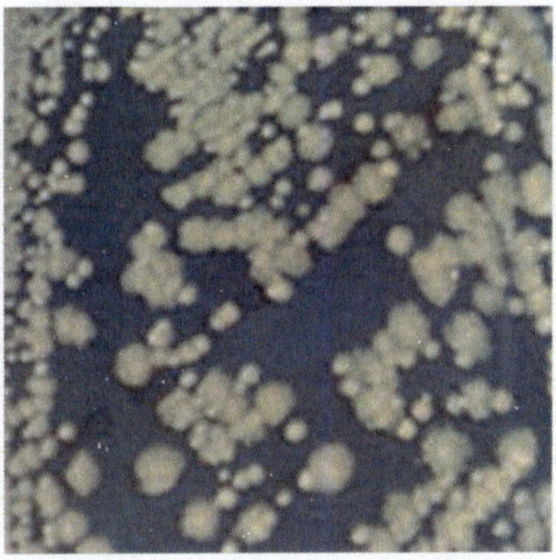

Fig. 2.79 Colonies of "Enterobacter cloacae," gastroscan.ru

produced by the Pseudomonas. Thus, helicopters from the description N.7 most likely have no relation to them. At that, natural experiments with health of people are considered improbable now.

Comment. The mechanism (hypothesis) of formation of gelatinous meteors as a result of duplication and development of colonies of bacteria such as Pseudomonas (or Algae), representing the coils, grids, thread packs covered by slime and glued by them the ice and hail particles arising in clouds on heterogeneous particles also brought from the ground is represented. Falling of large gelatinous meteors can occur during thunderstorms, rains, and downpours, and at influence on clouds of space dust pieces improving conditions of these meteors falling.

The given mechanism also shows a possible and effective way of these bacteria propagation over a surface of the earth by means of clouds. Moreover, this way can lead to propagation of pathogenic bacteria together with bacteria Pseudomonas, which survive in slime of the bacteria Pseudomonas. By this process, it is possible to explain unexpected epidemics of bird flu and unexpected flashes of infectious diseases in places far removed from the known centers.

This puts forward a question on preliminary measures against appearance of such bacteria in streams of air from hospital, medical institution, and microbiological laboratories burial of wastes. It is useful to inform people about this phenomena and possible harmful consequences of direct contact with gelatinous natural objects.

Some other structures

When talking about gelemeteors, they always remember another substance that falls out of long-lived luminous formations in front of people. These are "angel hair"-thin, spider-like threads that can cover entire cities. In the hands they melt quickly, in the open air-slower, and in a sealed container they are well preserved, and they can be brought to the laboratory. Here are some examples (Fig. 2.80).

1. Newspaper Komsomolskaya Pravda from 10.23.2013. On 10.17.1952, the sky was clear over the French town of Oloron, and the weather was excellent. Then hundreds of residents saw a whitish "cloud" of an ideal shape—a long narrow cylinder shrouded in a whitish haze, tilted at an angle. It moved slowly in a

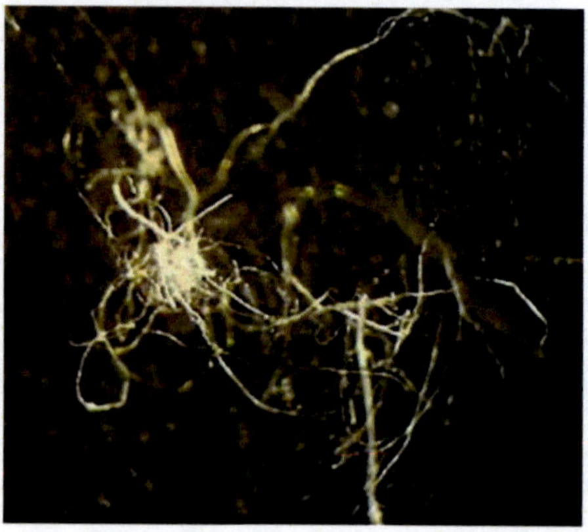

Fig. 2.80 Newspaper Komsomolskaya Pravda. 23.10. 2013. "Angel's hair" falling to the ground

straight line and left a white plume behind its upper part. The cylinder was not alone—dozens of balls were flying in front of it. To the naked eye, they seemed colorless, like balls of smoke. However, through binoculars, the headmaster of the school, Yves Prijan, discerned a red ball surrounded by a yellowish ring in the middle of each "tangle." There was a trail behind the squadron of objects, which slowly settled to the ground. For several hours, threads and lumps of matter hung on trees, wires, and roofs, as if millions of spiders had spun a web at once. Ten days later, exactly the same scene played out over another French town—Gaillac. A giant cylinder, tilted at an angle, accompanied by dozens of objects crossed the sky, covering streets and houses with thin threads.

2. (Tutt 2004) October 1962. Pape R. H. the captain of the ship "Roxburgsky Castle," bypassing the deck, found thin fibers resembling hair on the cables enclosing it. "Having called the first assistant, I pulled one of these threads from the pillars and found that it was very strong and elastic. I tried to break it, but I did not succeed immediately, i.e. the substance was much stronger than the web. After I held the thread in my hands for a few minutes, it melted into thin air—in other words, it simply disappeared. Raising our heads, we saw small cocoons of this substance slowly gliding from the sky, but we could not see anything that could be the source of the strange rain."
3. Newspaper Komsomolskaya Pravda from 2.10.2013. 4.05.1981 residents of the American city of Denville observed "cobwebs" falling from the sky. They clung to houses, wires, and trees. Everything began to look like "hundreds of years have passed." Local resident William Hammer looked up and noticed "something in the sky, spinning around the sun." Through the binoculars, he saw that metal disks were spinning in the sky, flying back and forth. Whitish material flowed from their bottoms in a continuous stream, covering everything around.

When the "angel's hair" was wound on a stick and sealed in a test tube, the analysis showed that it was a rather complex chemical compound with a predominance of boron, silicon, calcium, and magnesium.

Comment. Such objects are likely to have a man-made nature. Particles of fibrous structure are generated in a number of high-temperature processes. Elongated carbon fibers are formed during the catalytic decomposition of carbon-containing gases at temperatures from 600 to 1300 K. Such gases are used in many industrial processes, and the resulting filamentous particles can be quite widespread. For example, it is known that they are formed during the work of enterprises of the oil, petrochemical, and nuclear industries (Shreider 1986), where carbon fiber deposits negatively affect the efficiency of heat transfer in various installations. A special study of fibrous carbon particles formed during pyrometallurgical refining in copper-smelting furnaces showed (Shreider 1986) that fibers with a diameter of 10 to 2×10^3 nm and a length of 10^2 to 10^6 nm (1 mm) are formed, which are assembled into aggregates of 10–20 microns in size. Their small size does not allow such particles to claim to participate in the formation of macroscopic gel-like meteors, but they can be the cause of the appearance of "Angel Hair" in a number of cases related to the industrial activities of people.

2.10.8 Photos and Videos with BL

Reliable material evidence of the existence of BL (photos, videos, traces of its action) rarely fall into the hands of researchers. However, unlike the descriptions of witnesses, who, as a rule, have very little time to analyze the features of the observed phenomena, these documents allow to study these objects in calm conditions using available equipment and methods. Currently, due to the widespread use of mobile phones with video cameras, the probability of obtaining visual BL information has increased significantly. A detailed study of the video data allows to put certain reasonable questions to a particular BL model and check its "strength." This section presents an analysis of a small number of photos and videos of films with BL. From this, the following conclusions are likely to follow: (1) about the presence of an uncompensated electric charge of the BL; (2) about the presence of a corona-type spatial discharge on its surface; (3) about its radiation of radio waves in the wavelength range of 1–13 cm, in addition to visible light; (4) that the BL substance often consists of small elements. The conclusions obtained are also confirmed in comparison with other cases of the manifestation of BL properties such as exposure to window panes, metal ring objects, radios, etc.

The BL occasionally comes into the camera's field of view when shooting linear lightning discharges. Usually, during a thunderstorm, the photographer directs the camera to the area of the sky, where lightning discharges operate, and opens the shutter for a few minutes. After a linear discharge of lightning closes the shutter and rewinds the film. Sometimes during the exposure, the photographer leaves the room. One can imagine his amazement when, after developing the film, he, along with the expected images of linear lightning discharges, discovers several zigzag continuous lines on it.

As an example, (Singer 1971) presents two photographs obtained in this way. At the same time, he warns against false photographs, in which a solid trace may appear as a result of camera movement in the presence of a bright light source in its field of view, for example, a street lamp.

In his book, there are also two photographs of BL taken with a short exposure. On one of them (Singer 1971, Fig. 17) (see Fig. 2.81), a large oval is visible, the major axis of which is oriented vertically, the upper and lower parts of the oval are somewhat blurred, and the ratio of the lengths of the axes is 1.28:1. It is reported that the diameter of the ball was 35 cm and during recording the ball fell to the ground. Let's analyze the photo. You can independently find (Bychkov et al. 2010) that the length of the large axis of the oval is 35 cm × 1.28 = 44.8 cm, and the difference in the lengths of the axes is 9.8 cm. Such a distortion of the image of the ball could occur if the ball moved a distance of 4.9 cm during the exposure. The duration of exposure is not specified in the book, so we will take it equal to the average shutter speed of the camera 1/30 s. The speed of the ball movement in this case $vb = 4.9 \times 10–2 \times 30 = 1.47$ m/s, this is a typical speed of movement for a BL. The lower level of negative blackening in the upper and lower parts of the oval can easily be explained by the vertical movement of the ball. For some reason, everyone, including Singer, is talking about the oval, not the ball. In the second photo presented by Singer (Singer 1971, Fig. 18), you can see a photo of a BL with a diameter of 50 cm, accompanied by the emission of sparks.

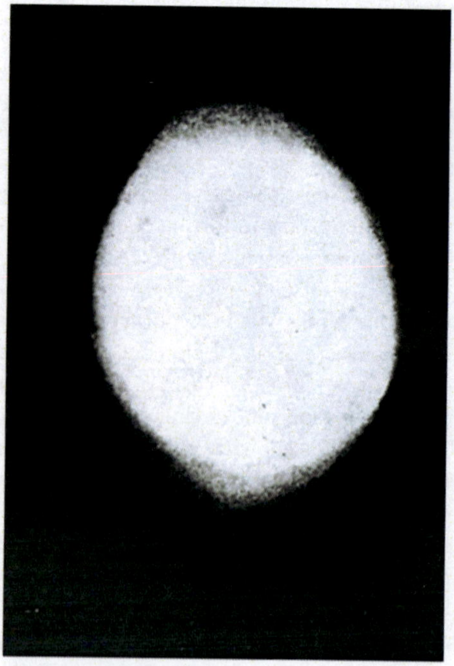

Fig. 2.81 A photograph taken during a thunderstorm in 1935. By Schneiderman (Singer 1971)

In 1960, the magazine "Ogonyok" published a photo taken in Sochi by I. Shagin during a summer night thunderstorm (see Fig. 2.82). Doctor of Technical Sciences I. S. Stekolnikov wrote in the comments that this photo was taken with a 5-min exposure, and a trace appeared from the sky in the form of a "rope with knots" left a descending with a diameter of 8–10 cm. "On the visible path, the Ball had an almost constant speed, judging by the uniform glow of the trace." The ball fell vertically about 5–7 m, after which it made "oscillatory movements sideways, up and down, showing itself near a kind of node. At these points, the BL became uncontrollable, as if it was losing its guiding force. On the last section of the way, having fallen below the top of the lighting mast and having passed a complex zigzag path, the BL then is 10–15 cm/s. He noted that "next to the BL track there are two completely identical, albeit lighter tracks, which apparently arose due to the reflection of light from the film inside the camera." Unfortunately, there were no further comments. There was no data on shooting conditions, type of camera, and film. It was not said whether the operator himself saw the falling bright ball, whether there were lightning strikes, etc. Fortunately, some of these questions (but not all) can be answered when analyzing a photo.

The discussed photo is presented in (Stenhoff 1999). The author expressed doubt about the authenticity of this photo from the point of view of the presence of many traces of a bright object in it. In his opinion, this trace could be formed by several

Fig. 2.82 Ball lightning near the Black Sea (Shagin and Stekolnikov1960)

different light sources and camera shifts during its positioning. This explanation cannot be taken seriously, because in this case it should be recognized that there are three lamps on the mast with different light intensities. The brightest of them had to be located not at the upper end of the pole, but at 2/3 of its height. Most likely, all the lamps were turned off during the thunderstorm. Thus, Stekolnikov's version, based on the reflection of light inside the camera, seems more plausible.

Taking this version for granted, the authors (Nikitin et al. 2006, 2007) gave an answer why the linear lightning track is not double and why the nodes of the additional images are shifted not horizontally, but at an angle of 55 to the horizon. These authors could not explain this phenomenon by overexposing the film with a near-field light source (which was the case), since this is not true for linear lightning, although the intensity of blackening of the film is comparable with both sources.

To answer these questions, the authors (Nikitin et al. 2006, 2007) conducted experiments with two bright sources when shooting with a Smena camera (one of the most common in the 1960s in the USSR) with an Industriar lens with a focal length of $F = 50$ mm, a sight at a distance of 5 m, and a lens. set the stop to f/3.5. A black-and-white "Svema" film with a sensitivity of 200 GOST units (ASA 200) was used. Analysis of the results (Nikitin et al. 2006, 2007; Bychkov and Nikitin 2014) allowed to conclude that when shooting, Shagin's camera was aimed at the place of linear lightning. Therefore, the shift of the additional image caused by light reflections inside the camera turned out to be small and only expanded the image of the lightning channel. It was found that additional images of BL tracks running sideways from the axis were shifted by a noticeable distance. Non-horizontal displacement of nodes on the route can be caused by the camera's tilt to the horizon. The experimental results (Nikitin et al. 2006, 2007) did not completely coincide with Shagin's photograph. In their case, the secondary lighter image was shifted towards the camera axis, and in the case of Shagin, the additional image was shifted in the opposite direction. If the shooting was carried out through a double window pane (although it is unlikely that there were double-glazed windows in Sochi and, moreover, they were closed in the summer), then an additional image of a bright object located to the left of the optical axis of the camera lens will be shifted to the axis of the camera lens in case the plane of the glass and the axis of the lens are perpendicular. Thus, additional images in Shagin's photo can be considered as having arisen due to multiple reflections of light by lenses or camera walls. Analysis of photographs in Fig. 2.81 showed that the distance to linear lightning was more than 13 km in diameter BL was equal to 3 cm, and the lengths of segments of a trace, counting from top to bottom, were 6.4, 5.8, and 5.6 M.

The authors, based on the model, positively charged BL and information about the distribution of charge in the lightning, and restored the charge of the BL. Nevertheless, it can be shown that a negatively charged BL could also be taken in this photo.

Ball lightning in Dolgoprudny

On the Internet in 2009 there was a video recording of an event that occurred near the city of Dolgoprudny in the Moscow region during a thunderstorm (YOUTUBE 1 2009; YOUTUBE 2 2009). Its author, Sergey Nikolaev, accidentally shot a spherical glowing yellow object with a purple aura, which most likely appeared in a thundercloud a few tens of seconds after the lightning discharge (see Fig. 2.83.). The object was captured by a Canon A 1000 camera in video mode. This camera has a 4 × zoom with an image stabilizer that changes the focal length from 6.2 to 24.8 mm and a maximum aperture of F2.5–F5.6. The sensor is a 1/2.3 in. (6.174.55 mm) CCD matrix and 10 effective megapixels. The resolution of the videotape is 640 × 480 pixels at a frequency of 30 frames per second. The author of the film kindly allowed us to use his film to study the properties of the object. The object was observed for 39 s. At the end of its service life, it split into two parts (see Fig. 2.84.), the decay of which lasted 1.7 s. The object was moving from left to right (from south to north) at a constant speed, rising up at an angle of 9.60 to the horizon.

The analysis of the terrain showed that the shooting was carried out from the loggia on the 7th floor of the house located at the address: Likhachevskoe highway, Dolgoprudny, Moscow region, Russia (Nikitin et al. 2013).

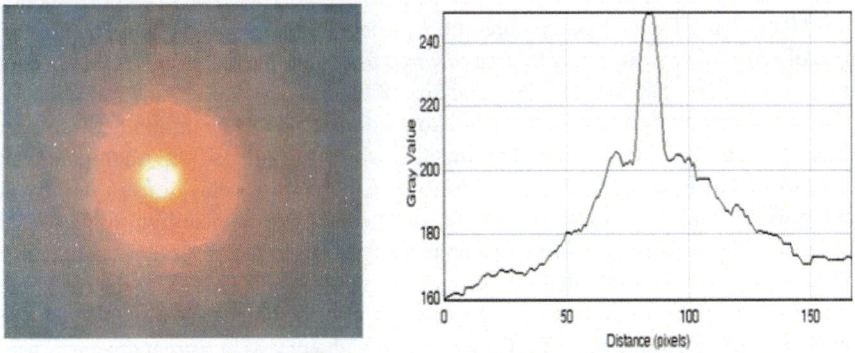

Fig. 2.83 View of the glowing object. On the right is the distribution of the brightness of the object along its radius

Fig. 2.84 A photo of two BLs surrounded by a dark halo

The object was observed to the east of the house. The film shows a 9-storey house, the wall closest to it is located at a distance of 100 m from the shooting location, and to the left of it are two 19-storey houses standing at a distance of 410 m. According to calculations, the standard height of the floor is approximately 3.2 m. The estimated height of the object above ground level, located at a distance of $l1 = 100$ m, is 68 m, and its projection on the plain, remote at $l2 = 410$ m, is 203 m. Subtracting from these dimensions the height of the point at which the operator was located, equal to 24.3 m, it was possible to find the values of the heights of the object above the horizontal shooting plane for two distances: $h1 = 68 - 24.3 = 43.7$ m (for $11 = 100$ m) and $h2 = 203 - 24.3 = 178.7$ m (for $12 = 410$ m). From here we find the value of the tangent of the angle α of the direction of the object relative to the horizontal $tn\ \alpha = h1/l1 = h2/l2 = 0.437$ ($\alpha = 23.60°$).

The first part of the film was shot with a small focal length. The diameter of the bright core of the object corresponds to 6 pixels of the photosensitive matrix, and the size of the luminous halo is 20 pixels. The height of the house floor (3.2 m), removed at $l1 = 100$ m, is displayed in the image with 20 pixels. Thus, one pixel corresponds to 0.16 microns. The size of the core of the projection of an object on a plane distant by a distance of $l1 = 100$ m is $0.16\ \text{m} \times 6 = 0.96$ m, and the angular size of the core is $\omega_c = 9.6 \cdot 10^{-3}$ rad $= 0.55°$. Angular size of the halo $\omega_h = 3.2 \cdot 10^{-2}$ rad $= 1.83°$.

It is assumed that the object was somehow connected with the lightning discharge. In the film (YOUTUBE 1 2009), 4 moments can be distinguished when a thunderclap is heard after a lightning flash. According to [10], the average delay time for the arrival of thunder is 9.62 ± 1.15 s. Based on this, the distance to the linear lightning is equal to 3250 ± 400 m. Since the luminous object was closer to the observer than the edge of the cloud, it is assumed that the distance to it did not exceed $L = 2000$ m. There is a large factor of arbitrariness in the estimation of distance (Nikitin et al. 2013); however, there are other facts confirming that the object was at a great distance from the observer. Taking the value $L = 2000$ m, it is found that the height of the object above ground level $H = L \cdot \sin 23.6° = 800$ m, the diameter of the core $Dc = L \times \omega_c = 19.2$ m, and the size of the halo $D_h = L \times \omega_h\ \ h = 64$ m. The angular displacement of the object in the horizontal direction during observation (39 s) was $\gamma = 28°$. From here, this distance is $G = 2 \cdot L \sin(\gamma/2) = 970$ m. Thus, the projection of the object's velocity onto the plain at a distance of $L = 2000$ m was $v = 970\ \text{m}/39\ \text{s} = 25$ m/s. A detailed discussion of the observation can be found in (Bychkov and Nikitin 2014).

Ball lightning in Mitino

On 27.07.2015, a unique event took place. BL was simultaneously registered by three observers located at different points and filming the object with a video camera in the phone. Knowledge of their coordinates and geography of the terrain, and determining the wind speed allowed in (Nikitin et al. 2018) to determine the distance to the BL, its diameter of 0.75 m and the speed of its movement at different points in time. The BL movement could be divided into three stages. At first, it was approaching the observers at wind speed. Then, for about 80 s, it moved at a height of 40–50 m inside a restricted zone with a width of 12 m and a length of 100 m. At the last stage, it rose at a speed of 5.7 m/s.

Some frames from video movies

(Kratko-News.com) 02/01/2018/0:00. Belaya. Columbia. 12. 30. 2017. A tourist recorded an unusual video from the camera of his mobile phone, Observation time is more than 20 s.

While filming the surroundings, a ball suddenly flew from his hotel room in the direction of the tourist's camera before it disappeared. Several people in this area have witnessed such an inexplicable observation. One of the commentators says that he has already seen a similar ball twice.

Since the uploader account does not show any violations such as fake or cgi-generated videos, we may consider that tourist captured a BL phenomenon (Fig. 2.85).

(https://finance.rambler.ru/other/47143122/?utm_content=finance_media&utm_medium=read_more&utm_source=copylink) 03.09. 21 Ball lightning in Yekaterinburg caught on video. Russia. Veronika Polynina.

A record of a rare natural phenomenon is published by the LifeNews telegram channel. The recording shows how a fireball forms in the sky, which burns up and disappears after a few seconds of flight (Fig. 2.86).

Fig. 2.85 Frames of BL in Columbia

Fig. 2.86 Frames of BL in Yekaterinburg

(Dijkhuis 2006) Dijkuis G. C. male, Ph.D. in Physics. 10.09. 2005. Heavy thunderstorm at a main street crossing the Rijswijk, the Netherlands.

Bystanders saw a bright-blue-white ball stuck to a tall lamppost for one minute before gradually extinguishing. Two eyewitnesses photographed this ball from nearby positions, one using her digital camera, and the other his cell phone. A month after the event the author took digital photos detailing the street crossing's complicated supports and wiring for streetlights, traffic lights, and electric power lines for trams. Comparison with corresponding fair weather photos locates the ball near a diagonal support bar attached to a tall support. A close-up photo indicates erosive damage on suspension wires near the lamppost, but no loose ends. By its brightness, color, fuzzy edges, and immobility near DC power lines, this event resembles a welding arc. By matching with similar welding arc photo the author arrived at 10–100 kW level for this BL event. For a self-supported lifetime of one minute, it would need storage of 06–6 MJ as internal energy. The comment of the observer F. Magdalens described the ball as slightly smaller than a soccer ball, and very bright, comparable with a photo flash radiating continuously (Fig. 2.87).

Comment. The information given by the eyewitness F. Magdalens and the estimate of BL energy store by Dijkuis G. C. give the estimate for this BL energy density as $0.7–7 \cdot 10^{9}$ J/m^3. That shows that the BL can be considered as high-energy object.

(https//ivrb.ru/07:39) 30.07.2021. St. Petersb. Russia. BL lightning got tangled in wires in St. Petersburg—video Author: Alexander Selivanov. On Thursday, a thunderstorm hits the Northern capital. BL hits the wires on one of the streets of the city. Observation time is more than 6 s during heavy rain (Fig. 2.88).

Comment. The phenomenon is similar to the previous one.

Comments to the section

These comments are based mainly on (Nikitin 2020). The use of digital video technology has greatly expanded the possibilities of studying such short-lived objects

Fig. 2.87 Digital photo of BL event by eye witness Mirjam Ruyter10.09. 2005. 15:19

Fig. 2.88 Frames of BL in St. Petersburg

as BL. A detailed analysis of the videos can significantly improve the reliability of determining its parameters. However, the use of new technology in some cases may be accompanied by the appearance of undesirable effects that must be taken into account when processing the results of observations.

A common circumstance for shooting BL is that it is almost always conducted under conditions of automatic exposure detection. In this case, the exposure level is determined not by the brightness of the glowing object, but by the brightness of the background. This leads to overexposure of the image of the object, which as a result makes it impossible to detect the details of its structure. In this respect, digital photography is completely inferior to film photography. If for the latter, thanks to the logarithmic law of blackening of the film, there is always hope to somehow correct the overexposed image, then for digital technology it is, in principle, impossible. The reason for this is that when the full charge level of the photosensitive matrix cell is reached, the further action of light only worsens the image due to the spreading of the charge to neighboring cells. In addition to the described difficulty, digital cameras can present other, more serious "surprises." The most trivial of them is the appearance of "mysterious" balls when shooting in a humid atmosphere. The reason for this effect is the shooting of raindrops or snowflakes illuminated by a flashbulb that happened to be near the camera. Another common defect in images of bright objects is the formation of black contours around them, Fig. 2.84. The appearance of these contours is the result of digital image processing by the camera computer. To obtain color information, an element of a 2 × 2 pixel photosensitive matrix is used, parts of which are covered with green, blue, and red filters. To get the correct color information, the original image is defocused using a special plate to a size of 2 × 2 pixels. After that, the image is processed using a special algorithm. This algorithm uses the "aggravation" program to reassemble information from a 2 × 2 pixel field into a 1 × 1 pixel field. This method is called "Unsharp Mask." This method is based on an optical illusion, when the eye will see the contour more sharply if you increase the brightness of the image in the light part near the border of the object and reduce the brightness in the dark part. If there is a bright object in the frame, then a dark rim appears around it. This procedure is also used in the Photoshop program to

Fig. 2.89 An image of a ball lightning shot in 2015 in Mitino

increase the clarity of digital images. The program compares the brightness values of neighboring pixels. If they are very different, then the program defines them as elements of the contour of the object and makes them even more contrasting. Pixels around light pixels become brighter, and pixels around dark pixels become darker (Andryushchenko and Kovalenko 2007).

Figure 2.89 shows an image of an object taken on July 27, 2015 in the Moscow District of Mitino using a smartphone NTS Desire 600 Dual Sum (Nikitin 2020). The image on the photosensitive matrix looks like a light spot surrounded by a black line. However, the observer said that he saw a round object, which he called the BL. In this regard, the question arises: what is more to believe—the testimony of an eyewitness or a "picture" created by a video camera? This question is not groundless, since when using methods of digital processing of "noisy" images (Gruzman et al. 2000; Huang 1979) you can sometimes find details in them that resemble, for example, drone propellers. Thus, when analyzing photo and video materials, it is always necessary to take into account the possibility of instrumental image distortions and take into account additional information received from observers of the event. Photographic materials should be accompanied by a description of the conditions of observation and shooting—the state of the weather, the events preceding the appearance of ball lightning, and its behavior in the period before the shooting. In addition, contact and conversation with the authors of photographs and videos can prevent the use of fakes and practical jokes, which are available in large numbers on the Internet.

2.10.9 On Phosphenes Formed as a Result of Energy Radiation from Linear Lightning Flashes

Material of this section is based on the work (Bychkov and Nikitin 2014). The influence of electromagnetic fields and phenomena on perception has been known for a long time. The effect of flash on perception was discussed in (Stakhanov IP 1996; Grigoriev et al. 1993), where it was shown that these effects are easily detected when analyzing a detailed interview of a BL observer.

Recently, there has been an opinion that BL is a kind of hallucination caused by the electromagnetic effect of linear lightning in a thunderstorm (Cooray et al. 2011) or the so-called "phosphene" hypothesis. According to the definition, "phosphenes" is a sensation of light caused by pressure on the eyelid of a closed eye or other mechanical or electrical interference with the visual system. At the same time, it was argued in (Keul et al. 2008): "According to eyewitness reports, in about 34% of cases, the duration of BL observation was less than 3 s, in about 24% of cases it lasted 3–5 s and in 19.5% of cases it lasted 5–10 s. It is currently unknown whether a single excitation due to a microsecond or such a long burst of energy radiation can cause visual sensations equal to the stated duration of a BL. However, some lightning bolts have a long duration and may contain a large number of subsequent reverse lightning strikes. Even in the case when the visual arousal from one flash is instantaneous, repeated arousals from subsequent shocks can be interpreted by the brain as a long continuous event." It has already been shown in (Dmitriev et al. 1986) that the "phosphenes" hypothesis has a low probability relative to the observation of BL.

Using BL statistics, we can roughly estimate this effect on BL observers. At the same time, we believe that all observers were in equal conditions and the same state of health (violation of this condition reduces the number of reliable observations). The probability (frequency or number of cases out of a hundred events) of observing a BL simultaneously with a linear lightning strike during a thunderstorm was (Cooray et al. 2011) fixed at $P1 = 0.106 \pm 0.007$ (1924 descriptions). The probability of observing $P2$ from a distance of 0–50 m, from which an electromagnetic effect can be caused, is $P2 = 0.87 \pm 0.052$ (Dmitriev et al. 1986), and from a more realistic distance of 0–10 m, which is true for both indoor and outdoor observations $P2 = 0.74 \pm 0.038$ (Cooray et al. 2011) (1441 observations).

Since these observations are uncorrelated, $P1$ and $P2$ are independent, then the total probability is $PT = P1 \pm P2$. Using the presented data, we obtain an upper estimate of the probability of this effect $PT1 \approx 0.092$ and $PT2 \approx 0.0784$.

It may also be recalled that the probability of observing $PTH = 0.51 \pm 0.016$ during thunderstorms is (Keul et al. 2008) $PTH = 0.51 \pm 0.016$ (1924 descriptions). Using the presented data, we obtain an estimate from above for the probability of this effect $PTTH1 \approx 0.047$ and $PTTH2 \approx 0.04$.

It is also possible to take into account the place of occurrence of the BL: in the linear lightning channel $\text{P3} = 0.101 \pm 0.019$, and at the point of impact of the BL $P3 = 0.11 \pm 0$. Here we do not know the definite distance between the BL and the observer.

In this case, $PT = P1 \times P3$, which gives $PT31 \approx 0.0101$ and $PT32 \approx 0.0117$.

Thus, using observational data from all distances from 0 to 50 m, we get the probability $PT = P1 \times P2 \times P3$, which gives $PT41 \approx 0.00679$ and $PT42 = 0.010$. And accounting the frequency of occurrence of BL during thunderstorms, we obtain the total probability $PTTH51 = 0.00346$ and $PTTH52 = 0.0051$.

Thus, it can be seen that the probability of the hypothesis about "phosphenes" is very small and in numbers this means that the estimate of the effect from above is approximately 3–50 "BL phosphenes" per 1000 observations. We also showed that the more detailed the data, the less significant the role of the "phosphene" effect becomes. Detailed interviews make it possible to highlight observations related to this phenomenon and take it into account in statistics. It should be emphasized that such a phenomenon associated with short-lived "flashes" without leaving material traces and effects on various objects is of little importance for the science of BL, whose attention is now focused on long-lived, large, high-energy, luminous, and non-luminous objects.

In conclusion of this section, we will consider the medical side of the danger of BL based on a detailed analysis of doctors (Dmitriev et al. 1986) BL can bring a variety of serious diseases. The most common lesion is burns (up to 12% of the body surface, including IV degree). Encephalopathy, brain contusion, traumatic paresis, neuritis, nystagmoid, a feeling of numbness, and weakness in the extremities are noted. In other cases, severe burns, bruises, brain concussion, and shock were detected. All the victims required medical assistance and long-term treatment.

The characteristics of BL factors obtained in the study of its impact on the environment are also presented there (Dmitriev 1967). The main factors are: thermal, radiation-chemical, electrical, radiation, chemical-physical, and air-toxic. The high temperature of the BL substance can lead to burns. Upon contact with the BL, charges recombine on the surfaces, so strong heating and burns can occur. Damage can also occur with chemical exposure to the substance of BL. The electric charge of the BL, although small, is quite enough for its effect. The lesion can be caused by the action of various radiations, primarily microwaves. In the event of a BL explosion, damage can occur as a result of a shock wave. Poisoning with toxic substances released by BL is also possible.

Burns can be caused by the action of many influencing factors of BL—thermal, radiation-chemical, electrical, direct radiation. The effect of the physico-chemical (explosive) factor can be explained by concussions and brain contusions, as well as shock states. Exposure to electrical and radiation factors can cause brain and heart damage, encephalopathy, traumatic paresis, neuritis, and nystagmoid. General weakness and concomitant diseases such as pneumonia may be associated with the action of toxic substances. All factors taken into account can lead to a fatal outcome.

References

Abakumov VI, Bikmukhametova AR, Bychkov VL, Safronenkov DA, Chernikov VA, Mikhailovskaya TO, Dolgoruky AM, Shvarov AP (2017) Interaction of long-lived luminous formations with the Earth's surface. In: Proceedings of 25th international conference on "electromag field and materials". Moscow, 24–25 November 2017. Infra Publishers, Moscow, pp 538–554

Abrahamson J, Bychkov AV, Bychkov VL (2002) Recently reported sightings of ball lightning. Philos Trans R Soc 360(1790):11–36

Akkuratov V (1991) The fiery, clever and cruel ball. Miracles Adventures Mosc 3:26–28

Alexeff I, Parameswaran SM, Thiyagarajan M, Grace M (2014) IEEE Trans Plasma Sci 32:1378

Amirov AKh, Bychkov VL (1991) Database on ball lightning for PC. In: Smirnov BM (ed) Sharovaya molnia (Ball Lightning). IVTAN Publishers, Moscow, pp 19–27
Amirov AKh, Bychkov VL (1994a) ANOVA of several parameters of the SKB data bank on ball lightning. Phys Scr 50:93–96
Amirov AKh, Bychkov VL (1994b) ANOVA of the parameters of ball lightning from the SKB data bank in correspondence to its location (indoors/outdoors). Phys Scr 50:588–590
Amirov AKh, Bychkov VL (2011) New data on the observation of ball lightning. In: Proceedings of 17th Russian conference on cold transmut nuclei chem elem and ball lightning. Dagomys. Sochi, 26 September–3 October 2010. Erzion Publishers, Moscow, pp 142–152
Andryushchenko IV, Kovalenko A (2007) 1000 and one question about digital photography. AST-Stalker Publishers, Moscow
Ardelyan NV, Bychkov VL, Bychkov DV, Kosmachevskii KV (2013) Electron-beam and non-self-maintained –driven plasmas for PAC. In: Matveev IB (ed) Plasma assisted combustion, gasification and pollution control, vol 1. Outskirts press, Denver, Colorado, pp 183–372
Ardelyan NV, Bychkov VL, Kosmachevskii KV (2017a) On electron attachment and detachment processes in dry air at low and moderate constant electric field. IEEE Trans Plasma Sci 45:3118–3124
Ardelyan N, Bychkov V, Belousov A, Kosmachevskii K (2017b) Processes of electron attachment and detachment in near breakdown conditions in air plasma. IOP J Phys: Conf Ser 927 N. 012012
Arsenyev VK (1951) In the mountains of Sikhote-Alin. State Publishing House Geogr liter, Moscow
Azhazha V (2013) Underwater UFOs. Veche Publishers, Moscow
Barry JD (1980) Ball lightning and bead lightning. Plenum Publishers, New York
Batygin A, Mosin I (1989) Visit of fiery lady. Newspaper Pravda, № 220, August 1989, p 8
Brand W (1923) Der Kugelblitz. Hamburg Verlag von Henry Grand, Hamburg
Brylev GB, Gashina SB, Evteev BF, Kamaldina II (1989) Characteristics of electrically active zones in stratified clouds. Gidrometeoizdat Publishers, Leningrad
Buckland L (2011). The real-life Blob: Is mysterious translucent jelly found in Cumbrian Fells from outer space? Mail Online. **Wednesday, Nov. 02.** http://www.dailymail.co.uk/home/search.html?s=&authornamef=Lucy+Buckland
Byalko AV (1983) Our planet - the Earth. Nauka Publishers, Moscow
Bychkov VL (2005a) Observations of ball lightning. In: Proceedings of 13th Russian conference on transmut nuclei chemical elements and ball lightning. Dagomys, Sochi, September 2005a. Erzion, Moscow, pp 237–245
Bychkov VL (2005b) The physical nature of gelatinous meteors 'Pwdre Ser' or 'Star Jelly'. Int J Meteorol 30 (N. 302):301–304
Bychkov VL (2006) On nature of gele-like meteors. Khimich Fiz 25 (N.3):108–111
Bychkov VL (2012) Star jelly what is it? Vestnik RAEN (N.1):96–101
Bychkov VL (2014) Atmospheric gelatinous meteors. In: Bychkov V, Golubkov G, Nikitin A (eds) The atmosphere and ionosphere: elementary processes, monitoring and ball lightning. Springer, Heidelberg, pp 369–384
Bychkov VL, Amirov AKh (2010) Data on airplanes interaction with ball lightning and other luminous objects of atmospheric electricity. Proc. 16-th Rus Conf Cold Nucl transmut Chem Elem and Ball Lightning. Dagomys, city of Sochi, June 1–8, 2009, Moscow, pp 64–80
Bychkov VL, Nikitin AI (2005) Observations of ball lightning. Materials 12-th Rus Conf on Cold Nucl transmut Chem Elem and Ball Lightning. Dagomys, city of Sochi, Sept 19-26 2004, Moscow 2005, pp 165–169
Bychkov DV, Bychkov VL (2008) Ball lightning: new observational data. In: Proceedings of 14th Russian conference on cold transmut nuclei chemical elements and ball lightning. Dagomys Sochi. October 1–8, 2006. Moscow 2008. (Proceedings of 13-th Russian conference on cold nuclear transmutation chemical elements and ball lightning. Dagomys, city of Sochi, 1–8 October 2006, Moscow 2008.) Erzion Publishers, Moscow, pp 142–151

Bychkov VL, Vaulin DN (2013) Observations of ball lightning. In: Materials of the 19th Russian conference on cold transmutation nuclei chemical elements and ball lightning. Krinitsa, Krasnodar Territory 3–10 September 2012. MATI Publishers, Moscow, pp 177–184

Bychkov VL, Nikitin AI (2014) Ball lightning: a new step in understanding. In: Bychkov VL, Golubkov GV, Nikitin AI (eds) The atmosphere and ionosphere. elementary processes, monitoring and ball lightning. Springer, Dordrecht, pp 201–367

Bychkov VL, Vaulin DN (2014) New observations of ball lightning. In: Proceedings of 20th Russian conference on cold transmutation nuclei chemical elements and ball lightning. Loo, Sochi, Krasnodar Territory, 29 September–6 October 2013. Publishing House of SIC-FTP-Ersion, Moscow, pp 111–121

Bychkov VL, Bychkov AV, Vasiliev MN, Klimov AI (1996) Evaluation of possible ball lightning energy by analysis of an event involving damage to a pot. J Meteorol UK 21:77–82

Bychkov AV, Bychkov VL, Abrahamson J (2002) On the energy characteristics of ball lightning. Philos Trans R Soc 360(1790):97–106

Bychkov VL, Bychkov AV, Bychkov DV (2003) Some new observations of ball lightning. In: Proceedings of 10th Russian conference on cold nuclear transmutation chemical elements and ball lightning. Dagomys, Sochi, 29 September–6 October 2002. Erzion Publishers, Moscow, pp 109–124

Bychkov VL, Bychkov VD, Sedov YuB (2004) Some new ball lightning observation data. In: Proceedings of 8th International Symposium Ball Lightning (ISBL04) 3–6 Aug 2004, Nat Centr Univ Chung-li, Taiwan, pp. 23–31

Bychkov VL, Bobkov SE, Stadnik SA (2006) On one type of luminous objects of Hessdalen. In: Proceedings of international project. Hessdalen Workshop. Bologna. Medicina, Italy, pp 182–187

Bychkov V, Nikitin A, Dijkhuis G (2010) Modern state of ball lightning investigations. In: Bychkov VL, Golubkov GV, Nikitin AI (eds) The atmosphere and ionosphere. dynamics, processes and monitoring. Springer, Dordrecht, pp 201–373

Bychkov VL, Abakumov VI, Bikmukhametova AR (2014) On the interaction of ball lightning with various objects. In: Proceedings 0th Russian conference on cold transmut nuclei of chemical elements and ball lightning. Loo, Sochi, Krasnodar Territory, 29 September–6 October 2013. Publishing House of SIC-FTP-Ersion, Moscow, pp 122–130

Bychkov VL, Gryaznov AYu, Nizovtsev VV (2015a) Observations of ball lightning, 2014. In: Proceeidngs of 21st Russian conference on cold transmutation nuclei of chemical elements and ball lightning, Dagomys. Sochi, 28 September–5 October 2014. MATI Publishers, Moscow. Pp 174–186

Bychkov VL, Nizovtsev VV, Toporov AA (2015b) Ball lightning. In: Abstracts of the 22nd Russian conference on cold transmutation nuclei chemical elements and ball lightning: Dagomys. Sochi. 27 September–4 October 2015b. Erzion Publishers, Moscow, p 29

Bychkov VL, Nikitin AI, Ivanenko IP, Nikitina TF, Velichko AM, Nosikov IA (2016a) Ball lightning passage through a glass without breaking it. J Atmos Solar-Terr Phys 150–151:69–76

Bychkov VL, Bikmukhametova AR, Nikitin AI (2017) Observations of ball lightning, 2016b. In: Proceedings of 23rd Russian conference on cold transmutation nuclei chemical elements and ball lightning. Dagomys, Sochi, Krasnodar Territory, 19–26 June 2016b, Moscow, MATI Publishers, Moscow, pp 176–189

Chernobrov VA (2003) Chronicle of UFO visits. Drofa Publishers, Moscow

Cooray V, Cooray G, Dwyer J (2011) On the possibility of phosphenes being generated by the energetic radiation from lightning flashes and thunderstorms. Phys Lett A 375:3704–3709

Corliss WR (1995) Handbook of unusual natural phenomena. Gramercy books, New York, Avenel

Corliss WR (1982) Lightning, auroras, nocturnal lights, and related luminous phenomena: A catalog of geophysical anomalies. Sourcebook Project, Glen Arm, MD

Covington AE (1970) Ball lightning. Nature 226:252

Data (2006) from the intern inform and analytical resource. Latvian Center for Res of Anomal Phenom "UFOlats"

Dijkhuis GC (2006) Site report and evaluation of Dutch ball lightning event recorded by two digital cameras. In: 9th international symposium on ball lightning (ISBL-06), 16–19 Aug. Eindhoven. Dijkhuis GC, Callebaut DK and Lu M eds The Netherlands, pp 39–42

Dmitriev MT (1967) The nature of ball lightning. Priroda 1967(1):98

Dmitriev MT, Bakhtin BI, Martynov VI (1981) Investigation of the thermal factor of ball lightning. Zhur Tekh Fiz 51:2567–2572

Dmitriev MT, Lakschin AM, Morozov SS (1986) Specific features of a ball lightning injury. Ortop Travmatol Protezirovanie 11:66–67

Dmitriev MT, Bakhtin BI, Lipatov AI (1985) Characteristics of ball lightningfrom results of natural observations. In: Sisin AN (ed) Academy of medical sciences USSR. SRI of General and Municipal Hygiene, Moscow

Doe RK, Keul AG, Bychkov VL (2009) An analysis of ball lightning-aircraft incidents. AGU fall meeting, 14–18 December 2009, San Francisco, California, USA

Doyle L (1970) Flight into the past. Nauka Publishers, Moscow

Egely G (1987) Hungarian ball lightning observation. Central Research Institute for Physics of the Hungarian Academy of Sciences KFKI-1987-10/D, Budapest

Enomoto Y, Hashimoto T (1990) Emission of charged particles from indentation fracture of rocks. Nature 346:641

Faye H (1890) Sur les boules de feu on globes electriques du tornado de Saint-Claude, d'apres le rapport de M. Cadanet. Compt Rend Hebd Acad Sci 111

Fedorov AA (ed) (1977a) Life of plants, vol 1. Prosvesheniye, Moscow

Fedorov AA (ed) (1977b) Life of plants, vol 3. Prosvesheniye, Moscow

Fedorov AA (ed) (1979) Flora of the European part of the USSR, vol 4. Nauka Publ. Leningrad

Fierani D (2014) Crop circles. BINOM Publishers, Moscow

Ford RA (2002) Homemade lightning. McGraw Hill, New York

Grigoriev AI (1990) Fire killers. Debut Publishers, Yaroslavl

Grigoriev AI (2006) Ball lightning. P.G. Demidov Yaroslavl state University Publishers, Yaroslavl

Grigoriev AI, Gershenzon N, Gokhberg N (1988) Rep USSR Acad Sci 300:1087–1089

Grigoriev AI, Grigorieva ID, Shiryaeva SO (1993) Observations of ball lightning and their analysis In: Smirnov BM (ed) Plasma chemistry, vol 17. Energoatomizdat Publishers, Moscow, pp 218–249

Gromyko AO (2004a) New information about ball lightning-prerequisites for synthesis. Phys Math Sci. Fundam Res 6:11

Gromyko AO (2004b) About the nature of ball lightning. Phys Math Sci Fundam Res 3:62

Gruzman IS, Kirichuk VS, Kosykh VP, Peretyagin GI, Spector AA (2000) Digital image processing in information systems. NSTU Publishing House, Novosibirsk

Hrgian AKh (1976) Physical meteorology. In: Meteorology and climatology, vol 3. Results of sci and technol VINITI Publishers, Moscow

Huang TS (ed) (1979) Picture processing and digital filtering. Springer, Berlin

Imyanitov I, Tikhy D (1980) Beyond the laws of science. Atomizdat Publishers, Moscow

Keul AG (2021.)A brief history of ball lightning observations by scientists. Hist Geo Space Sci 12:43–56

Keul A, Sauseng P, Diendorfer G (2008) Ball lightning – an electromagnetic hallucination? Int J Meteorol 33:89–95

Klimov AI (2008) Analysis and modeling of new reports on plasmoid observation in atmosphere and under sea water surface. In: Proceedings of 10th international symposium on ball lightning (ISBL-08). Kaliningrad, Russia, pp 76–83

Kogan-Beletskii GI (1957) Priroda 4:71–73

Kolosovsky OA (1981) Investigation of ball lightning trace on the window pane. ZTF 51:856–858

Kozlov BN (1978) Maximum energy liberation by ball lightning. Sov Phys Dokl 23:41–42

Kroit GR (1955) Science about colloids, vol 1. Inostrannaya literature Publishers, Moscow

Landau LD, Lifshits EM (1982) Electrodynamics of continuous media. Nauka Publishers, Moscow

Landau LD, Lifshitz EM (1988) Hydrodynamics. Nauka Publishers, Moscow

Leonov RA (1965) Enigma of ball lightning. Nauka Publishers, Moscow

Likhosherstnykh GV (1983) 138 approaches to the riddle of nature. Tekhnika Molodezhi. No. 3:38–39
Likhosherstnykh GV (1994) Abnormal cases of encounters with ball lightning. Ball lightning in the laboratory (in Rus). Khimiya Publishers, Moscow, pp 218–223
Lindow SE, Arny DC, Upper CD (1982) Bacteria ice nucleation. Plant Physiol 70:1084–1089
MacGorman DR, Rust WD (1998) The electrical nature of storms. Oxford University Press, Oxford
Mashkovich MD (1969) Electrical properties of inorganic dielectrics in microwave range. Sovetskoe radio Publishers, Moscow
McKenny HT (1910) Pwdre ser. Nature 83:492–494
Menzel DH (1962) Flying saucers (Russian translation). Harvard University Press, Cambridge
Mikhailov K (1998) Fortress for a chick. Nauka I Zhizn (Science and life), vol 3.
Mosby's Medical Dictionary, 8th edn. Elsevier, New York (2009)
National Geographic Television (2010) Paranatural blood rain and star jelly SATRip.
Nikitin AI (2020) The main properties of ball lightning as the basis for creating its model. Bull Rus Acad Nat Sci 1:25–34
Nikitin AI, Velichko AM, Vnukov AV, Nikitina TF (2007) Estimation of ball lightning parameters based on the analysis of its photograph. Khim Fiz 26:80–89
Nikitin AI, Velichko AM, Vnukov AV, Nikitina TF (2006) Evaluation of the characteristics of ball lightning based on the analysis of its photographs. In: Proceedings of 9th international symposium on ball lightning (ISBL06). Eindhoven, The Netherlands, pp 148–156
Nikitin AI, Nikitina TF, Velichko AM (2013) Analysis of video filmed June 3, 2009 in town Dolgoprudny, Russia. In: Proceedings of 10th international conference: wave electrodynamics of conducting liquid. Long-lived plasma objects and poorly investigated forms of natural electric discharges in atmosphere. Yaroslavl, Russia pp 149–165
Nikitin AI, Bychkov VL, Nikitina TF, Velichko AM (2014) High energy ball lightning observations. IEEE Trans Plasma Sci 42:3906–3911. https://doi.org/10.1109/TPS.2014.2359235
Nikitin AI, Velichko AM, Nikitina TF, Stepanov IG (2018) Analysis of the unique case of ball lightning observation in Mitino, the northwest district of Moscow. J Atmos Solar-Terr Phys 179
Nikolaev GV (1999) Scientific vacuum. NGO Rital Publishers, Tomsk
Nikolaev GV (2001) Secrets of electromagnetism. LLC "Banner of Peace". Publishers, Tomsk
Novoseltsev VN (1996) Optical illusions when observing extended atmospheric traces. Priroda 5:14–25
Ohtsuki YH, Ofuruton H (1989) Nature of ball lightning in Japan. In: Ohtsuki YH (ed) Science if ball lightning. World Scientific Publishers, Singapore, New Jersey, pp 31–57
Olkhovatov A (2008) The Tunguska phenomenon of 1908. Moscow. Binomial. Laboratory of knowledge Publishers, Moscow. (Part of the materials of the book is taken from Camille Flammarion. Atmosphere. Generally understandable meteorology. A complete description of air phenomena on the globe. (1900) Book 6 - Electricity. Thunder and lightning. Translated from the French by EA Predtechensky and DA Koropchevsky. SPB. Publication of the PV Lukovnikov bookstore)
Papkov SP (1974) Gelatinous state of polymers. Khimia Publichers, Moscow
Petrisko P (1998). http://www.sightings.com/health/mysteryskygel.htm
Platov YuV, Rubtsov VV (1999) UFOs and modern science. Nauka Publishers, Moscow
Pudovkin AK (1999) Ball lightning inSiberian academic city. Uspekhi Fiz Nauk 166:1253–1254
Pulinets SA, Hegai VV, Boyarchuk KA, Lomonosov A (1998) Uspekhi Fiz Nauk 168:582
Rainey PB (2007) Unity from conflict. Nature 446:616
Rakov VA, Uman MA (2003) Lightning: physics and effects. Cambridge University Press, Cambridge
Reiter R (1992) Phenomena in atmospheric and environmental electricity. Elsevier, Amsterdam
Rogers RR (1976) A short course in cloud physics. Pergamon Press, Oxford
Sattler B, Puxbaum H, Psenner R (2001) Bacterial growth in supercooled cloud droplets. Geoph Res Lett 28:239–242
Seckbech J (ed) (2007) Algae and cyanobacteria in extreme environments. Springer, Dordrech
Shagin I, Stekolnikov IS (1960) Thunderstorm near the Black Sea. Magazine Ogonek 20:34

Shelkunov G (2001) Ball lightning: observation and analysis of traces. Nauka i Zhizn 10:52–53
Shreider DR (1986) Heterogeneous chemistry of the atmosphere. Gidrometeoizdat Publishers, Leningrad
Singer S (1971) The nature of ball lightning. Plenum Publishers, New York
Skryagin LN (2000) Secrets of the flying Dutchman. Veche Publishers, Moscow, pp 178–187
Slyusarev NM (1989) Lightning and wood, deposited. VINITI Publishers, Moscow
Smirnov BM (1988) The problem of ball lightning. Nauka, Moscow
Stakhanov IP (1985) On the physical nature of ball lightning. Energoatomizdat, Moscow
Stakhanov IP (1996) About physical nature of ball lightning. Nauchnyi Mir, Moscow
Stenhoff M (1999) Ball lightning. An unsolved problem in atmospheric physics. Kluwer Academic/Plenum Publishers, New York
Stepanov SI (1990) About the energy of ball lightning. Zhur Tekh Fiz 60:214–216
Strand E (1993) Project Hessdalen, instrumentation of possible BL. In: Progress in ball lightning research. Proceedings of interdisciplinary congress Vizotum, 20–22 September 1993. Salzburg. Austria, pp 102–106
Tar D (2006a) Observation of lightning ball – a new phenomenological description of the phenomenon (Part 1). In: Proceedinsg of 9th international symposium on ball lightning (ISBL-06). Eindhoven, the Netherlands, pp 223–226
Tar D (2006b) Lightning ball and ball lightning: two different phenomena – a new phenomenological description of the phenomenon (Part II). In: Proceedinsg of 9th international symposium on ball lightning (ISBL-06). Eindhoven, the Netherlands, pp 227–233
Tarasenko GV (2008) The formation of the dynamo effect and its role in the structure of the planet Earth. The connection of the surface structures of the Earth's crust with the deep ones. In: Materials of 14th international conference on Petrozavodsk, 27–31 October 2008, pp 252–255
Tareev BM (1973) Physics of dielectric materials. Energia Publishing House, Moscow
Tekhnika molodezhy (1982) Attention: ball lightning. 5:46–47
Tsarev IV (1995) It cannot be. No 5:43
Tsarev I, Tsareva I (1982) The secret of witch circles. RIPOL classics, Moscow
Turner DJ (1997) The interaction of ball lightning with glass window panes. J Meteorol UK 22:52–64
Tutt K (2004) Mysterious phenomena of nature. Book Club Publishing house, Kharkov
Viostrukhin (1985) Exactly at 4.10… the newspaper "Trud (Labor)" Moscow
Wittman A (1971) In support of physical explanation of ball lightning. Nature 232:625
Yablokov M (2005) Inzhener N 1322 (#2):22
YOUTUBE 1 (2009) Storm, hurricane, hailstorm and fireball, Russia, Moscow region, city Dolgoprudny. http://www.youtube.com/watch?v=e5nceNOX8vE
YOUTUBE 2 (2009) Thunderstorm in Dolgoprudny, Russia, Moscow region. http://www.youtube.com/watch?v=vMQEpG1xL_A
Zacher CA, Studying lightning effects at the northern forest fire laboratory. Nat Notes 44, N.3:28
Zou Y-S (1995) Some physical considerations for unusual atmospheric lights observed in Norway. Phys Scr 52:726–730

Chapter 3
Analysis of Observations

Abstract This chapter is devoted to the statistical analysis of ball lightning observations. Statistical analysis of observations by the simplest mathematical methods is discussed. Modern methods of statistical analysis of BL are discussed also. Examples of correlation and variance analysis are given. The analysis of the dependence of the lifetime of the BL on its diameter, as well as the influence of thunderstorm atmospheric conditions on the properties of the BL is given.

Keywords Observations · Statistical analysis · Simple mathematical methods · Modern statistical approaches

3.1 Data Collection

In order to specify what we mean by the term ball lightning (BL), consider the following observations.

Nikitin A.I., a physicist, a researcher of ball lightning, was made by him at the age of eight.

Observer Nikitin A.I., male, physicist DS, 66 years old at the time of the survey in 2005. Observation in the village of Belaya Glina of the Krasnodar Territory, Russia. 1948, Mid-August, about 20.00 LT

It was a hot summer day without rain. I was walking along a rural street, to my left, there were pillars 3–4 m high, on the upper horizontal crossbar of which four pairs of steel telephone wires were stretched on insulators. To my right at a height of about 15' was the sun setting in the clouds.

Suddenly, with my peripheral vision to my left, I noticed a red ball with a diameter of about 15 cm. The ball was above the wires, the minimum distance to it was about 4 m. The ball had a sharp border, it shone evenly, and could be seen even against the blue sky. The glow was not dazzling, it was possible to look at the ball calmly. The shape of the ball was perfectly spherical, it did not rotate, and its surface was smooth. There were no protrusions on it. The ball was moving slowly (at a constant speed of about 1 m/s) over the wires in the direction of my initial movement. After I saw the ball, I froze in place and could only follow it with my eyes. The ball moved along the middle line of the wires, keeping above them at a height of 30–40 cm.

V. L. Bychkov, *Natural and Artificial Ball Lightning in the Earth's Atmosphere*, Springer Atmospheric Sciences, https://doi.org/10.1007/978-3-031-07861-3_3

During the observation, the altitude remained constant. The ball was in my field of vision for about a minute, after that it disappeared behind the foliage of trees that obscured the upper part of the pillars from me. The ball moved in absolute silence. After it disappeared, for at least 10 min I did not hear anything like the sound of an explosion from the direction where the ball went.

At that time, neither my parents nor I had heard anything about ball lightning. I realized that it was BL when I got older.

Now another example is the observation of N.N. Lutsenko (born in 1912), a hydrobiologist from letters to I.P. Stakhanov in 1975. Time 1932 August, 17–18.00 LT, the resort of Sernovodye Chechen-Ingush ASSR, Russia, 40 km from the city of Grozny.

Within 10–15 min before the thunderstorm, at a distance of about 200 m on a grassy slope, glowing balls began to appear, which moved up the slope (slope 20°) and disappeared into the grass. The width of the area where the balls appeared is about 50 m. The balls passed a distance of 50–75 m. They moved with a slight crackle and disappeared with a slight click. About 20 balls were observed at the same time. They sailed over the grass to a hot sulfur spring. Some balls "splashed" and sparkled when they disappeared.

For example, I.P. Stakhanov identified the following properties of this observation: the object is a ball; its size is 20–30 cm; the distance to the observer is 200–300 m; the distance traveled by the object is 50–75 m. Color: orange, blue, green, yellow, almost white; radiation power compared to an incandescent lamp is 500–1000 Watts; the balls quietly went out before a severe thunderstorm.

Obviously, after reviewing these observations, it is impossible to say anything specific about the properties of these objects. In addition, significant features and the number of objects dropped out of the second observation during the processing attempt.

Nevertheless, we can enter a definition: ball lightning is a luminous formation in the air (or some other medium), observed for several seconds or longer. This formation most often has a spherical shape, is not glued to any walls, and does not noticeably change its size during observation.

To get an answer to the question about the nature of BL, detailed observations of BL are constantly being collected, which would help to find out and model its structure. These studies include the works of Arago (1859), Brandt [2], McNally (1966), Rayle (1966), Dmitriev (1985), Stakhanov (1985), Grigoriev (1990), Grigoriev et al. (1993), Keul and Schwarzenbacher (1989), Keul (1993), Likhosherstnykh (1994), Egeli (1987), Stenhoff (1999), Abrahamson and Bychkov (2002, 1991) , etc. These data differ from each other and characterize different properties of the observed object. Often the differences are caused by conceptual representations.

Therefore, Stakhanov (1985) built a model of ball lightning with a small energy reserve, so it is not possible to find observations of high-energy BL in his books.

Let us focus on the issues of collecting information about BL. As a rule, collecting information is a routine job, because the observer is very reluctant to talk about the phenomenon, embarrassed by his prejudices, emotions, and experiences, and distrustful of his observations. It is extremely difficult to conduct a survey of the observer a second time. Most rarely and reluctantly, representatives of technical professions and scientists talk about it. It can be difficult to analyze their observations because they try to give their interpretation of the phenomenon in the story or tie it to a ready-made theory.

In general, BL from the point of view of social psychology is a "hidden phenomenon" (Keul and Schwarzenbacher 1989; Keul 1993), and therefore it is not possible to obtain a reliable estimate of the frequency of its occurrence, with the exception of small geographical areas. For example, A. Kehl (private communication) specifically conducted a study of the frequency of observations of BL by eyewitnesses in the Montafon Valley (Austria) for 60 years and obtained an observation frequency of 9.0×10^{-10} km^{-2}c^{-1}, which turned out to be 20 times higher than previously accepted. Let us translate it into the probability of BL observing by one observer who has the opportunity to observe BL over an area of 25 km^2 annually for 60 years, 4 months a year, 6 h a day (we limited ourselves to the number of months and observation time typical for BL in Europe). After making calculations, we get the frequency value of 0.12, which is quite small, confirms the small number of observations of BL in Europe. Therefore, an essential role in the preparation of materials for analysis is played by a questionnaire. It, on the one hand, should be sufficiently detailed to identify all the most characteristic features of the object and, on the other hand, not very long, so that the observer can fill it out at one time without losing interest in discussing the observation. A similar questionnaire of 13 questions was proposed by I.P. Stakhanov, which served as the basis for correspondence with readers of the journal Nauka i Zhizn "Science and Life" in 1975 and found out details on a more detailed questionnaire (Stakhanov and Lopatnikov 1975). I.P. Stakhanov received more than 1,500 questionnaires. In 1992, B.M. Smirnov made an attempt to attract readers of the journal Nauka i Zhizn "Science and Life" to write observations of BL again (Smirnov 1992a, b), but only about 15 questionnaires were received for a questionnaire of 30 questions. A similar collection of information on observations was carried out by Grigoriev (1990) and Egeli (1987). Data on the observation of BL in aircraft were obtained from the Main Meteorological Center of the USSR Air Force by letter of Prof. B.M. Smirnov (Bychkov and Nikitin 2014).

The questionnaire on the observation of BL by the Austrian psychologist Ph.D. Keul contains 46 questions (Keul 1993). In his opinion, questionnaires consisting of more than 50 questions are psychologically unjustified. The questionnaire should consist of questions that help formalize observations. Based on our experience, we present our questionnaire (see below).

Most likely, for purely conceptual reasons, the data on the observation of BL at sea and underwater (Skryagin 2000; Azhazha 2013; Olkhovatov 2008), and in the air over them fell out of BL consideration. Observations of BL in complex geophysical, seismic, and tectonic conditions (Skryagin 2000; Azhazha 2013; Olkhovatov 2008; Tsarev and Tsareva 2005; Ohtsuki and Ofuruton 1989) have fallen out, which are often presented in UFO observations (Tsarev and Tsareva 2005; Menzel 1962; Chernobrov 2003). They are easily separated from observations of spacecraft artifacts (Platov and Rubtsov 1999) by both shape and typical times of appearance and destruction. In particular, Prof. M.T. Dmitriev and DS A.I. Nikitin attributed large

UFOs with a long lifetime to the BL. These thoughts were not taken seriously by scientists who were not engaged in collecting and analyzing observations until a well-documented description of BL with a diameter of 100 m appeared in Abrahamson et al. (2002, Bychkov and Nikitin (2014).

In order to organize the characteristics of the BL, Brandt compiled a list of BL properties without "geophysical", "aviation" and "marine" BL. Based on data collections from Brandt (1923) and discussed in Bychkov et al. (2010) one can make the following conclusions about general properties of BL:

1. The thunderstorm in the majority of cases is the cause of BL appearance, however, BL can be observed in fair weather.
2. One can most often observe a single BL, however, there are cases when two or more number of balls were observed simultaneously.
3. BL were observed on a plain, in mountains, and near seas.
4. BL is formed not only during thunderstorms but also during hurricanes and tornados.
5. Maximum of BL observation number in Central Europe falls in summer months—from May to August, this correlates with thunderstorm appearance frequency. BL most often is observed from 14 to 20 h during daytime, but a period of most thunderstorm activity falls during the period of 12–18 h. It means that BL more often appears at the end of the thunderstorm.
6. BL form in majority of cases is spherical, sometimes it is a hollow sphere, and sometimes, it is compared with a soap bubble. Sometimes one observes deviations from the spherical form: BL has the appearance of an egg, an ellipsoid, a flame tip, a bar, or a pole.
7. BL contour is usually clearly outlined, but sometimes it seems to be blurred, surrounded by a glowing gas. During its motion, one can sometimes observe projections on its surface that protrude on different sides. Sometimes one can see a halo around BL. Its color in majority of cases is different from those of the ball.
8. Sometimes one can observe an internal structure of a ball: decrease in light intensity from center to an edge; the ball is dark red in the center and brighter to the edge; a transparent ball, with a blue flame in the center.
9. Sometimes sparks, flame bunches, and beams of approximately 50 cm long fly from the surface of a ball.
10. BL sometimes changes its form: it elongates going through a hole, it is transformed from a "fire snake" to a ball and vice versa, a ball little by little elongates transforming to thin weakly glowing strip.
11. Average diameter of BL is 20 cm, and its size range is from 1–2 cm to 27 m. BL size during its lifetime can be constant, increase or decrease.
12. BL in a number of cases was rotating or rolled over a ground.
13. Main colors of BL are white, white-blue, dark-blue, yellow-green, yellow–red and red, at that, the last one meets more often than others. Changing of the ball color usually takes place before an explosion of the ball: it becomes white

from red-yellow; the ball being yellow before an explosion becomes red-blue; the white central part of a ball becomes red before an explosion.

14. There are reports about observations of black BL as if consisting of a dense smoke.
15. Light irradiated by BL is characterized by a range from a dim to a dazzling bright. Observers often say that it was bright but not blinding. Brightness of BL often increases before its disappearance. There are cases when luminescence brightness of different BL parts was different.
16. Sometimes BL leaves a smoky trace behind itself or is covered by a smoke or vapor. The smell of the smoke observers usually compares with those of burning sulfur or burning powder. A vapor they describe as the orange gas, which can be identified as nitrous oxide N_2O. In other cases, the smell was identified as the smell of ozone, especially when BL was observed by people belonging to science.
17. In most cases, there are no material traces after BL explosion.
18. Observers found themselves at close distance to BL or having touched it say in the vast majority of cases about heat feeling absence from a heat going from it. Sometimes on the contrary observers report heat irradiation and even cases of singing or inflammations of objects by BL. Effects with heat release most often take place at BL explosions. It is usually difficult to separate consequences of BL explosions and effects caused by a stroke of usual linear lightning.
19. Encounter BL usually leaves an indelible impression in an observer's memory. Most often, it causes a feeling of danger and fear.
20. BL at close contact with an observer's body can cause temporary loss of consciousness or paralysis or even death. There are many cases of animal deaths at BL explosion. A "mechanism" of breather affection is similar to impact of usual linear lightning.
21. There are data about mechanical work executing by BL: stone spreading, transportation of people and animals in space, making holes in thick walls, in wooden doors or window glasses. BL is capable to make a deep pit in a ground, to make a trench in earth, to throw a tile from a roof, to swing a church bell. It can evaporate wine, oil and water in a earthenware leaving the vessel safe; it can tear a shirt to pieces; it can instantly melt down a piece of ore or gold; and it can tear away a ring from a finger.
22. BL can impact electrical devices: makes to ring electrical bells, causes glowing of bulbs; its explosion is accompanied by melting down of safety fuses, and burnout of light bulbs.
23. BL sometimes is a sound source. This sound is compared with a whistle of rising rocket or a shell. Crackling noise sometimes accompanies BL motion. St. Elms fires give a similar set of sounds, but the sound generated by BL proves to be much louder. Probably the sound presence one can connect with electric charge loss.
24. In a large number of cases the ball dropped out of the sky is observed. BL sometimes reach earth surface, but sometimes their fall is stopped at a height of 1–15 m, and they start moving horizontally. There are cases when BL on

the contrary was rising up to the cloud from earth surface. BL sometimes stops and stays motionless in air. BL way is never straight but the cases when it falls down from a cloud to earth along the vertical trajectory.

25. BL moves along a complex trajectory in closed premises or near buildings. It can jump aside as the elastic object at contacts with walls or a ground.
26. In premises it often moves round a circle approaching walls and a floor nor closer than one meter. BL flying in a room through a window or a door usually moves to an opposite wall. Sometimes it leaves the room along the same way by which it flew in. This can be caused by electric field variation in the room or by a complicated airflow gas-dynamics in premises with respect to location of intake and air vent holes.
27. Sharp change of BL motion direction can be caused by air flows or an impact of electric conductors. However, sometimes air motion does not visibly impact it; and it moves upwind. Features of BL motion can be explained by its attraction to metallic objects; and its tendency to move through chimneys can be connected with an electrical conductivity of a smoke.
28. BL motion way in some cases lays along electric conductors: it moves along telephone, telegraph and electric cables, wires of lightning rod groundings, drainpipes and tree trunks.
29. There were marked cases when BL was moving along surfaces of people and animals leaving skin burns. However more often it demonstrates a tendency to escape contacts with people, animals and good conductors.
30. BL can penetrate through keyholes, slits and cracks. In rare cases it can penetrate glasses leaving it intact; and sometimes it makes a fist-size orifice.
31. BL during its lifetime can cover a substantial distance of several hundreds of meters.
32. BL moves with a constant velocity only in rare cases, usually it moves accelerating and decelerating. BL can be separated in two groups with respect to their motion velocity. The first one consists of BL moving from clouds to earth or between the clouds. In average their velocity is of several tens meters per second. The second group consists of BL observed in premises or outdoors near earth. Their average velocity is 2 m/s.
33. An existence BL time is from one second to five minutes and more. In majority of cases one can observe BL with the observation time below 5 s. Shortly-lived BL were observed inside premises and in open air, but the long-lived ones were observed only outdoors.
34. A stroke of usual linear lightning often precedes an appearance of BL. There are observations when BL was originated in the linear lightning channel. However, observers often say nothing about BL appearance connection with usual lightning. The more, there are cases when BL had appeared not directly after the linear lightning stroke but several minutes after and in the absence of linear lightning.
35. BL often disappears silently and does not leave any traces. Its disappearance can take place in air and also at touching of some object, ground, or water. BL disappearance in some cases is accompanied by a sizzle and burst of flame

tips or sparks. BL life in many cases is finished by an explosion. An explosion sound sometimes is described as a gunshot or a rifle shot. The explosion is often accompanied by a bright light flash, lightning, pieces of a matter, and sparks fly out of BL.

36. BL sometimes divides into several separate balls which can continue their existence or go out. BL division can take place in air and at a collision with some object. Balls of a smaller size practically do not differ from the initial ball.
37. There is a description of the case when BL on the contrary was gathered of smaller zero-sized balls. At that, they were moving towards a cable or a wire that came nearer and created a luminescent mass.
38. There are rare observation cases of BL pairs located one over another and connected by a "thread" consisting of small luminescent beads.

Besides conclusions made based on Brand's descriptions one can formulate additional BL features:

39. BL can lose a part of its matter at impact with an obstacle, but it does not interrupt its existence as a single whole (Br.14, Br.36, Br.92). (The mark (Br.) means the book of Brandt, and a numeral means a number of a report from this book).
40. Natural tissues (cotton, cloth) become fragile or tear to pieces after touching BL. Traces of "burns" left on a tissue disappear after washing (Br.3, Br.95).
41. Tinted tissue becomes colorless at the contact with BL (Br.95).
42. BL as a linear lightning can strike a tree. At that, a tree trunk can be split, and people being under the tree can be killed (Br.33, Br.146, Br.149).
43. There is a report that St. Elms fires had appeared at mast tops 2 min after BL explosion in water at a distance of 2.5 m from the sailing-ship side (Br.47).
44. There is a detailed description of circular holes in a glass left by BL (Br.95). "Their diameter as a rule 8 cm, the fracture was even, not the star-shaped, smooth on the inner surface and exhibited a small sharp edge on the outer surface. In certain cases, there was a series of concentric wave patterns around the circular hole".
45. An electric discharge had occurred between BL and the grounded electric wire at BL coming to it. This caused melting of wires (Br.136).
46. Sometimes disappearance of BL is accompanied by a loud noise; however, sparks do not fly from it (Br.171). This resembles a collapse of evacuated sphere.

Below we present additional BL features from Bychkov et al. (2010).

47. BL can magnetize steel objects located at a close distance from them.
48. BL can penetrate through a metal mesh.
49. BL is a source of radio emission at frequencies of 0.3–3 MHz.
50. Ozone and nitrogen dioxide are formed in the trace of BL with a delay of about a minute.
51. BL can penetrate into all-metal aircraft. At the same time, it usually moves in the center of the cabin at a constant altitude at a speed of 1–2 m/s in the direction from the cockpit to the tail of the aircraft.

52. BL does not detect visible magnetic action at a distance of more than 0.5 m.
53. By moving around ring-type metal objects with a diameter of several millimeters to several centimeters or exploding near them, the BL can quickly vaporize these objects.
54. There are cases when it is possible to estimate the energy density of BL, which reaches 2.4×10^{12} J/m^3. BL can emit an energy value of 2.3×10^9 J.
55. BL energy absorbed by water can be transferred through a dielectric wall (wood).
56. In the presence of a magnetized object, the BL can also become a magnet and be attracted to it.
57. The magnetic features of the BL can suddenly disappear even in the presence of a magnetized object.
58. The BL is able to hold objects weighing about one kilogram or more near it and lift them into the air.
59. BL can cause symptoms of radiation sickness in humans.
60. BL can have a high electrical potential and an electric charge.
61. The explosion of BL can cause the appearance of a very large induction electromotive force in the surrounding wires.
62. BL in the course of its existence can emit electromagnetic waves in the radio range.
63. BL can be a source of hard ultraviolet and soft X-rays.
64. BL may appear on wires with electrical overvoltages.
65. The BL has a shell inside in which the movement of small elements can be observed.
66. The BL can change its shape: it can transform its shape from a ribbon to a ball and from a ball to a ribbon.
67. The BL "feels" a change in the external electric field; it is sensitive to the movement of objects (people, cars).
68. It is possible to observe the interaction of individual elements in a group of BLs: two BLs move at a certain distance from each other, and three BLs create an isosceles triangle.
69. The BL is capable of lifting and carrying a heavy object, for example, a stone, a steel fork, or a sheet of iron.
70. In an explosion, BL can reduce iron oxides to metal.
71. BL can transmit energy at a distance, most likely in the form of electromagnetic radiation. This causes a reaction of electrical devices (bells, lamps) or may lead to the melting of wires, sand, or glass.
72. BL is capable of making holes or depressions in walls, metal plates, and glasses.
73. BL can penetrate through the glass without damaging it.
74. Getting into the water tank, the BL can heat the water to a boil and evaporate it. The energy density of BL, estimated in this case, is 3.4×10^{10} J/m^3.
75. BL may have a significant uncompensated electric charge.
76. BL can be a source of powerful electromagnetic radiation that causes resonance of currents in annular conductors with a diameter from 1 mm to 2 cm.
77. Visible BL radiation can take the form of a narrow beam.

78. BL can get into an all-metal aircraft through the cockpit glass, air fairings, or through holes in the fuselage. Once in the cabin of the aircraft, it usually moves slowly between the rows of seats in the direction from the nose of the aircraft to the tail. It can leave the tail of the plane silently or by explosion, leaving a hole.
79. The BL shell (cover) is a complex system that may consist of a solid inner layer and a liquid outer layer or vice versa.
80. The flight of a BL over a field with green shoots can leave circles on the field.
81. The appearance of BL in areas with seismic and volcanic activity correlates with the times of activity (for and aftershocks).
82. During seismic activity, the dimensions of the BL can reach 100 m and much more.
83. The lifetime of BL during seismic activity can reach several minutes or more.
84. BL can stimulate the appearance of holes with the realization of the impact on the elemental composition of the soil.
85. BL may consist of complex soil compounds.

Below we present a working questionnaire, which we use for the collection of BL data.

Questionnaire for the survey of ball lightning observers.

1. Surname, first name, patronymic of the observer.
2. Gender of the observer.
3. Year of birth.
4. Specialty.
5. Position.
6. Age at the time of observation.
7. Home address, phone number, or E-mail (for possible correspondence in order to clarify the details of surveillance).
8. Date of observation (including year, month, hours, and minutes). If you do not remember exactly, then specify the interval of years.
9. Season (summer, autumn, winter, spring).
10. Time of day. If you do not remember exactly, then specify one of the specified intervals:3–6 h—early morning, 6–9 h—morning, 9–12 h—before noon, 12–15 h—noon, 15–18 h—after noon, 18–21 h—evening, 21–24 h—before midnight, 0–3 h—after midnight.
11. Place of observation. Specify the region, district, locality, or distance to the nearest locality.
12. Specify where the object was located: indoors or outdoors (outdoors).
13. Type of place of observation. For example: an apartment on the 5th floor, a kitchen, a field, a forest, on a mountain, under a mountain, in an airplane cabin, in a steamer cabin, in a mine.
14. What was the shape of the object?

    The object can have any shape, including possible shapes: ball, ellipsoid, pear-shaped, ring, spiral, disk, pancake, ribbon, thread, oblong, and irregular. If the shape was incorrect or did not have the appearance specified in the list, then describe in detail the shape of the object.

15. Estimate the diameter of the object. If you don't remember exactly, then specify one of the specified intervals (in centimeters): less 1, 1–2, 3–5, 5–7, 7–10, 10–15, 15–20, 20–25, 25–30, 30–40, 40–50, 50–70, 70–100, and more than 100 cm. If the size of the object was more than 100 cm, then describe the object in detail.
16. How different were the longitudinal and transverse diameters of the object? Specify one of the following answers: practically did not differ, differed by 20–30%, differed by 1.5–2 times, differed by 2 times. If the object had an elongated shape, then specify in which direction: from bottom to top, in the direction of movement, etc.
17. Was the border of the object sharp or blurry?
18. Did the object have a core? If it was, then specify its dimensions (as in No. 14).
19. Did the object have a halo? If there was, then specify its dimensions (as in No. 14).
20. What was the glow of the object?
    Here are the brightness standards: "candle", "kerosene lamp"—less than 10 watts; "dim"—10–20 watts; "transparent", translucent—less than 20 watts; "dim"—20–50 watts; "moon"—50–100 watts; "bright"—100–200 watts; "very bright"—200 watts, and above. If you can compare the brightness of an object with another standard, then describe this standard.
21. What was the smallest distance from you to the object (in meters)? If you don't remember exactly, then specify one of the specified intervals: less 0.5, 0.5–1, 1–1.5, 1.5–2, 2–2.5, 2.5–3, 3–3.5, 3.5–4, 4–5, 5–7, 7–10, 10–15, 15–20, 20–25, 25–30, 30–40, 40–50, 50–70, 70–100, and more than 100 m.
22. What distance did the object travel during the observation (in meters). For example: 1–3, 3–5, 5–7, 7–10, 10–20, 20–30, 30–50, 50–100, 100–200, 200–300, 300–400, 400–500, and more than 500 m.
23. What, in your opinion, was the maximum and minimum speed of the object? For example: minimum—1 m/s, maximum—3 m/s.
24. How long have you been watching the object? If you don't remember exactly, then specify one of the specified intervals in seconds: less 0.5, 0.5–1, 1–1.5, 1.5–2, 2–3, 3–4, 4–5, 5–7, 7–10, 10–15, 15–20, 20–30, 30–40, 40–60, 60–90, 90–120, 120–180, and more than 180 s.

    If you have observed the object for more than 180 s, then specify the observation time.
25. At what distance from the earth was the object? (For example, as in No. 21).
26. Specify the color of the object. To do this, use the specified spectral colors and their combinations: white, red, orange, yellow, green, blue, and purple.
27. Specify the color of the object's core.
28. Specify the color of the object's halo.

29. Did the object have a structure. For example: the object consisted of threads, burning dots, sparks, rays emanated from it, the surface was twigs, the object was a jelly-like formation, the object was a ball of threads, cotton wool, etc.
30. Specify if the object was a double, triple, etc. group of interconnected objects.
31. Did you feel the heat from the ball lightning?
32. Did the size of the object change during the observation? If yes, then specify how many times?
33. Did the color of the object change during the observation? If yes, then indicate which one?
34. Have you seen the moment of the formation of the object? If so, describe this moment in detail.
35. How did the object disappear? For example: exploded, went out, fell apart, and went out of sight. If none of the methods corresponds to observation, then describe this process.
36. If the process was a complex sequence of events, then describe this process.
37. If the object exploded, then indicate at what distance from you (in meters). If you don't remember exactly, then specify one of the specified intervals: less 1, 1–2, 2–4, 4–6, 6–8, 8–10, 10–15, 15–20, 20–30, 25–30, 30–40, 40–50, 50–70, 70–100, 100–150, 150–200, 200–300, 300–500, and more than 500 m.
38. Were there objects of an electrical circuit near the object? If yes, specify which ones. For example: an electrical outlet (serviceable or faulty), bare wires, transformer booth, etc.
39. Did the appearance of the object affect the operation of radio, television receivers, or electrical appliances? Specify the type of interference and the type of devices.
40. What were the weather conditions at the time of observation? Choose one of the listed answer options in each of the three groups:

    A. Clear, cloudy, rain, fog, snow.
    B. There was a thunderstorm, there was no thunderstorm.
    C. The wind is weak, the wind is moderate, the wind is strong, a hurricane.

41. Specify the wind direction.
42. How did the object move? Choose one of the listed answer options in each of the groups:

    A. Up, down, horizontally.
    B. Smoothly, in leaps.
    C. Rotated around its axis, did not rotate. G. Moved along the movement of the wind or against it.

43. Describe the trajectory of the object.
44. Have you observed the ignition of objects when they interact with the object?
45. Was the appearance of the object accompanied by the onset of fires, earthquakes, and other natural disasters? If so, describe in detail the sequence of events.

46. Was the interaction of people with the object accompanied by their injury or death?
47. If you can, then draw the object of observation.
48. Has anyone observed the object with you? If yes, then specify the data about the observer.
49. Describe your observation in the form of a story.

This questionnaire is useful when conducting a survey of observers. It should be answered that with the development of computer technology, the culture of letter exchange has been lost. Therefore, we have to use information about the observation of BL from the Internet, which is incomplete and insufficient. However, there are many videos and photos of BL, with which additional work is required to separate reliable information from fake, a separate section above is devoted to these issues.

3.2 Analysis of Observations by the Simplest Methods

Currently, several collections of observational data of BL have been collected, on their basis, the average characteristics of ball lightning, presented, for example, McNally (1966), Rayle (1966), Stakhanov (1985), Grigoriev et al. (1993), Egely (1987), Keul and Schwarzenbacher (1989), Abrahamson et al. (2002) have been identified. They concern the distribution of BL by size, by distances to the observer, the distribution of ball lightning by colors, by seasons and time of day; the characteristics of the birth and death of these objects. To date, several thousand observations have been accumulated, on the basis of which it is possible to compile a table of average parameters of ball lightning. Below, we present histograms of the main parameters from the Stakhanov-Kel-Bychkov computer Databank of BL, (SKB) which was created on the basis of data from I.P. Stakhanov (about 1500 observation events), A. Kel (about 150 observation events) and V.L. Bychkov (about 150 observation events) in 1991 (Amirov and Bychkov 1991; Amirov et al. 1995). The databank existed until 1997, when I.G. Stakhanova, for unclear reasons, forbade the use of I.P. Stakhanov's collection data obtained by him during correspondence with readers in the framework of contacts with the journal Nauka i Zhizn ("Science and Life"). The letters themselves are unknown where currently stored; this does not allow restoring the databank on up-to-date level. All our work with the databank were having indications of I.P. Stakhanov.

3.2.1 Parameters

Note the features of storing information about numerical parameters stored in the database. It consists of the fact that the observer, as a rule, gives an estimate of the parameter, i.e., indicates the range of values in which the parameter falls. For example, when estimating the diameter, the answer is more likely to be 20–25 cm than 24 cm. When using the SKB, the average value between these limits was used without taking into account the width of the range. Another feature of the numerical parameters is their low accuracy. Errors in determining the parameter value are determined both by the peculiarities of the perception process (Meshkov and Matveev 1989) and by the unusual conditions of observations of the BL. In the description of the observations of BL, the emotional state of the observer is very often emphasized—surprise, fright, etc. Estimates of the values of the estimates indicate that they can reach 50% in an untrained observer. This is also evidenced, in our opinion, by the correlation coefficients between the size of the BL and the distance to it, which are noticeably different from zero.

3.2.2 Lifetime

The lifetime of the BL is one of the difficult to estimate parameters. The observer rarely records the moment of occurrence of the BL, often the description refers to an "unexpected occurrence", or that the BL came into view. The death of a BL often occurs outside the observer's field of view, it simply disappears from it, hiding behind other objects. It often happens that, having disappeared from the field of observation, it soon dies, for example, with the sound of an explosion. Only then, there is both the time of observation and the time of its life. In statistical analysis, the SKB uses the observation time as the "lifetime". Although these values correlate with each other, they do not coincide. This can lead to an uncontrolled error during statistical processing.

3.2.3 Equivalent Electrical Power of Radiation

A method for comparing the power of BL radiation with the radiation of an incandescent light bulb was proposed by Stakhanov (1985). This parameter may have a large systematic error because, firstly, the error in estimating the radiation power is large, and, secondly, because comparison with other light sources may give a different result for the electrical power. For example, luminescent ones have a radiation efficiency of 5–6 times greater than incandescent ones with the same electrical power. Since the mechanism of BL radiation is unknown, and existing theoretical models show the possibility of several mechanisms, including non-quilibrium ones, the comparison with an incandescent electric lamp is very conditional. Another feature is that the produced incandescent electric lamps have discrete power values of—40, 60, 100 watts, etc. Because of this, the probability of getting a value of 80 watts in the description is less than the value of 100 watts. Thus, the exact value of the equivalent electrical power of radiation has a large error, but as a relative value, it is used in the SKB to find out the trends of BL radiation.

3.2.4 Color

As a rule, a BL observer indicates several colors for an object. At the same time, it is not possible to determine which color is the main one and which is a side color from the available descriptions. In statistical analysis, when using SKB, the primary color is considered to be the first one indicated, and the secondary colors are the remaining colors. In addition, the observer often reports conditional colors, such as "metallic", "golden", "fiery", etc. In statistical analysis, such descriptions were ignored in most cases. Shades of color ("bright", "dazzling", and "dirty color") were not taken into account but were used to verify the estimate of the radiation power given by the observer.

3.2.5 Geography of Observations

Most of the observations in the SKB database took place in the middle latitudes of the northern hemisphere on the territory of the former Soviet Union and Austria.

Below, we present average data obtained on the basis of SKB Databank (Amirov and Bychkov 1991; Amirov et al. 1995) (Figs. 3.1, 3.2, 3.3, 3.4, 3.5, 3.6, 3.7, 3.8, 3.9, 3.10, and 3.11).

In Table 3.1, we present the average parameters of BL, based on the summation of data from these collections collected in the Northern Hemisphere, mainly in Europe (Grigoriev et al. 1993; Smirnov 1992a, b) and usually used for theoretical modeling

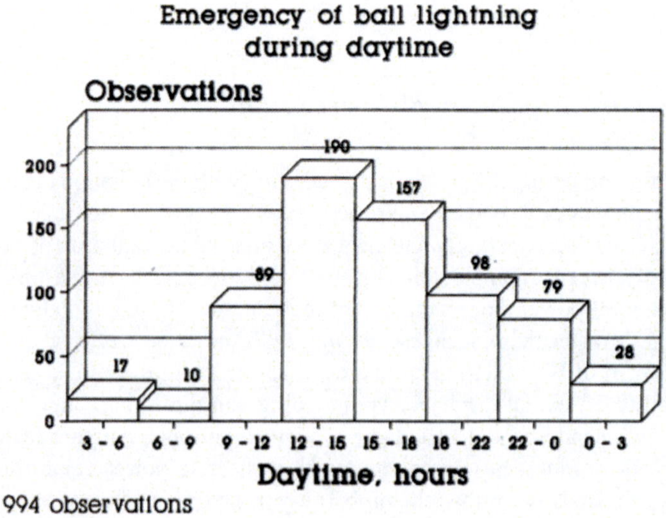

Fig. 3.1 Distribution of the number of observations of BL by observation time during the day (intervals are in hours)

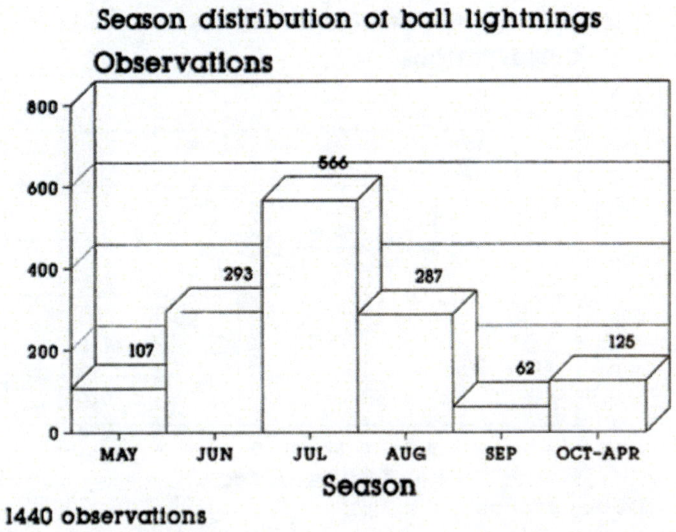

Fig. 3.2 Distribution of the number of BL observations by month (from left to right: May, June, July, August, September, October-April)

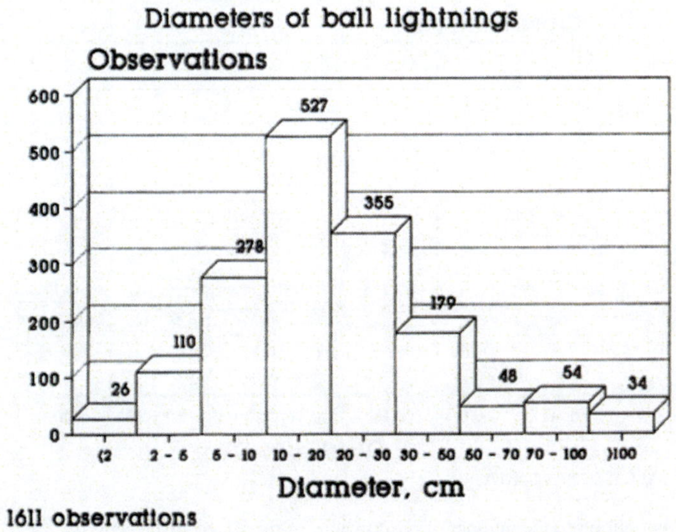

Fig. 3.3 Distribution of the number of BL observations by diameter (in cm)

of BL. However, they do not account for anomaly and rare BL features which make them to be unreliable. In Table 3.1 estimates of the internal energy of the BL were presented above based on observational data on the interaction of the BL with various objects. The most complete analysis of observations of ball lightning parameters

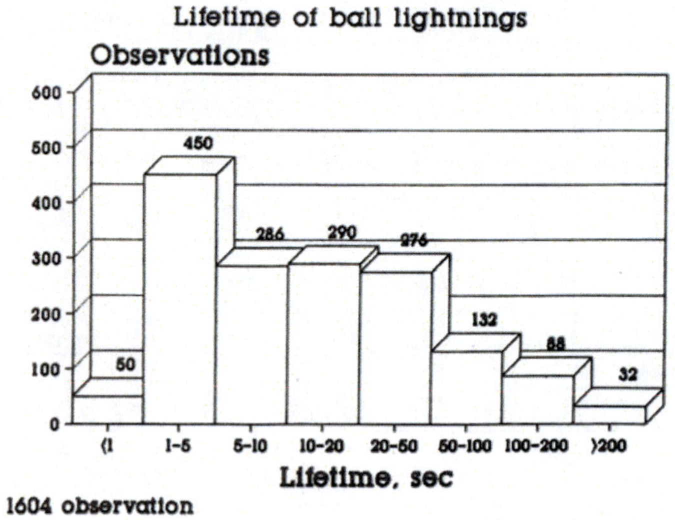

Fig. 3.4 Distribution of the number of BLs by the duration of the object observation (in seconds)

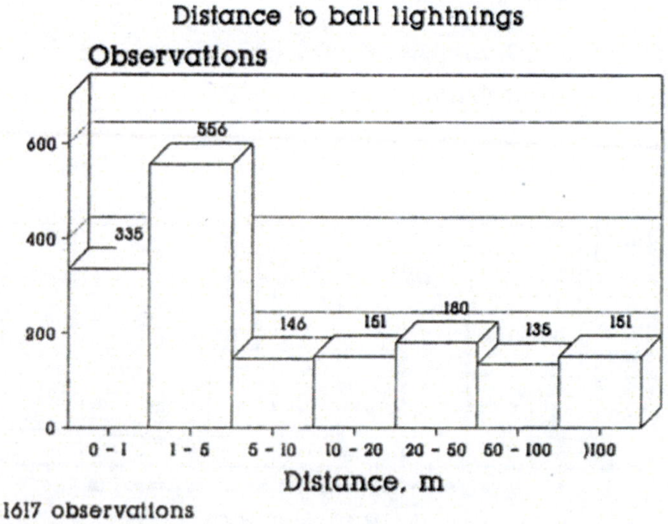

Fig. 3.5 Distribution of the number of observations of the BL by distance to it

based on histograms was done in Stakhanov (1985), so in essence, the main work on the analysis of the observational properties of BL using a simple analysis was completed in Stakhanov (1985), Grigoriev et al. (1993). It should be noted, that the most unusual manifestations of ball lightning, such as abnormally high energy, the passage of BL through holes and through various materials, etc., have not been

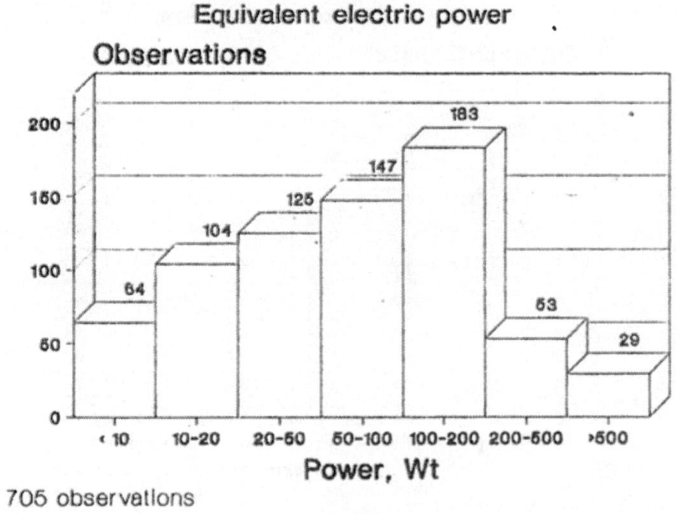

Fig. 3.6 Distribution of the number of BLs by radiation power in comparison with an incandescent lamp, W

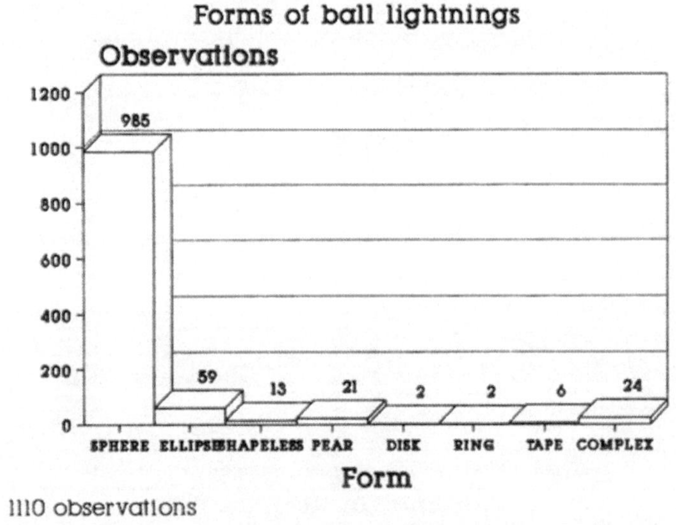

Fig. 3.7 Distribution of the number of BLs observations by shape (from left to right: sphere, ellipsoid, shapeless, pear-shaped, disk, ring, ribbon, complex or shape change)

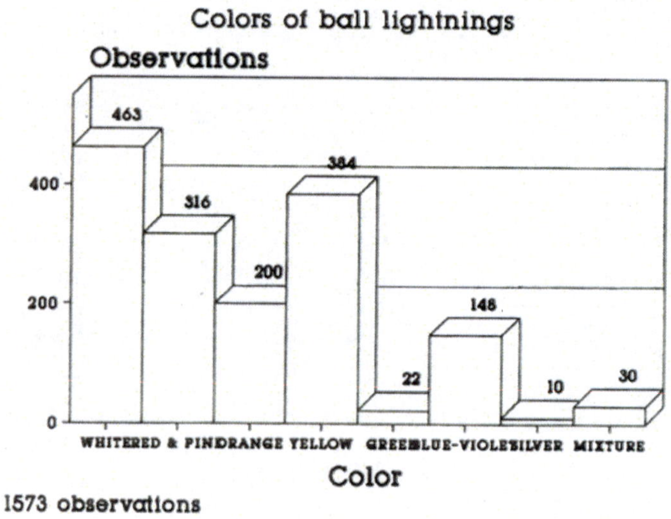

Fig. 3.8 Distribution of the number of observations of BLs by color (from left to right: white, red + pink, orange, yellow, green, blue + purple, silver, mixture)

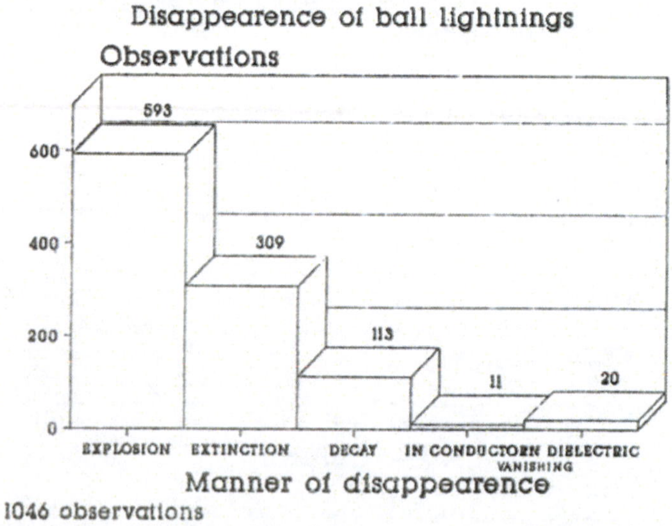

Fig. 3.9 Distribution of the number of observations of BLs by a way of disappearance

sufficiently studied. Therefore, the collection of observational data and their analysis is of great interest.

It is important to note that the dependencies presented above relate to conditions typical of Europe in the absence of tectonic activity. Data from Japanese researchers (Ohtsuki and Ofuruton 1989). In the country where high seismic activity is observed,

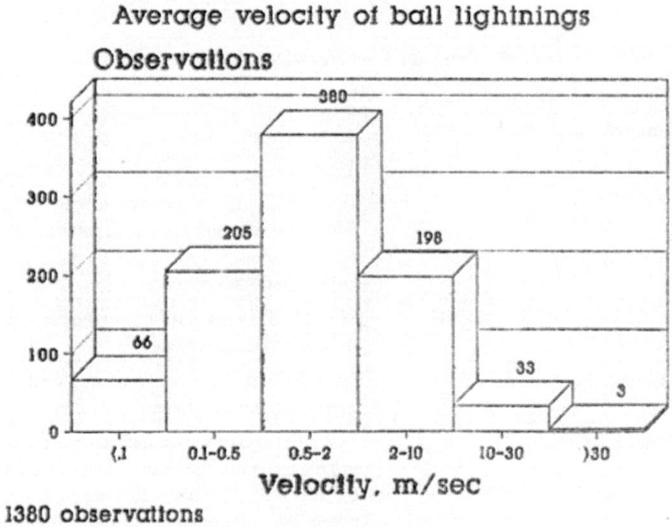

Fig. 3.10 Distribution of the number of observations of BLs by velocity (m/s)

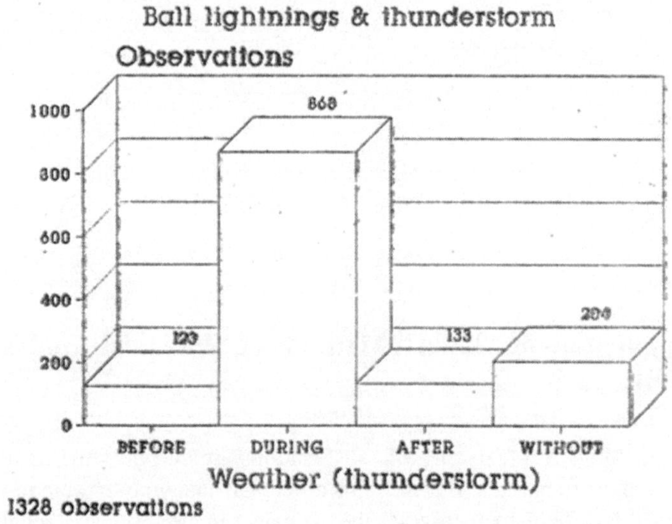

Fig. 3.11 Distribution of the number of observations of BLs in relation to thunderstorm conditions (from left to right: before a thunderstorm, during a thunderstorm, after a thunderstorm, and in the absence of a thunderstorm.)

Table 3.1 Average parameters of BL

	Parameter, unit of measurement	Parameter values
1	Probability of spherical shape, %	91
2	**Diameter, cm**	23 ± 5
3	Lifetime, s	4–16
4	Color	White, red, orange, yellow—about 20%, each. Blue, purple—about 11%. A mixture of colors—about 9% Green—about 1.5%
5	Correlation with atmospheric Electricity	80% of BL in the northern hemisphere is observed in thunderstorm weather
6	Seasonality	Over 80% of BL is observed in the summer months in the Northern Hemisphere
7	Death (disappearance)	In about 60% of cases, the existence of a BL ends in an explosion. In about 34% of cases, the existence of BL ends with a slow extinction. In about 7% of cases, the existence of BL ends in disintegration into parts
8	Luminous flux, lm	1500 ± 30
9	Light output, lm/W	0.6 × 10 ± 0.5
10	Probability of observation of km^{-2} s^{-1}	9.0×10^{-10}

it is shown that for conditions different from European conditions, the dependencies may be different and lead to the need for a broader analysis of the observational data of the BL (Fig. 3.12).

3.3 Application of Modern Methods of Statistical Analysis of BL

Statistical methods for analyzing BL observations are currently the main ones for determining their properties. Because this object has many observable properties—parameters, it is difficult to distinguish the main ones in the absence of reliable experimental information or a reasonable model. Until recently, research has been limited to only the simplest statistical methods. These include estimates of the frequency of manifestations of BL parameters and obtaining the correlation coefficient according to (Pearson) (Stakhanov 1985; Grigoriev et al. 1993) and various modifications of these methods (Grigoriev and Grigorieva 1989; Smirnov and Strizhev 1994), often insufficiently mathematically justified. At the same time, only numerically determined parameters (for example, diameter, lifetime, and equivalent radiation power) are used to identify the properties of BL, and qualitative descriptions of BL properties remain practically beyond the attention of researchers. At the same time, statistical

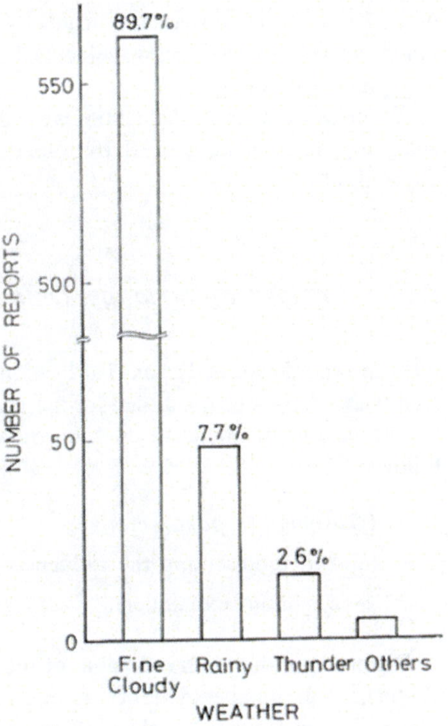

Fig. 3.12 The number of observations of BL depending on weather conditions in Japan (Ohtsuki and Ofuruton 1989) (from left to right: clear and cloudy weather, rain, thunderstorm, other conditions)

methods developed in the middle of the twentieth century allow to involve such information.

This became obvious from our papers (Amirov and Bychkov 1994–1999), some of the results of which we will present below. A.Kh. Amirov supervised the work on calculations and the use of statistical methods.

3.3.1 Statistical Studies of BL Color

The color of the BL is one of the observed parameters that is most difficult to interpret. From the point of view of statistics, two tasks can be formulated as

- Are the processes that determine the color of the BL also essential for other BL parameters, such as diameter and lifetime or radiation power.
- What factors influence the color (or perception of color) of the BL.

We will consider the answer to the first question using the variance analysis (ANOVA) procedure, which allows analyzing the influence of qualitative factors on the observed parameters in conditions when the relationship between the factor and the observed parameter cannot be determined. To do this, we chose such observations of the BL, when the observer indicates the color and diameter (or lifetime) in pairs. For the analysis of variance, we selected the five most common colors—blue, orange, red, white, and yellow.

To answer the second question, we will use the correlation analysis procedure, analyzing the distribution of BL by color depending on various factors that can affect the color of BL.

3.3.2 Analysis of Variance. Data Processing Procedure

As is known Johnson and Lyon (1981), the analysis of variance can be used to identify factors that cannot be quantified for the observed parameter.

The parametric model used Johnson and Lyon (1981) can be represented as follows:

$$\begin{aligned}&(Observable\ parameter) =\\&(Term,\ \text{representing the influence of the corresponding factor})\\&+\ (\text{a random remainder}).\end{aligned} \tag{3.1}$$

Denote Y_{ij}—the observed value of the parameter, i—the number of the group (factor), j—the number of the observation in the group. At the same time, $i = 1, ..., I$, where I is the number of selected groups; $j = 1, \ldots, n_i$, where n_i is the number of observations in the group. Then the parametric model (3.1) can be written as $Y_{ij} = G_{ij} + \varepsilon_{ij}$, where G_{ij} is the term representing the influence of the studied factor, ε_{ij} is the random remainder, and its average value $\overline{\varepsilon}_{ij} = 0$.

We also denote $\overline{Y} = N^{-1}\sum_{i=1}^{I}\sum_{j=1}^{n_i} Y_{ij}$.

- the average for all observations (here $N = \sum_{i=1}^{I} n_i$—the total number of observations),

$$\overline{Y}_i = n_i^{-1}\sum_{j=1}^{n_i} Y_{ij}$$

- the average within the group.

It is known from the theory of mathematical statistics (Kramer 1975) that group averages $\overline{Y}_i$ are unbiased and consistent estimates of the average. If the hypothesis is accepted that the studied factor affects the value of the parameter, then the average for all observations $\overline{Y}$ should statistically differ from the group averages $\overline{Y}_i$.

Based on the accepted model (3.1), for the sum of the squares of deviations from the mean, we can write

$$S^2 = \sum_{i=1}^{I}\sum_{j=1}^{n_i}(Y_{ij} - \overline{Y})^2 = \sum_{i=1}^{I}\sum_{j=1}^{n_i}(Y_{ij} - \overline{Y}_i)^2 + \sum_{j=1}^{n_i}(\overline{Y}_i - \overline{Y})^2 = S_R^2 + S_A^2,$$

where S_R^2—the sum of the squares of deviations within groups, S_A^2—the sum of the squares of deviations between groups. Value

$$R = \frac{\overline{S}_R^2}{\overline{S}_A^2} = \frac{\overline{S}_R^2/(I-1)}{\overline{S}_A^2/(N-1)}$$

- the ratio of the mean square of deviations from the mean between groups to the mean square of deviations from the mean within groups—has a F (Fischer)-distribution (Johnson and Lyon 1981). If our hypothesis is correct and the studied factor is significant, then

$$R > F_{I-1,N-1,1-\alpha}.$$

Here, $F_{I-1,N-1,1-\alpha}$ is the percentile of the F-distribution with $I-1$, $N-1$ degrees of freedom and significance level α. Otherwise, the qualitative factor is not significant. The significance level α determines the range of values $(R-\Delta R,\ R+\Delta R)$ in which, with a probability of α at given degrees of freedom, the value R would fall if the sample were random, i.e., the studied factor was a random (insignificant) way of dividing the data.

3.3.3 *The Results of the Analysis of Variance for the BL Color Factor*

(i) *The color of the BL—the diameter of the BL.* We investigated the influence of the processes that determine the color of BL by identifying five main colors of BL—white, red, orange, yellow, and blue. Observations with indications of blue and violet BL were attributed to the blue color. Thus, the number of factors $I = 5$. In the case of diameters, we obtained the number of observations in groups: $n_1 = 154$ (blue BL), $n_2 = 213$ (orange), $n_3 = 199$ (red), $n_4 = 375$ (white), and $n_5 = 321$ (yellow).
The total number of observations is $N = 1262$. The average value of diameters by groups: $\overline{Y}_1 = 21.184$ cm, $\overline{Y}_2 = 27.805$ cm, $\overline{Y}_3 = 24.469$ cm, $\overline{Y}_4 = 24.2639$ cm, $\overline{Y}_5 = 26.924$ cm.
The average square of deviations from the mean between groups $\overline{S}_A^2 = 1325.93$, and the average square of deviations from the mean within groups $\overline{S}_R^2 = 614.36$. Thus $R = 2.16$.

Percentile of the F-distribution with 4.1257 degrees of freedom at the usual level of confidence in mathematical statistics $\alpha = 0.05$ $F_{4,\infty,0.95} = 2.37$ (Korn and Korn 1974. As you can see, $R < F_{4,\infty,\,0.95}$.

The obtained result indicates that the hypothesis about the significance of the factor determining the color of the BL on the diameter of the BL should be discarded. ***The color factor is not significant for the diameter of the BL.***

(ii) *The color of the BL—the lifetime of the BL.* The number of factors $I = 5$. In the case of diameters, we obtained the number of observations in groups: $n_1 = 148$ (blue BL), $n_2 = 210$ (orange), $n_3 = 201$ (red), $n_4 = 371$ (white), and $n_5 = 317$ (yellow). The total number of observations is $N = 1247$. The average value of lifetimes by groups: $\overline{Y}_1 = 28.264$ s, $\overline{Y}_2 = 26.686$ s, $\overline{Y}_3 = 26.291$ s, $\overline{Y}_4 = 29.737$ s, and $\overline{Y}_5 = 24.629$ s. The average value of the CMM lifetime for all observations $\overline{Y} = 27.194$ s.

The average square of deviations between groups $\overline{S}_A^2 = 1238.02$, and the average square of deviations from the average within groups $\overline{S}_R^2 = 2106.75$. Thus, $R = 0.59$.

The percentile of the F-distribution with 4.1243 degrees of freedom at the usual level of confidence in mathematical statistics $\alpha = 0.05$ $F_{4,\infty,\,0.95} = 2.37$. As can be seen $R < F_{4,\infty,\,0.95}$.

The result obtained indicates that the hypothesis about the significance of the influence of the factor determining the color of the BL for the lifetime of the BL should be discarded. ***The color factor is not significant for the lifetime of the BL.***

Thus, the color of the BL (or the processes that determine it) is not significant for the lifetime and diameter of the BL.

One can assume that the processes that determine the color of the BL are electric discharge processes. In this case, the color of the BL is determined both by the electrical discharge processes themselves on the surface of the BL, and by the chemical composition of both the BL itself and the air layer surrounding the BL. The presence of observations when the rainbow color of the BL is reported may indicate the scattering of radiation (both incident light and discharges) on the particles of the BL surface, similar to interference in thin films: in this case, the color factor is not significant for either the diameter or the lifetime of the object.

3.3.4 Correlation Analysis of BL Color Distributions

Let us now turn to the task of identifying the factors that determine the color (or perception of color) of the BL. Our statistical procedure is based on the hypothesis that if there is a certain factor that significantly affects the color of the BL (and therefore the distribution of BL by colors), then the distribution of BL by colors for such a factor will poorly correlate with distributions for non-essential factors. This means that the correlation coefficient r of the BL color distribution for this factor

Table 3.2 Sample size for various factors used in correlation analysis

Sampling factor	Number of observations
Indoors	1136
Outdoors	553
Disappearing with an explosion	584
Disappearing with extinction	375
Disappearance with decay	118
Observation in June	289
Observation in July	613
Observation in August	329
Observation in September	96
Observation in the morning	70
Observation during the day	939
Observation in the evening	418
Night observation	143
Observation during a thunderstorm	949
Observation without thunderstorms	315
Observation before a thunderstorm	176
Observation in clear weather	192
Observation after a thunderstorm	152

and the non-essential factor will differ from the correlation coefficient of the color distributions for non-essential factors.

We have identified various factors that could affect both the perception of BL color and BL radiation. Among them is the position in space, relative to thunderstorm manifestations, from the time of day, from the month of observation, from the BL way of death. A total of 18 factors were identified (see Table 3.2).

For each of the factors, we calculated the distribution of BL by color. The six most common BL colors were selected (see above): red, yellow, white, orange, blue (including cyan and purple), and green. Pearson correlation coefficients were determined between all distributions (Kramer 1975):

$$r_{XY} = \frac{\sum_i (X_i - \overline{X})\cdot \sum_i (Y_i - \overline{Y})}{\sqrt{\sum_i (X_i - \overline{X})^2 \cdot \sum_i (Y_i - \overline{Y})^2}}, \tag{3.2}$$

where X_i, Y_i, are the frequency of manifestation of the i- th color for the factors X and Y, respectively, $\overline{X}$, $\overline{Y}$ are the average frequency of the BL color for the factors X and Y ($\overline{X} = n_X^{-1} \sum_i X_i$, $\overline{Y} = n_Y^{-1} \sum_i Y_i$, where n_X, n_Y—the total number of observations for each factor). The obtained values of the correlation coefficients r_{XY} for BL distributions in the case of different factors are presented in Table 3.2 (see Fig. 3.13). As can be seen from this table, the correlation coefficients have a noticeable spread.

Fig. 3.13 Table 3.2 Correlation matrix for BL distribution for different factors, which can influence the BL color or its perception. Designation A—indoors, B—outdoors, C—explosion, D—extinction, E—decay, F—June, G—July, H—August, I—September, J—morning, K—day, L—evening, M—night, N—during thunderstorm, O—without thunderstorm, P—before thunderstorm, Q—clear weather, R—after thunderstorm

	A	B	C	D	E	F	G	H	I	J	K	L	M	N	O	P	Q	R
A	1.000																	
B	0.973	1.000																
C	0.951	0.973	1.000															
D	0.960	0.984	0.953	1.000														
E	0.844	0.934	0.871	0.891	1.000													
F	0.956	0.984	0.993	0.962	0.892	1.000												
G	0.944	0.979	0.988	0.936	0.920	0.990	1.000											
H	0.988	0.983	0.986	0.962	0.867	0.986	0.979	1.000										
I	0.266	0.362	0.304	0.480	0.324	0.360	0.254	0.267	1.000									
J	0.955	0.966	0.976	0.913	0.893	0.974	0.992	0.982	0.149	1.000								
K	0.987	0.988	0.972	0.955	0.903	0.976	0.981	0.993	0.229	0.987	1.000							
L	0.949	0.985	0.994	0.955	0.920	0.992	0.997	0.983	0.291	0.984	0.980	1.000						
M	0.746	0.842	0.909	0.815	0.829	0.881	0.899	0.834	0.275	0.860	0.816	0.911	1.000					
N	0.965	0.998	0.979	0.978	0.932	0.992	0.985	0.982	0.375	0.969	0.984	0.989	0.856	1.000				
O	0.945	0.981	0.975	0.969	0.923	0.964	0.969	0.968	0.303	0.955	0.969	0.984	0.900	0.975	1.000			
P	0.899	0.797	0.785	0.838	0.569	0.767	0.727	0.850	0.217	0.757	0.825	0.752	0.520	0.773	0.791	1.000		
Q	0.938	0.991	0.960	0.959	0.971	0.973	0.982	0.959	0.340	0.963	0.974	0.984	0.869	0.991	0.976	0.713	1.000	
R	0.967	0.932	0.956	0.947	0.744	0.947	0.912	0.970	0.342	0.913	0.939	0.927	0.776	0.930	0.921	0.917	0.880	1.000

Let us turn to the question of the statistical significance of the observed variations, i.e. consider the question of the ratio of the influence on the color of the factors we have identified in comparison with other uncontrolled factors. To do this, similarly to Bychkov et al. (1993), the databank was randomly divided into two parts, for each of which, a distribution of BL by color was obtained. The $\boldsymbol{r}_{\text{rand}}$ correlation coefficient between these random distributions was determined by the formula (3.2). It was $\boldsymbol{r}_{\text{rand}} = 0.987$. As an error in determining the correlation coefficient, we cannot use, as, for example, in the case of the average value, the value of the standard error $d(r) = n^{-1/2}(1-r^2)$, because, unlike the average, the correlation coefficient does not have a normal distribution (Kramer 1975). According to the theory of mathematical statistics (Kramer 1975), the value of Z is associated with the correlation coefficient for the general population ρ, calculated by the formula

$$Z = \text{Arcth}(\rho) = \frac{1}{2}\ln\left(\frac{1+\rho}{1-\rho}\right) \tag{3.3}$$

is distributed normally at moderate values of n, (where n is the number of observations for which the correlation coefficient is determined). In this case, the average value of Z is equal to

$$\overline{Z} = \frac{1}{2}\ln\left(\frac{1+\rho}{1-\rho}\right) + \frac{\rho}{2(n-1)} \tag{3.4}$$

with a standard deviation. Consequently, the Z-transformation for the correlation coefficient for two random samples from the general population of observations of BL (Kramer 1975) lies within (k_1, k_2), where $k_{1,2}$ are determined from the equality:

$$k_{1,2} = \frac{1}{2}\ln\left(\frac{1+r}{1-r}\right) + \frac{\lambda_\alpha}{\sqrt{n-3}}, \tag{3.5}$$

were λ_α is the quantile of the normal distribution of the level $\alpha\%$. To find α the percentage confidence intervals for the correlation coefficient between two random samples from the general population of BL observations, it is necessary to solve with respect to ρ the equation

$$k_{1,2} = \frac{1}{2}\ln\left(\frac{1+\rho}{1-\rho}\right) + \frac{\rho}{2(n-1)}. \tag{3.6}$$

Substituting the obtained $\boldsymbol{r}_{\text{rand}}$, value into Eq. (3.3), we find the values $k_{1,2}$ for the usual confidence level in statistics $\alpha = 95\%$ (Kramer 1975, for which the quantile is $\lambda_{0,95} = 1.65$ (Korn and Korn 1974. Knowing the values of $k_{1,2}$, we solve the Eq. (3.3) to obtain α—percent confidence intervals for ρ. We obtain that for two random samples, the correlation coefficient with a probability of 95% lies in the range of values $0.90278 \leq \rho \leq 1$.

As can be seen from Table 3.2, the correlation coefficients between the overwhelming number of samples made by us lie in the specified interval. This means that all samples whose correlation coefficients fall within the interval $(0.90278 \div 1)$, can be considered random.

Correlations associated with observations of BL in September, observations at night and before a thunderstorm, as well as correlations with decaying BL fall out of it.

Let us note two more points that are important. When defining $\boldsymbol{r}_{\text{rand}}$, we split the database in half, so we got a sample of the volume, (here N-is the number of records in the database). It is clear that if the volume of the random sample is equal n, then $\lim_{n\to\infty} r_{rand} = 1$.

At the same time, the samples n_X for exploding and observed BLs before a thunderstorm are significantly less $1/2N$ half of the records in the database (Amirov and Bychkov 1991). That is, it would be more correct to make a random sample n_{rand} of the volume ~n_X, as for factor X. At the same time, it is clear that the correlation coefficient for a smaller sample $\left|r_{rand}^{n_X}\right| \leq \left|r_{rand}^{1/2N}\right|$, and therefore its confidence interval $\rho_1 \leq \rho \leq \rho_2$ will be larger at the same confidence level α. That is, the probability of getting into the confidence interval of those samples that did not fall into it at $n_{rand} = 1/2N$.

The second point is related to the fact that we can carry out the procedure we have done to determine the confidence interval for any correlation coefficient. If the correlation coefficient is close to the boundary of the confidence interval $\rho_1 \leq \rho \leq \rho_2$ (and this is the case for decaying BL), then the confidence interval is obtained from the Eqs. (3.2–3.6) for all will ρ_X intersect with the interval $\rho_1 \leq \rho \leq \rho_2$, and therefore the probability of coincidence ρ_X and ρ will be different from zero. This probability is equal to

$$P_\alpha = \sqrt{\frac{n-3}{2\pi}} \int_{Arcth(0.90278)}^{Arcth(\rho_2^X)} \exp\left(-\frac{(n-3)\cdot(x-\overline{Z}_X)^2}{2}\right) dx.$$

Here, $\overline{Z}_X$ the average value for a normally distributed quantity Z_X with a standard deviation of $(n-3)^{-1/2}$, obtained by substituting the quantity into Eq. (3.6), is the upper bound $r_X Arcth(\rho_2^X)$ of the confidence interval for ρ_X for factor X.

Consider those factors for which the correlation coefficient was outside the confidence interval $(0.90278 \div 1)$ of the correlation coefficient for a random sample.

September. The loss of correlation coefficients for September can be explained by insufficient statistics—the total number of BL observations in September is 86 cases, which is almost 3 times smaller than, for example, in August, and 11 times smaller than with a random sample (see Table 3.1).

Observation at night and before a thunderstorm. The result with the loss of correlation coefficients during night observations and before a thunderstorm can be explained by the complexity of the light conditions for observing the BL and the peculiarity of light perception in these conditions. It is obvious that the light conditions of observation are very difficult, especially in cities with a large number of outdoor lighting devices. Light conditions before a thunderstorm are also difficult when the intensity of illumination can change dramatically. Under these conditions, observations of luminous objects are associated with adaptive processes of physiological optics such as light adaptation, dark adaptation, and color adaptation (Meshkov and Matveev 1989). Under these conditions, a person's adaptive time for color perception is equal to several minutes (Meshkov and Matveev 1989). From this point of view, it can be assumed that the loss of the correlation coefficient for BL observers at night and before a thunderstorm is the result of the observer's adaptive ability in difficult light conditions.

Decaying BL. As if observers of the BL in September should pay attention to a smaller number of observations decaying BL compared with exploding (less than 5 times) and extinction (less than 3 times) BL (see Table 3.2).

At the same time, the observers of the BL death as decay, allow us to draw an important conclusion (Amirov and Bychkov 1995a) that such BL has on the surface a layer, the destruction of which is perceived as decay. In the case of the formation of such a layer, its emission and reflective properties should differ from the properties of the BL substance that comprises its core.

3.3.5 Factors Affecting the Observed Parameters of the BL

Let us use the variance analysis procedure in order to identify which factors influence the parameters of the BL. From this point of view, two factors are interesting.

The place of observation of the BL—inside or outside the room. The connection of BL with atmospheric electricity and thunderstorm manifestations is well known. At the same time, it is known that BLs are often observed in the absence of thunderstorms (see above), and according to Ohtsuki and Ofuruton (1989) up to 70% of BL observations, in Japan occur without thunderstorms (see above). In closed rooms, where BLs are often observed, the effect of atmospheric electricity of the formed BLs should be different from outdoors. Inside and outside the room, there may be different concentrations of substances that form the basis of BL. Therefore, the factor of the place of observation of the CMM may be significant from the point of view of the distribution of CMM parameters.

An interesting question is the problem of the relationship between factors or processes that determine the BL way of death and the parameters of the BL.

Analysis of Variance for the Factor of the Place of Observation of the BL Outside and Inside the Room

To carry out the dispersion analysis, we selected in the database those cases of BL observations when the observer indicated both the studied factor—the position in space, and the studied parameter—the diameter, lifetime, or equivalent radiation power. The variance analysis procedure is similar to that used to analyze the significance of the color factor for the BL parameters (see above).

(i) *The position of the BL in space—its diameter.* The number of factors $I = 2$, the number of observations in the group $N_1 = 572$ (indoors) $N_2 = 1124$ (outdoors). The total number of observations is $N = 1696$. Thus, the number of degrees of freedom is 1694. The average value of diameters by groups: $\overline{Y}_1 = 15.61$ cm, and $\overline{Y}_2 = 30.38$ cm. The average value for all observations is $\overline{Y} = 25.40$ cm.

(ii) *The position of the BL in the space—BL lifetime.* The number of factors $I = 2$, the number of observations in the group $N_1 = 571$ (indoors) $N_2 = 1097$ (outdoors). The total number of observations is $N = 1668$. Thus, the number of degrees of freedom is 1666. The average value of diameters by groups: 18.96 s, 35.27 s. The average value for all observations is $\overline{Y} = 29.69$ s. The average square of deviations between groups is $\overline{S}_A^2 = 99{,}902$, and the average square within groups is $\overline{S}_R^2 = 2411.6$.
Thus, $R = 41.43$. The percentile of the F-distribution with 1.1666 degrees of freedom at the confidence level $\alpha = 0.001\,F_{1,\infty,0.999} = 10.83$ (Korn and Korn 1974). As you can see, $R > F_{1,\infty,0.999}$.
The obtained result indicates that the hypothesis about the influence of the observation location on the lifetime of the BL is significant.

(iii) *The position of the BL in space—the equivalent radiation power.* The number of factors $I = 2$, the number of observations in the group $N_1 = 333$ (indoors) and $N_2 = 455$ (outdoors). The total number of observations is $N = 788$. Thus, the number of degrees of freedom is 786. The average value of diameters in groups: $\overline{Y}_1 = 87.57, \overline{Y}_2 =$ W, 167.12 W. The average value for all observations is $\overline{Y} = 133.51$ watts. The average square of deviations between groups is $\overline{S}_A^2 = 1{,}216{,}203$, and the average square within groups is $\overline{S}_R^2 = 28{,}107$.
Thus, $R = 43.27$. The percentile of the F-distribution with 1.788 degrees of freedom at the confidence level $\alpha = = 0.001\,F_{1,\infty,0.999} = 10.83$ (Korn and Korn 1974). As you can see, $R > F_{1,\infty,0.999}$.
The obtained result indicates that the hypothesis about the influence of the observation location for the equivalent power of the BL radiation is significant.

Thus, the results of the dispersion analysis indicate that such a qualitative factor as the place of observation of the BL is essential for all the observed parameters of the BL investigated by us—diameter, lifetime, and radiation power. As can be seen from Figs. 3.14, 3.15, 3.16, where histograms of distributions of BL parameters are

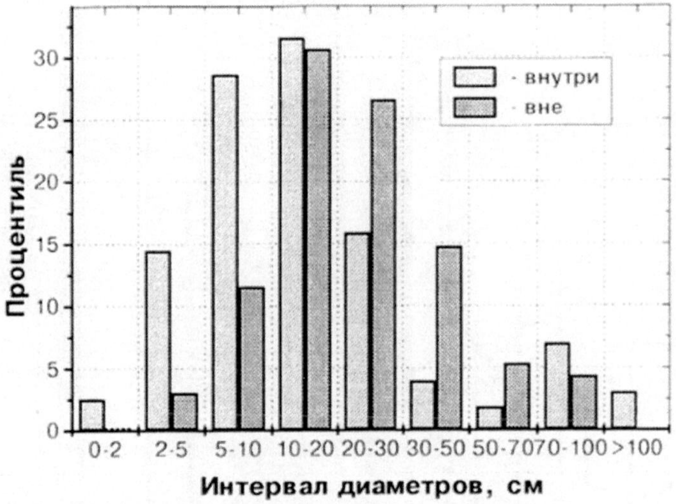

Fig. 3.14 Histogram of the distributions of the BL by diameter depending on the position of the BL in space (indoors or outdoors). Vertical axis—percentile, horizontal axis—interval of diameters. Indoors—white, outdoors—gray

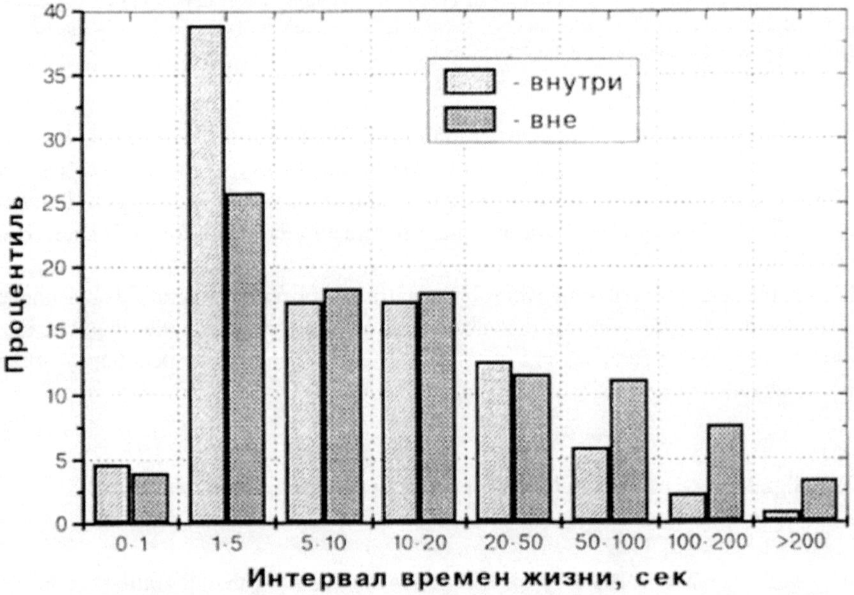

Fig. 3.15 Histogram of the distributions of BL by lifetime depending on lifetime of the BL in space (indoors or outdoors). Vertical axis—percentile, horizontal axis—interval of diameters. Indoors—white, outdoors—gray

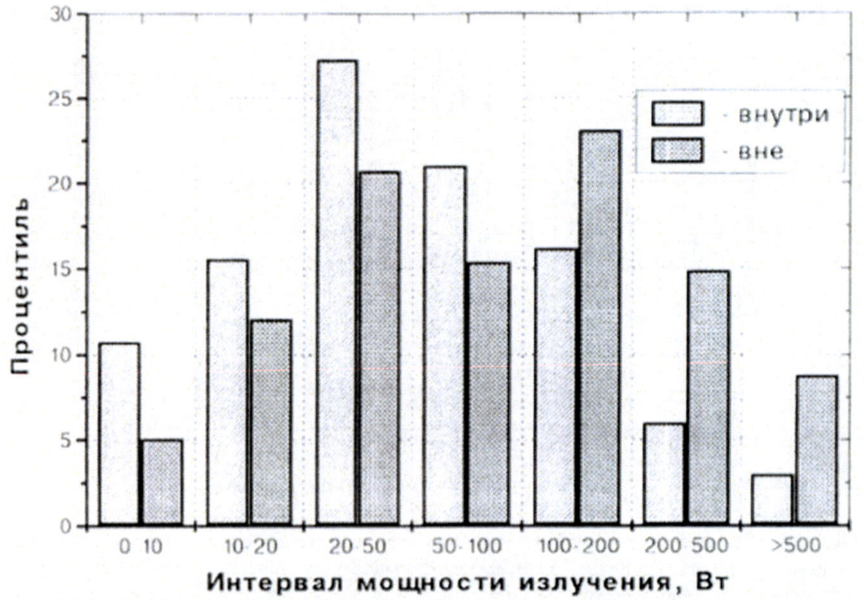

Fig. 3.16 Histogram of the distributions of the BL by radiation power depending on the power of the BL in space (indoors or outdoors). Vertical axis—percentile, horizontal axis—interval of radiation power. Indoors—white, outdoors—gray

presented, BLs have different distributions and dependences on the place of observation of the object. Possible reasons for the result may be psychological peculiarities of perception (Fress and Piaget 1978); the secondary nature of BL observed indoors in relation to BL observed outdoors; the difference in the conditions of occurrence and death of BL inside and outside the premises. In Fress and Piaget (1978) distortions in the perception of dimensions by an untrained observer depending on distance were noted. Since the possible distances to the BL indoors are smaller than outside, the results of the differences in Fig. 3.14 can be explained by the peculiarity of the BL observation (Figs. 3.14, 3.15, 3.16, and 3.17).

The Significance of the Factors Determining the Method of Death of BL. Analysis of Variance

It is well known that BL dies in three ways—exploding, disintegrating, or going out. Let us consider with the help of variance analysis whether the processes that determine the ways of the death of the BL are significant for the parameters of the BL. For the analysis of variance, those descriptions were selected from the database when the observer simultaneously reported the parameter of the BL and the way of its death.

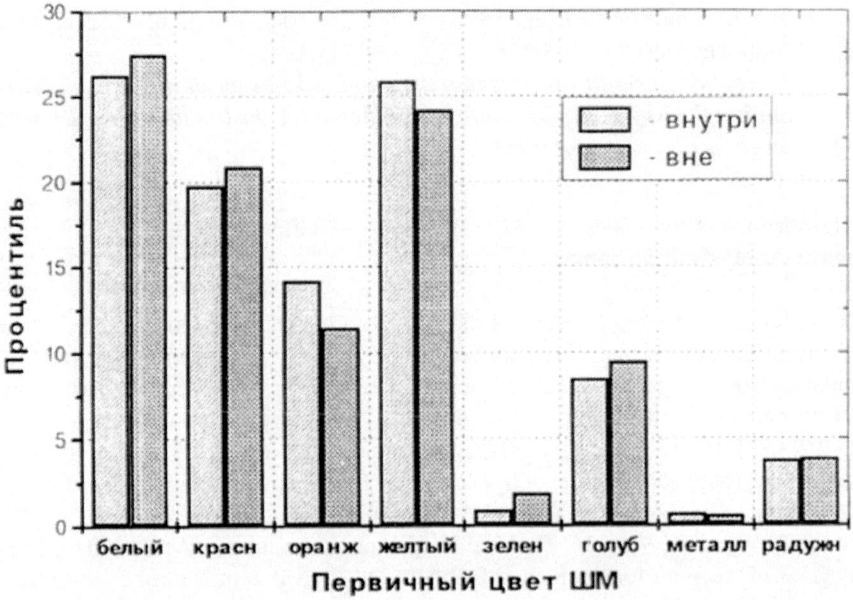

Fig. 3.17 Histogram of BL primary color distributions depending on the position of the BL in space (inside or outside the room). (indoors or outdoors). Vertical axis—percentile, horizontal axis—interval of primary colors. Indoors—white, outdoors—gray (left to right: white, red, orange, yellow, green, blue, metallic, rainbow)

(i) *The way of death of the BL—the diameter of the BL.* The number of studied factors $I = 3$, the number of observations in the group $n_1 = 600$ (explosion of BL), $n_2 = 351$ (extinction of BL), and $n_3 = 123$ (decay of BL). The total number of observations is $N = 1074$. The average value of diameters by groups: $\overline{Y}_1 = 22.26$ cm, $\overline{Y}_2 = 28.52$ cm, and $\overline{Y}_3 = 28.70$ cm. The average value for all observations is $\overline{Y} = 25.04$ cm. The mean square of deviations from the mean between groups is $\overline{S}_A^2 = 5267.03$. The mean square of deviations from the mean within groups is $\overline{S}_R^2 = 549.71$. Thus, $R = 9.58$.
The percentile of the F-distribution with 2 degrees of freedom at the confidence level $\alpha = 0.001$ $F_{2,\infty,0.999} = 6.91$ [46]. As you can see$R > F_{2,\infty,0.999}$.
The obtained result suggests that the qualitative factor of the process determining the way of BL death is significant for the diameter of the BL with an error probability of 0.1%.

(ii) *The method of death of the BL—the lifetime of the BL.* The number of studied factors $I = 3$, the number of observations in the group $n_1 = 593$ (explosion of BL), $n_2 = 337$ (extinction of BL), $n_3 = 122$ (decay of BL). The total number of observations is $N = 1048$. The average value of lifetimes by groups: $\overline{Y}_1 = 21.48$ s, $\overline{Y}_2 = 33.12$ s, $\overline{Y}_3 = 33.78$ s. The average value for all observations is $\overline{Y} = 23.64$ s. The average square of deviations from the mean between groups is $\overline{S}_A^2 = 18{,}050.17$, $\overline{S}_R^2 = 2105.221$. Thus, $R = 8.57$.

The percentile of the F-distribution with 2 and 1048 degrees of freedom at the confidence level $\alpha = 0.001\, F_{2,\,\infty,\,0.999} = 6.91$. As you can see, $R > F_{2,\,\infty,\,0.999}$. ***The obtained result suggests that the qualitative factor of the process determining the way of the BL death is significant for the lifetime of the BL with an error probability of 0.1%.***

Significance of the Diameter of the BL for Its Lifetime. Results of the Analysis of Variance

We investigated the significance of diameter as a qualitative factor for the lifetime of the BL as a parameter. Within the framework of the parametric model of variance analysis (Johnson and Lyon 1981) described above, we have divided the observations of BL into groups depending on their diameter. The following groups were selected: 0–2, 2–5, 5–10, 10–20, 20–30, 30–50, 50–70, 70–100, and over 100 cm. Thus, we have nine groups (or 8 degrees of freedom). The number of observations in the groups turned out to be: $N_1 = 26$, $N_2 = 128$, $N_3 = 306$, $N_4 = 565$, $N_5 = 386$, $N_6 = 187$, $N_7 = 70$, $N_8 = 56$, $N_9 = 36$. The total number of observations is $N = 1760$ (or 1751 degrees of freedom for the Fisher distribution). The value of R turned out to be $R = 15.06$. The value of the F-distribution with 8 and 1751 degrees of freedom and the usual confidence level for statistics $\alpha = 0.95$ equals $F_{8,\,\infty,\,0.95} = 1.94$ (Korn and Korn 1974). Thus, $R > F_{8,\,\infty,\,0.95}$, and the diameter factor turns out to be significant for the lifetime (as a parameter) for all BL.

At the same time, as we found out, the lifetime and diameter depend on the way of death of the BL. Let us now perform a similar analysis of variance independently for each way of BL destruction—explosion, extinction, and decay.

(i) Exploding BL. The number of observations in the group turned out to be: $N_1 = 16$, $N_2 = 32$, $N_3 = 110$, $N_4 = 204$, $N_5 = 125$, $N_6 = 52$, $N_7 = 12$, $N_8 = 216$, $N_9 = 9$. The total number of observations is $N = 575$ (or 566 degrees of freedom for the Fisher distribution). The value of R turned out to be $R = 2.84$. The value of the F-distribution with 8 and 566 degrees of freedom and the usual confidence level for statistics $\alpha = 0.95$ равно $F_{8,\,\infty,\,0.95} = 1.94$. Thus, $R > F_{8,\,\infty,\,0.95}$, and the diameter factor turns out to be significant for the lifetime (as a parameter) for exploding BL.

(ii) Decaying BL. The number of observations in the group turned out to be: $N_1 = 1$, $N_2 = 14$, $N_3 = 12$, $N_4 = 36$, $N_5 = 23$, $N_6 = 20$, $N_7 = 6$, $N_8 = 5$, $N_9 = 4$. The total number of observations is $N = 575$ (or 112 degrees of freedom for the Fisher distribution). The value of R turned out to be $R = 20.25$. The value of the F-distribution with 8 and 566 degrees of freedom and the usual confidence level for statistics $\alpha = 0.95$ равно $F_{8,\,120,\,0.95} = 2.02$. Thus, $> F_{8,\,120,\,0.95}$, and the diameter factor turns out to be significant for the lifetime (as a parameter) for decaying BL.

(iii) Extinguished BL. The number of observations in the group turned out to be: $N_1 = 3$, $N_2 = 214$, $N_3 = 44$, $N_4 = 93$, $N_5 = 73$, $N_6 = 32$, $N_7 = 29$, $N_8 = 11$, $N_9 = 9$. The total number of observations is $N = 318$ (or 309 degrees of freedom for the Fisher distribution). The value of R turned out to be $R = 1.609$. The value of the F-distribution with 8 and 309 degrees of freedom and the usual confidence level for statistics $\alpha = 0.95$ is $F_{8,\infty,0.95} = 1.94$. Thus, $R < F_{8,\infty,0.95}$, and the diameter factor turns out to be insignificant for the lifetime (as a parameter) for extinguished BL.

Our results mean that there is a difference between BL depending on the way their death. This allows us to take a fresh look at the question of the ratio of the lifetime of a BL τ from its size R_s. To identify the desired dependence, it is necessary to take into account another factor—the loss of BL energy. At the same time, the energy loss of the BL as a function of the diameter should be different functions depending on the way of death of the BL. The studies performed by the linear regression method taking into account the BL radiation factor ($W\tau$ -radiation power multiplied by the BL lifetime) in [39] showed that $W \cdot \tau \sim R_s^{0.14}$ in the case of explosion and extinction; and $W \cdot \tau \sim R_s^{0.43}$ in the case of BL decay.

The interpretation of these results requires the development of a model of the substance and its structure, of which BL are composed.

Dependence of the Lifetime of the BL on Its Diameter

For the first time, the analysis of the relationship between the lifetime of the BL and its diameter was made in Grigorjev et al. (1988) based on a correlation analysis of the collection of observations of the BL by Professor A.I. Grigoriev. It was shown that the lifetime of BL increases with an increase in their diameters. However, this analysis did not take into account the various ways in which BL disappeared. In addition, it was important to compare the results of processing observations of data (Grigorjev et al. 1988) with data from the SKB database (SKB database).

As the lifetime of the BL in statistical analysis, we used the observation time of the BL specified in the database of the SKB. This correlates with the lifetime of the BL but is not always equal to it. Observers very rarely record the moments of the appearance and disappearance of BL in the same description. Nevertheless, based on our previous experience, we used the values of the observation time as the lifetime of the BL, although this may introduce some errors in the analysis.

Statistical analysis

Here, we apply nonparametric regression methods (Härdle 1990). The following numbers of life diameters of pairs were selected in the database of the SKB, which were indicated in the same observation. 112 in the case of the disappearance of the BL as a result of decay, 773 in the case of the disappearance of the BL as a result of explosion and 296 in the case of the extinction of the BL.

To obtain a more accurate averaged dependence of the lifetime of the BL on its diameter, we need not only to use the averaging procedure for a given diameter but also to take into account the values of the lifetime at neighboring points $d_i \pm \Delta d$. Such a procedure of local averaging and smoothing makes it possible to reduce the influence of random errors in estimating the characteristics of BL. If we have a set of n pairs of values $\{X, Y\}$, then the averaged and smoothed values of the parameter y at point x can be written as Handbook of Applicable Mathematics (1986):

$$m(x) = n^{-1} \sum_{i=1}^{n} W_{nl}(x) \cdot Y_i. \tag{3.7}$$

Here, $m(x)$ is the average, smoothed over the vectors $\{X, Y\}$ value of the parameter y at point x, $W_{nl}(x)$ is the weight of the i-th observation for point x, $Y_i - i$—is the i-th value of the parameter at point x. As can be seen from the formula (3.7), the value $m(x)$ may depend on the entire vector $\{X, Y\}$.

For the processing procedure according to (3.7), we use the Nadaray-Watson kernel smoothing estimate (Handbook of Applicable Mathematics 1986): for which, the value $m(x)$ has the form

$$m(x) = \frac{n^{-1} \sum_{i=1}^{n} K_h(x - x_i) \cdot Y_i}{n^{-1} \sum_{i=1}^{n} K_h(x - x_i)}. \tag{3.8}$$

Here, $K_h(x)$ is a smoothing kernel with a scale factor h. As a kernel, we used a parabolic type function, the so-called Epanechnikov kernel (Handbook of Applicable Mathematics 1986),

$$K(u) = 0.75\,(1 - u^2) \cdot I(|u| \le 1) \tag{3.9}$$

and function I

$$I(|u|) = \begin{cases} 1, & |u| \le 1 \\ 0, & |u| > 1. \end{cases} \tag{3.10}$$

The value of u in the case of the Nadaray-Watson estimate $u = (x - x_i)/h$.

Since in the case of BL observations, the scattering of diameter values reaches several orders of magnitude, the normalization procedure was used during averaging (Handbook of Applicable Mathematics 1986). Instead of the diameter values, we inserted the values of their logarithms in (3.7)–(3.10). Thus, in our case, x_i is the value of the logarithm $\lg(d_i)$ of the diameter of the BL for the i-th observation, Y_i is the lifetime t_i for the i-th observation.

From formulas (3.7)–(3.10), it can be seen that when $h \to 0$ $m(x)$ tends to the average value of Y at the point x. This means that $h \to 0$ after averaging according to (3.7)–(3.9) at the points d_i, we get the average lifetime value t_{di} for each diameter d indicated by the observers. When $h \to \infty$ we get a simple straight line $y = \text{const}$, where const will be equal to the average value of the lifetime for all observations const $\equiv n^{-1} \sum_i t_{di}$. The question of choosing the averaging parameter h becomes essential. We assumed that the error in estimating the diameter of the BL could reach 50% (Grigoriev and Grigorieva 1989). Therefore, for h, when processing logarithms, we took the value $h = 0.2$. The latter means that averaging is performed for values that differ by 50%, although the weight of the boundary points, according to (3.10), tends to zero. As can be seen from Figs. 3.18, 3.19, and 3.20, the obtained functions are not sufficiently smoothed due to an insufficiently large number of sample values. Another likely way to choose the value of h is to find the minimum of differences, i.e., it is necessary to solve the equation

$$\frac{d}{dh}\sum_i [m(x_i) - Y_i]^2 = 0.$$

The ranges of confidence intervals for the smoothed function *m(x)* at the confidence level were estimated in accordance with Härdle (1990):

$$m(x) \pm q_\alpha \Lambda \sigma(x) \Big/ \left(nh \sum_{i=1}^{n} K_h(x - x_i)\right)^{1/2} \tag{3.11}$$

Here, q_α is the quantile as a percentage of the normal distribution, Λ is a constant depending on the smoothing main function ($\Lambda = 0.6$ for the Epanechnikov kernel, and $\sigma(x)$ is the variance determined by the equation (Härdle 1990):

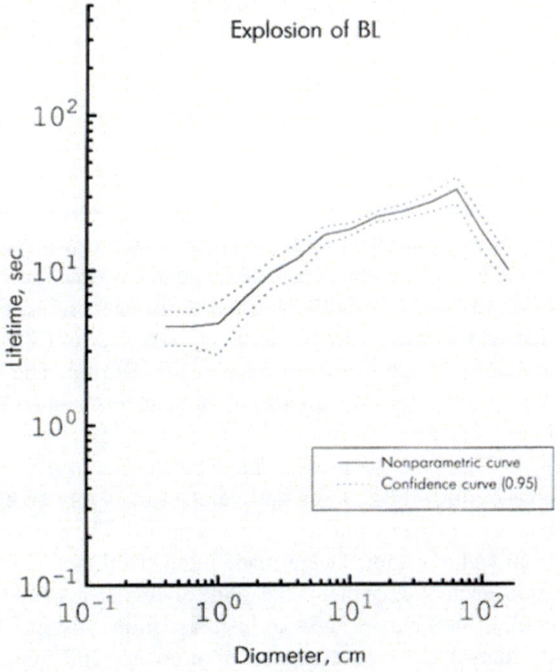

Fig. 3.18 Data on the lifetime of the BL depending on the diameter of the BL in the event of an explosion of the BL. Solid line; $h = 0.2$ in Eq. (3.8); dotted line: the confidence lifetime range for $h = 0.2$ at the confidence level $\alpha = 0.95$, according to (3.11)

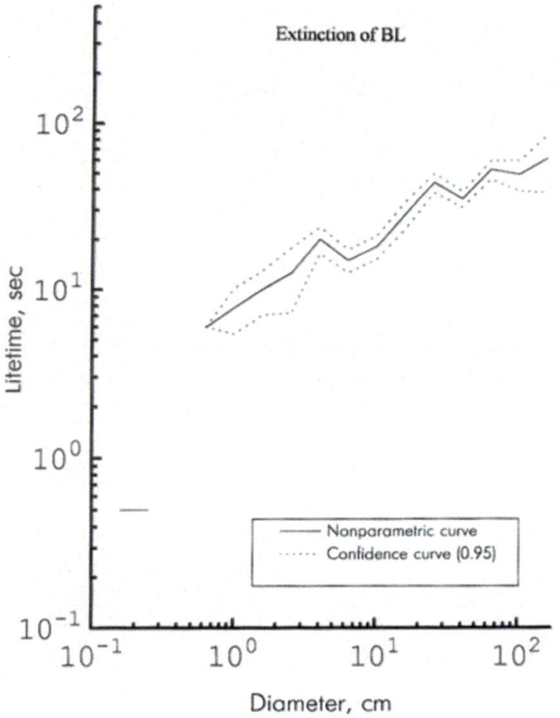

Fig. 3.19 Data on the lifetime of the BL depending on the diameter of the BL in case of extinction of the BL. Solid line; $h = 0.2$ in Eq. (3.8); dotted line: the confidence lifetime range for $h = 0.2$ at the confidence level α = = 0.95, according to Eq. (3.11)

$$\sigma^2(x) = n^{-1}\sum_{i=1}^{n} W_{hi}(x)[Y_i - m(x)]^2 \tag{3.12}$$

Figures 3.18, 3.19, and 3.20 present the Nadaray–Watson smoothed estimated dependences for the lifetime of the diameter for various methods of the disappearance of the BL (explosion, extinction, and decay) and their confidence intervals obtained using (3.7)–(3.11), with the usual statistical confidence level $\alpha = 0.95$. One can see that the lifetime of the BL increases in the case of BL that disappear during extinction and decay in the diameter range of 0.5–100 cm. The lifetime of BLs that disappear during an explosion increases with diameters smaller than 60 cm and decreases with larger diameters.

In Fig. 3.21, we present the dependences of the lifetime diameter in the case of considering all BL, when the average values for two parameters in this interval were taken, and the error in this case is standard. The number of observations of the object from which samples were made in our work and Grigorjev et al. (1988) was about 2000, each. We see that in the range of diameters of 1–100 cm, both our and Grigorjev et al. (1988) curves show an increase in the lifetime with the size of the BL. Note that they decrease at diameters of more than 200 cm (this is not shown in the figure).

The results of nonparametric regression analysis show that this effect is caused by the death of the BL in an explosion, the number of data which is the largest in the

Fig. 3.20 Data on the lifetime of the BL depending on the diameter of the BL in the case of BL decay. Solid line $h = 0.2$ in Eq. (3.8); dotted line: the confidence lifetime range for $h = 0.2$ at the confidence level $\alpha = 0.95$, according to Eq. (3.11)

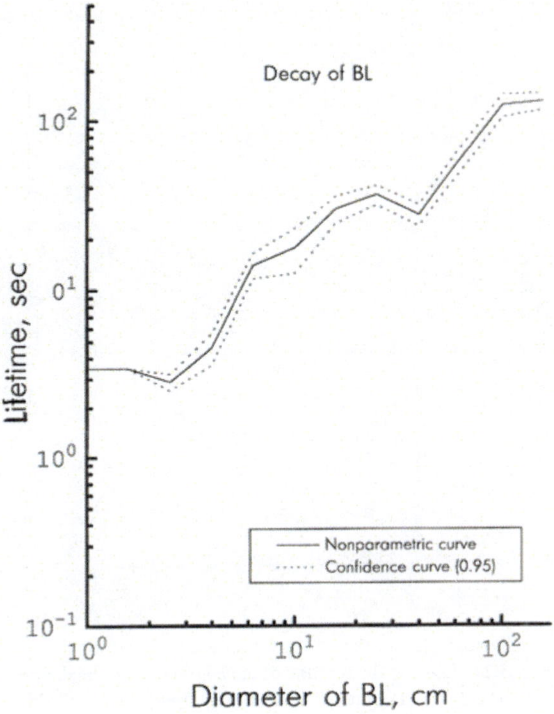

sample. A comparison of our data and data (Grigorjev et al. 1988), i.e., data from different collections shows that the spread of observational data can reach 1.5–2 times. This also shows that it is necessary to increase the number of observations in collections in order to obtain smoother approximations.

The Influence of Thunderstorm Atmospheric Conditions on the Properties of BL

We also note the works (Amirov and Bychkov 1996, 1997). They attempt to use rank correlations to investigate the influence of thunderstorm atmospheric conditions on the lifetime (observation time) of the BL and the diameter of the BL based on the analysis of observations from the database of the SKB. For the analysis, the BL was ranked by size; weather conditions were ranked in relation to humidity correlated with atmospheric charge, thunderstorm conditions were ranked in relation to the electric field strength of the surface layer of the atmosphere (the relationship is based on fragmentary data on humidity and the electric field strength of the atmosphere, which in principle may be different). The conducted statistical analysis showed that under conditions of increasing humidity and electric field strength, the lifetime and

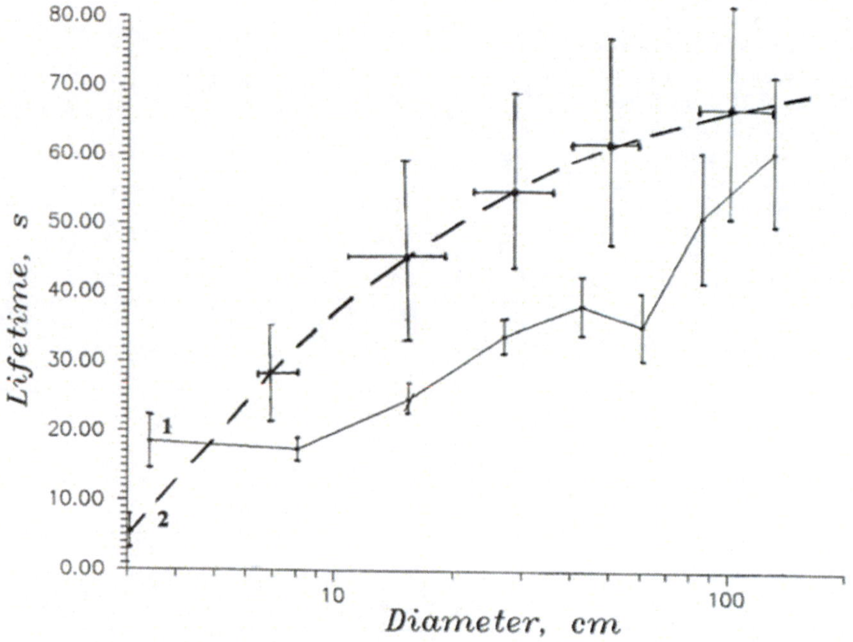

Fig. 3.21 Data on the lifetime of the BL relative to the diameter of the BL for averaged data values. (1) our results; (2) Grigoriev's results (Grigorjev et al. 1988)

diameter of the BL decrease. The interpretation of the results obtained depends on the BL model.

Notes to the Section

The section discusses the application of modern statistical methods to the study of BL properties. these are methods of parametric and nonparametric analysis of variance, rank correlation analysis from works (Amirov and Bychkov 1991, 1995; Amirov et al. 1995; Amirov and Bychkov 1995–1999).

When interpreting the results obtained, the data of physiological optics and psychology of perception are taken into account. Statistical relationships between various properties of BL are determined.

The presence of a statistical relationship between the parameters of the BL and the way of its death is shown. The distribution functions of the parameters of the BL radiation, diameter, lifetime, and radiation power depend on the way of death. The distribution function of the BL parameters cannot be described by a log-normal distribution, and the question of the nature of the distribution function remains open.

Studies have shown that there is not enough information in the created databanks to analyze a number of properties of BL, in particular decay, luminescence in difficult lighting conditions, structure, rheology, probability of occurrence not in the summer months, etc., and at least a twofold increase in the number of observations in the databanks is necessary.

For a correct and unambiguous analysis of observations, many data on the physiology of perception and physiological optics are currently lacking. Their consideration significantly affects the analysis of the observation results. Without combining the efforts of psychologists and physicists in determining the influence of psychological factors on the reliability of descriptions, significant progress in this area cannot be expected.

References

Abrahamson J, Bychkov AV, Bychkov VL (2002) Recently reported sightings of ball lightning. Phil Trans Roy Soc 360(1790):11–36

Amirov AKh, Bychkov VL (1994a) ANOVA of several parameters of the SKB Data Bank on ball lightning. Phys Scr 50:93–96

Amirov AKh, Bychkov VL (1994b) ANOVA of the parameters of ball lightning from the SKB Data Bank in correspondence to its Location (Indoors/Outdoors). Phys Scr 50:588–590

AKh, Amirov, Bychkov VL (1995a) Correlation analysis of ball lightning distribution over colors of the SKB Data Bank. Phys Scr 52:222–224

Amirov AKh, Bychkov VL (1995b) Ball lightning diameter-lifetime statistical analysis the SKB Data Bank. Phys Scr 51:413–416

Amirov AKh, Bychkov VL (1995c) A study of the distribution function of the parameters of ball lightning. J Meteorol 20(199):159–164

Amirov AKh, Bychkov VL (1996) Influence of atmospheric thunderstorm conditions on the life time of ball lightning. Phys Scr 53:252–255

Amirov AKh, Bychkov VL (1997) Influence of thunderstorm atmospheric conditions on the properties of ball lightning. Tech Phys 42:430–435

Amirov AKh, Bychkov VL (1999) On reducing the influence of data inaccuracy when analysing ball lightning observations. J Meteorol UK 24(237):95–101

Amirov AKh, Bychkov VL, Bobkov SE (1998) On the dependence lifetime-diameter for ball lightnings. Phys Scr 58:56–60

Amirov AKh, Bychkov VL, Strizhev AYu (1995) Principles of creating and processing data banks. J Meteorol 20(197):85–93

Amirov AKh, Bychkov VL (1991) Database on ball Lightning for PC. Sharovaya molnia. (Ball Lightning). Ed. prof. B.M.Smirnov. IVTAN Publishers, Moscow. pp 19–27

Arago F (1859) Thunder and lightning. St. Petersburg Published, Translated from French

Azhazha V (2013) Underwater UFOs. Veche Publishers, Moscow

Barry JD (1980) Ball lightning and bead lightning. Plenum Publishers, New York

Brandt W (1923) Der Kugelblitz. Hamburg, Henry grand. (English translation is available in Internet)

Bychkov V, Nikitin A, Dijkhuis G (2010) Modern state of ball lightning investigations. In: Bychkov VL, Golubkov GV, Nikitin AI (eds) The atmosphere and Ionosphere. Dynamics, processes and monitoring. Springer, Dordrecht, pp 201–373

Bychkov VL, Nikitin AI (2014) Ball lightning: a new step in understanding. In: Bychkov VL, Golubkov GV, Nikitin AI (eds) The atmosphere and Ionosphere. Elementary processes, monitoring and ball lightning. Springer, Dordrecht, pp 201–367
Bychkov VL, Smirnov BM, Stridjev AYu (1993) Analysis of the Russian-Austrian ball lightning data banks. J Meteorol UK 18(178):113–120.
Chernobrov VA (2003) Chronicle of UFO visits. Drofa Publishers, Moscow
Dmitriev MT, Bakhtin BI, Lipatov AI (1985) Characteristics of ball lightningfrom results of natural observations. Academy of Medical Sciences USSR, Sisin AN ed. SRI of General and Municipal Hygiene, Moscow
Egely G (1987) Hungarian ball lightning observation. Budapest: Centr Res Inst Phys Hung Acad Sci KFKI-1987–10/D
Fress P, Piaget J (1978) Experimental psychology. 6. Progress Publishers, Moscow
Grigoriev AI (1990) Fire killers. Debut Publishers, Yaroslavl
Grigoriev AI (2010) Ball lightning. Demidov Yaroslavl state University Publishers, Yaroslavl, P.G
Grigoriev AI, Grigorieva ID (1989) Correlation dependencies between some properties of ball lightning. Zhur Tekh Fiz 59:79–87
Grigoriev AI, Grigorieva ID, Shiryaeva SO (1993) Observations of ball lightning and their analysis. In: Smirnov BM (ed) Plasma chemistry. Energoatomizdat Publishers, Moscow, vol 17, pp 218–249
Grigorjev AI, Grigorjeva ID, Shirjaeva SO (1988) Statistical analysis of the ball lightning properties. In: Prof. Y-H Otsuki (ed) Science of ball lightning. World Scientific, Singapore pp 88–134
Handbook of Applicable Mathematics (1986) W. Ledermann Chief ed. 4. John Wiley & Sons, Chichester
Härdle W (1990) Applied nonparametric regression. Cambridge University, Cambridge
Johnson N, Lyon F (1981) Statistics and experiment planning in engineering and science. Mir Publishers, Moscow, Methods of experiment planning
Keul AG (1993) Ball lightning—a case for physics and technology. Progress in ball lightning research. In: Proceedings Interdisciplinary Congress Vizotum. Salzburg. Sept. 20–22, pp 36–40
Keul A.G., Schwarzenbacher K. (1989) Phenomenological and psychological analysis of 150 Austrian ball lightning reports. In: Science of ball lightning. Ed. Prof. Y.-H. Ohtsuki. World Scientific. Singapore. N.Jersey. London. Hong Kong, pp 58–80.
Korn G, Korn T (1974) Handbook of Mathematics for researchers and engineers. Nauka Publishers, Moscow
Kramer G (1975) Mathematical methods of statistics. Mir Publishers, Moscow
Likhosherstnykh GV (1994) Abnormal cases of encounters with ball lightning. Ball lightning in the laboratory (in Rus). Publishers Moscow, Khimiya, pp 218–223
McNally JR Jr (1966) Preliminary report on the ball lightning. O.R.NL-3938. Oak-Ridge Nat. Lab
Menzel DH (1962) Flying saucers. Harvard University Press, Russian translation, Moscow
Meshkov VV and Matveev AB (1989) Fundamentals of lighting engineering. Part 2. Physiological optics and colorimetry. Energoatomizdat Publishers, Moscow
Ohtsuki YH, Ofuruton H (1989). Nature of ball lightning in Japan. In: Ohtsuki YH (ed)Science if ball lightning. World Scientific Publishers, Singapore, New Jersey, pp 31–57
Olkhovatov A (2008) The Tunguska phenomenon of 1908. Laboratory of knowledge Publishers, Moscow, Moscow. Binomial
Platov YuV, Rubtsov VV (1999) UFOs and modern science. Nauka Publishers, Moscow
Rayle WD (1966) Ball lightning characteristics. NASA Tech. Note. NASA-TN-D- 3188
Singer S (1971) The nature of ball lightning. Plenum Publishers, New York
Skryagin LN (2000) Secrets of the flying Dutchman. Veche Publishers, Moscow, pp 178–187

Smirnov BM (1992a) Experiment "ball lightning". Nauka i Zhizn. "Science and Life." No 9:35–39
Smirnov BM (1992b) Observational properties of ball lightning. Uspekhi Fiz Nauk 162:43–81
Smirnov BM, Strizhev AJ (1994) Analysis of observational ball lightning by correlation methods. Phys Scr 50:606–608
Stakhanov IP (1985) On the physical nature of ball lightning. Energoatomizdat, Moscow
Stakhanov I, Lopatnikov S (1975) Ball lightning: riddles remain. Nauka i Zhizn. "Science and Life." No 12:86–95
Stenhoff M (1999) Ball lightning. An unsolved problem in atmospheric physics. Kluwer Academic/Plenum Publishers, New York
Tsarev I, Tsareva I (2005) The Secret of the witch circles. Ripoll Classic Publishers, Moscow

Chapter 4
Experimental Modeling of BL and LLO

Abstract This chapter is devoted to the analysis and description of experimental modeling of ball lightning and long-lived luminous formations. The early experiments of the nineteenth and twentieth centuries on the purposeful creation of BL are described. Modern modeling of BL and LLO is discussed, namely, with the help of high-frequency discharges in the air, under electric discharge conditions. The production of artificial plasmodynamic formations and artificial BL in erosive discharges is described. Data on experimental modeling of the interaction of BL with glasses are presented.

Keywords Experiments on ball lightning · Artificial objects · Plasma jets · High-frequency discharges · Capillary discharges · Electrical discharges · Interaction with glasses

4.1 Introduction

Over the years of studying BL, attempts have been made repeatedly to create it experimentally in the laboratory. One can divide these experiments into two groups. The first is an experimental verification of the conclusions of the BL models. The second is attempts to "invent" new forms of electrical discharges, in which conditions for the creation of "fiery spheres" are automatically generated, similar to those realized in nature. This separation is largely conditional since both approaches have common features: the desire to reproduce nature and somehow guess the "technology" that it uses to create BL.

A distinctive feature of these objects is an anomalously large time of their existence τ_l (luminescence time) compared to the time of their generation by a sourceτ_g, $\tau_l \gg \tau_g$, or typical of the time of recombination of charged particles in the plasma $\tau_{rec} = \frac{1}{\alpha_{rec} N^-}$ (where α_{rec}—effective recombination coefficient of charged particles, N^- the concentration of charged particles) formed by the same source, and $\tau_l \gg \tau_{rec}$. Their other feature is their autonomy, that is, the ability to move freely in space while maintaining their shape, size, and color for a time comparable to the time of existence. According to these properties, artificial BL or LLO (plasmoid) is

V. L. Bychkov, *Natural and Artificial Ball Lightning in the Earth's Atmosphere*, Springer Atmospheric Sciences, https://doi.org/10.1007/978-3-031-07861-3_4

similar to natural BL. They differ from the lights of St. Elmo precisely by the property of autonomy. The St. Elmo lights are usually attached to charged surfaces or charged droplets. This property, in particular, is absent in plasma objects generated and supported by microwave radiation, see below.

In our consideration in this chapter, we will limit ourselves mainly to discussing experiments in which the autonomy of objects is manifested at or near atmospheric pressure (300–760 Torr), and the experimental implementation of the theoretical model is understandable. Therefore, we will practically not discuss experiments and models in which only the long-term existence of plasma is investigated or long-lived objects at low pressures.

Long before this was confirmed by statistical analysis of observations, researchers intuitively knew that the birth of the BL is somehow connected with the discharge of linear lightning. Therefore, it was natural to assume that it could be obtained by means of an electric discharge in the air or other natural environment, which models the properties of lightning discharges.

According to (Ford 2002), the appearance of BL in the discharge can contribute to the presence of water vapor in the air, dust, or soot; the presence of inhomogeneities in the bending current conductors; a very slow increase in the electric field strength to the breakdown potential; the presence of a shock or sound wave, for example, thunder.

Experiments on the generation of luminous spheres using an electric discharge began with the creation of the first electrical devices, the first electrical capacitors (Leiden bank), and the first chemical current sources (galvanic batteries and accumulators).

The first European scientist who purposefully tried to create a BL was R. L. G. Plante. He considered BL to be a special, primary form of a stable state of electrical energy. Linear lightning is simply a combination of BL. A description of G. Plante's experiments on the generation of luminous balls (see Fig. 4.1) during the passage of direct (Planté 1891) or alternating current can be found in books (Leonov 1965; Singer 1971; Imyanitov and Tikhiy 1980; Barry 1980). There were two types of

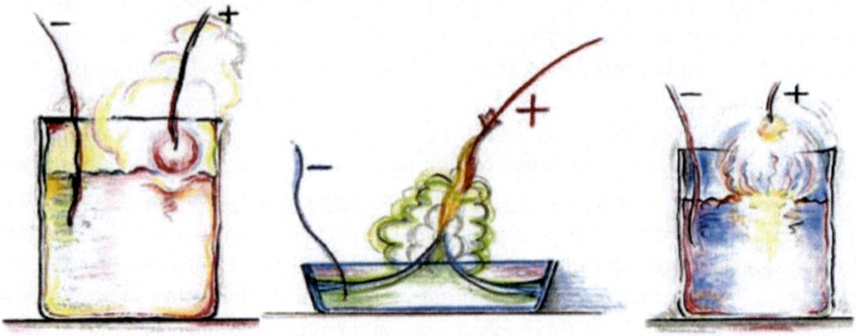

Fig. 4.1 Schemes of Plante's experiments Planté G (1891)

current sources that made it possible to do such experiments. Some were intended to produce very strong, in kiloamperes, currents at a voltage of 4–7 V. Others gave a high pulse voltage of up to 10 kV at currents of tens of amperes. In any case, Plante received a unique combination of current, voltage, and time of their action.

The first experiment in which a ball with the properties of BL was accidentally obtained was done in 1757 by the naturalist Arden in the presence of Vm. Constable, Esq. (Ford 2002). In experiments, he used a Leiden vessel made of kronglass with a volume of 4 gallons (18 L), which was connected to a high-voltage generator through a copper core located in a glass insulating tube. After the researcher, turning the handle of the electric machine, made "about 100 or 150 revolutions," in the lower part of the jar he "saw a fireball very similar to a red-hot iron bullet, a full three-quarters of an inch in diameter. It turned around its axis and went up a glass tube in which there was a copper wire." The researcher continued to rotate the handle of the machine. The "fireball continued to rotate around its axis and climb up the glass tube until it reached the top of the main conductor. There it turned around his axis for a while, and then gradually descended to the bottom of the jar, turning around its axis, as it did when lifting." "But soon after, a very bright flash was noticed; there was a strong explosion, and a strong smell of sulfur began to be felt throughout the room; a round hole was cut in the side of the jar more than three quarters of an inch (2 cm) in diameter and from two to three inches (5–7.6 cm) below the top lid, as if it had been cut by a diamond." The researchers replaced a perforated Leyden jar, but the effect was not repeated.

4.2 Early Experiments of the Nineteenth and Twentieth Centuries on the Purposeful Creation of BL

In 1899, N. Tesla, in experiments using discharges created by a high-frequency generator with an electric coil, realized RF discharges at which luminous balls were obtained (Tesla 1978), reached several centimeters.

In this case, the high-voltage electrode was covered with insulation made of natural rubber and soot (Corum and Corum 1989). Therefore, metal particles from the electrode, soot, and polymer, could participate in the creation of the BL structure. In (Corum and Corum 1989), the authors of the experiments, the Corums, decided to repeat the experiments of N. Tesla in a smaller installation, discharges generated by a Tesla generator at a frequency of 67 kHz were used to produce luminous spheres, the average power supplied to a high-voltage electrode was 3.2 kW. It was found experimentally that during the development of discharges, a large number of carbon particles and vaporized metal particles are formed from the electrode in a small near-electrode region. In the same area, there are strong electric fields (voltage up to 10–20 kV/cm) and magnetic fields, and the formed particles heat up quickly. The spheres appeared next to the high-voltage electrode. It seemed "out of nothing" since they were not visible in the previous frame of the video recording; i.e. the time of their formation was less than 40 ms.

The spheres had different colors, including red, yellow, blue, and white; their lifetime reached 2 s, and a diameter of 1–5 cm. Often at the end of their existence, they exploded with a loud explosion, which indicated an energy reserve greater than 0.2 J/cm^3. The best results were obtained in those experiments when the high-voltage electrode was coated with wax or charred wood.

Similar experiments were conducted by V. V. Kopeikin (2011) in an installation similar to that of Corums. He also obtained glowing formations against the background of discharge streamers.

The cases of the formation of fiery spheres on American submarines had a great resonance and were tested by Silberg in 1963 and Dijkhuis in 1965.

The balls occurred when the relay with silver contacts was disconnected when a strong current passed through it. The green fiery sphere separated from the contacts, and "floated" around the room for several seconds. "In 1947, experiments were conducted in Philadelphia on the submarine "Cattle". The sphere was created in the engine room at a DC voltage of 260 V and a current strength of 156 kA, with a peak power value of 40 MW. Since the shutdown time for mechanical type interrupters is 0.1–0.01 s, the energy of the formation of a luminous formation—a "plasmoid" was 0.4–4.0 MJ. The diameter of the sphere turned out to be 10–15 cm."

In the eighties of the last century, G. S. Dijkhuis undertook experiments to reproduce the events described by Silberg. To do this, he used a set of underwater batteries with a capacity of 30 MW. A short circuit was created. At a current exceeding 150 kA, orange fire spheres with a diameter of 10 cm with a lifetime of 1 s were obtained, scattered along parabolic trajectories from the place of disconnection of relay contacts.

Now one can say that the experiments of N. Tesla, K. L. Corum and J. F. Corum, and G. S. Dijkhuis were successful in realization of BL analogues.

The approach to modeling a BL went in two directions: the development of ideas about it as an independent object, and about an electric discharge in the air. The most impressive experiments on the generation of high-frequency discharges in the air were implemented in 1941 by G. I. Babat in besieged Leningrad, for its discussion see (Leonov 1965). Babat used a generator with a frequency of 60 MHz and a power of 20 kW. The energy of the generator was fed into a glass cylinder with a capacity of 5 L through its walls in the form of an alternating magnetic field. First, the cylinder was pumped out to a pressure of 0.1–5 Torr. When the high-frequency generator was turned on, a ring glowing red appeared in the cylinder. After the tap was opened and air-filled the cylinder, the ring began to shrink into a ball and turn into a fiery sphere. As the pressure increased, its color first turned purple, then green, and finally, it became a blinding white. Fiery arrows shot out from the surface of the sphere, in it was possible to distinguish concentric zones of different colors and different intensities of glow. When the generator was turned off, the sphere went out, and a dim green glowing fog remained in the cylinder.

4.3 High-Frequency Discharges in the Air

Purposeful experiments on obtaining luminous balls were carried out by P. L. Kapitsa. Above we talked about Babat's experiments with high-frequency discharge in the air. However, he did not connect them with the BL in any way. The hypothesis that the BL is powered by the energy of radio frequency radiation of a thunderstorm was put forward by P. L. Kapitsa in 1955 (Kapitsa 1955). He was one of the few scientists who tested his idea in practice. Kapitsa and the researchers managed to realize a high-frequency counteracted discharge in an atmosphere of helium, argon, carbon dioxide, and air at a pressure from 10 Torrs to several atmospheres (Kapitsa 1969). The discharge was ignited in a resonator with a spherical cavity operating according to the H011 wave type, the wavelength of the radio emission was 19 cm, and the power supplied to the resonator was in the range from 0.6 to 6.4 kW. At low energies, the discharge had the shape of an oval with blurred boundaries. However, with increasing power, it took the form of a cord, reached the upper wall of the resonator, and took off, being at some distance from it. The discharge became more stable if the gas was rotating, while the temperature of its central part was about 6000 K. Despite such a high temperature, the heat exchange of the hot region with the wall occurred insignificantly due to the appearance of a double layer of charges at its boundary. The plasma formations obtained by Kapitsa do indeed bear a faint resemblance to BL. However, the discharge in the air is maintained only with a constant supply of significant radio emission power to it. When the RF generator is switched off, the discharge goes out.

Japanese scientists (Ofuruton and Ohtsuki 1989) managed to obtain LLO, objects similar to glowing spheres, similar to the plasmoids studied by Kapitsa. LLO in air at atmospheric pressure appeared inside a resonator with a cylindrical cavity with a diameter of 16.1 cm and a length of 37 cm when a continuous stream of radio emission with a power of 5 kW was supplied at a frequency of 2.45 GHz (λ = 12.24 cm). Four types of plasmoid spheres were discovered. LLO spheres of the first type were stable and only occasionally showed a tendency to split into two spheres. The formation of plasma spheres of the second type began with a bright flash lasting 0.2 s. After that, the sphere was divided into two spheres; these spheres began to move in different directions in the resonator. The discharge of the third type had the shape of a flame and was located at the junction with the waveguide, which supplies radiation to the resonator. Sometimes, this flame was drawn into the waveguide. If a ceramic plate with a thickness of 3 mm was located between the waveguide and the resonator, then the plasma formation passed through the plate without destroying or melting it. The fourth type of plasma formations had the form of a flame flowing out of the resonator into free space through a hole made by it in an aluminum foil covering the end of the resonator. This "plasma flame" continued to exist for 1–2 s after the magnetron was turned off.

Luminous plasma formations similar to Kapitsa plasmoids and Ofuruton and Ohtsuki plasma spheres were also obtained by the authors (Zhiltsov et al. 1995).

Later, Ofuruton and his colleagues learned how to obtain fiery spheres not in a waveguide, but in the open air (Ohtsuki and Ofuruton 1991; Ofuruton et al. 1997). The experiments used a magnetron continuously emitting power up to 5 kW at a frequency of 2.45 GHz. Radio radiation was released into the air through an open waveguide. Two copper electrodes with a diameter of 3 mm were located near the waveguide; the gap between them was 3 mm. The distance from the spark gap to the waveguide section could vary from zero to 9 cm. A capacitor with a capacity from 0.2 to 10 μF, charged to 8–10 kV, was discharged through electrodes. When turned on, magnetron glowing balls appeared in the air after the discharge was turned off; they had a spherical or ellipsoidal shape. At distances less than 3 cm from the spark gap to the waveguide, the plasma sphere usually moved horizontally and passed through the waveguide. At distances of more than 5 cm, the LLO moved upwards. The color of the sphere changed from red to white. After the start of the discharge, the sphere gradually decreased in size and disappeared after 0.5 s. The spheres appeared with 100% probability at high discharge power and/or high microwave radiation power.

As one can see, if it was possible to find processes that ensure the continuous output of high-frequency power of several kilowatts in some allocated area of space, then the "Kapitsa BL" would become a real fact. As for the shape constancy factor, it seems to be an unsolvable problem for this LLO. However, the second part of the problem—the concentration of energy in a certain place—can be solved thanks to the interference of radio waves. In this regard, "Anderson localization" is an intensively researched task; it is the possibility of a local increase in the intensity of the high-frequency field due to the summation of the amplitudes of waves reflected from randomly located objects (Anderson 1958).

K. Tanaka and M. Tanaka (1997) suggested that due to the "localization of Anderson" in the air, conditions could be created for the formation of a fireball in the form of plasma spheres obtained by Otsuki and Ofuruton. The authors (Kamogawa et al. 1999) performed calculations of the electric field strength with a frequency of 2.45 GHz in a rectangular waveguide for the case when a number of ceramic plates with a thickness of 2.25 mm are located in it. Two types of one-dimensional disordered systems of plate arrangement were investigated: a quasi-periodic system of the Fibonacci series type and a fractal system of the Cantor lattice type. In both cases, according to the authors' calculations, the intensity of the high-frequency field should increase by more than 100 times in limited sections of the waveguide. Further work of Japanese scientists was devoted to experimental verification of the results found (Ofuruton et al. 1999; Kamogawa et al. 2004). When using ceramic plates with a thickness of 2.25 mm, the dimensions of which were equal to the dimensions of the waveguide (109 × 55 mm), the transmission and reflected power were measured at a power of 1–2 kW of microwave radiation. The plates were placed according to the Fibonacci series from the first to the sixth generations (in the latter case, 22 plates were used). During the transition from 4 to 5 and from 5 to 6 generations, the sound of air breakdown in the waveguide could be heard, but the plasma discharge did not appear. Since it is unlikely that there may be conditions in nature when the intensity of the high-frequency field of external sources exceeds the value of air breakdown

($\geq 3 \cdot 10^6$ V/m) or there will be a local decrease in pressure that contributes to the ignition of the discharge.

Studies of discharge ignition processes using passive initiators are of interest. Metal needles, multi-needle metal brooms, metal-dielectric plates, flame, and aerosol can serve as initiators (Brovkin et al. 1994; Shibkov et al. 1994; Mitsuk and Mitsuk 1994a, b). Using an initiator made of wire with a diameter of 1 mm, the authors (Brovkin et al. 1994) obtained a low-threshold spherical high-frequency discharge at a frequency of 7 GHz. It existed at energy flows of 10–100 MW/cm^2 in the air pressure range from 3 to 110 Tor. The maximum diameter of the sphere was equal to the wavelength of the radiation, the minimum was a quarter of the wavelength. As the pressure increased, the discharge took the form of a torch, and the intensity of its glow increased dramatically. At atmospheric pressure, the discharge was maintained with a power of several hundred watts. Note that luminescent objects obtained by (Brovkin et al. 1994) cannot be considered artificial BL since there were realized at a time constant source of excitation.

In (Shibkov et al. 1994), a short pulse of powerful high-frequency radiation was used to ignite the discharge and the plasma was maintained by the second pulse of small amplitude, unable to independently cause a gas breakdown.

Chukanov K. (2008) obtained glowing spheres in a quartz flask with a diameter of 10 cm placed in an industrial microwave oven with a power of 2.2 kW. The flask could be pumped out by a pre-vacuum pump and filled with various gases. It had two tungsten electrodes placed against the inner wall. An alternating voltage (60 Hz) of 1500 V at a current of 10 A could be applied to the electrodes. In addition, the electron beam could be inserted into the flask in the radial direction (from the periphery to the center). First, the flask was pumped out. When the microwave was turned on, a discharge was ignited in it. As the gas was fed into the flask, the brightness of the glow increased. The maximum gas pressure was 1 atm. When the electron gun was turned on, the glow of the gas became very bright. When an alternating current was applied to the electrodes, the plasma began to glow like a 12 kW lamp. The discharge burned most easily in air, oxygen, nitrogen, and carbon dioxide. In hydrogen, helium, and water vapor, the discharge was not ignited. Plasma balls of Chukanov cannot be considered as artificial BL since they were realized at time constant source of excitation. Estimations of energy in these experiments required special attention to the influence of inductive component of the circuit, about which there was no information in (Chukanov 2008).

A special group of experiments consists of experiments in which combustion processes in a high-frequency field were investigated. R. Golka (1994) obtained LLO by placing a 5-cm burning candle in a household microwave oven. During the time it was turned on in the camera, spheres similar to fire spheres were flying, but not very bright. The authors (Mitsuk and Mitsuk 1994a, b) also investigated the interaction of a high-frequency discharge with a candle flame. They found that a diffusion-type electrodeless discharge occurs in the area of a burning candle due to the formation of a cavity with a reduced gas density.

In (Dikhtyar and Jerby 2006), a technique for implementing fiery spheres from hot spots of molten matter is described. In the experiments, a rectangular waveguide with a length of 20 cm and a cross section of 8.6×4.3 cm^2 was used, the open end of which was blocked by a mirror of blades with a cut-off, and at the other end, 0.6 kW power was supplied from a magnetron operating at a frequency of 2.45 GHz. A hole was drilled in the center of the wide side of the waveguide at a distance of 5 cm from its open edge, into which a metal drill could be inserted, reaching the inner surface of the lower side of the waveguide. This configuration was called by the authors (Dikhtyar and Jerby 2006) a "microwave drill". In the lower part of the waveguide, a substrate of silicon, glass, germanium, alumina, or basalt was placed, having the shape of a plate, volumetric, or in the form of a powder. At first, the drill led to contact with the substrate, thereby concentrating the release of microwave power at the point of contact. A hot spot appeared, and about a second after that, the melting of the substance took place. "After that, the microwave drill bit was extracted and torn away from the molten spot, at the same time pulling out a condensed, almost vaporized drop from the substrate. A jet of fire is ejected from the molten hot spot, forming a pillar of fire… Soon after, the pillar of fire rises, breaks away from the hot spot and forms a fireball in the air. A limited fireball flies up to the metal ceiling, where it is slightly compressed to a Gaussian-like shape. The floating fireball looks like a thick heavy vapor, almost liquid, glowing yellow–red; it rises into the air and shivers elastically like a jellyfish. A floating fireball (~15 cm^3 in volume) can remain stable as long as the microwave power is turned on, and it continues to glow for another 30–40 ms after the microwave power is turned off." It was possible to observe the bounces of the fire sphere during collisions with walls, the rotation of the fire column, and the movement of the sphere in the waveguide towards the energy source. Upon contact with the Teflon partition, the sphere left burn marks on it, and its contact with a glass plate 2 mm thick led to its melting, cracking, and destruction of the fiery sphere.

4.4 Experiments on BL Obtaining Under Electric Discharge Conditions

To date, several areas of research and the creation of luminous balls in electric discharge conditions have been identified.

The first direction includes the creation of balls from the plasma of metal vapors, when the contacts of powerful electric batteries are closed (Stakhanov 1985; Dijkhuis 1981), Golka R. K. Jr (1994) accompanied by the oxidation of metal atoms, as well as the generation of balls when metal electrical wires are closed, both in air and in water (Dijkhuis 1981; Golka 1994).

The size of the glowing balls that appeared when the battery contacts were opened reached several centimeters, and the lifetime was tenth of a second. The brightness temperature corresponded to the melting and evaporation temperature of the metal.

Luminous balls with a diameter of 2–4 mm and a lifetime of 2–5 s were obtained (Golka 1994) when the circuit was short-circuited by two electrodes in water. The balls resembled drops of molten metal burning in the air. After cooling, they were 2 mm metal spheres with traces of oxides on the surface.

In (Urutskoev et al. 2000), when metal foils exploded in cuvettes with water over a device containing these cuvettes and covered with a vacuum seal, luminous balls of up to 10–15 cm in size and a lifetime of up to 5 ms appeared. The energy invested in eight cuvettes, with titanium foils of 50 microns × 1 cm × 5 cm, was 50 kJ, and the pulse time was 60 microseconds. The photographs presented in the article show that as the observation time increases, the purely diffuse glow of the balls changes to a glow filled with luminous particles. This allows us to make an assumption about the condensation processes that occur in the plasma of metal vapors that have penetrated through the seal and formed a luminous area.

The second direction is the formation of luminous balls with a lifetime of several milliseconds to several seconds when using pulsed or brush discharges in mixtures of air with organic gases or organic aerosol (Cawood and Patterson 1931; Ohtsuki and Ofuruton 1990; Powell and Filkenstein 1970), or during the destruction of metal electrode material (Powell and Filkenstein 1970) in the afterglow of the discharge.

One of the first works devoted to the appearance of luminous balls in the discharge is (Cawood and Patterson 1931). A brush discharge was ignited in 4 m^3 of air when aerosol particles p-xylene-azo-β-naphtol were sprayed in the discharge at the rate of 15 mg/m^3. The observed discharge color was greenish-blue. Then, in the center of the chamber, a sphere with a diameter of 20 cm was formed of a bright red color, which was possessed by large particles forming a sphere. Analysis of these particles under a microscope showed that these particles are long chains of small particles, up to a thousand pieces each.

Barry's book (Barry 1980) describes the author's experiments when propane was added to a pulsed discharge (with a voltage between the electrodes of 10 kV and a pulse energy of ~250 J) in the air at atmospheric pressure. Its volume concentration was 1.4–1.8%, which is less than the propane concentration required for its ignition. After that, a yellow-green ball appeared in the volume, which existed for 1–2 s.

Singer's book (Singer 1971) describes Nauer's experiments (Nauer 1953) to produce glowing balls during a gas discharge in the atmosphere of hydrogen and hydrocarbons such as methane, propane, and gasoline, with a concentration slightly below the limit required for their ignition. The greatest effect was observed when using gasoline. Bright glowing balls appeared even with insignificant traces of gasoline in the discharge chamber. When explaining the experiments in this work, it was suggested that the glow occurs on a clump of threads. (Gasoline emits a large number of light conductive lumps of soot during combustion (Wojciechowski and Wojciechowski 1986)). However, the glow itself was considered a result of the organic matter burning. At the same time, glowing spheres were also observed when the discharge chamber was filled with an inert gas with small additives of hydrocarbons.

In (Wojciechowski and Wojciechowski 1986), it was possible to observe the glow of a red ball with a diameter of 5 mm, flying between two electrodes, one of which was a metal grid, and the second was a rod. The electrode gap with a potential difference of 40 kV was watered from above with artificial rain. A shapeless sheet suddenly separated from the grid, which turned into a regular red ball for 15 mm; continuing to move along the power line, the ball silently disappeared before reaching the second electrode. The ball, according to the authors of the article, arose on the basis of a lump of aerosol, since the smoke generator was tested at the site of the experiment the day before. The glow was explained as follows: the appearance of conductivity of various filaments and shells floating in the air, when wetted with water, even if they were formed from non-conductive materials, leads to filament heating at the points of contact of the filaments and shells or the appearance of gas discharges between the filaments at their points of contact. The appearance of BL was also explained as a result of the glow of tangles of carbonized organic filaments.

In the work of Otsuki (Ohtsuki and Ofuruton 1990), which ideologically continued the work (Barry 1980) describes the author's experiments when propane was added to a pulsed discharge, with a voltage, a pulse discharge was created in the air between copper electrodes located at a distance of 3 mm, while the discharge voltage varied in the range of 8–10 kV, and the pulse energy was $E = CU^2/2$ ~350 J. The discharge chamber had dimensions of 150 × 120 × 120 × 150 cm. When 2.1% ethane was added to the air, a red ball with a diameter of up to ~4 cm and a lifetime of up to 0.3 s was observed. When 100 cm^3 of cotton fibers were added, a white object ~3 cm in diameter was observed in the afterglow of the discharge for 0.8 s. In a mixture of 2.7% ethane, 100 cm^3 of cotton fibers processed to fine particles, and the appearance of luminous objects was observed four times. Once the diameter of such an object was 5 cm, and its lifetime was 2 s after the discharge pulse was turned off.

The results of early experiments with artificial balls in discharges with organic additives were interpreted (Cawood and Patterson 1931; Singer 1971) as being directly related to natural BL, experiments with metal vapors were also seriously discussed in this regard.

Later, in (Smirnov 1988a, b), when analyzing the long afterglow of a high-frequency discharge in air (with a frequency of $7.5 \cdot 10^7$ Hz), which reached 1 s in the experiment (Powell and Filkenstein 1970), it was concluded that fractal clusters formed in metal vapors entering the discharge from the electrodes play an essential role in this glow. A fractal cluster is an aggregate in which the mass grows with size according to the law (Smirnov 1988a, b)

$$m = m_0 \cdot \left(2R_{sp}/r_0\right)^D,$$

where m_0 and r_0 are the mass and size of the smallest link, and R_{sp} is the radius of the cluster (structure). D is the fractal dimension, in particular for organic polymers (such as C_2H_2 chains) $D \approx 2.5 \pm 0.3$, and for inorganic (such as SiO_2 chains) $D \approx 1.75 \pm 0.2$ (Smirnov 1988a, b).

Artificial balls were also obtained in (Slyusarev 1990) during the breakdown of birch wood in a discharge at atmospheric pressure when a piece of wood was located between the electrodes. Red-blue balls with a diameter of 5–10 mm (two cases of 450–500 breakdown acts) came out of wood, on the surface of which there were light traces of lignin. Such spots have anomalies of electrical conductivity, in them, the electrical conductivity decreases by 15–20 orders of magnitude. In these areas, under the influence of an electric field, the most effective effect on lignin occurs—the release of gases, the separation of hydroxyl groups, a decrease in crystallinity, and, consequently, the possibility of polymerization and the formation of a heterogeneous structure. In (Slyusarev 1990), the idea of lignin (i.e., a high-molecular organic substance) as the basis for the formation of natural BL was expressed.

Experiments on the closure of electrical circuits with electrodes coated with charcoal or a layer of water (Ford 2002) should be attributed in the same direction. They produced LLO with a size of 9–12 mm and a lifetime from 30 ms to 0.8 s.

Based on the work on the appearance of glowing balls during the interaction of plasma with particles of organic polymers, a ball lightning model was proposed in (Bychkov 1994), where the BL framework was formed from chains of organic macromolecules (this idea was put forward earlier than the similar idea of (Abrahamson and Dinniss 2000)).

The third direction can be attributed to the formation of structures in the conditions of evaporation of materials in the presence of an electric field when the formation of the structure was recorded in the first place, and nothing was reported about the glow.

One of the first works from this direction is the work of the Podmoshensky group (Aleksandrov et al. 1982a). During the explosion of thin wires and foils made of tungsten at a pressure of 100–760 Torr and the contribution of energy in a pulse of up to 50 J (the explosion was realized by discharge from a capacitor), the formation of filamentous conglomerates with a transverse size of microfibers in a conglomerate of 20–30 nm was observed. In (Aleksandrov et al. 1982b), the results of these experiments are associated with the possibility of the formation of BL from such aerosol particles.

The formation of aerogel structures from fractal aggregates was studied in (Lushnikov et al. 1990a, b). Under the action of radiation from a ruby laser with a pulse energy of ~70 J and a duration of 10^{-3} s. The formation of fractal aggregates with a mass of ~7 g/m^3 was observed on the surface of titanium in an applied electric field with a strength of $100–1.5 \cdot 10^3$ V/cm. They were formed during steam condensation with the formation of small singlets (1–10 nm in size) and their subsequent coagulation, with the formation of a fractal aggregate with a fractal dimension $D = 1.82$. After that, a macroscopic structure was formed, visible to the eye. The authors come to the conclusion that such a structure can serve as a framework for BL, for example, from particles arising from the impact of linear lightning to obstacles.

The paper (Mikhailov et al. 1995) is devoted to the peculiarities of the growth of fractal clusters in the gas phase. It shows that fractal aggregates of metals can be formed during the burning of metals (in particular magnesium) both in arc discharge and open air. At the same time, at high temperatures $T > 1000$ K, the ionization state of the medium has a noticeable effect on the structure of growing aggregates.

The papers of Abrahamson (Abrahamson and Dinniss 2000; Abrahamson 2002) turned out to be important works for understanding the possible nature of BL. In (Abrahamson and Dinniss 2000), it was suggested that the origin of the BL is associated with a linear lightning strike into the ground. The so-called fulgurite areas appear at the site of the lightning strike, where the lightning energy is released. These areas have shapes such as tubes up to several meters long and several centimeters wide, or areas with lateral channels. In them, silica is reduced to metallic silicon, while chains of atoms are formed, and those that fall into the air create a tangle of chains of atoms or balls of the type predicted by Smirnov (Smirnov 1988a, b). Studies (Abrahamson 2002) were devoted to the emergence of a spherical ball of silicon from Si nanoparticles. Experiments have been carried out to study the impacts of discharge current pulses into the soil at an energy input of E ~ $4.5 \cdot 10^4$ J. It was assumed that silicon oxide particles SiO_2 were present in the soil. The creation of Si/SiO/SiC nanoparticles occurs at the place of energy release during the interaction of the discharge with the organic components of the soil, where the dissociation of organic substances to carbon atoms occurred $SiO_2 + 2\,C \rightarrow Si + 2\,CO$. When silicon vapors enter the atmosphere, they condense and nanoscale silicon structures are formed. Under the influence of an electric field, these structures, as in (Smirnov 1988a, b), are assembled into large aggregates that begin to glow due to heating in the silicon oxidation reaction $Si + O_2 \rightarrow SiO_2$. In (Abrahamson 2002), experiments were undertaken to simulate a lightning strike into the ground. In this case, a circuit with a capacity of 204 μF was used at a voltage of 20 kV, the discharge current passed through a moistened soil layer with a thickness of 3 mm. At the same time, the energy released in the soil was 110 kJ and a charge of 1.3–3.4 C passed through the soil. The reaction products were collected and investigated. With the help of a scanning electron microscope, filaments of fibers with a diameter of 100 nm and a length of up to 7 microns were detected. Also, several spheres with a diameter of 2 microns were found and chains with a length of 25–120 nm were found, consisting of balls with an average size of 25 nm. Such spheres with a diameter of 2 microns consisted mainly of silicon, and Al, Mg, Ca, Na, and Fe (components of soil minerals) acted as impurities. In (Abrahamson 2002), the formation of isolated aerogel macrospheres was not observed, i.e., direct experiments to obtain luminous did not lead to the production of luminous vortex spheres. This was explained by the generation of a strong shock wave during discharge of the discharge and, accordingly, by strong gas disturbances near the discharge and the possible destruction of the spheres. Therefore, in (Abrahamson 2002), it was assumed that in nature, vapors from the cavity formed by lightning come out after the termination of the thunderstorm discharge. In this case, the vortex is formed from charged nanoparticles. In Figs. 4.2 and 4.3, the scheme of the experiment and the resulting luminous formations are presented.

After the appearance of these works, several studies were published on the combustion and explosion of porous silicon with the formation of plasmoids with a radiance temperature above 2000 °C, among which were the following works (Lazarouk et al. 2006; Paiva et al. 2007, 2010; Stephan and Massey 2008; Dikhtyar and Jerby 2006).

Studies (Lazarouk et al. 2006) were devoted to the explosion and combustion of nanostructured silicon in the presence and absence of hydrogen on its surface. Spherical plasma structures formed in these processes were observed. Layers of porous silicon were used as a source of silicon nanoparticles. They were formed by electrochemical anodizing of p-type monocrystalline silicon with a resistivity of 12 cm in a 48% aqueous solution of hydrofluoric acid at a constant current density of

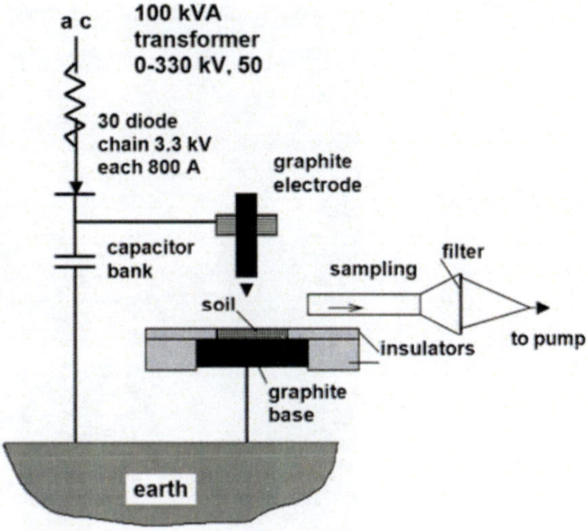

Fig. 4.2 Schematic of the experiment (Abrahamson 2002)

50 mA/cm^2. After that, the samples were placed in a 10% aqueous solution of KNO_3 and dried at ~60 °C. The explosion and combustion of porous silicon were initiated thermally, mechanically, electrically, and optically. Thermal initiation was carried out with contact heating up to 900 °C. Mechanical initiation was carried out by scratching (or piercing) the surface of porous silicon with a metal needle. Electrical initiation was carried out by passing an electric current pulse through the structure. Optical initiation was carried out by laser radiation with a power of 40 MW/cm^2. There were no differences in the explosion patterns regarding the method of initiation. Explosions and combustion of porous silicon took place at room temperature with controlled humidity of 50, 70, and 90%.

The theory (Abrahamson and Dinniss 2000; Abrahamson 2002) was also tested in (Paiva et al. 2007, 2010). At the same time, the evaporation of small pieces of high-purity silicon wafers was carried out by an electric arc at normal atmospheric pressure. The authors managed to create glowing fireballs that had a long (seconds) lifetime and demonstrated several properties common to natural BL. As the lower electrode and the tungsten (or graphite) upper electrode, Si elements with a diameter of 5 cm, doped with p-type, and a thickness of 350 microns × 50 microns, located on a flat steel plate with a thickness of 1000 × 1000 mm^2 with a thickness of 5 mm were used. The upper electrode had a diameter of 4 mm and a length of 30 cm. The voltage on it varied in the range from 20 to 25 V and the current—from 100 to 140 A. The upper electrode was movable. The whole procedure of the electrodes and the short circuit of the sample took from 1 to 2 s. The experiments were carried out at a temperature of 29 °C and a relative humidity of 70%.

Video frames show many small hot glowing fragments flying in all directions during the discharge. Glowing fireballs resembling BL clearly behaved. Their diameter was in the range of 1–4 cm and much larger than that of a typical fragment. Figure 4.4 shows photos of these objects.

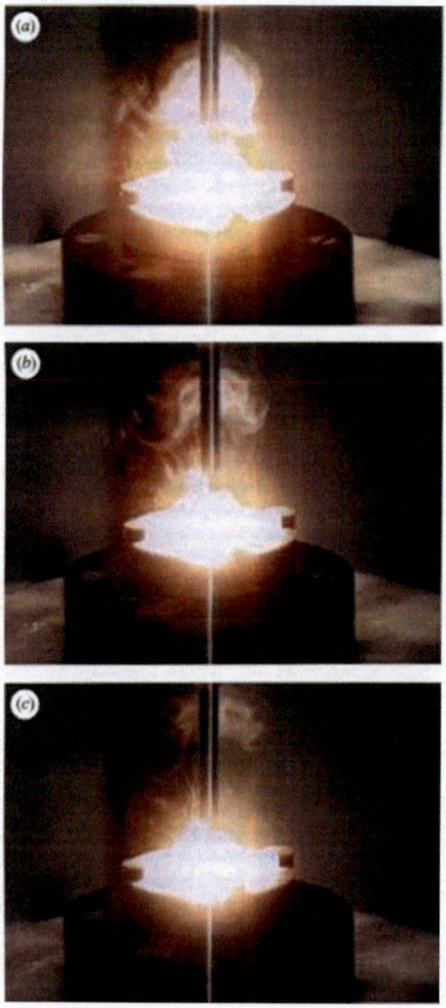

Fig. 4.3 Glowing objects in the experiment (Abrahamson 2002)

Their lifetime was up to 8 s, while the fragments cooled in 1 s. The fireballs had properties that were similar to the properties of BL in nature. The experimental statistics showed about 1 event out of 30 trials. In these experiments, no free-floating fireballs were obtained. To compare the results obtained with the results of experiments using an electrode other than Si, similar experiments were carried out with the replacement of the Si plate with: Al and Cu foil, salt water, dry and wet SiO_2 microspheres (the type used for sandblasting), moistened wood and a mixture of moistened wood and SiO_2 microspheres. None of these materials resulted in glowing fireballs with behavior observed when pieces of silicon wafers were deposited.

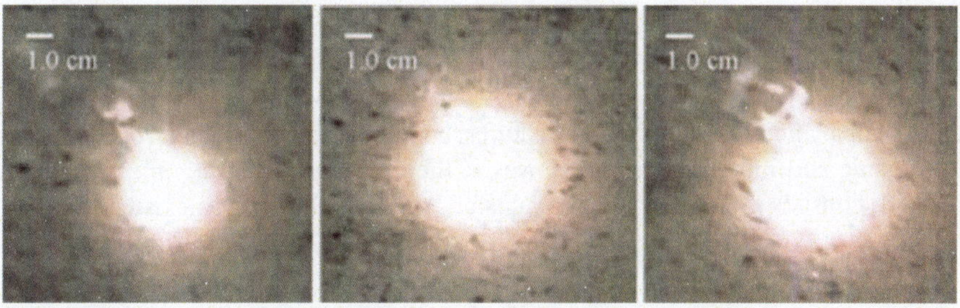

Fig. 4.4 Sequential frames showing the obtained objects with smoke emanating from the surface (Paiva et al. 2007). The time interval between frames is 80 ms

In (Dikhtyar and Jerby 2006), see above, it was reported that fireballs were generated by exposure to silicon and other solid materials, such as glass and alumina, by concentrated microwaves. Thanks to the melting and evaporation of the material, the researchers were able to form a column, and then a fireball. The fireball did not live for a long time—it glowed only about 30–40 ms after the microwave source was turned off.

The work (Stephan and Massey 2008) was devoted to experiments and continuation of experiments on burning silicon and other materials using an electric arc. It has been shown that the luminous objects described in (Paiva et al. 2007, 2010) apparently consist of burning droplets of molten silicon; burning was initiated by the high temperature of the electric arc. At the same time, the oxidation of molten silicon was inhibited by the presence of a layer of liquid SiO_2. Experiments (Paiva et al. 2007, 2010; Stephan and Massey 2008) and our own experiments (see below) confirm the hypothesis (Abrahamson and Dinniss 2000) that at high temperatures of silicon oxidation, a glowing ball can form during a reaction that lasts several seconds. However, the details of this process differ greatly from the hypothesis (Abrahamson and Dinniss 2000) since it involves not a cloud of silicon nanoparticles, but macro-sized structures.

All of the above works were devoted to the combustion of materials that may appear in the recovery processes under the influence of plasma. In this regard, they represent a continuation of experiments on burning materials, in different conditions from those considered in (Emelin et al. 1997, 2012) (see below) under the influence of plasma and an electric field. Doped Si is a convenient material for such experiments. Combustion is not surprising. Note that the amount of SiO_2 in the earth's soil is about 60%, so the process of silica excitation can actually occur under the influence of a linear lightning strike into the soil, and this can lead to the appearance of one of the possible types of BL.

In conclusion in this section, it should be noted that the formation of various kinds of structures occurs in the plasma of RF discharges with the participation of organic components (Yasuda 1985). Thus, (Vinogradov et al. 1983), during the formation of thin films in the freon-318 discharge at a pressure of $P = 25–75$ Pa, the formation of structures up to 3 mm in size took place. They resembled intertwined chains of 10–20 microns thickness.

Thus, experiments with gas discharges show the possibility of the formation in plasma of artificial BL, luminous structures, conglomerates, and fractal clusters, both of organic and inorganic origin. However, the difference in these objects with the real BL consists in the absence of explosions of them during their existence.

4.5 Plasmodynamic Formation of BL and LLO

In the second half of the twentieth century, in the course of plasma studies of various types of discharges, the concept of "long-lived luminous objects" (LLO) appeared, which for the first time, apparently, was introduced at the Department of Physical Electronics of the Physics Faculty of M. V. Lomonosov Moscow State University in works devoted to the creation of long-lived luminous objects (LLO) in plasma, see, for example (Ershov et al. 1990; Alexandrov et al. 1990; Aleksandrov et al. 1999), "long-lived plasma formations" (LPO) (Avramenko et al. 1990a, b, c), or plasmoids of various types (Pankova et al 1999; Bychkov et al. 1999).

These objects were investigated in connection with the creation of particularly bright radiation sources, issues of drag reduction of aircraft (Pankova et al. 1994), the effect of plasma on various materials, and plasma chemistry.

* Experiments that have attracted the attention of researchers over the past 20 years are connected with using of plasma jets. Experiments on the creation of plasmoids were carried out using pulsed plasma injectors of various types with a range of energy invested in the plasmoid, 100 J–100 kJ (Ershov et al. 1990; Alexandrov et al. 1990, 1999). The lifetimes of plasma structures significantly (by several orders of magnitude) exceeded the typical times of plasma creation by energy sources, as well as the time of plasma decay. In these experiments, during pulsed plasma injection into the air, plasma structures of various shapes were formed—spheroidal (at a specific energy into the plasma torch channel $Q > 1\ kJ/cm^3$), toroidal (at a specific energy into the plasma torch channel $Q \sim 0.5\ kJ/cm^3$), cylindrical (Ershov et al. 1990; Avramenko et al. 1990a, b; Avramenko et al. 1994; Pankova et al. 1994, 1999; Bychkov et al. 1994, 1999; Gridin and Klimov 1999), etc., with typical dimensions of 10–40 cm and a lifetime of up to 0.1–1 s. At the same time, the power supply time varied from 10 microseconds to 10 ms. The typical lifetime was up to 50 microseconds, and the typical temperature value in the initial stage of the plasmoid's existence was 7000–10,000 K. The initial electron concentration was 10^{16}–$10^{17}\ cm^{-3}$. According to (Ershov et al. 1990; Alexandrov et al. 1999), during pulsed jet injection, there is a hierarchy of times: the time of energy introduced into the discharge τ, the time of formation of the characteristic structure τ_f by the jet, and the time of development of instability of the flow τ_i. When $\tau = \tau_{f,}$ the barrel-shaped or mushroom-shaped structures of the jet turn over time into a sphere or torus. At $\tau_f = \tau_i$, Kelvin-Helmholtz instability develops, also leading to the formation of a toroidal vortex.

Plasma emission spectra and their temporal evolution were studied in (Ershov et al. 1990; Aleksandrov et al. 1990, 1999). It is shown that at the late stages of the existence of luminous balls, their radiation spectrum corresponds to the radiation of solid carbon particles and metal oxide particles, as well as the radiation spectrum of burning organic materials. It should be noted that these objects cannot belong to natural BL due to their short lifetime.

Experiments in Vladimir State University. Since 1971, experiments have been conducted at Vladimir State University (Russia) to obtain electrical discharges with a possibly high current strength (1997, 2007). The last option of the power plant is a system with inductive energy storage (Kunin et al. 1990), in the store, there is an electromagnet with an open U-shaped core made of transformer steel weighing 22 tons. Fifteen coils of several parallel bundles of copper wire are wound on a core with a total cross section of 4000 mm2. The inductance of the drive is 6.5×10^{-4} g. With a supply current of 30 kA, energy of about 500 kJ is stored in it. The duration of the current pulse when removing energy from the storage device can be changed from 70 to 120 ms. This setup works as follows. The inductance of the drive is connected to the rectifier unit via a switch. After the accumulation of magnetic energy, the storage device is connected to the load using the second switch, and after 70 ms after that, the first switch is turned off. An additional cut-off current is generated in the operating circuit. This current passes through a round metal diaphragm with a thickness of 56 microns. One electrode is connected to the peripheral part of the diaphragm, and the second is connected to its center. As a result, the current flows along the disk radii. At the moment of the diaphragm explosion, the currents flowing in it and in the lead wires create a toroidal magnetic field. The ions seem to be "frozen" in this field. Convection electric currents appear in an electrically neutral plasma. These currents take the form of a torus and then the form of a sphere (Kunin et al. 1990; Kunin et al. 1997a, b, 1999). Figure 4.5 shows the time dependences of the current strength I, the voltage drop U during an electric explosion, and the radiation flux P recorded by a photometer at different points in the evolution of plasma formation at an energy input of ~30 kJ (Fig. 4.6).

The transformation of the torus into a sphere occurs in 0.25 s, while the magnetic field fades in 0.4 s. The generated luminous object has the shape of a spheroid with a sharp border. After that, there is a diffuse decay of electromagnetic energy from thermal to optical, and the object slowly fades away. Sometimes, when the surface of the sphere is locally destroyed, "the flash energy attenuates, including a beam with characteristics close to those of a laser beam."

It is revealed that the radiation is polarized by 1.5–5.5%, and the plane of polarization is parallel to the plane of the diaphragm (Kunin and Furov 1990). The authors (Kunin and Pleshivtsev 1999) believe that these phenomena can be explained by the accumulation of electromagnetic radiation energy in the inner region of plasma formation, which occurs due to the reflection of light from the inner walls of the toroid caused by an increase in the concentration of free electrons in the direction of the vortex radius. Data on the development of previously conducted experiments are presented in (Kunin et al. 2006a, b; Kunin and Furov 2006; Furov 2019). In these works, an LLO with a shape close (plasma vortex) to spherical, 30–35 cm in size,

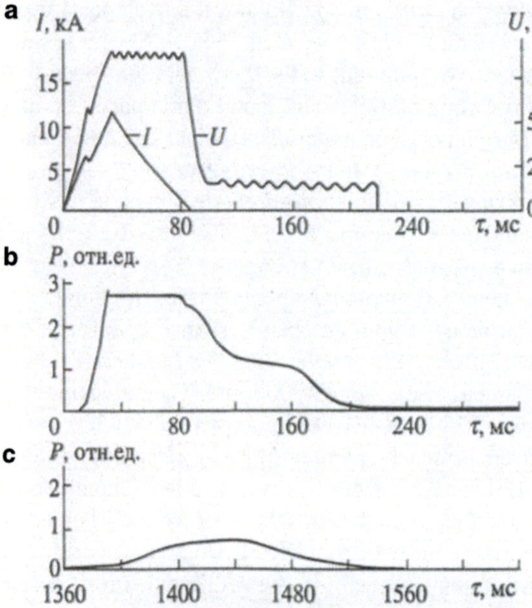

Fig. 4.5 Time dependences of the current strength I and voltage drop U during an electric explosion (**a**) and the radiation flux P recorded by the photometer at different moments of the evolution of plasma formation (**b, c**) (the introduced energy is ~30 kJ)

Fig. 4.6 Photo of a long-lived plasma formation

with a lifetime of about 2 s was formed when a round metal diaphragm exploded with a powerful current pulse (I_{max} ~ 10–12 kA, U ~ 80 V, pulse time ~80 ms) with an energy of the order of 10 kJ.

In our opinion, hydrodynamic instability also developed in this work, which, like (Ershov et al. 1990; Aleksandrov et al. 1999), led to the appearance of a ball, and the glow was determined by the radiation of eroded and condensed metal particles of the diaphragm.

The parameters of the obtained LLO correspond to the parameter of the average BL (see above) with an energy density of 0.1–0.17 J/cm^3, which is very small, compared to the energy density of real BL. In addition, these objects died as a result of extinction without explosions typical for BL. Nevertheless, the LLO data can be considered experimental in terms of lifetime and size analogues of natural BL.

Experiments of Alexeff et al. This author obtained plasmoids in the form of heated air spheres kept from expanding by atmospheric air pressure. They came to the conclusion that heating a gas in a volume of several cubic centimeters by a powerful spark discharge leads to an increase in pressure in it to several thousand atmospheres, which causes the rapid destruction of the sphere by a supersonic shock wave. As a result, the lifetime of the heated sphere does not exceed one millisecond. In this regard, they chose the mode of slow heating of the gas volume. Six high-voltage transformers (60 Hz, 120 V/15 kV, 60 mA) were used, the secondary windings of which were switched on in parallel through spark arresters, exciting a small volume of air. Transformers were periodically switched on for 0.5 s and then turned off for 0.5 s. Luminous formations rose up from the discharge zone that existed from 0.25 s to 0.5 s. It should be noted that these objects cannot belong to natural BL due to their short lifetime.

Experiments of Shvilkin et al. The authors (Kuz'min et al. 2003; 2006) obtained glowing formations during the explosion of 0.5 mm diameter wires made of copper, tungsten, iron, titanium, and nickel in a small plastic cuvette filled with a mixture of organic material and water. There was a hole in the ditch through which the products of the explosion could enter the atmosphere. When a current of 5–20 A passed through the wire, it evaporated and excited the gas that passed through the hole. At the same time, a luminous sphere with a diameter of 5–6 cm was formed. The sphere glowed for 2–3 s. Then the sphere flattened and gradually turned into a non-luminous toroidal ring. It rose up—first quickly, then slowly, and its size increased from 5 to 50 cm. According to the authors, the spherical glow could not be maintained by recombination of charges, since the characteristic time of this process is very short. Most likely, burning of dispersed particles of wire material in the air was the cause of the glow. It should be noted that these objects cannot belong to natural BL due to their low energy and lifetime.

Experiments of Vlasov et al. A.N. Vlasov (Vlasov 1997, 1999, 2006a, b, 2008b), Vlasov and Kolesnikov (2008) believed that the BL is a toroidal current layer induced inside a vortex gas ring. A toroidal vortex can form when linear lightning strikes the ground.

To simulate this process, Vlasov (Vlasov and Vlasov A 2006) undertook experiments on the induction of a strong toroidal magnetic field when a current pulse passes through a long wire coil wound in the form of a torus. The wound coil was made of copper wire with a diameter of 0.5 mm. Through it, with the help of an electronic key, a capacitor bank with a capacity of 45 mF was discharged and charged to a voltage of 300 V. At the same time, an electric explosion of the coil occurred. A shapeless or pear-shaped body of 15–20 cm in size quickly rose up from the explosion site, it shone for about 0.3 s. Before the collapse, the plasmoids assumed a toroidal shape. Long-lived plasmoids with large sizes did not appear during the explosion of a wire spiral that was not rolled into a torus. In the experiments (Vlasov et al. 2015) the INGIR-Mega-15 installation was used, which is a generator of single submillisecond pulses (up to 0.5 ms) with a peak current of up to 35 kA. The charge voltage of the capacitor bank was 450 V both at a current of 35 kA and at 400 kA. With the electric explosion of 4 spirals of 16 turns installed toroidally (size 2 cm, Fig. 4.7a), a plasma LLO was formed with a diameter of about 6 cm (Fig. 4.7b). The typical lifetime of such LLO was 0.5 s and the best result was 1.6 s (Fig. 4.8).

The energy density of luminous objects is estimated at 70–80 J/cm^3, which brings the energy of the obtained objects closer to the energy of low-energy BL. The nature of their movement does not correspond to the description of natural BL. Most likely, burning of dispersed particles of wire material in the air was the cause of the glow, as in experiments (Alexeff et al. 2005; Kuz'min et al. 2003). These objects died as a result

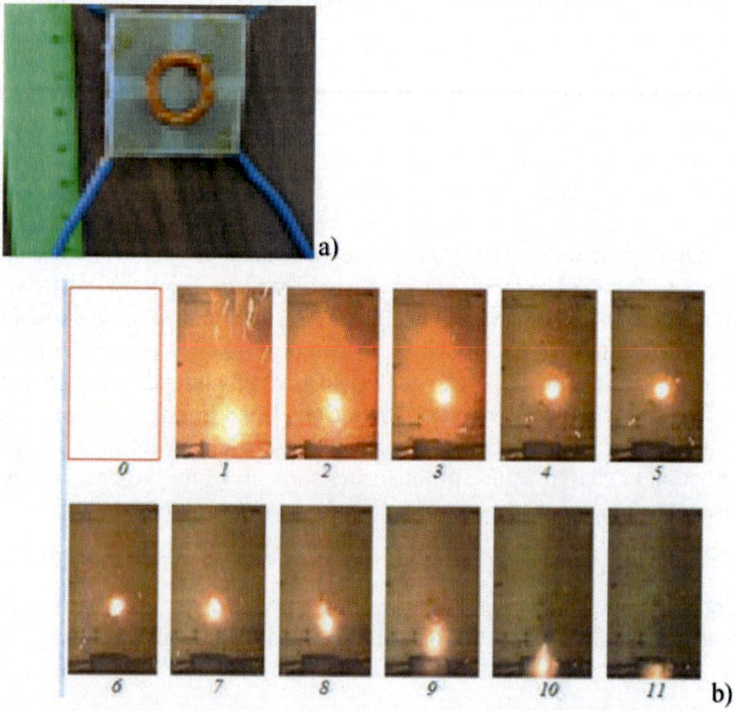

Fig. 4.7 Spirals (**a**) and video frames of electric explosion (**b**, **c**, **d**) with a sequence of 1/15 s

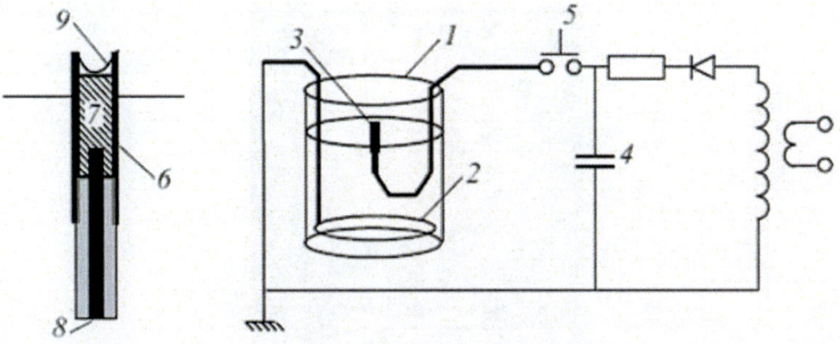

Fig. 4.8 Installation for obtaining LLO. 1—polyethylene container. 2—ring electrode. 3—the central electrode. 4—capacitor bank with a capacity of 0.6 mF. 5—discharger. 6—quartz tube. 7—carbon or metal electrode. 8—copper bus. 9—a drop of water or a substance injected into the discharge

of extinction without explosions typical for BL. Nevertheless, the obtained LLO in terms of lifetime and energy reserve can be considered experimental analogues of natural BL.

Gatchina discharge of Shabanov. In 2000 in the city of Gatchina near St. Petersburg (Russia) a new type of electric discharge called "Gatchina discharge" was discovered by its inventors (Shabanov 2002; Shabanov and Zherebtsov 2003, 2004a, b; Emelin et al. 2002; Shabanov and Sokolovskyi 2005; Shabanov et al. 2006a, b, c, 2008a, b). They consider the erosive discharge ("Avramenko flow") to be the prototype of this discharge, but, unlike the latter, they can increase the size and lifetime of a luminous object by two orders of magnitude (Shabanov and Zherebtsov 2003; Emelin et al. 2002).

The basis of the installation for the discharge is a 0.6 mF rechargeable battery, charged to a voltage of 5.5–6.0 kV. A plastic cylindrical vessel with water is used. The grounded anode is a ring of wire lying at the bottom of the vessel. The thickness of the water layer is usually 15 cm. As a cathode, a graphite electrode with a diameter of 6 mm is used, protruding above the water surface to a height of 2–3 mm. The side surface of the electrode is insulated from water by a quartz tube. With the help of a manual spark gap, the capacitance voltage is applied to the cathode. A sliding discharge develops on the surface of the water, and from an electrode protruding above the water, a luminous stream is ejected upwards. The spark gap is switched off at the 80th millisecond (at the same time, the residual battery voltage is 3 kV), and the luminous flux separating from the electrode turns into a luminous object. From the 60th to the 100th millisecond, it has the shape of a jelly-like body, and sometimes takes the form of an ideal sphere (see Fig. 4.9). The object rises up and disintegrates. The decay starts at the 200th millisecond, and the decay itself lasts about 400 ms. (The videos show that during the decay of a spheroidal object turns into a toroidal one.) (Fig. 4.10).

Using electric probes, the authors (Emelin et al. 2002; Shabanov and Zherebtsov 2004a, b; Shabanov and Sokolovskyi 2005; Shabanov et al. 2006a, b, c, 2008a, b)

Fig. 4.9 Photo of the "Gatchina" LLO

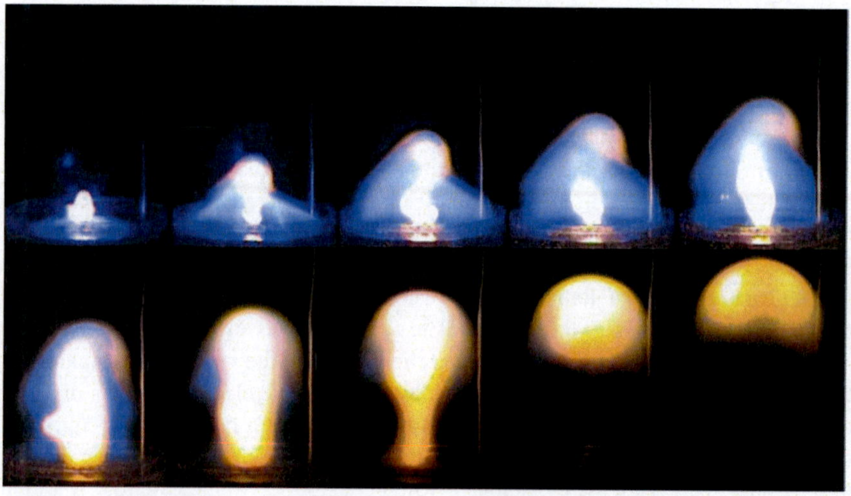

Fig. 4.10 Video-shooting of the discharge and the afterglow (Emelin et al. 2008)

found that the object carries an uncompensated negative charge of 2 10–7 S located on its shell. At the same time, the potential of the shell exceeds 10 kV. The plasmoid did not affect the dielectrics but melted thin wires. In (Shabanov et al. 2006b), it is reported that when melting a wire with a thickness of 0.1 mm, a ball with a diameter of 0.5 mm was formed. The authors believe that "the main energy of ball lightning is stored in the form of an electric field created by an electronic sphere, and this energy has a potential character" (Shabanov et al. 2006b). One can estimate the energy of the "Gatchina LLO". Let the diameter of the charged sphere be $D = 13$ cm, and its charge is $Q = 2$ -7· 10^{-7} C. The potential of the sphere is $U = Q/2\ \pi\varepsilon_0 D$ and the energy of the electric field $W_e = Q^2/4\ \pi\varepsilon_0 D$. For the accepted values of D and Q, we get $U = 27.7$ kV and $W_e = 2.8\ 10^{-3}$ J. Now let's determine the energy required to melt the nichrome wire. The specific heat capacity of nichrome is $Cp = 0.46$ kJ/kg·K, its density is $\rho_N = 8.4\ 10^3$ kg/m^3, and the melting point is $T_m = 1390$ °C (Physical values 1991). The formation of a molten ball with a diameter d requires energy $W_m = \pi \cdot d^3 \rho_N\ Cp\Delta T/6$. (Here ΔT is the difference between the melting point and room temperature). To form a ball with a diameter of 0.2 mm, an energy of $2.2 \cdot 10^{-2}$ J is required, and a ball with a diameter of 0.5 mm can be formed at an energy cost of $3.5 \cdot 10^{-1}$ J. This energy is 10–100 times greater than that of the electrical energy of the plasmoid. However, in (Shabanov et al. 2008a, b), the authors report that immediately after the discharge ceases, the charge of the plasmoid is $Q_0 = 4 \cdot 10^{-6}$ C, and, accordingly, we find that the energy of its electric field is 1.1 J. At the same time, according to the authors' observations, in the initial stage of existence (within $\tau \sim 100$–200 ms), the plasmoid intensively loses charge due to a corona discharge. Apparently, this can explain that by the time it rises to the height at which the probe is located (when the wire melts), the charge of the plasmoid decreases by 99%. Hence, the average corona discharge current is $Ic = Q_0/\tau = 20$–$40\ \mu$A.

For a sphere with a charge Q, the current determined by the air conductivity λ_a is equal to $I_a = \lambda_a\ Q/\varepsilon_0$ (Physical values 1991). Thus, $I_a = 1.6 \cdot 10^{-2}\ \mu$A at $\lambda_a = 3{,}5\ 10^{-14}\ \Omega$/m (dry air conductivity) (Nikitin 2006), and $Q = 4 \cdot 10^{-6}$ C. For the Gatchina LOO, the Ic current exceeds Ia by 2000 times. Apparently, this is due to the high conductivity of the channel ("leader") connecting the emerging luminous object with the cathode of the installation. The inconsistency of the results of measuring the laboratory energy of the LLO obtained as a result of melting conductors and evaluating its electrical energy indicates that there must be other energy sources in it.

In the luminous formations obtained in Shabanov's experiments, the electric charge is located on the surface of the shell. Therefore, it cannot cause the appearance of a force that creates surface tension. This force could appear if the charge was in a volume limited by the shell (Nikitin 2006). Therefore, the sharp boundary of the plasmoid most likely arises due to the vortex motion of the volume of heated gas rising upwards.

The creators of a new type of discharge (Shabanov and Sokolovskyi 2005) report that if the capacitive discharge is broken up to the 80th millisecond when a voltage of more than 3 kV is present on the capacitor bank, the second sphere takes off from the copper electrode of the rod, closes and disconnects the discharge circuit. The color of both spheres is approximately the same. Based on this, the authors (Shabanov and Sokolovskyi 2005) concluded that water in the installation plays only the role of variable resistance. This conclusion was experimentally confirmed in (Emelin et al. 2006). It is shown that the current decreases faster than the discharge voltage, while the voltage on the water surface at a point located 1 cm from the cathode begins to decrease by 50 ms earlier than at a point 7 cm away from the cathode. In addition, it was found that luminous formations similar to those observed with the standard "water" discharge configuration appear during a short-term short circuit

Fig. 4.11 LLO attained during the experiment (Kogut et al. 2008)

and disconnection of electrodes made of coal, steel, copper, and aluminum or when a current pulse passes through glass with a metal coating (Emelin et al. 2006, 2008; Emelin and Pirozerski 2006). In Fig. 4.10, one can see video-shooting of the discharge in experiments (Emelin et al. 2008).

Successful experiments with Gtchina discharges with a demonstration of spherical vortex were undertaken in (Kogut et al. 2008) see Fig. 4.11.

In 2007, the "Gatchina discharge" was investigated by German scientists (Juettner et al. 2007; Versteegh et al. 2008). Their installation was carried out according to the scheme (Shabanov 2002). A cylindrical vessel made of glass or plastic with a diameter and height of 25 cm was used, filled with tap water, or distilled water with salt (NaCl or $CaCl_2$) was used. A copper ring (anode) was placed at the bottom of the vessel, and the central electrode (cathode) protruded several millimeters above the surface of the water. A capacitor bank with a capacity of 1 mF, charged to a voltage of 4.8 kV, was connected to the anode through a spark gap, which caused a discharge current from 10 to 100 A. 100–150 ms after the current break. At the 30th millisecond after the interruption of the current, a spherical object with a clear boundary was formed. With a large exposure of photographing, it looked like a bright uniformly glowing sphere, but with a small exposure, a teardrop-shaped bright core was observed in the image, the size of which was two times smaller than the diameter of the sphere surrounded by a weakly luminescent coating. Usually, an energy of about 8 kJ was discharged into the discharge.

Calorimetry showed that the energy spent on heating and evaporation of water was about 5 kJ and the remaining 3 kJ were spent on the formation of a plasma sphere. About 0.7 kJ (23%) of the plasmoid's energy was spent on radiation. The temperature at the time $t = 235$ ms from the beginning of the discharge was equal to 900 K, and at $t = 260$ ms, it was $T = 600$ K. At $t = 200$ ms, the thermocouple detected a temperature above 1300 K. The floating potential of the rising plasma sphere was measured using the probe. It turned out to be positive. From 200 to 300 ms, it was equal to about 0.5 V, and by 650 ms, it decreased to 0.05 V. This indicates the presence of positive charges in the sphere, which is also confirmed by the presence of CaII and MgII ions in the emission lines spectrum. The radiation spectrum consists of lines of elements with low excitation energy, while there are no lines with an energy of more than 5 eV. Therefore, the electron temperature in the sphere barely exceeds 1 eV. Based on the measurement of the broadening values of the Stark lines of hydrogen, LiI and CuI, it was found that the electron density at the initial stage was about 10^{20}–10^{22} m^{-3} (from t = 0 to t = 75 ms), but in the generated plasmoid it decreased to 10^{14}–10^{15} m^{-3}. The temperature of the electrons after a rapid initial drop from 5000 to 4000 K gradually decreased to 2000 K during the subsequent existence of the plasmoid. The molecular band CaOH was the predominant source of visible radiation. Thus, it can be assumed that the source of radiation energy is chemical energy stored in the dissociation products of water. Measurements of the deflection angles of the laser beam when passing through the volume of the sphere showed that the average density of particles in it is about eight times less than the density of air, which implies that the temperature of the substance of the sphere is $T = 8 \times 273$ K = 2200 K (Versteegh 2008). It can be seen that the conclusions of the authors of the works (Juettner et al. 2007; Versteegh et al. 2008) differ somewhat from the conclusions of the creators of the "Gatchina discharge": the temperature of the sphere turned out to be higher than previously assumed, the charge of the sphere turned out to be smaller.

When discussing the nature of the Gatchina plasmoid, special attention was not paid to its gas-dynamic features, and it was not discussed in the works cited above, therefore, the purpose of this section is to discuss the possible gas-dynamic nature of the observed objects. The question of plasma processes is not considered at this stage of research. In the works of Emelin, the gas-dynamic features of Gatchina discharge were noted for the first time. In 2014, in the work (Bychkov et al. 2014), a possibility of the gas-dynamic nature of the "plasmoids" of the Gatchina discharge is analyzed. The analysis was carried out using two models for the formation of heated gas flows over the discharge region. Thus, gas heating by a volumetric heat source in the area of the central electrode and gas heating by a hot jet created above the surface of the discharge region and having a maximum velocity in the area above the central electrode are considered. A better agreement has been established between the results obtained in the second model for creating an aerodynamic vortex and experimental data (Shabanov 2002).

In the same year, ultrasonic sounding of the Gatchina plasmoid was carried out in (Stepanov 2014). Assuming that the plasmoid is formed from water vapor, the temperature of the plasmoid is calculated from the ultrasound velocity. It is shown that its temperature during its lifetime decreases from 2800 to 600–700 K. The temperature decrease is due to heat transfer and mixing with the surrounding air. In the photographs in (Stepanov 2014) perpendicular to the motion of the plasmoid, its vortex nature is recorded.

In (Stelmashuk and Hoffer 2017), the evolution over time of "Gatchina plasmoids" generated by a millisecond atmospheric discharge in contact with water was investigated. The authors claim that an initial electrical breakdown in a cathode ceramic tube filled with a conductive liquid is necessary for further plasmoid formation. It generates a shock wave that leads to the release of a conductive liquid. The resulting expansion of water vapor and bubbles with the addition of weekly ionized plasma serves as the main source of matter for the formation of a plasmoid. The electric arc that occurs after the breakdown is transformed into a plasma jet. This jet accelerates the plasma to the plasmoid.

In (Bychkov et al. 2018), a gas-dynamic simulation of a "plasmoid" created in an experiment by a pulsed discharge was carried out. Experimentally measured data were used as input data in the model. The qualitative agreement between the behavior of the "plasmoid" observed in the experiment and the dynamics of the model object has been established. The temperature distribution obtained as a result of the simulation is also consistent with experimental data (Stepanov 2014), where the gas temperature of the order of 2000 K is optically recorded. In Figs. 4.12, 4.13, 4.14, and 4.15, one can see the comparison of experimental results and computations of temperature.

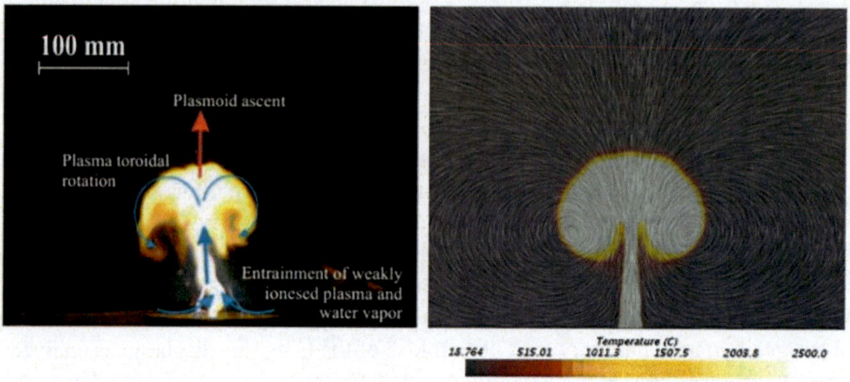

a *b*

Fig. 4.12 **a** is a schematic diagram of the motion of the media (Stelmashuk and Hoffer 2017), **b** is the direction of motion of the medium obtained as a result of the calculation (Bychkov et al. 2018)

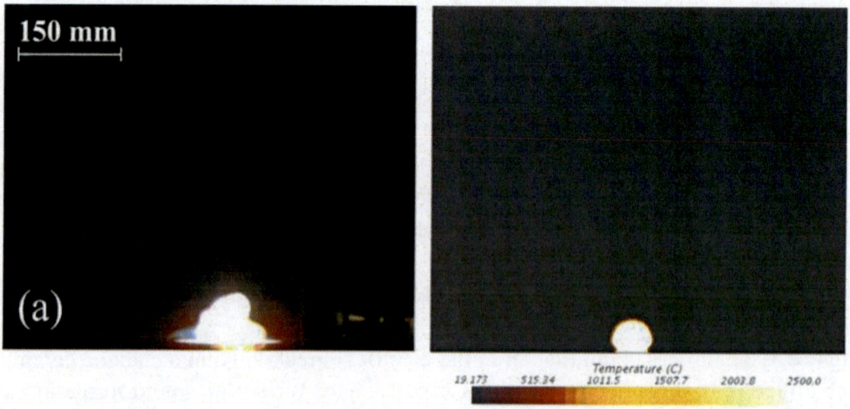

a *b*

Fig. 4.13 Temperature distribution at time 33.418 ms: **a**—experiment, **b**—calculations

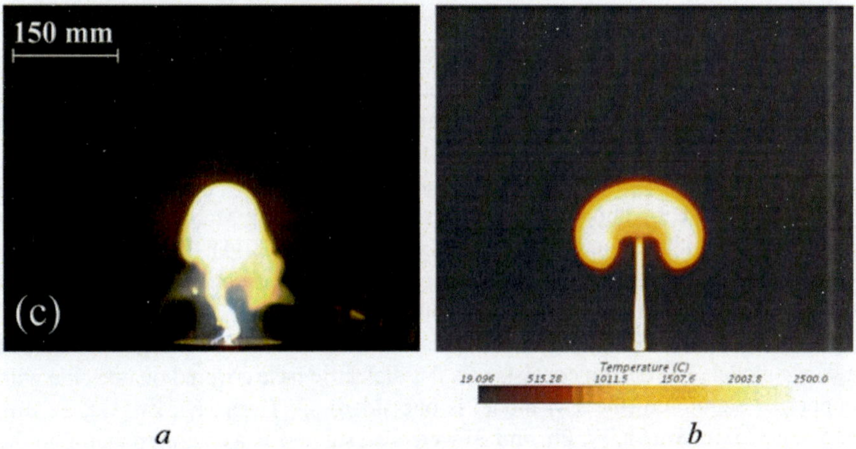

Fig. 4.14 Temperature distribution at time 110.987 ms: **a**—experiment, **b**—calculations

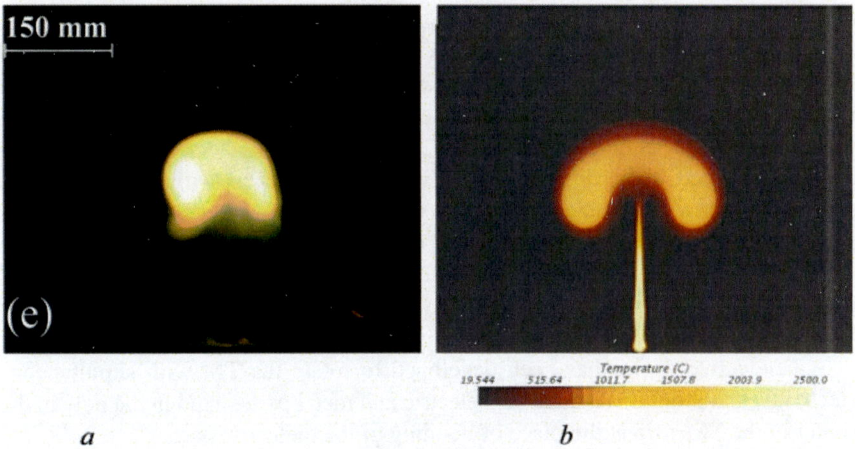

Fig. 4.15 Temperature distribution at time 170.938 ms: **a**—experiment, **b**—calculations

The calculations showed agreement between the dynamics of the behavior of plasma formation obtained as a result of the calculation and recorded during the experiment (Stelmashuk and Hoffer 2017). In addition, they are consistent with the results of the experiment (Versteegh et al. 2008) where the gas temperature of the order of 2000 K is optically recorded, which indicates the gas-dynamic nature of the hot gas rise. An incomplete coincidence indicates the need to take into account in the model (Bychkov et al. 2018) gas-discharge processes of energy release over the surface of the discharge region, leading to the formation of a hot jet and kinetics of elementary processes in created plasma.

It should be noted that the gas-dynamic properties of the "Gatchina plasmoid" dropped out of the discussion in the review paper (Shabanov 2019) where the author considers this plasmoid to be only an electric formation, which he identifies with the BL. In fact, the lifetime, size, glow and interaction features of this plasmoid make it to be an analogue of the BL, but it is due to its macroscopic vortex properties.

Shakhparonov experiments with undirected contours. The issue of artificial production of BL by undirected contours in the form of a Mobius tape (TM) was investigated in (Shakhparonov 1994; Evstigneev et al. 2011). The manufacturing technology of TM, the ways of connecting them to a power source, and other details for conducting discharge experiments with TM are given in (Shakhparonov 1994). In short, they are as follows. The basic design for obtaining BL analogues is the TM electric model. The design consists of a flat dielectric base covered on one side with copper or aluminum (the TM model is one-sided). C, Ti, Fe, Ni, Zn, Ga, Zr, Nb, Mo, Ag, Cd, In, Sn, Ta, W, Pb, and Bi were also studied as a coating material in the form of films sprayed in vacuum or glued foils. Of the listed materials, metals that conduct electric current well are the most applicable: Cu, Al, and Ag. Dielectrics of both organic and inorganic origin, for example, glass, were used for the dielectric base. Adhesives of organic origin, as well as epoxy resins, were used for gluing foils.

For the manufacture of TM, a blank is made from a dielectric coated with metal foil. The dielectric blank should be several millimeters wider and shorter than the metal foil blank. We rotated one of the ends of the dielectric strip relative to the other clockwise or counterclockwise by 2π. When turning clockwise, a left-screw TM was obtained, and a right-screw TM was obtained. Then, without changing the position of the twist, the ends of the strip were brought together and glued together with each other. Next, a metal conductor was glued. During vacuum deposition on the dielectric, the film thickness should not exceed 800–1000 nm; during chemical metallization, the film thickness can reach 1–3 microns. The maximum thickness of the glued foil is 50 microns.

A conductive device has been developed to rotate the TM with simultaneous electrical excitation. It is a node of dielectric and metal parts, fixed in the hole in the center of the TM strip at the place of bonding of the dielectric base.

Before switching on the TM as a load in the power circuit, the TM is processed with a high-frequency current using a Tesla generator, for no more than 60 s, as well as sinusoidal and pulse voltages with an amplitude of up to 10 V. Signal modulation was also used. Then the RF current generator was turned off and the TM prepared in this way was included in a power circuit with parameters 220 V; 50 Hz, 6–10 A. TM switching circuits can be either single-phase or three-phase, see Fig. 4.16. In (Shakhparonov 1994), only a single-phase switching circuit with a right-torsion LM was used. Figure 4.17a-b shows the discharges on undirected contours in the form of a Mobius tape (TM).

In (Shakhparonov 1994; Evstigneev et al. 2011), the main attention is paid to the study of the electrodynamics of TM radiation. It has specific properties and can be used in applications for processing various materials. However, almost no space is

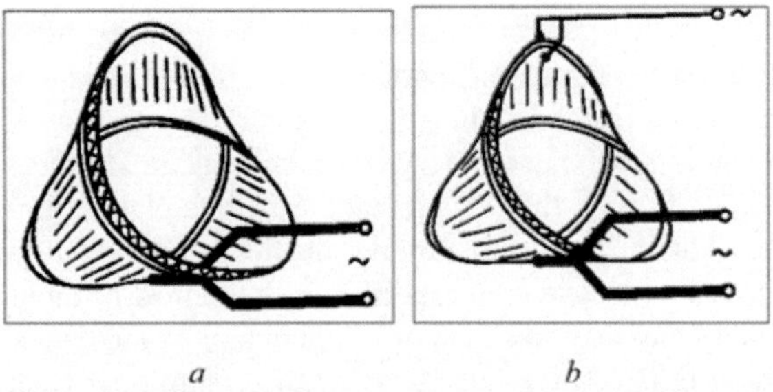

Fig. 4.16 TM switching circuits from left to right: single-phase, then three-phase

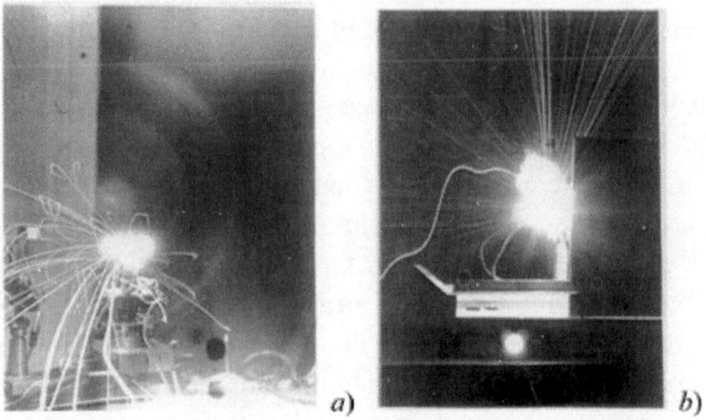

Fig. 4.17 **a** Integral photo of a discharge on a rotating TM. In the upper part of the photo, there is a luminous formation in the form of LLO. At the bottom of the photo, on the right, there are gray and black objects. **b** Integral photo of the discharge on a non-rotating TM. The luminous area with rays is the discharge area. Part of the frame on the right is covered by an opaque screen. A luminous spherical formation in the form of a BL with two protrusions is located in the lower part of the photo

given to the properties of luminous and black objects. One can only guess whether the LOO was obtained as a result of the electrodynamic and plasma interaction of the realized discharges into the air or whether it was the result of an explosion of TM followed by the formation of soot.

4.6 Erosive Discharge

4.6.1 Pulsed Electric Discharge in a Channel with Evaporating Walls

In the early 90s of the twentieth century, the interest of researchers was attracted by a pulsed electric discharge in a channel with evaporating walls (erosive discharge). The plasma stream that is ejected from the channel has a density close to the density of air, respectively, with low temperature and low radiation intensity, high energy density, and a large concentration of charged particles. This flow burns holes in metal plates but leaves the paper untouched and tends to maintain integrity when colliding with an obstacle. According to the authors (Avramenko et al. 1994), this set of properties is typical for BL, as a result, they stated that the erosive plasma flow is a laboratory analogue of BL. This analogy looks rather artificial (and should cause an active protest among those who have seen real BL) since the flow of moving plasma does not resemble BL in any way in shape or in the nature of movement. Nevertheless, the analogy stated in (Ershov et al. 1994) was unconditionally accepted by other researchers.

In a typical installation, capacitors from 630 to 2.5 mF charged to a voltage of 450 V are used to generate erosive plasma flows. Capacitors are discharged through a channel with a diameter of 1 mm and a length of 5 mm drilled into the insulator plate (organic glass is most often used). To maintain the laminar flow of the evaporated substance from the channel, it is necessary that the rate of increase in the current does not exceed 105 A/s, at high speeds the flow becomes turbulent. To create the desired shape of the current pulse, inductance is included in the discharge circuit. The typical length of the jet at atmospheric pressure is about 10–40 cm, and its speed is 60–100 m/s. The flow consists of a thin central part (core) with a diameter of 1–2 mm and a shell with a diameter of 6–10 mm. The continuous spectrum of the radiation of the stream is caused by the luminescence of soot particles, lines of hydrogen, carbon, copper, sodium, and calcium radiation atoms, as well as molecular bands C_2 and CN are also observed. In general, the plasma formation turns out to be electrically neutral, the excess negative charge in the core does not exceed 10^{-3} of the charge that passed through the discharge channel. The maximum floating potential of the plasma is 12 V. The radiation power of the plasma formation in the wavelength range from 0.2 μm to 10 μm, apparently, is about 30 watts, so no more than 1% of the energy stored in it (50) J) goes away during its existence. The temperature of the neutral gas in the stream turned out to be in the range of 1000–6000 K and higher (Ershov et al. 1994), and the temperature of the ions was determined as $6.5 \cdot 10^3$ K. The plasma flow is able to pass through a hole in the textolite plate. If the diameter of the hole is small, then only the "core" passes through it, and the shell is "removed".

In Klimov et al. (2006), the interaction of the end of an erosive discharge plasma jet with an incoming supersonic air flow was investigated. It was revealed that no shock wave is formed in front of the plasmoid, which is why the drag force drops by 20%. The plasmoid is a source of radio waves with a wavelength of 0.5–2 cm and amplifies laser radiation at wavelengths of 0.63 μm and 10.6 μm. It was shown in Klimov (2003) and Klimov et al. (2006) that a cluster plasmoid obtained in an erosion discharge carries a positive charge. The floating potential of the plasmoid was equal to 100 V. This value is 100 times greater than the expected potential of a decaying plasma with an electron temperature of 10^4 K.

The development of erosive discharges allowed to create a number of laboratory long-lived formations with a lifetime of about 1–10 ms. The main advantage of these discharges is the ability to supply energy much more than 30 MJ/m^3 into the channel, which is comparable to the energy density possessed by plasma regions of linear lightning. A nonequilibrium highly excited plasma is created inside a pulsed plasma generator under conditions of a powerful capillary discharge. When a discharge occurs in a channel with a diameter of 1–2 mm with walls made of polymer material, a lightning discharge is simulated. Here, we present the data obtained in experiments with erosive or capillary discharges when it was possible to separate the observed phenomena from burning.

4.6.2 *Characteristics of Capillary Plasmatron Plasma Jets*

When using different plasma-forming materials for capillaries, the outer regions of the plasma (outside the capillary) differed from each other in appearance and size. The plasma was created using capillaries made of plexiglass (PMMA), paraffin, a mixture of wax and rosin, as well as a mixture of paraffin with rosin and ground wood (particle size 100–300 microns) (composition 1:1:3—volume parts).

The typical volt-ampere characteristic of this plasma torch with a capillary made of a mixture of rosin, paraffin, and ground wood is shown in Fig. 4.18a, and in Fig. 4.18b, the volt-ampere characteristic of a plexiglass capillary is presented. From these volt-ampere characteristics, it can be seen that the current pulse (upper curve) and voltage (lower curve) have a triangular shape in time. When using a plexiglass capillary, the duration of the discharge pulse is less by $\tau \sim 1.5$ ms. The following explanation can be given for this fact: in the case of a capillary made of rosin, paraffin, ground wood, as well as their mixtures, the evaporation of materials from the capillary surface takes longer than when using a plexiglass capillary. The melting point of paraffin, wax, and rosin is T ~ 45, 64, and 68 °C, respectively, while plexiglass has a melting point of 120–130 °C. This leads to the fact that when the capillary cools, paraffin, wax, and rosin vapors form in the cooling capillary longer than in the case of PMMA,

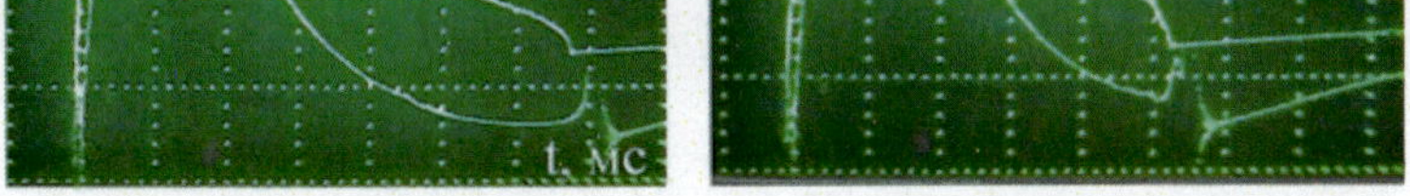

a b

Fig. 4.18 Typical volt-ampere characteristics of a capillary plasma generator. **a**—capillary made of a mixture of rosin, paraffin, and ground wood. **b**—capillary from PMMA

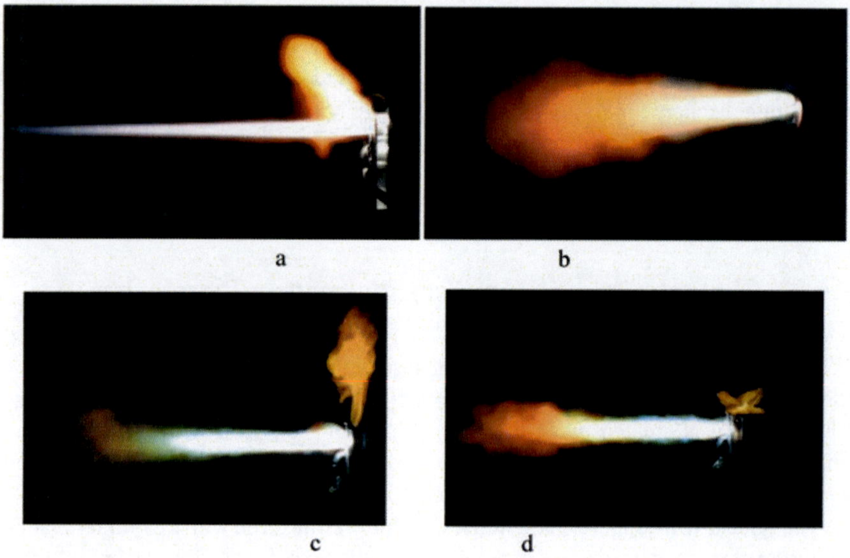

Fig. 4.19 Typical view of capillary plasma torch jets when using capillaries made of various materials. **a**—PMMA, **b**—paraffin, **c**—rosin + wax, **d**—paraffin with rosin and ground wood

forming a conductive channel, which in turn leads to an increase in the duration of the current pulse.

A typical type of jet is shown in Fig. 4.19a–d, respectively. The characteristic length of the luminous region was 10–12 cm, the thickness in the case of PMMA was 3–5 mm, and in the case of using other capillaries it increased to 15–40 mm (see Fig. 4.19c and d). At the exit from the capillary, in addition to the main area of discharge plasma, the presence of a yellow area is visible, which, as shown by the analysis of literature apparently corresponds to burning of acetylene (C_2H_2) molecules. They are formed by when paraffins are heated. As the video showed, the duration of the combustion process of eroded, organic materials coming out of the capillary of a complex composition reaches 30 ms.

Temperature measurements and the distribution of the electron concentration in the plasma jet created using this plasma generator with a capillary from PMMA were carried out by Ershov and Shibkov in 1994. Figure 4.20 shows the axial (Fig. 4.20a) and radial (Fig. 4.20b) distributions of the electron concentration in the plasma jet. Figure 4.6.14 shows the axial and radial distributions of the gas temperature in the plasma jet (Fig. 4.21).

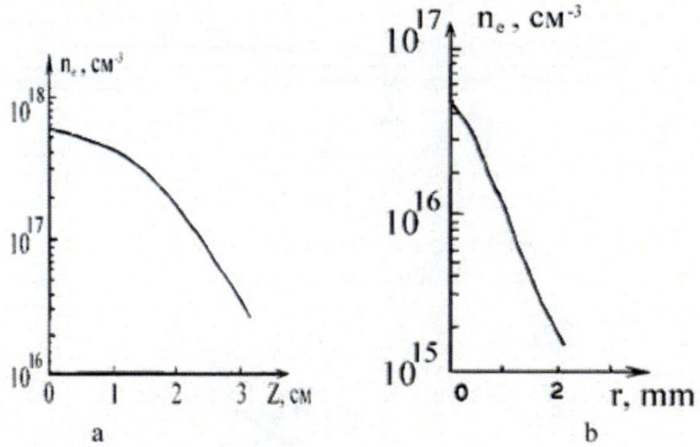

Fig. 4.20 Electron concentration distributions in a plasma jet (Ershov and Shibkov): **a**—axial; **b**—radial

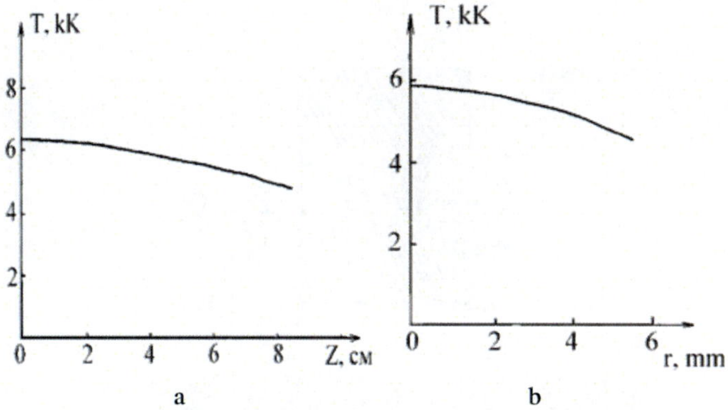

Fig. 4.21 Temperature distributions in the plasma jet. **a**—axial, **b**—radial

4.6.3 Experiments with Erosive Discharge with a Capillary Made of Wax

Using the idea of the interaction of the discharge with polymer materials that are closest in properties to natural ones, a plasma generator with paraffin (or wax) walls was used in (Bychkov et al. 1994 and Pancova in 2006). The scheme of the experiment is shown in Fig. 4.22.

The energy supplied per pulse to the discharge was 5.6 kJ and the pulse duration was 30 ms. At the exit from the capillary, a luminous object was formed in the air of atmospheric pressure. It had the form of a plasma jet 40 cm long and 2–3 cm wide, or in the form of a structured cylinder with an inner bright white core with a diameter of 1 cm, a blown shell of purple color, 2–3 cm in size, located around, see Fig. 4.19. The lifetime of the radiation was l ~ 0.1–1 s, which is abnormally large for an object of plasma.

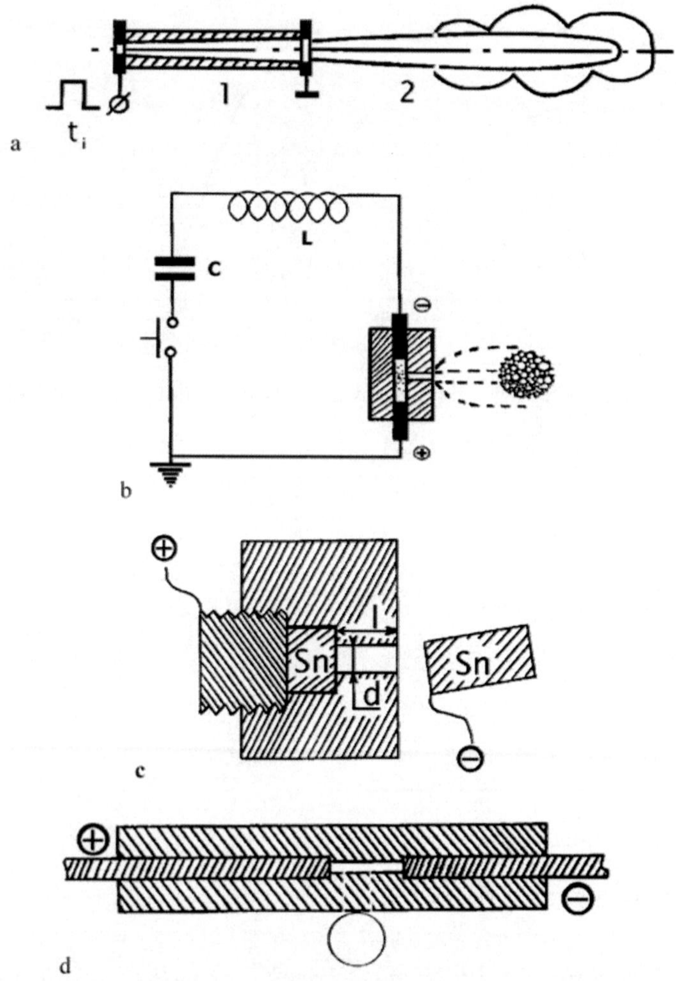

Fig. 4.22 Used in experiments plasma generators

The object had the following properties: it could make holes in thin metal films, which is possible with an object energy of at least 10 J, it could bounce elastically off dielectric materials, it could penetrate through small holes as an autonomous object, and this resembled a BL.

As a result of a series of experiments, it was revealed that the mass of the polymer material that left the walls in one shot was about 10 mg. In some experiments, after the end of the glow, a trace resembling smoke remained, consisting of dust particles, including thin filaments ~ 1 cm long and 1–10 μm in diameter.

When the object interacted with liquid nitrogen, the entire surface of the nitrogen cell was covered with a multitude of fluffy polymer threads, which then rolled into small tangles and sank to the bottom of the cell. At the end of the experiment, such formations could be removed from the cell and analyzed using a microscope (Bychkov et al. 1996).

Several shots were fired into the surface of liquid nitrogen, water, a metal ball, and crystal NaCl (see Fig. 4.24) for analysis in a transmission electron microscope of the JEM-100C type and in a scanning electron microscope of the JSM-U3 type. Studies in SEM (Bychkov et al. 1996) have shown the following: when an object interacts with liquid nitrogen, polymer wax structures appear on its surface; when the object interacted with the water surface, a thin film of wax with a thickness of 0.1–1 mμ appeared on it, the presence of carbon filaments and intertwined filamentous structures with a thickness of 0.1–10 microns and a length of up to 200 microns appeared on the surface of the metal ball (Fig. 4.23).

The experiments have shown that the obtained objects are similar in some properties to natural BL. They found that organic materials can appear in the ionized region in various forms, from dielectric hydrocarbon filaments to conductive pyrographite agglomerates, depending on the cooling rate (in liquid nitrogen, water, on a metal ball, NaCl crystal). They confirm the assumption of the theories of BL as polymer composites that during the lifetime of BL can change its properties from dielectric to conductive.

In (Pancova 2006), the high-power hydrocarbon plasmoid was generated at the outlet of the capillary erosive discharge in the atmosphere. The fireballs of spherical form were created. Their colors were from orange to white. They appeared from the glass tubes of different diameters, which were located over these plasmoids. The inner walls of the tube were covered with the organic polymer like wax. Not every

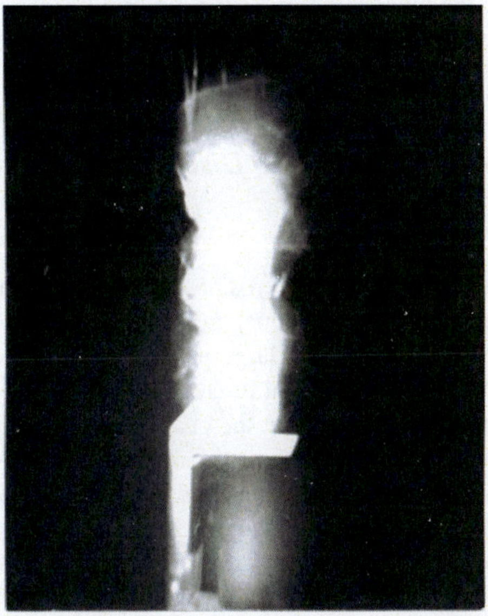

Fig. 4.23 LLO from an erosive discharge in a wax channel

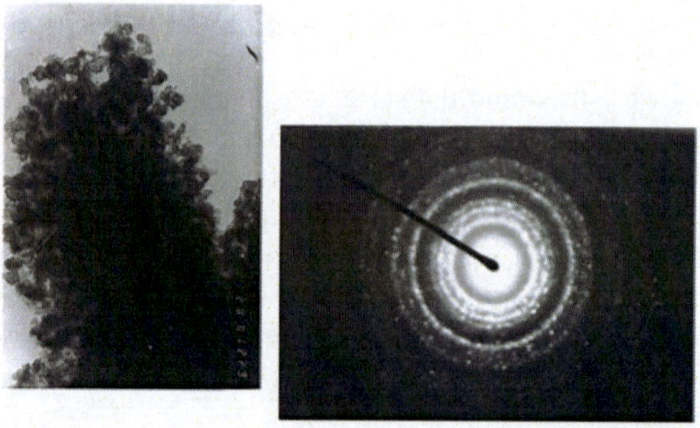

Fig. 4.24 Local diffraction analysis of films created on the surface of the NaCl crystal in JEM-100C showed that they consist of hollow spherical pyrographite particles with an average diameter of 1 micron, which form large microporous aggregates, see Fig. 4.24a and their electronogram in Fig. 4.24b

selected material for coverage breeds a spherical fireball. The interaction with some kinds of organic coverage breeds a spherical fireball. The interaction with some kinds of organic coverage gave only luminous volume of air. But the experimental fact was the following: no coverage inside the glass tube- no fireball at its opposite outlet. The luminous ball has a structure like the tangle of inhomogeneous threads. It was generated due to the interaction of the plasmoid surface, radiation, and gas dynamics with the organic coverage of the tube. The fireball is attached to the tube. Its lifetime was about 0.5 s and the pulse duration of the plasmoid generation was 30 ms. The stability of its appearance was high, almost at each generation of the plasmoid.

4.6.4 Experiments with Erosive Discharge with a Capillary from PMMA

The development of experiments with erosion discharge was carried out in (Emelin et al. 1997). Using the idea that the polymerization process of macromolecules occurs much faster with the participation of large particles as polymerization centers, in experiments small particles of charcoal or cotton wool (about 10 units, the size of each of them ~0.1 mm) were placed in the chamber. The channel (1–2 mm wide, see Fig. 4.22b) had walls made of polymethylmethacrylate (PMMA). A discharge with an input energy of 1–3 kJ led to the appearance of one or more independent formations. They flew up to a height of 0.5–1.5 m and fell like snowflakes. They glowed pink, had an irregular shape, and size 5–6 mm larger than the size of the channel. The glow time was 5–6 s before it went out, it was concentrated in the center of the object. No trace of smoke. The remains of these objects are shown in

Fig. 4.25 Polymer structure

Fig. 4.25. They had a white-yellow highly porous structure, sometimes consisting of bundles of fibers or individual long strands.

There were no traces of charring, and the density of the material was about 50 kg/m^3. Conductivity at direct current was practically absent. After sufficient deformation-compression and stretching, the object regained its shape, which indicates a high elasticity of the material, characteristic only of polymers. Under suboptimal discharge conditions, the object consisted of parts of charcoal and polymer with significantly less elasticity. Transmission electron microscopy showed that the frame of the object consists of a modified PMMA.

Getting this object onto cotton wool led to the ignition of cotton wool. This can easily be explained by the fact that the release of volatile substances from PMMA under normal conditions begins at T = 246–354 °C (and for polymer materials obtained in plasma, it becomes higher (Yasuda 1985), and the ignition of cotton wool occurs at T = 228 °C.

The results of these experiments qualitatively confirmed the possibility of the formation of a polymer-composite structure of BL as a result of polymerization, aggregation, and charging in plasma or thunderstorm conditions.

To obtain fractal charged objects, the authors in the group of S.E. Emelin (1997) placed activated carbon or lumps of cotton wool in a discharge chamber. When discharged, parts up to 5 mm in size flew out of the channel, which glowed for about 6 s with a red light. Under a microscope, a white porous structure with short fibers could be seen on the surface of the parts; their density was about 5 × 10–2 g/cm^3. The authors (Emelin et al. 1997) placed metal particles with a size of 50–100 μm in the channel of a plasma generator made of organic glass. At the same time, at the outlet of the plasma generator at a distance of 30–100 cm from the injector, an explosive formation of bright areas of 10–30 cm in size with a brightness of about 0.1 s occurred. When the plasma flow interacted with tin wire, yellow-red balls with a diameter of 3–5 mm appeared. Falling on the plastic surface, they bounced to the side, leaving burn marks on it 3–5 mm in diameter. The cooled surface of the balls turned out to be covered with tin oxide threads up to 50 microns long. In addition to the balls, there were fractal structures up to 3 cm long and several millimeters wide with a fractal dimension of $D = 1.8 \pm 0.2$. Within a few minutes they were very elastic, and then they became brittle and quickly collapsed.

4.6.5 Experiments with Polymer Materials

The schematic diagram of the used plasma generator is shown in Fig. 4.26 in experiments of Bychkov and Bychkov (Bychkov et al. 2004) it is a circuit including a switch (2), a capacitance (1), and a discharge device (3), which includes flat electrodes separated by a dielectric with a capillary, see Fig. 4.27. This capillary plasma torch has the following characteristics: the duration of the current pulse is 6 ms, the pulse energy is 200 J, the voltage at the discharge interval is 300–340 V, and the maximum value of the current in the pulse is 200–230 A (Fig. 4.28).

The values of the resistances of the voltage divider used to remove the voltage dependence on the discharge interval from time: R1—62 kOhm and R2—1.2 kOhm.

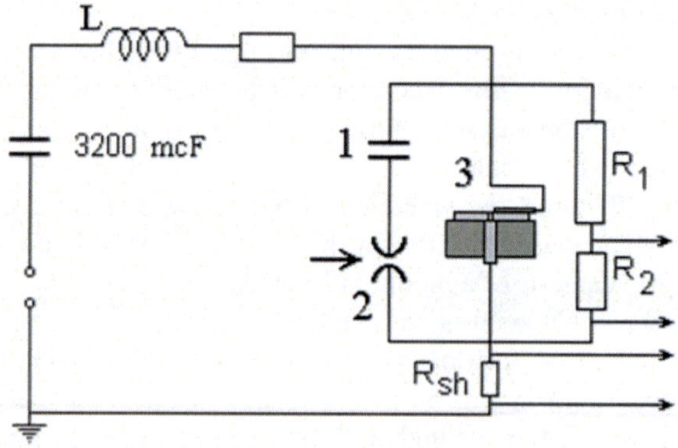

Fig. 4.26 General scheme of the capillary plasma generator. 1—the initiating capacity, 2—the switch that creates the initiating breakdown, 3—the capillary. R1, R2 are the resistances of the voltage divider

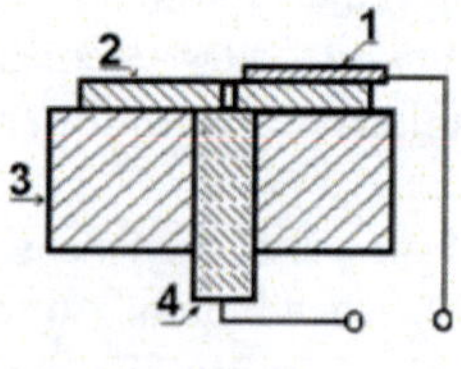

Fig. 4.27 Erosive plasma generator. 1, 4—Electrodes; 2—dielectric plate with capillary (discharge chamber); 3—plasma generator bed (made of plexiglass)

Fig. 4.28 Discharge chamber of capillary plasma torch, 1—capillaries

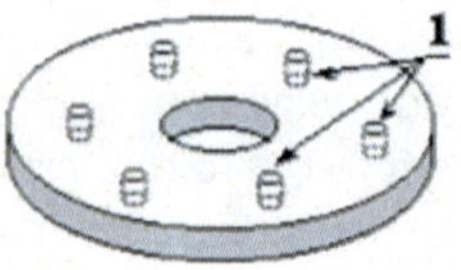

The resistance of the shunt Rsh is 0.013 ohms. The total storage capacity of C = 3200 μF.

To obtain the discharge, a capillary made of various polymer materials (see below) with a diameter of 1–2 mm and a length of 3–4 mm was used. The lower electrode (4) in Fig. 4.27 was made of an alloy of tungsten with copper, the upper electrode was copper, and the remaining elements of the plasma generator design (with the exception of the capillary) were made of plexiglass.

The launch of the plasma torch and the recording oscilloscope was carried out by synchro pulses coming from the pulse generator G5-15. The discharge was photographed in integral mode with the aperture open using neutral filters. Video shooting was carried out in continuous mode with video cameras with a frame duration of 33 ms and 41 ms. The current waveforms at the interelectrode gap were recorded using an ohmic shunt with a resistance of 0.013 ohms.

The Effect of Plasmatron Plasma on Polymer Materials

In these experiments, the interaction of a plasma jet with various samples was studied. The experimental scheme is shown in Fig. 4.29. A sample with which the plasma jet interacted was placed in front of the plasma torch nozzle at a distance of 3–10 cm. In various experiments, cotton wool, nylon wool, dielectric and conductive plates coated with wax or paraffin 3 mm thick, aluminum foil (50–100 μm thick), and paper (80 g/m^2 density) were used as samples.

When the plasma jet interacted with a cotton wool sample (a lump with a diameter of 3 cm and a density of 0.12 g/cm^3), this sample ignited at a distance of up to 7 cm from the plasma generator nozzle, for a time not exceeding 30 ms.

When the plasma jet interacted with a nylon wool sample (a lump with a diameter of 3 cm), the latter only acquired a yellowish trace of slightly charred fibers.

The fact that the interaction of nylon wool and cotton wool with the plasma of the jet is of a different nature can be explained if we take into account that, as shown in (Drysdale 1990), the yield of volatile substances during thermal decomposition of polymer materials in the case of structures with a high degree of reticularity, to which nylon can be attributed, forms a non-volatile, charred mass when heated. This significantly reduces the supply of oxygen to the flame.

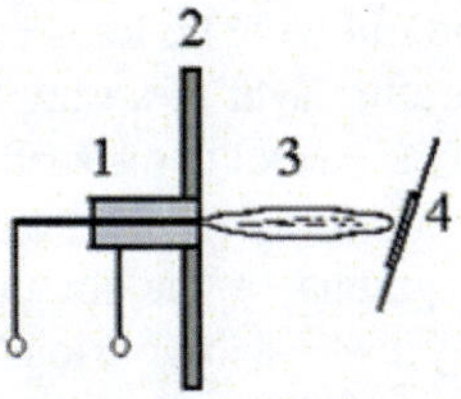

Fig. 4.29 Scheme of experiments using a plasma generator: 1—plasma generator, 2—bed, 3—plasma jet, 4—sample on a substrate

Table 4.1 Characteristics of some polymer materials (Drysdale 1990) (Reference book 1976)

Substance	Density, 10^3 kg/m^3	Heat capacity kJ/(kg·K)	Thermal conductivity W/(m·K)	Heat of combustion kJ/kg	Melting point, K
Nylon	1.2	1.4	0.4	31.9.10^3	523–533
Cellulose (cotton)	0.8	1.3		16.1.10^3	
Wood (birch, spruce, common pine)	0.4–0.5	1.3		16.1.10^3	

Table 4.2 Temperatures, °C, at which the specified vapor pressure (VP, Torr) of organic substances is set (Reference book 1976)

substance	formula	VP, 60	VP, 100	VP, 200	VP, 400	VP, 760	T, °C melting
Stearic acid	$C_{18}H_{38}O_2$	275.5	291	316.5	343	370	69.3
Stearin Aldehyde	$C_{18}H_{38}O$	244	260	285	313	342.5	63.5

Unlike nylon, when cellulose is heated, a large release of volatile substances occurs, which actively participate in the burning process (Kislitsyn 1990). In Tables 4.1 and 4.2, we present some data on organic materials necessary for the analysis of experiments.

The next group of experiments was aimed at explaining the paradoxical properties described in (Avramenko et al. 1994), where during the interaction of a jet from an erosive plasma torch with walls made of polymethylmethacrylate (PMMA) with plates made of aluminum, titanium, and tantalum (0.05–0.2 mm thick), the formation of round holes with a diameter of 6–10 mm was observed, while with the same contact with sheets of paper, LLO only occasionally left a charred trace.

The experiments were carried out according to the scheme shown in Fig. 4.29. Samples of aluminum foil and paper were located at a distance of 4–7 cm from the nozzle of the plasma torch. During the experiment, the plasma torch jet, when interacting with aluminum foil (100 microns thick), burned holes in it with a diameter of ~1 mm, but when a layer of wax with a thickness of over 0.7 mm was applied to the foil surface, the holes in the foil were no longer burned. For the evaporation of an experimentally observed paraffin layer of a sample with dimensions of 40 40 × 40 × 0.4 mm at a distance from the ejector to the surface of ~5 cm, the estimate gives an energy of $W \approx 100$ J.

Figure 4.30 presents photographs of the interaction of the plasma jet with aluminum foil coated with a layer of wax of various thicknesses. In Fig. 4.30a and with a thickness of the wax layer is 0.3 mm, the photo shows that under these conditions a hole is formed, and plasma (characterized by white glow) and wax vapor

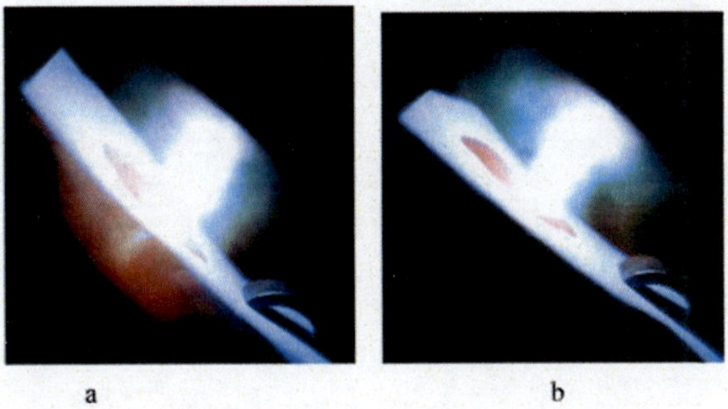

Fig. 4.30 Interaction of a plasma jet with aluminum foil samples coated with wax

(characterized by red glow) pass through the formed hole. In Fig. 4.30b, the thickness of the wax layer is 0.8 mm and the plasma jet no longer creates holes in the foil.

The following explanation can be given for this fact. When the plasma jet interacts with the polymer surface, thermal evaporation of the polymer material occurs. At the same time, the saturated vapor pressure, as can be seen in Table 4.2 increases. For example, we use data on stearic acid and stearic aldehyde, which are part of paraffins and waxes. This leads to an increase in the concentration of molecules in the wax mixture with air.

When the plasma jet interacted with wax or paraffin samples with a thickness of more than 2 mm, the surface of the samples was melted in the absence of other surface deformations, while the thickness of the sample decreased by 0.4–0.5 mm. The results of the interaction of paraffin and wax with a plasma jet when using different types of substrates (dielectric and conductive) had no visual differences.

During the interaction of a sheet of paper (density $\rho = 80$ g/m^2) with a plasma jet, holes with a diameter of 1–2 mm and breaks up to 5 cm long, having a gas-dynamic nature, were formed in the sheet. These experiments show that the paradoxical properties of erosive plasma described in (Avramenko et al. 1994) have a thermal and gas-dynamic nature, as in Ershov et al. (1994). In this work, the penetration of metal samples with the help of erosive discharge plasma was investigated. Based on electron microscopy of the interaction region, the thermal nature of the hole appearance was established.

In a number of experiments, a cuvette with molten wax was placed directly under the nozzle of the plasma generator, Fig. 4.31 (melting temperature, which is ~60–80 °C for various types of waxes (Figs. 4.32 and 4.33).

The plasma jet ignited the contents of the cuvette. Glowing areas appeared several times, rising, as the video showed, to a height of h ~25 cm in a time of t ~ 0.2 s.

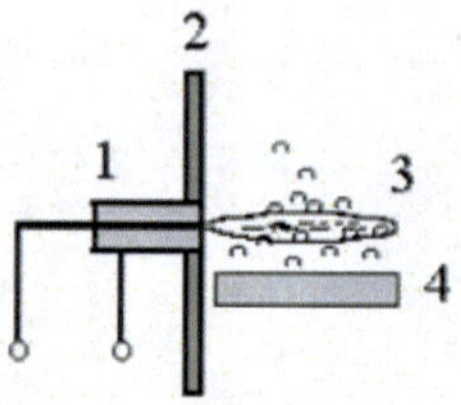

Fig. 4.31 The scheme of conducting experiments in the presence of a cuvette that creates wax vapors: 1—plasma torch, 2—bed, 3—plasma jet, 4—cuvettes with heated wax

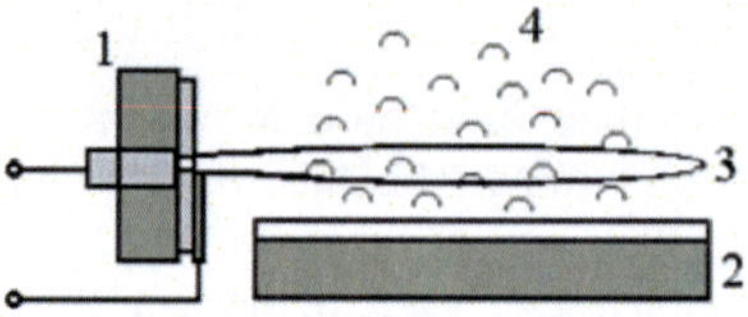

Fig. 4.32 The scheme of the experiment using an erosive plasma generator. 1—plasma generator, 2—a dish with evaporating substance, 3—a plasma jet, 4—a vapor

4.6.6 Production of Glowing Spheres

The scheme of experiments using a capillary plasma torch is shown in Fig. 4.33. In the cuvette (2), the wax was brought to a boil, after which the cuvette was placed at a distance of 1–2 cm from the capillary of the plasma generator. Burning of a light fraction of wax vapors as a result of interaction with a plasma jet is shown in Fig. 4.6.16. In this case, the jet was formed using a capillary made of PMMA with a length of 3 mm and a diameter of 1 mm. It can be seen from the figures that after ignition, the burning area rises to a height of 15–20 cm with the burning area separating from the surface of the cuvette with molten wax. This process lasted up to 0.4 s.

The typical time and height of the floating of the luminous area allow us to estimate the temperature of the gas contained in it (see below), which turns out to be of the order of $T \sim 2000$ K. This indicates (Ravich 1961) the complete combustion of paraffin vapors. Assuming that when paraffin is heated, its decomposition to methane occurs, it is possible to estimate the ratio of molecular components above the surface of the heated paraffin: CH4:O2:N2≈1:1:8.

The results of this experiment deserve special attention since such luminous spheres with a lifetime of $t \leq 1$ s have been repeatedly observed in natural conditions, namely, in forests during thunderstorms in hot weather (Taylor 1974). In this case, external factors, such as high temperature and lack of wind, could lead to a locally increased content of organic vapors in the air near the earth's surface. Since the diffusion rate of the organic polymer particles that comprises these vapors in the air is significantly lower than that of oxygen and nitrogen molecules ($D_{org} \sim D_{O2}(M_{O2}/M_{org})^{1/2}$) ($D_{org}$, D_{O2}—are coefficients of diffusion of organic and oxygen

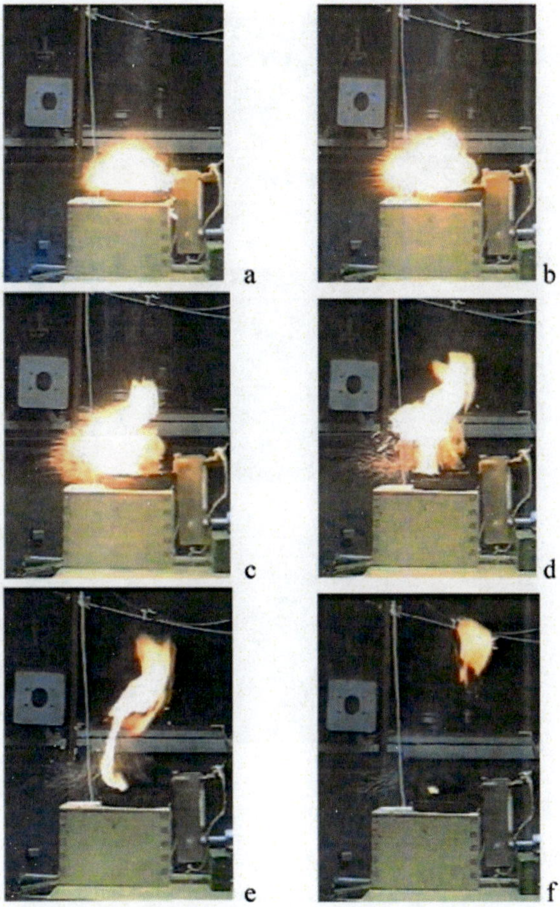

Fig. 4.33 Burning of a light fraction of wax vapors as a result of interaction with a plasma jet

molecules, respectively, M_{O2}, M_{org} their masses) local accumulation could occur near the source (for example, the tree that releases them). Therefore, this experiment, in essence, simulated the process of vapor ignition during the interaction of a linear lightning discharge with the subsequent flowing-up of a burning area. These results explain the nature of the formation of such spheres.

4.6.7 Obtaining of Fireballs with a Polymer Structure

With the help of a capillary plasma generator, experiments were carried out aimed at creating LLO, the characteristics of which would correspond as best as possible to the characteristics of BL observed in nature, namely, a lifetime of 1 s or more at a size of 1 cm or more.

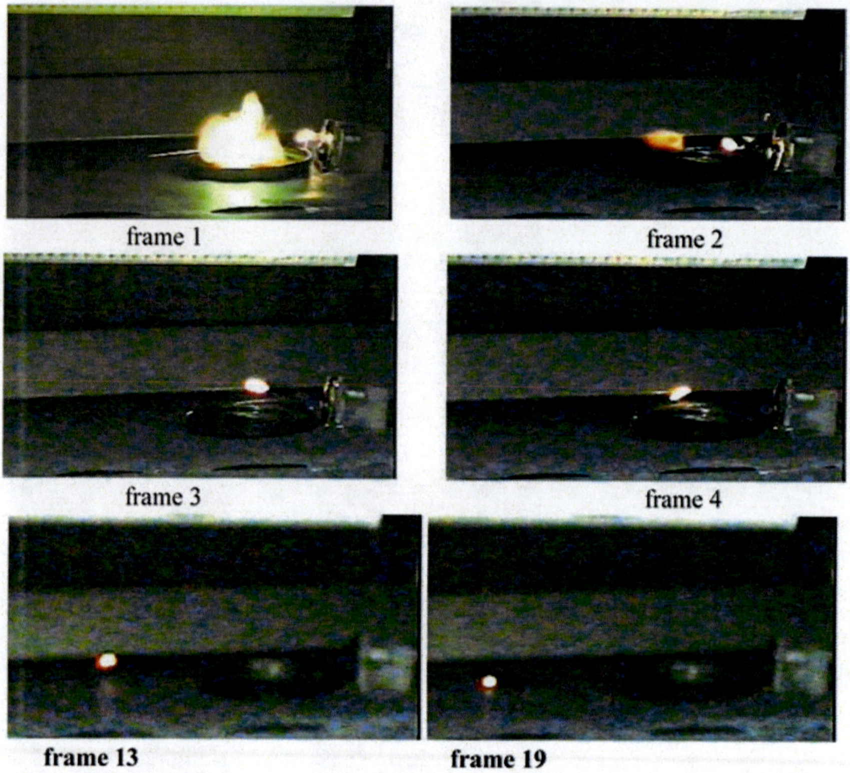

Fig. 4.34 Time dynamics of typical LLOs created with the help of a capillary plasma generator

The experiments, as a result of which LLO were obtained, were carried out according to two schemes. The first scheme is shown in Fig. 4.34. The second scheme differs from the first in that it lacks a cuvette that creates wax vapors. The presence or absence of hot vapors did not have a noticeable effect on the formation of LLO.

Figure 4.34 shows the typical LLO created during this series of experiments. The sequence of images was taken using a video camera with 41 ms duration of one frame. By the number of frames, it was possible to estimate the lifetime created by the LLO. Figure 4.34 shows the sequence of frames: 1–4, 13, 19 from the total number of frames 29. The 0-th frame corresponds to the discharge pulse, it is usually highlighted due to the limited capabilities of the video camera. It should therefore be noted that, as follows from the videos, the appearance of LLO from the channel occurs, as a rule, ~40 ms or more after the end of the main discharge pulse.

The lifetime of the presented LLO was ~1.2 s. The main condition for the appearance of DSO during this series of experiments was the choice of certain components from a mixture of which discharge chambers (capillaries) were made. The capillary dimensions remained the same: 3–3.5 mm in length and 1 mm in diameter.

During the research, various types of waxes, paraffins, and ground wood, as well as their mixtures, were tried. It was not possible to create a capillary in a rosin plate, due to its fragility.

Only one set of components consisting of pure medical paraffin (p), rosin (k), and sowed wood (d) with an average grain size of 1 × 0.3 × 0.3 mm led to the appearance of LLO in these experiments, while the volume proportions of the mixture in the manufacture of the capillary were maintained as follows: p:k:d = 1:1:3. When using finer sowed wood, the formation of LLO was not recorded. Pine wood was used in our experiments. Based on the work Konev (1977), it was concluded that the use of this type of wood is preferable due to the high content of lignin in it since we assumed that polymerization processes of a new structure are possible when plasma is exposed to lignin.

To make plates with capillaries, paraffin and rosin were melted in a cuvette on the fire of an alcohol burner, after which sowed wood was added to their mixture. The plate samples were formed in Teflon form.

During these experiments, the appearance of LLO was recorded, as a rule, in three out of four discharge pulses (with a single use of a capillary to create a jet). With repeated use of one capillary during these experiments, it was impossible to predict the possibility of the appearance of LLO as a result of each subsequent discharge pulse, but the appearance of LLO was repeatedly observed after 20 and 30 discharges, while the diameter of the capillary could already reach 2–2.5 mm. With not very frequent discharges (1 discharge in 5 s), the capillary retained its original diameter of 1 mm during the first 5–7 discharge pulses.

Recall that the appearance of LLO occurred with a typical time delay of t ~ 40 ms with respect to the discharge pulse, which was easily estimated using video shooting. It was not possible to estimate the energy input in the capillary during the time after the current pulse before the appearance of the LLO from the channel, due to the design features of the generator. Nevertheless, the fact that energy is released in the channel at this time is indicated by a sufficiently high initial velocity of LLO (3–5 m/s), which allows to estimate the energy in the capillary channel $E \leq 0.4$–0.6 J.

Almost all the observed LLO during the time of its existence made repeated collisions with the surface of the experimental installation covered with black photographic paper, when deciphering the video recording it was clear that the LLO experienced almost absolutely elastic collisions. The rate of departure of the LLO from the capillary determined by the video recording was from 3 to 5 m/s. Some of the observed LLO gradually decreased in size from 5–7 mm in diameter to 2–3 mm. Sometimes, during frame-by-frame analysis of the video, it was clear that large LLOs were breaking up into smaller ones.

In several cases, after the LLO was extinguished, remnants of its structure could be found on the surface of the experimental installation, which were black particles up to 2 mm in size. These residues were examined using the SCAN-240-Cambridge electron microscope at various magnifications. Figure 4.35 presents a number of the most interesting images obtained using this electron microscope.

Also, with the help of the electron microscope, the material from which the discharge chambers (capillaries) were made was examined. Enlarged fragments of

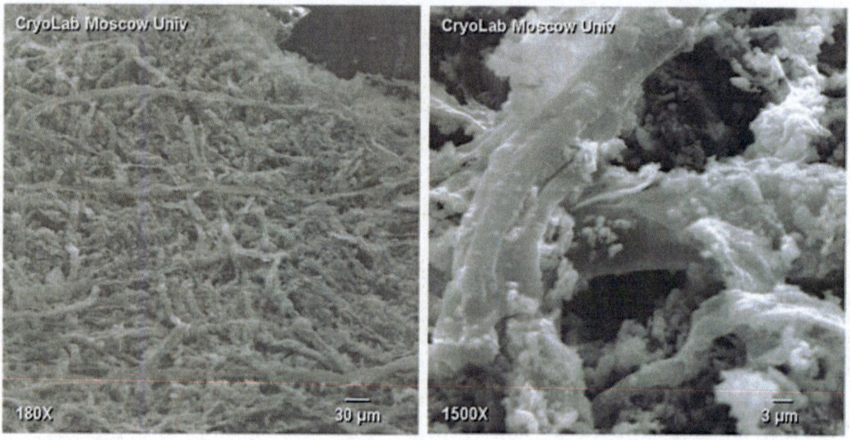

Fig. 4.35 Enlarged photos of the remains of the LLO

the source material are shown in Fig. 4.35. They are continuous polymer formations with various inclusions. The individual images clearly show structures in the form of large filaments or in the form of fibers up to 20 μm thick, which have a cellulose nature. Large white clusters, which may correspond to lignin or dust particles, are clearly visible in photos with high magnification (up to 3 μm).

Figure 4.35 shows enlarged photographs of fragments of the remains of the LLO. Compared with the source material Fig. 4.36 one can see the almost complete disappearance of the surfaces of smooth-homogeneous material. All surfaces at magnification up to 2 μm are porous materials consisting of intertwined threads of cellulose nature with clearly distinguishable clusters or soot particles between them. It can be assumed that during the exposure, the paraffin or rosin material burned, which led to the visualization of sowed wood.

The nature of the LLO observed in the experiment can be the combustion of large polymer clusters appearing in the channel due to polymerization of the wall material or due to the separation of fragments of material in the capillary and their association into clusters inside the channel. In this case, the heating of the surface of the particles can lead to the formation of a burning film of gases escaping from their surface. The flame front value for polymer materials of natural origin (Konev 1977) reaches several (5–8) widths of thin samples, with a lifetime for the flame stage of combustion $t \sim 1.5 \cdot d^2$, s, where d is the initial diameter of the sample ($d < 5$ mm).

Using the formulas from Konev (1977), on combustion of wood, one can show that materials with a density of spruce, pine, or birch (which are close in terms of the parameters of our mixture with $\rho \sim 0.3$–$0.5 \cdot 10^3$ kg/m^3) can float under the action of the pushing force of cold air acting on the area occupied by the luminous object.

Thus, one of the types of natural BL with a lifetime of $t < 20$ s and a diameter of up to 2 cm can be polymer-burning particles formed when linear lightning strikes trees.

In this case, the flame front, which is much larger in size than the particle size, will be perceived by observers as the effective size of the BL. At the same time, this flame front will determine the ability of the particles to levitate, due to the Archimedean force.

4.6.8 Experiments on the Interaction of an Erosive Discharge Jet with a Metal Sample

Another series of experiments (Emelin et al. 1997) was carried out with the discharge installation shown in Fig. 4.22c. Objects were formed by the interaction of an erosive plasma jet with low-melting metals. The channel length was l = 3–5 mm, diameter d = 2–5 mm, inductance L = 0.5 MH, and characteristic resistance of the discharge circuit Rx = (L /C) 1/2 = 0.8 ohms. A cylinder made of metal (for example, tin) with a diameter of 8 mm was located near the outlet of the gas discharge tube. At the initial storage voltage U0 = 800–1500 V, the average specific energy was ~80–150 MJ/m^3. The erosive jet burned grooves in the tin, sometimes with periodically varying width and depth, like a corrugated tube, from which brightly glowing yellowish-red balls with a diameter of about 1–2 mm or less were ejected laterally.

When these balls hit a solid horizontal surface, they bounce up to 50 times to a height of several centimeters, which is accompanied by a gradual decrease in the height of the rebound, as well as a decrease in the diameter of the balls with constant brightness. The glow usually goes out abruptly. The remains of the object were an irregular sphere with a diameter of 0.2–0.5 mm.

The surface of the autonomous formation, which showed considerable activity (high and long jumps), consisted of alternating sections of pure metal and a highly developed dark surface on which one could see "twigs" and various long thin polymer fibers. The dark parts are a dark structure (see Fig. 4.37) formed by intersecting forms of fibers, the surface of which is covered with a "carpet" of light polymer formations. The remains of an inactive object have a typical tin surface that has melted and hardened.

When using a tin inner electrode with the widest and shortest channel and at voltages in the upper part of the specified range, the "smoke" trail of autonomous formations flying through the air became non-dispersing, the trail first represented a continuous, twisting, and rapidly thickening thread, and then broke up into Sects. 3–5 cm long, which narrowed to about 1 cm with a diameter of 0.1–0.2 mm. The highly elastic structure observed in the first minute was replaced by an extremely fragile structure within 24 h.

A relaxed structure was observed under the microscope, see Fig. 4.38a, and its TEM in Fig. 4.38b, resembled a low-density aerogel consisting of chains and clumps of particles approximately 0.1–1 μm in size. The fractal dimension D of this structure was equal to $D = 1.8 \pm 0.2$, which is typical for the cluster-cluster aggregation of Brownian particles, which can be used to describe condensation processes.

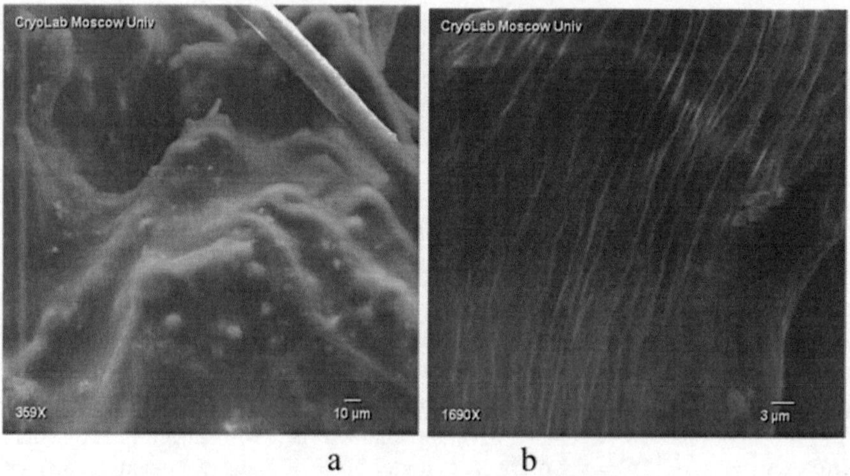

Fig. 4.36 Enlarged photos of fragments of the material used for the manufacture of capillaries

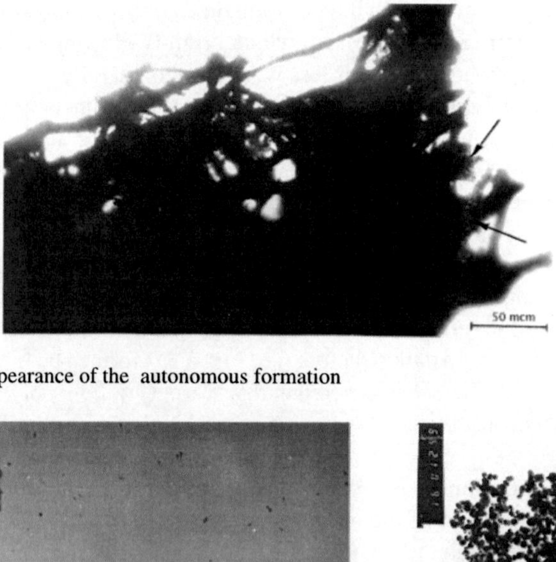

Fig. 4.37 An appearance of the autonomous formation

Fig. 4.38 A "smoke" trail of autonomous formations

The electronogram showed that the trace consists of SnO_2 particles. That is, during the formation of traces, tin particles were oxidized and a dielectric structure was formed.

To some extent, this experiment is similar to the experiments of Abrahamson (Abrahamson and Dinniss 2000; Abrahamson 2002), since in it a high-energy discharge interacts with metal, vaporizes it and fractal structures arise. However, no luminous structures with nanometer particles were observed.

In the case of (Abrahamson 2002), soil with silicon in its composition was used. It is well known that micron-sized silica filaments appear at temperatures of ~600–1000 K in chemical reactions involving $SiCl_4$, SiH_4, crystalline silica with a density of 1.97 g/cm^3 is formed in the gas phase during the oxidation of silica at a temperature of 1200–1400 °C and appears as paper-type films. It is unstable at temperatures above 1400 °C. Thus, we rather expect the formation of such structures in experiments (Abrahamson and Dinniss 2000; Abrahamson 2002), and we consider the nanostructures detected in this experiment as clusters of $Si(OH)_4$ monosilicon acid, which were previously in the soil, which is a colloidal solution, and were extracted due to a gas-dynamic disturbance caused by the interaction of the discharge with the soil.

This series of experiments has shown that long-lived structures of metallic and metallic oxide nature can be formed by the interaction of high-energy plasma with metal objects. This shows that some of the observations of BL can be caused by the interaction of linear lightning with metal objects with the destruction of their surfaces and the formation of luminous spheres.

4.6.9 Experiments on Erosion Discharge from a Slit

A closed discharge gap is optimal from the point of view of optimizing the supplied energy, but then it is difficult for an autonomous discharge to exit. This occurs only as a result of spontaneous damage to the discharge gap caused by the penetration of the dielectric wall of the channel by the object (Emelin et al. 1997). Some of the energy is lost by the autonomous formation and the correct combination of a large number of parameters is required. As a result, some relatively large objects up to 10 mm were observed only a few times.

One of the designs of the closed discharge gap consisted of a polyethylene tube with a cross section of 8 mm × 1.5 mm, into which steel rod electrodes with a thread of 2 mm were screwed, see Fig. 4.22d.

The length of the channel between the electrodes was 4–6 mm, the initial voltage was about 1.8 kV, the current cut-off voltage was 1–1.2 kV, the limiting resistance was 100 ohms, the inductance was 2.3 MH, the discharge energy was about 0.9 kJ, and the energy density in the channel volume was about 85 MJ /m^3.

When the discharge was successfully created, air and a small part of the plasma passed through the gap between the short electrode and the tube sealing the discharge volume. The main part of the plasma interacted with the vaporized polymer and metal and formed a medium that formed a gap in the side of the channel, through which the object almost silently flew out. The hole on the inside had the shape of a slit, which as a result transformed into the shape of a dumbbell about 1 mm long on the outside. A brightly glowing yellow ball with a diameter of less than 10 mm initially flew horizontally at a speed of several meters per second or less.

After moving about 2 m, one of these autonomous balls with a diameter of 6–7 mm hit a wooden wall and exploded, leaving traces of soot in the form of a downward expanding fan. The sound of the explosion resembled a high loud click. In another case, the discharge occurred in two stages with a spontaneous delay of half a second at a voltage of 1.5 kV. At the first stage, three autonomous formations of different sizes flew horizontally, skirting the throttle coil, and silently disappeared at the painted steel wall of the apparatus. At the second stage, a lot of small autonomous entities emerged that faced the same point. After the first stage, the resulting hole was completely closed, which was observed several times.

In the presented experiments, both molten and partially evaporated metal and polymer formed objects that could be considered glowing balls, this experiment is similar to the experiment with jets with a wax capillary, where evaporation and destruction of the polymer wall occurred.

The glow, lifetime, and explosion of the described object indicate that it is the closest to natural BL.

The development of this experiment was carried out in (Emelin et al. 2012). Its purpose was to continue the studies described above and to verify the model (Bychkov 2010) according to which the formation of BL occurs as a result of the impact of linear lightning on the components of the Earth's soil with the formation of a reservoir area having a strong oxide coating filled with metal powder.

To clarify the role of the core and shell in the formation of the properties of a luminous ball, the following experiment was conducted. In a polymer tube with a length of 160 mm and a diameter of 15.8 × 10 mm, an electric discharge was realized in a gap of 20 mm between the ends of steel electrodes screwed into a tube with a thread of 10.5 × 1 mm. The tube with electrodes was placed in the hole of a steel casing with a thread of 16 × 1 mm and a length of 140 mm, consisting of two halves, tightened by 8 bolts M14, see Fig. 4.39.

To protect the tube wall from the breakthrough and jetting of discharge products, a hole with a diameter of 3–3.5 mm inside the tube and 14 mm in diameter in its outer part was made in the middle of the casing cover between the ends of the electrodes. The storage capacity of 3.9 mF can be charged up to 5 kV and discharged through a choke with an inductance of 20 MCG. Current and radiation sensors are allowed to determine the moment of discharge products release relative to the moment of termination of the discharge current.

To create objects based on aerogel (type 1), 0.5 g of basalt filaments were distributed through the discharge chamber, and the battery was charged to 1.8 kV.

To create objects with a solid core (type 2), basalt filaments were crushed in a mortar and pressed into the anode, and the battery was charged to 2 kV.

In the third case (type 3), 1 g of basalt filaments and 0.5 g of polymer fibers were grinded into powder in a mortar and the chamber was filled, and a steel needle with

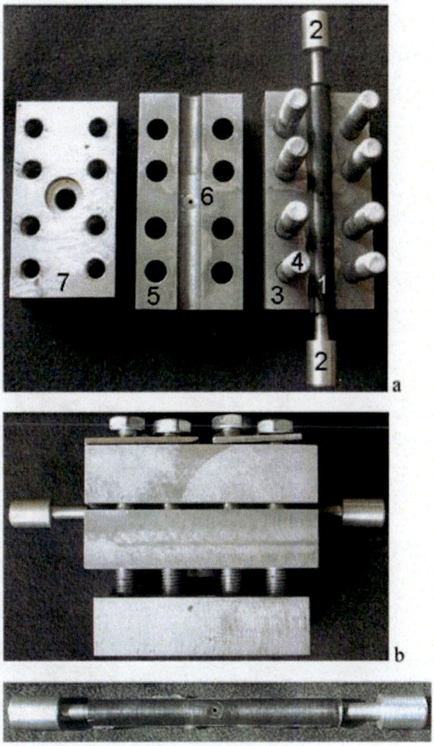

Fig. 4.39 **a** The discharge device. 1—polymer tube. 2—electrodes. 3—side of the steel casing. 4—mounting bolts. 5—side of the inner steel plug. 6—the inner hole. 7—reinforcing cover with an outer hole; **b** discharge tube in the casing; **c** discharge tube

a length of 19 mm and a diameter of 0.6 mm was inserted on the axis. The battery was charged to 2 kV.

The video was recorded using a SONY HDR-HC9 video camera in DV mode of 200 fields per second 720 × 576 pixels. In all cases, after the discharge was realized, a supersonic plasma jet first appeared, which formed a spherical plasmoid in the braking zone. The luminescent formations flying out of the spark gap following the plasma flow had a low velocity not exceeding 25 m/s.

Objects of the first type with a diameter of less than 1 cm had a lifetime of ~0.35 s. Their brightness and speed decreased faster at the beginning and end of life than in the middle. The trajectory of their movement could contain turns and accelerations. Their structural residue was a tangle of black fibers.

Objects of the second type did not reduce the brightness. With an object diameter of about 1 cm, it begins to grow avalanche-like in about 10–20 ms until the moment of explosion. The diameter of the object increases significantly.

Figure 4.39 (b-d) shows a sequence of frames of this object's movement with an explosion at the 130th millisecond of its lifetime.

Part of the substance remains in the discharge chamber. Figure 4.41a, b shows images of a piece of this substance stuck in the front of the outlet, obtained by electron microscopy. Figure 4.40a, b shows that the shell was realized during the discharge pulse, it consists of cavities formed during the boiling of basalt.

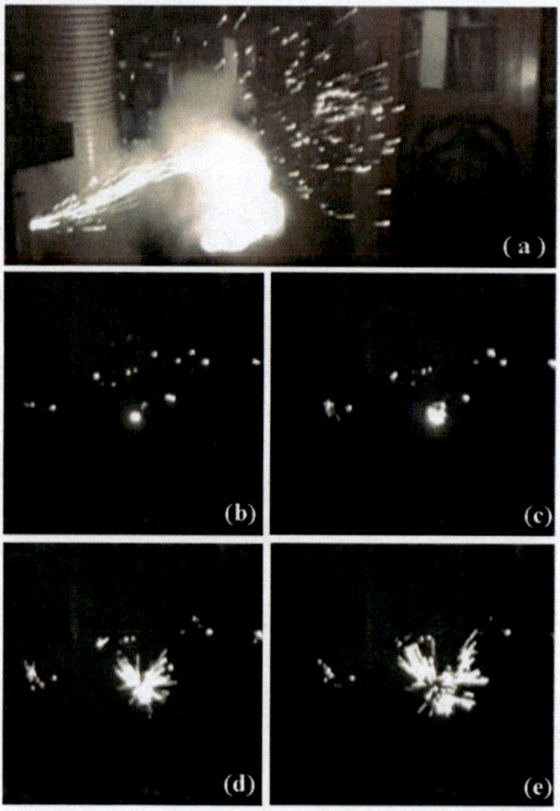

Fig. 4.40 Dynamics of the departure of luminous particles from the nozzle of the plasma generator

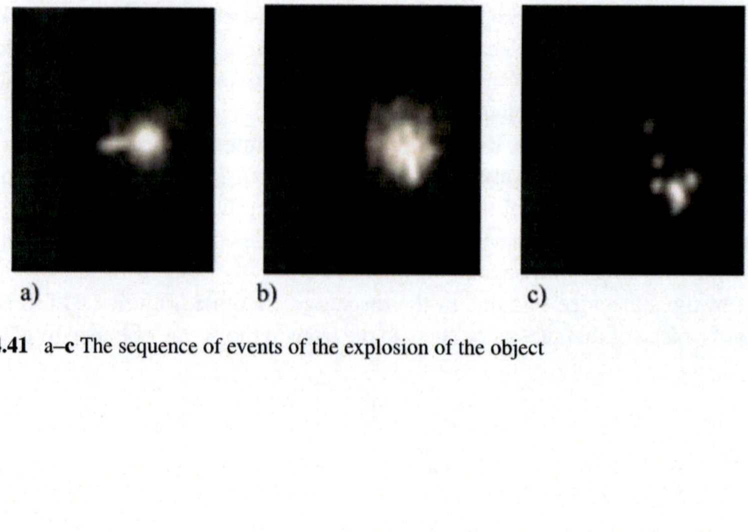

Fig. 4.41 a–c The sequence of events of the explosion of the object

The behavior of objects of the third type was significantly different from the one described above. One of them had a bright glow at the start from the discharge device, about 10 ms after discharge it was moving at a speed of about 25 m/s. During the first 60 ms, its speed stabilized at a value of about 5 m/s, and the glow decreased to poorly visible on the video. During the next 150 ms, the object, moving along a horizontal trajectory without changing speed and brightness, turned 90 degrees to the right and passed along the device. After that, its glow began to intensify, it abruptly stopped, the object bounced slightly, throwing out sparks, hung for 100 ms, and then began to fall vertically with a gradual fading of the glow. The total time of its existence before the glow was extinguished was about 0.6 s.

The observed phenomena can be given the following qualitative explanation.

Objects of the first type are "aerogel", created on the basis of basalt and polymer, containing particles of electrode materials and destroyed basalt, including the size of nm. These particles provide Gorenje plasma-excited hydrocarbons and their combustion.

The significantly reduced period of luminescence of objects with a core and shell (type 2), coinciding with the period of horizontal movement at a low speed, is most likely due to the surface nature of the core surface burning and the action of the air-gel shell, limiting the flow of air to the surface of the core after its formation. The destruction of the shell under external influence changes the combustion conditions of the core and can lead to its explosion.

The research of the third type of object requires its continuation. In Fig. 4.42 we present a photo of the structure of a ball formed inside the tube.

In this section, it is shown that when using some solid materials and powders in a gas discharge, a complex sequence of chemical and physical processes takes place with the creation of heterostructural artificial BL. It is very important that such BL can have a core of a complex material (powder, vapor particles) and a shell of a solid material (oxide, polymer, or aerogel), as proposed by (Bychkov 2010). This can lead to a complex nature of movement, fire, and explosion of such an object. In real conditions, the presence of sand particles, basalt, alumina, rust, etc., can lead to rapid ignition when plasma appears, therefore the combustion of such substances in plasma conditions requires careful investigation. The analysis of these processes in plasma can lead to an understanding of the physical nature of the real BL.

We have to note that the objects created in these experiments mostly resemble real BL since their lifetime often finishes with explosion. So they can be considered analogue of real BL.

4.6.10 Experiments on the Interaction of an Erosion Discharge Jet with Substances

Experiments on the Interaction of an Erosion Discharge Jet with an Alloy

The experiments carried out in (Emelin et al. 1997, 2012) confirmed the formation of luminous balls under discharge conditions in specially designed tubes filled with various organic and inorganic materials (such as basalt containing Si). The lifetime of these objects ranged from fractions of a second to a second, and the diameter ranged from a few millimeters to a centimeter. Some of them ended in an explosion or disintegration, which allegedly confirmed the theory (Bychkov 2010). However, microscopy of the remains of these objects showed that the internal structure of the objects consists of pores and cavities as if they arose during the boiling process. This led to a rethinking (Emelin et al. 1997, 2012) where explosions of experimentally obtained luminous spheres and experiments on the creation of "hollow" metal spheres were observed during the explosion of titanium foil in a liquid. These experiments have shown the important role of the processes of heating, melting, and evaporation of the material inside the formed object. It should be noted that the important role of steam in LLO and BL was discussed in (Bychkov 2014), where it was shown that the energy of BL filled with water vapor is comparable to the high energy of BL observed in nature, and its energy density can be greater than 10^9 J/m^3. The important role of the BL shell was indicated in Abrahamson 2002 where the observer saw how the BL disintegrated in flaps. The important role of the metal shell in BL was discovered in Gromyko (2004), where the destruction of the natural BL was accompanied by the appearance of metal spheres with a diameter of 0.1–1.2 mm. In addition, a number of observations (Stakhanov 1985) claimed that the BL was formed when lightning struck metal objects. In this regard, we present the results of experiments (Bychkov et al. 2015) using a capillary erosive discharge, the plasma jet of which interacts with the evaporating material and forms a long-lived luminous formation. These experiments also simulate a linear lightning strike into a metal object and attempts to obtain a glowing ball with a solid or liquid shell filled with steam.

Experiments (Bychkov et al. 2015) were carried out using a capillary plasma torch. The schematic diagram of the plasma torch is shown in Fig. 4.43. Above the outlet (capillary), from which the plasma jet is ejected into the air, there is a metal wire, see Fig. 4.43a, b, which is used to form LLO when plasma interacts with the metal of the wire.

The maximum current value was 100–150 A. Total storage capacity C = 3200 μF. To realize the discharge, a plexiglass capillary with a diameter of 1–2 mm and a length of 3–4 mm was used. The lower electrode (4) in Fig. 4.43a is made of an alloy of tungsten with copper, the upper electrode is made of copper, and the remaining structural elements of the plasma generator are made of plexiglass.

Video shooting was carried out in continuous mode using a video camera with a frame duration of 33 ms. The typical length of the plasma jet from the capillary was 10–12 cm. The energy stored in the generator was estimated as $Wn = (C \cdot U^2)/2$,

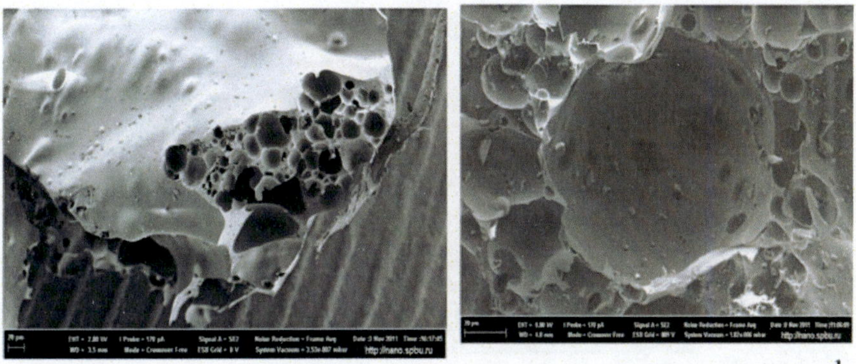

a b

Fig. 4.42 **a, b**—images of the surface of the substance (scale 20 μm)

Fig. 4.43 **a** 1,4—electrodes, 2—dielectric plate, 3—plasma generator rack, 5—metal wire. **b** Discharge jet with a plexiglass capillary

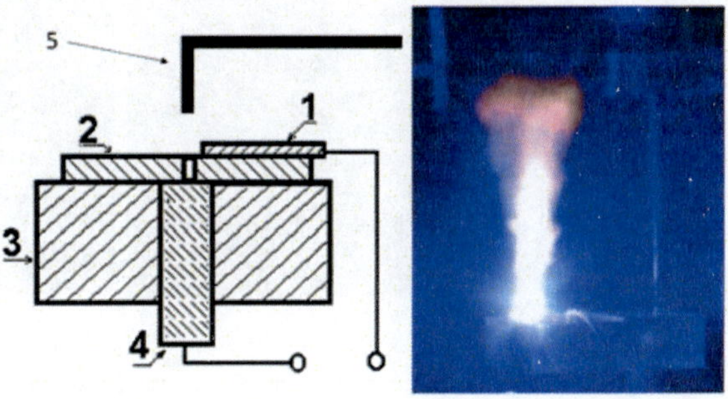

a b

where C = 7 mF is the storage capacity, U = 300 V is the voltage on the generator, therefore, Wn = 315 J. The energy entering the jet is usually 50–60% of the energy supplied to the generator, i.e., J. The typical appearance of the device is shown in Fig. 4.43b. The typical discharge time was 10–12 ms, and the maximum time was about 15 ms.

Experiments were carried out on the interaction of a plasma jet with an alloy wire consisting of about 70% Sn, 20% Pb, and 10% organic additives. The lower part of the wire (~0.3–0.5 mm long) heats up and melts, and glowing objects fly along with the plasma jet. The frames of these objects are shown in Fig. 4.44a–e.

The film frames of these objects are shown in Fig. 4.44a–d. Figure 4.44a corresponds to the category in which the LLO appeared, the first frame. In Fig. 4.44b, one can see the appearance of two LLOs in the second frame. The following frames show only one object (of the first two), shot on frames 21 and 31. Later it disappeared. The lifetime of such objects was about 1–1.5 s. The apparent diameter of the object was up to 5 mm.

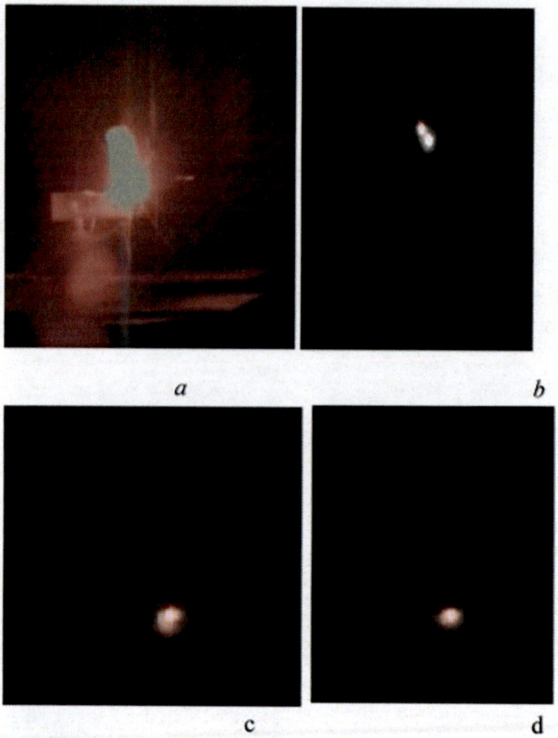

Fig. 4.44 a–d The appearance of the LLO. a—the appearance of the jet, b the 2nd frame, c—the 21st frame, e—the 31st frame

To study the structure of the luminous spheres around the experimental setup, a sheet of white paper was spread out, and the LLO, flying out of the plasma jet, fell on the paper, leaving traces, see Fig. 4.45a, b

In the case of Fig. 4.45a on the paper, there was a trace in the form of stars. This allowed to make the assumption that the explosion of the LLO from the alloy and

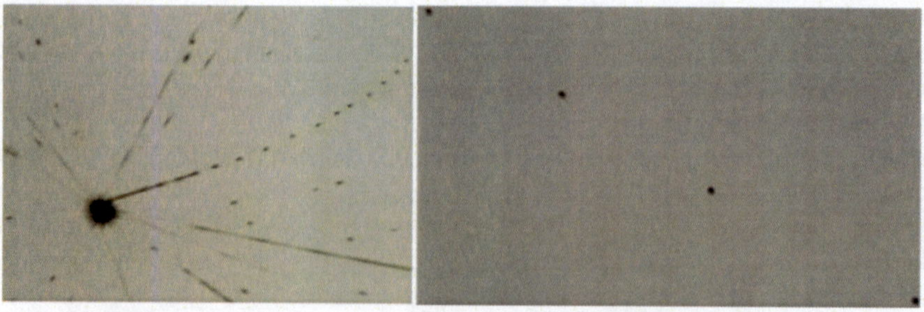

Fig. 4.45 Traces of LLO

the destruction of its material occurred upon contact with the paper. This could be realized when the surface of the LLO breaks during a fall with the release of steam from its core. In the case of Fig. 4.45b, the marks resembled the usual spots left when a drop of liquid fell; the object did not explode. He "jumped" on the paper, leaving traces of 20 or more in the series. The traces of these objects did not exceed 2 mm in diameter, and their average value was 0.5–0.8 mm. The size of the "stars" Fig. 4.45a reached 5–6 mm.

Experiments were also carried out in which LLO fell into a cuvette with water. Some of them drowned in it. Some of them appeared on the surface of the water. The average radius of these floating spheres was 1 mm. When trying to pull the sphere out of the ditch, it collapsed. A metal ball with a diameter of ~600 μm was found inside the shell. Figure 4.46 a shows a photograph of two balls of two "hollow" spheres made with the help of SEM, and Fig. 4.46b shows the shell of a "hollow" sphere.

Fig. 4.46 Photos of the sphere. **a** inner ball, **b** shell

From these photos, it can be seen that the diameter of the ball was about 600–640 μm, and the thickness of the shell was about 8–12 μm. Mass spectrometric analysis of the material of the balls and shells was also carried out. The composition of the ball was O: Sn: Pb = 14.95: 67.59: 17.34%. Shell composition: O: Sn: Pb = 18.95: 54.71: 26.64%. From these data, it can be concluded that the shell and the ball (core) consist of the same substances. However, there is more oxygen in the shell. This indicates the oxidation processes during the formation of the shell.

The appearance of a metal ball in a "hollow" sphere when falling into water indicates the condensation of steam inside the ball when it cools. This confirms the conclusion (Bychkov 2014) about the possibility of the formation of objects filled with metal vapor during the interaction of plasma with metal objects.

So, in experiments (Bychkov et al. 2015), LLOs filled with metal vapors were obtained. The experiments were performed using capillary discharge plasma acting on an Sn alloy wire. At the same time, it was possible to obtain two types of LLO: the first were drops of molten metal, and the second were spheres filled with metal vapors. Their lifetime was about 1–1.5 s, and their visual diameter was up to 5 mm.

The creation of steam-filled, so-called "hollow" spheres is consistent with the results of (Emelin et al. 1997, 2012) and confirms the theory (Bychkov 2014) about the role of steam in the composition of long-lived luminous objects.

The results of this work confirm the hypothesis (Bychkov 2014), according to which some types of natural BL may have a shell of solid or liquid type material filled with steam of this material.

The question of the mechanism of shell formation remains unclear, however, the results of this experiment show that metal oxidation can occur during its formation.

We have to note that the objects created in these experiments mostly resemble real BL since their lifetime often finishes with explosion. So they can be considered analogue of real BL.

Similar experiments were made in the work (Kirko et al. 2014a). In it was used an experimental setup consisting of a discharger connected with a power source and a plexiglass support, on which a metallic foil was placed. A discharger was made of a dielectric plate (material plexiglass, thickness 3–4 mm) with a capillary of diameter 1–2 mm. The power source was the capacitor battery (charge voltage 200–250 V, energy 100–270 J). Discharge current was 100–150 A at a pulse duration of 6–10 ms. During the discharger switching, the interaction of capillary discharge jet with a metallic foil (material: aluminum, copper, brass, thickness 0.05–0.1 mm) took place. Thus, in the foil, the hole with dimension 0.5–1.5 cm was burnt through. The mass of evaporated metal was 1–3 mg. The touch length by interaction process decreased from 10–15 cm to 5–7 cm, and its form transformed from cylindrical (diameter 2–3 cm) to irregular. The radiation from the region of touch and metallic foil interaction was investigated by spectrometer Ava Spec 2048 (spectral region 200–1100 nm, resolution 0.3 nm). In the radiation spectra, atomic lines of the metallic foil element and atomic and molecular lines of the capillary material and air were fixed. At the final interaction stage, the origin of luminous regions (FB-fireball) starting from the range of touch and foil contact was observed. The best results were obtained using aluminum foil. Their motion trajectories were mainly parabolic. The FB motion in space was from some cm to 1 m. The calculated mean FB motion velocity was in the range of 0.5–10 m/s. The radiation color was usually white with a little dark blue shade. In film frames around the main white color luminescence region, a white-gray train or a dark blue halo sometimes were observed.

Note that in difference to previous experiments the inputted energy was larger by about two times, it allowed to obtain FB on a basis of aluminum.

Development of Work on Erosive Discharges with Tin Alloy Electrodes

Consider the research described above is the work (Pirozersky et al. 2014). Experiments were carried out in it with the effect of an erosive discharge on a copper-tin solder wire. Several types of LLO with a lifetime of up to 8 s were obtained. It was concluded that during the formation of these objects, the energy invested in the discharge is transferred to the energy of nonequilibrium excitations of the LLO components. Elucidation of the nature of these excitations and the mechanism of energy conversion requires further research.

In the work (Emelin et al. 1997) cited above, several methods of reproducible production of LLO were proposed in the interaction of highly nonequilibrium erosive plasma with metals and polymers. The resulting LLO had a long time of autonomous existence and could "float" in the atmosphere, explode and burn thin metal foils. At the same time, their characteristics improved quantitatively and qualitatively as the specific energy input into the plasma of the erosive discharge increased, which reached tens of eV/atom. At the same time, the specific energy of the LLO itself was not directly measured but was estimated indirectly by the energy input to the discharge and the mass of the substance involved in the discharge. The efficiency of energy input into the discharge plasma, as well as the proportion of energy transferred from the plasma to the LLO themselves, remained unknown. LLO of one of the types discovered and investigated in (Emelin et al. 1997) occurred when a jet of erosive plasma from a capillary discharger made of polymethylmethacrylate was exposed to a tin anode. These LLOs had a spherical shape, diameter ~1–2 mm and glowed with a bright yellow-reddish light, after falling to the surface they jumped and fell down again up to several dozen times. They were followed by "tails" of aerogel, which over time did not dissipate in the atmosphere like smoke, but on the contrary, compressed into thin threads. Examination of the aerogel under a microscope showed the presence of a mesh structure consisting of chains and agglomerates of light gray and dark particles measuring about 1 micron in diameter.

In order to clarify the necessity of the presence and role of polymers in the formation of LLO, another method for obtaining such objects was proposed in Pirozersky et al. (2014), in which a capillary spark gap is not used, and there was no entry of polymer material into the erosive plasma. The objects were created using an erosive discharge between tin alloy electrodes and were similar in almost all properties to the LLO from the work (Emelin et al. 1997) described above. In Pirozersky et al. (2014), using high-speed videography, studies of the dynamics of objects were carried out, and the interaction of objects with various targets was studied. The structure of the aerogel formed in the active phase of the LLO existence in a wide range of length scales has been studied by optical and electron microscopy methods. It is shown that the aerogel is a fractal network of tin oxide nanoparticles SnO_2 connected by filamentous nanocrystals (nanowiskers), also consisting of SnO_2. The characteristic sizes of the nanoparticles were 35–90 nm, some of them had clearly defined faces and edges, which indicates their crystal structure. The diameters of typical nanowiskers were in the range of 10–40 nm, while their length could reach 1 micron or more. When LLO is obtained using the method proposed in Pirozersky et al. 2014, the apparent specific energy input into the discharge, calculated as the ratio of the electrical energy received from the capacitive storage to the decrease in the mass of the electrodes, is about 20 eV/atom. This value significantly exceeds the maximum possible specific energy density of LLO due to the chemical and thermal mechanism.

Therefore, questions naturally arise about what proportion of the energy of the accumulator is actually transferred to the discharge plasma, what is the value of the specific energy of the LLO itself, and whether there are energy accumulation mechanisms in the LLO that differ from chemical and thermal. These issues require further study.

Conclusions to the section.

Experiments with erosive and capillary discharges leading to the production of LLO with high energy and long lifetime are considered. In experiments with metals and composite materials (a mixture of Si with organic materials), LOOs are realized. The existence of which sometimes ends with an explosion. According to the totality of their properties, with the exception of their small size and the absence of an electric charge, they can be considered analogues of BL.

4.7 Electric Discharges in Liquid

Studies of electrical discharges in liquids began mainly in the 90 s of the twentieth century. They were stimulated by repeated observations of luminous objects near ships in the seas and oceans. We talked about them in Chap. 2. At the same time, the nature of these objects remains unknown. In addition, the conditions for the creation of luminous objects in a liquid may differ fundamentally from the conditions for their creation in a gas environment. Conducted experiments (Klimov et al. 1994; Dimitrov et al. 1994) showed that varieties of LLO can appear in pure liquefied gases (nitrogen, argon, helium, etc.) or in their saturated vapors with the help of various sources of excitation: electric discharges, jets, UV illumination, etc. Promising media turned out to be water, saturated water vapor, and liquid nitrogen. A high degree of excitation of the electronically excited states of molecules occurs in them. Large relaxation times of metastable molecules. The low temperature of the cryogenic liquid slows down relaxation processes and promotes condensation of excited states. Let us consider the results of some of the works on this topic.

4.7.1 To reproduce the conditions for the occurrence of fireballs on Silberg PA submarines (1965), Golka conducted experiments using a 150 kW transformer providing a current of 10 kA at a voltage of 15 V at a frequency of 60 Hz. In his experiments, (Golka 1994) produced a small LLO. When a copper cable with a thickness of 2.5 cm and an aluminum plate with a thickness of 3–6 mm were short-circuited under water, LLOs appeared on the surface of the water in the form of fireballs with a diameter of 6 mm. Hissing, they slid across the surface. Sometimes glowing spheres of irregular shape rose in the air, leaving spiral smoky traces. Such spheres could be obtained without water, but in this case, they dispersed at much higher speeds, and it was difficult to observe them. These spheres were glowing with a bright white light; the reason for this could be slow burning of aluminum. Yellow fireballs appeared when the iron electrodes came into contact. Their glow lasted 2–5 s, and at the end of their life, their size decreased. The cooled residuals were metal spheres with a diameter of about 2 mm, coated with oxides in the form of small hairs.

4.7.1 Experiments (Klimov et al. 1994, 2006; Dimitrov et al. 1994; Kirko et al. 2014) Devoted to the Appearance of LLO in Cryogenic Liquids

In experiments of Klimov in 1994, studies were conducted on the generation of compact long-lived excited formations (CLO) in liquid nitrogen and its vapors using various excitation sources at the installation, see Fig. 4.47.

It consisted of: a cuvette made of foam with a size of $6 \times 26 \times 35$ cm^3, a pulsed plasma torch of an erosive type, a powerful-pulsed UV source; electric spark arresters; a powerful pulsed laser ($\lambda = 1.04$ μm), a microwave interferometer, a PEA. Electric probes and a camera with neutral optical filters were used as diagnostic tools.

During the experiment on the excitation of liquid nitrogen using a pulsed plasma torch, three types of CLO were registered in it: CLO_1—liquid nitrogen and vapors uniformly faintly glowing throughout the cuvette; CLO_2—bright spherical plasmoids with a diameter of up to 5 cm; CLO_3—dark spheroids with a diameter of 1–2 cm.

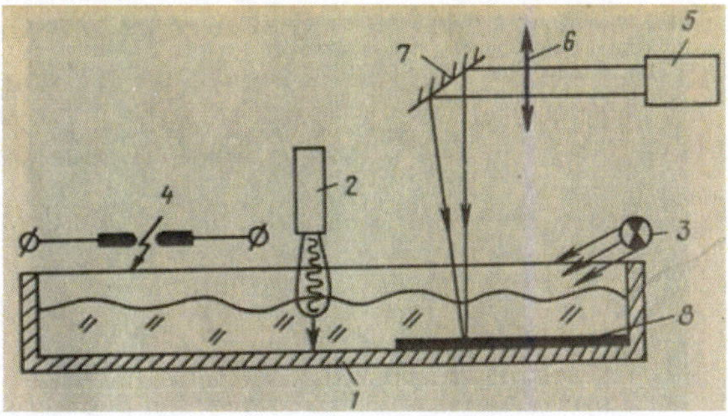

Fig. 4.47 Scheme of the experimental setup. 1—cuvette for liquid nitrogen, 2—pulsed plasma torch, 3—pulsed ultraviolet lamp, 4—spark gap; 5—pulsed laser ($\lambda =$ 1.04 μm), lens, 7—mirror, 8—target

The illumination time of CLO_1- was 10–20 s, CLO_2- 20–40 s. The lifetime of ClO_3 did not exceed 1–2 s. The color of CLO_1 was light blue and the color of CLO_2 changed during evolution from bright white to rich purple. The color of CLO_3 was dirty gray. The energy stored in CLO_2 could reach 100–200 J. Note that CLO_2 and CLO_3 appeared irregularly in the experiment.

CLO_2 and CLO_3 were also formed in nitrogen vapor. In this case, CLO_2 had a distinct halo, the outer border of which reached 10–20 cm.

All three types of CLO were also well obtained using electrical dischargers located near the surface of liquid nitrogen. In addition, CLO_1 was obtained using intense UV radiation, and the intensity of its glow was higher than in the case of a pulsed plasma torch.

However, it was not possible to get CLO_2 and CLO_3 in this way. Based on these data, a conclusion was made about the various mechanisms of the effect of plasma jets, discharges, and external UV radiation on liquid nitrogen.

In the work (Klimov et al. 2006), "strange traces" of luminous particles created by a pulsed generator of erosive plasma in an external electric field were observed. The size of the particles was 10–100 microns, they moved at a speed of 20–200 km/s, had an electric charge, and, upon impact with the target, emitted soft X-ray radiation directed forward. The characteristic glow time of these particles (~1 s) significantly exceeded the glow time of the cluster plasma at the output of the plasma generator (~10 ms). Luminous particles flew out of a capillary dielectric channel with a diameter of 1–2 mm with walls of organic glass, paraffin, or wax. They could be collected in a cuvette with liquid nitrogen. The analysis showed that they consist of pyrolytic graphite. In other experiments, aluminum powder with a grain size of 10–100 microns was mixed into the flow coming out of the channel. Burning of these particles occurred at the exit of the nozzle. They turned out to be negatively charged and sometimes exploded due to Coulomb's forces. In the spectrum of burning aluminum particles, along with the Al and AlO lines, Na, Li, and K lines were observed, and the latter appeared with a delay of more than 10 ms. The radiation in the quasi-continuum in the spectra was amplified approximately during the same time, which corresponded to the equilibrium radiation of heated solid-state particles.

It should be noted that there is no direct analogy of the obtained objects with BL, however, they may have common mechanisms of formation and existence with BL.

4.7.3 Kirko and Dimitrov group in 1994 and 2014 conducted sequential studies of the formation of luminous objects in cryogenic liquids. In experiments (Kirko et al. 2014b), the following studies have been conducted. The scheme of experimental setup is represented in Fig. 4.48. Liquid nitrogen was placed in a thick wall vessel (1).

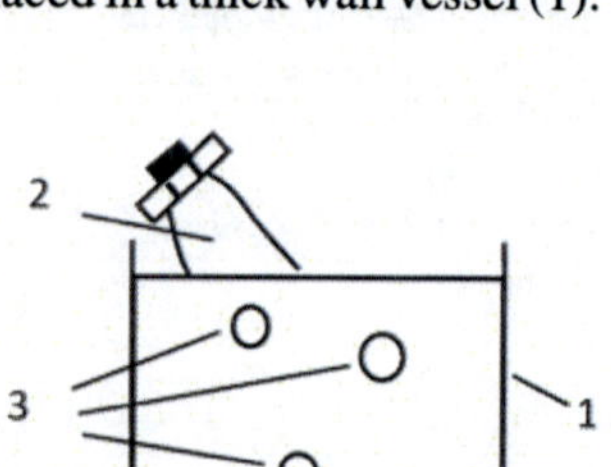

Fig. 4.48 Scheme of the experimental device: 1—a vessel with liquid nitrogen, 2—capillary discharge, 3—luminous regions

An impulse discharge (capillary (2) or arc) was placed near liquid nitrogen surface (at a distance of 2–4 cm). Besides, the arc discharge may be switched on inside a liquid nitrogen volume. Visual observations were realized in a dark room. For photo and filming, Nikon 1G (time distribution 1 ms) was used. After the discharge switching on inside the liquid nitrogen vessel, a volume luminescence was observed, which was weakened sufficiently rapidly. Then in liquid nitrogen volume luminous regions (CLO) (Fig. 4.48) of spherical, stretched, and irregular forms appeared. CLO dimensions were 0.1–3 mm in volumes $(3–5)·10^2$ cm^3 and to 0.5–2 cm in volumes $(2–5)·10^3$ cm^3. Its lifetime was in the range of 10–40 s. The color of luminescence was dark blue (blue, violet), rarely yellow. The spectrum of liquid nitrogen volume luminescence and CLO luminescence was in a range of 380–560 nm with a maximum wave length of 430 nm. CLO moved mainly along the volume due to the acting of boiling liquid flows. After collisions with a vessel wall, CLO continued to exist. At action of a strong metallic pivot, the CLO disappeared. Also, CLO disappeared if one was trying to take out it from the liquid nitrogen volume. In our experiments, the presence of strong electrical and magnetic fields of CLOs was not registered, this may mean that they have a compensated structure. Kirko group in 2014 conducted sequential studies of the formation of luminous objects in cryogenic liquids. The scheme of experimental setup is represented in Fig. 4.48. Liquid nitrogen was placed in a thick wall vessel (1). An impulse discharge (capillary (2) or arc) was placed near liquid nitrogen surface (at a distance of 2–4 cm). Besides, the arc discharge may be switched on inside a liquid nitrogen volume. Visual observations were realized in a dark room. For photo and filming, Nikon 1G (time distribution 1 ms) was used. After the discharge switching on inside the liquid nitrogen vessel, a volume luminescence was observed, which was weakened sufficiently rapidly. Then in liquid nitrogen volume luminous regions (CLO) (Fig. 4.48) of spherical, stretched, and irregular forms appeared. CLO dimensions were 0.1–3 mm in volumes $(3–5)·10^2$ cm^3 and to 0.5–2 cm in volumes $(2–5)·10^3$ cm^3. Its lifetime was in the range of 10–40 s. The color of luminescence was dark blue (blue, violet), rarely yellow. The spectrum of liquid nitrogen volume luminescence and CLO luminescence was in a range of 380–560 nm with a maximum wave length of 430 nm. CLO moved mainly along the volume due to the acting of boiling liquid flows. After collisions with a vessel wall, CLO continued to exist. At action of a strong metallic pivot, the CLO disappeared. Also, CLO disappeared if one was trying to take out it from the liquid nitrogen volume. In our experiments, the presence of strong electrical and magnetic fields of CLOs was not registered, this may mean that they have a compensated structure.

A condition explanation of the origin of ball luminous regions acting is an important moment of research. It was found that the significant fact is the presence of ultraviolet radiation in exiting discharges in regions with 300–360 nm.

The excitation process has a threshold and appears when energy exceeds a value $E \approx 20$ mJ (pulse duration $\Delta t = 5–10$ ms). Other main parameters are material, dimensions (volume value), and a form of liquid nitrogen vessel. The best results were obtained in aluminum vessels, this, perhaps, is explained by the high reflecting

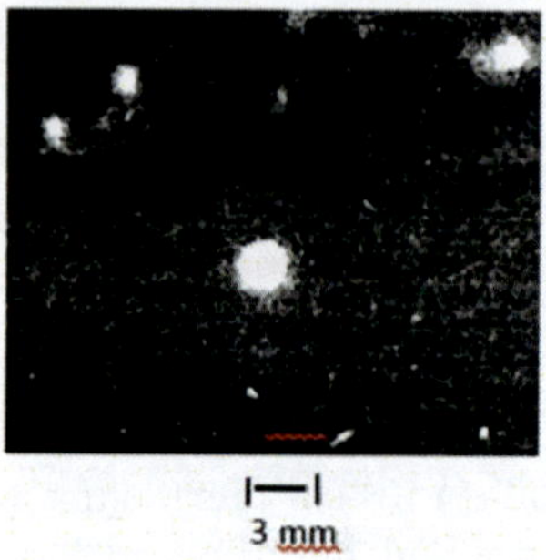

Fig. 4.49 Photo of luminous regions: ball luminescence near vessel bottom (exposition time $\Delta t = 10$ s)

characteristics of aluminum for UV radiation. At the same time, formation of the spherical regions was observed also in brass and still vessels. A minimum volume of vessel ($V \leq 2{\cdot}10^2$ cm^3) exists, below which CLO appearance was not observed. As volume increases, the effect strengthening is the main tendency. We fixed an influence of little metal grit (dimensions 0.05–0.5 mm) in liquid nitrogen volume, for example, drops of molten metal cooling of arc discharge arising. These metal particles can act as original condensation centers during CLO origin. The main region of LR appearance is a liquid nitrogen layer placed near a vessel bottom (Fig. 4.49).

4.7.4. "Golubnichiy's experiments". In the work of Golubnichiy in 1990, the results of relaxation of the expanding plasma formed by high-voltage discharge in water are described. To obtain a discharge lasting 2 microseconds, a Marx generator with a voltage of 60 kV and energy stored in capacitors equal to 30 J was used. The discharge chamber filled with water had a volume of 1.5 dm^3. The discharge was monitored through windows located on the axis of the camera. Under the influence of the discharge, a pulsating cavity with a diameter of up to 2.5 cm was formed in the water. It was revealed that sometimes the plasma in the cavity did not disappear, but broke up into separate luminous objects with a size from 0.1 to 1 mm. Their number in different experiments ranged from several units to several dozen. Some objects passed through the walls of the cavity into the liquid and existed in it for about 0.5 ms. The study of the luminescence spectrum of objects showed that when the wavelength changes from 550 to 750 nm, an increase in the intensity of the glow is observed, while the course of the dependence approximately coincides with the type of the short-wave wing of the blackbody spectrum at T = 2850 K. However, in the wavelength range from 750 to 900 nm, the radiation intensity drops sharply. There are no lines and bands of atoms and molecules in the radiation spectrum. The authors believe that the glow cannot be caused by the radiation of heated solid particles, since photometric measurements of images have shown that glowing objects are optically transparent. Later, the researchers included a device in the installation that allows the decomposition products of water plasma to be released into the air (Golubnichiy et al. 2006a, b). When objects emerged from the water, their size increased 3–5 times (from 1 to 5 mm), and the glow time reached 0.1 s. Sometimes, it was possible to observe how spherical objects were formed during about 300 microseconds from a luminous cloud consisting of the decomposition products of water plasma. These objects were destroyed when passing through the hot air zone (T = 800–1000 °C), and, conversely, condensation of objects was observed on the surface of the plate cooled to the temperature of liquid nitrogen. Sometimes objects exploded, while the speed of the scattering parts exceeded 50 km/s. Based on the observation of the behavior of these objects, the authors (Golubnichiy et al. 2006b) suggested that they can be porous loose bodies, the "skeleton" of which consists of unusual metastable compositions of hydrogen and oxygen. During the transition from the metastable to the ground state, energy is released.

4.7.5. Experiments of T. Matsumoto. Matsumoto in 1997 and 1999 conducted studies on spark discharge in tap water and in electrolyte solutions. In one modification of its installation, a wire was used as an electrode; it was made of lead, cadmium, iron, or gold with a diameter of 1 mm. It was immersed with one end in a solution of KOH (1.5 mol/l) or cesium carbonate Cs_2CO_3 (0.6 mol/l) to a depth of 3 mm. As another electrode, a 25 × 25 × 25 mm box assembled from polished copper plates was used. A box with a wire inserted into it stood at the bottom of a Petri dish, in which there was an 8 mm thick layer of electrolyte. Alternating current pulses with a voltage of 120 V (50 Hz) and a duration of 20 ms were applied between the electrodes, the time between the pulses was 20–30 s. Traces of a round shape with a diameter of 20–200 microns were found on the copper plates. The same traces appeared in films with nuclear emulsion, which were located around the discharge zone. Matsumoto believes that microscopic BL left these traces. In his opinion, they are clusters of highly compressed hydrogen atoms. The binding of atoms in such a cluster is provided by the exchange forces of attraction of degenerate electrons. These clusters are able to pass through dielectric plates, along thin copper wires, jump away from obstacles and explode at the end of life.

4.7.6. "Urutskoev's experiments". In (Urutskoev et al. 2000), are described the results of studies of an electric explosion of metal foils with a thickness of 50 microns and a size of 1 cm × 5 cm located in a sealed tank filled with water. Eight tanks were used, into which the discharge energy of 50 kJ was introduced at a pulse duration of 60–150 microseconds. Spheres with a diameter of 10–15 cm appeared above the tanks during discharge, they burned for 5 ms. The photographs of the frame-by-frame shooting of the process show that the initially diffuse glow of the spheres changes to the radiation of bright microparticles filling their volume. This can be explained by the process of condensation of metal vapors in plasma that has penetrated into the atmosphere through the destroyed vacuum sealing of the tank.

4.7.7. "Nesterovich's experiments." One of the promising ways for changing the physical properties of materials is the method of electric spark action, which can be implemented, in particular, by forming a periodic (quasi-periodic) high-voltage discharge in a liquid stream (PRP) using a voltage source with a falling volt-ampere characteristic. In method of Bogdanovich in 1996 and Nesterovich in 1999–2000 the use of a liquid jet makes it possible to simultaneously implement the functions of a current-carrying electrode and a current switch, with the help of which sequences of electric discharge pulses are formed with a duration of about several nanoseconds with a pulse packet repetition rate of up to several tens of Hz. Under conditions of a fast-flowing electric discharge in the near-electrode space and on the surface of a flat electrode, conditions with high intensity of electric and magnetic fields, as well as powerful shock-wave effects, are created. In the course of their work to determine the optimal treatment modes for materials of various compositions, attention was drawn to the appearance of structural changes in materials that are uncharacteristic of other methods of exposure (laser, ion-plasma, electronic, etc.). These changes consist, in particular, in the presence of extended formations of various shapes and sizes on the surface of the treated samples.

Studies of the electrical and optical characteristics of the discharge were carried out by two methods: the method of obtaining a two-dimensional image and the method of obtaining a stereo image using high-speed cameras. A high-speed filming camera SCS-1 M was used with a shooting speed from 150 to 4000 frames per second. To obtain a stereo image, the shooting was carried out with two synchronized cameras located at an angle of 90 degrees with respect to each other.

The obtained results allowed to establish the following properties of the discharge. The discharge occurs and begins to develop on the outer surface of the jet in a space saturated with water vapor, ozone, and other decomposition products of water and the atmosphere. Under certain conditions, formations may occur in the near-electrode space, identified by the authors presumably with fractal clusters, which may have both a spherical shape and the shape of a polyhedron. These formations can be translucent and have a brightness comparable to the main discharge. Sometimes it was possible to identify their granular structure. In some cases, luminous current channels from the main electrode to these formations were observed. The decoding of the film allowed to assert that the time of their existence is longer than the time of the current pulse, and is at least 0.2–0.3 s for spherical objects, and more than 2–3 s for polyhedra. In addition, it was noted that the formation of polyhedra (having almost the same size) occurs more intensively with increasing discharge power.

During the discharge, objects of a more complex shape are also observed, which have dimensions of 2–4 mm and can exist much longer than the discharge period (up to 5 min), transforming during this time. In some cases, torus-shaped objects have a fairly uniform structure externally and at the same time reach sizes of 4–10 mm in diameter. It is possible to form luminous formations on the surface, and above them single and paired objects with a clearly defined geometry.

Comment to the section.

In conclusion to the section, it should be noted that in the experiments considered, several types of LLO were obtained under different conditions, different shapes, and lifetimes. The possibility of their implementation in experiments confirms that the observation of luminous formations in the seas and oceans is not an illusion and requires additional research and observations. With the exception of LLO received by Golka, there is no simple analogy of the obtained objects with BL, however, they may have common mechanisms of formation and existence with BL in highly nonequilibrium and anomalous geotectonic conditions.

4.8 Experiments on Modeling of BL Features

Along with the attempts described above to obtain artificial BL in experiments on modeling its individual properties, using processes known in engineering and engineering are no less interesting. In this simulation, there is a problem whether we understand this property of the BL correctly. With a successful model, new data about the subject appears.

4.8.1 Corona Discharge for Modeling

If one imagines a BL as an object with a charged surface or a charged ball, then it will show features. As is known (Landau and Lifshitz 1982) the electric field affects the behavior of the charged surface of the liquid. It attracts the surface of the liquid against gravity and surface tension, while the movement of the disturbed surface of the liquid in a homogeneous electric field causes the development of electrostatic instabilities. Under the action of strong homogeneous field on the charged liquid is observed instability of Tonks—Frenkel, which is realized under the condition of (Landau and Lifshitz 1982)

$$E^2 > 2\sqrt{\rho \cdot g \cdot \alpha}/(\varepsilon\,\varepsilon_0), \tag{4.8.1}$$

where E is the electric field strength at the surface of the liquid, ε_0 the dielectric constant of air, α—surface tension of the liquid, ρ—its density. This instability may explain the appearance of columns and jets.

Another effect may be associated with the appearance of the pondermotive force near the charged surface. As is known, a pondermotive force acts on a dielectric sample placed into the inhomogeneous electric field, the density of which is described by the formula (Aleksandrov et al. 2011)

$$f = \frac{\varepsilon_1 \varepsilon_0 - \varepsilon_0}{2} \operatorname{grad} E^2, \tag{4.8.2}$$

where ε_1, ε_0—are the dielectric constants of the dielectric sample and vacuum, respectively. One can see from the formula that the direction of the force acting on the dielectric does not depend on the direction of the field; it is always directed toward the maximum increase in the electric field strength. In this case, the dielectric moves to the region of the greatest electric field strength. In our case, it creates a column. For example, for charged sphere, this force is

$$f = \frac{\varepsilon_1 \varepsilon_0 - \varepsilon_0}{2} Q^2/(4\pi\,\varepsilon\,\varepsilon_0)^2 \cdot -\frac{4}{r^5} e_r,$$

where Q—is the electric charge of the sphere, r—is the distance from the center of the sphere, e_r—is a unity vector directed normally to the surface. This formula shows that the pondermotive force attracts dielectric particles to the sphere. Most evidently, these effects can be demonstrated in experiments with corona discharges over liquids and dispersed media.

Structures Over Charged Surface in Corona Discharge with Liquid or Disperse Electrode

The schematics of the experimental setup are shown in Fig. 1a and the photos of the experimental implementation are in Fig. 4.50.

The installation consists of a cuvette filled with liquid (water, alcohol, kerosene, etc.) and an electrical circuit. The upper electrode or a set of electrodes with a diameter of 0.9 mm (tip radius 0.2 mm) or 2 mm (tip radius 0.4 mm) was located at a height of 5–15 mm above the surface of the liquid. The electrodes were under positive or negative potential.

Experiments of Aleksandrov and Bychkov have shown that in the case of a corona discharge, the electric field under the needle is strongly inhomogeneous. At high electric fields, the development of Tonks-Frenkel instability takes place. Figure 4.51 shows structures that appear under the action of an electric field over the surface of butyl glycol when it is charged in a corona discharge.

Figure 4.51 clearly shows the structures above the liquid surface: columns, funnels, and cylindrical waves. Figure 4.52 shows a photograph of the impact of a multi-needle corona discharge on the surface of a clay powder. The formation of structures in the inhomogeneous electric field of the corona discharge is visible.

It should be noted that similar manifestations occur in the case of the burning of thin wires covered with a porous layer of Al_2O_3. The appearance of jets occurs due to a violation of the structure of the oxide shell as a result of cracking and tearing of the oxide shell by metal vapors and diffusion of the latter into the gas phase.

In real BL with the participation of charged surfaces and combustible materials, the appearance of jets and structures can go simultaneously.

4.8.2 Levitation of Corona-Forming Spheres

Experiments on the observation of the levitation of corona-forming spheres in the region of a flat capacitor are described in (Aleksandrov et al. 1990). The sphere with a diameter of 12 cm was made of a steel micro-wire with a thickness of 20 microns, the weight of the ball was 0.03 g. A vertically directed electric field was created between two parallel metal disks with a diameter of 120 cm, located at a distance of 100 cm from each other. A negative voltage of 20–160 kV was applied to the upper

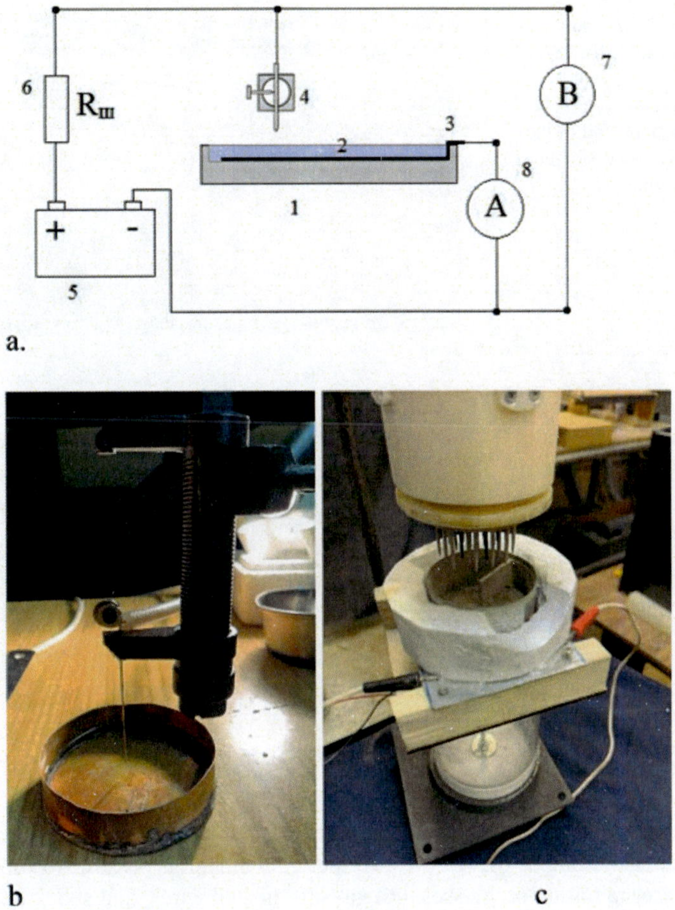

Fig. 4.50 Realization of the corona discharge. **a** The scheme of the experimental installation. 1—the cuvette into which the liquid is poured 2, 3—the second electrode (non-corona), 4—the first electrode—corona, 5—a high voltage source, 6—Rh—ballast discharge resistance to limit the current in the discharge circuit, 7—voltmeter to measure the voltage drop on the corona discharge, 8—ammeter to measure the current strength of the corona discharge. **b** A photo of a single needle discharge over a cuvette with liquid. **c** A photo of a lot of needle discharge over a cuvette with powder. Under the cuvette, there is a device for mixing the powder

electrode, the lower electrode was grounded. At voltages above 60 kV, a corona appeared on the surface of the sphere, which oscillated in a vertical direction without touching the electrodes or hung at a certain height between the electrodes. When the uniformity of the field was distorted, it was possible to observe the tendency of the sphere to be drawn into the region of a stronger field. When entering several (2 or 3) spheres into the capacitor, they were lined up along one vertical axis and, connected by a common ion current, performed a group flight in a horizontal plane.

a b

Fig. 4.51 **a, b** Columns, funnel, and waves above the surface of butyl glycol. Negative corona. The distance from the end of the needle to the liquid surface is 7 mm. The voltage is 10 kV

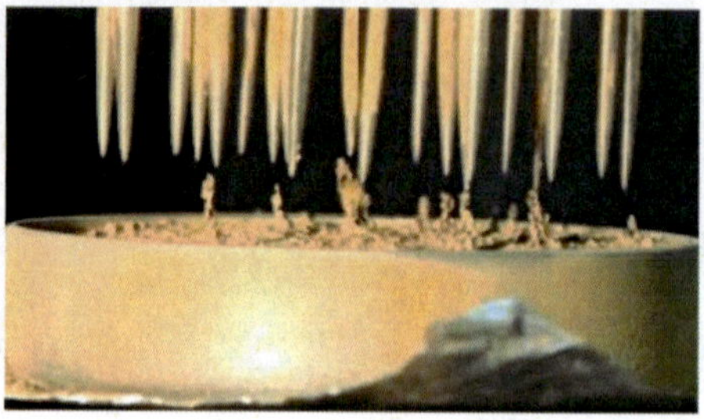

Fig. 4.52 Columns of clay powder in a multi-needle corona

4.8.3 Experimental Modeling of BL Interaction with Glasses

Experiments with modeling the interaction of BL with glasses are of interest for general understanding of BL features. At the same time, it is possible to simulate the effect of laser radiation on glasses, hot gas jets, heated wires, and shock devices.

One can think that the most likely mechanism for the formation of holes in the glasses in the case of their excitation by BL was their destruction due to local heating. This phenomenon has long been established.

Kolosovsky in 1981 conducted experiments on the formation of holes in glasses by heating with a CO_2 laser with a power of 500 watts. The glass was heated until a hole began to appear, after which the laser was turned off. When the glass was cooled, a round crack appeared, which captured the area of the molten glass. A disk with a molten hole bounded by a circular crack fell out of the glass spontaneously or after it was hit. The edges of the holes were shaped like a cone, with a larger diameter on the side facing the laser. At a lower power, the glass did not melt but softened. At the same time, multiple cracks appeared during the cooling process, and "the resulting hole had an irregular shape with sharp fragmented edges."

Similar results were obtained by Kolosovsky when glass was heated by the flame of a gas jet. It was found that the diameter of the hole is 1.4 times larger than the diameter of the heating area. To form a 3.5 cm diameter hole in a 2.5 mm thick glass, it was necessary to spend about 20 kJ of energy on heating the glass.

Valuable results obtained by Kolosovsky concerned the requirements for the power released by the BL and the mechanism of the formation of holes in the glasses. However, the simulation results cannot be considered fully consistent with the results of the natural impact of BL on glass, since in some cases, as shown above, not only the glass with the hole remained intact but also the disk fell out of it.

In this regard, in (Nikitin et al. 2006), see about this (Bychkov et al. 2014), experiments were conducted on local heating of glass using a flat spiral through which an electric current passed. Spirals of nichrome wire with a diameter of 0.7 mm were used. Their outer diameter was 40 mm, the distance between the coils was 3 mm, and the spirals had 3, 4, or 5 turns. The spiral was laid on a glass surface, and a ceramic washer with a diameter of 40 mm and a thickness of 4 mm was placed on top of it. The spiral washer was pressed against the glass surface by a brass cylinder weighing 400 g. Alternating current passed through the spiral, causing the spiral to heat up to red or yellow. The surface temperature of the glass was measured with a chromel-copel thermocouple at a distance of 3–4 mm from the outer spiral edge; to protect it from spiral radiation, it was covered with a thin asbestos sheet.

Figure 4.53 shows the result of heating the glass plate with a current of 12.2 A at a voltage of 16.4 V, its dimensions were 121 × 15 cm, thickness–3.7 mm. The plate was damaged after 37 s of heating. Two fan-shaped cracks were formed; they started from opposite sides of the spiral area of the heated diameter. The central glass part, which has an approximately ellipsoidal shape, remained unchanged.

The dimensions of the ellipsoid axes were 31 and 40 mm. The central fragment of glass has the shape of a truncated cone with an angle of about 50°, with the base of

Fig. 4.53 The result of heating the glass plate with a current of 12.2 A at a voltage of 16.4 V; its dimensions are 12 × 15 cm, thickness 3.7 mm. The plate was damaged after 37 s of heating

the cone facing the surface heated by a spiral. The temperature of the glass, measured before its destruction, was about 400 °C. The type of glass damage resembling that indicated was observed when the power (feeding the spiral) changed from 150 to 400 W, the size of the glass plate changed from 15 × 15 cm to 50 × 50 cm, and the thickness of the glass changed from 2.5 to 3 mm, and when the glass was fixed in the frame state. The average value of the energy that had to be input into the spiral turned out to be equal to 18 ± 6 kJ.

A completely different type of glass damage from the one described above is shown in Fig. 4.54. In this case, the crack core accidentally turned out to be at the edge of the glass located to the left of the heated area. The crack reached the heating area 11 s after the start of heating the glass in a spiral, which was powered by 260 watts. The rest of the glass remained intact throughout the heating process. After 5 min, the current turned off and the glass began to cool down. In the process of cooling the glass, the following cracks were formed: an annular crack, whose diameter (50 mm) is 1.25 times larger than the diameter of the outer spiral, and a right crack, the beginning of which is tangential to the annular one, and cracks on the circular disk. The edges of the hole and disk have the shape of a cone, the cone is wider on the heating side, and its angle is 4–6. Note that energy exceeding 80 kJ was applied to the glass during heating, which led to softening of the glass surface (spiral coils soldered into the glass).

Since, when analyzing unusual holes in glasses, it is often believed that the cause of their appearance may be the impact of a flying object into the glass, an experiment was conducted on punching glasses with metal-headed arrows or lead bullets from an air pistol. Usually, the glass was positioned horizontally and the flying object fell on it vertically from top to bottom. Wooden arrows with a length of 340 mm were used, their metal heads in the form of a hemisphere had a mass of 16 and 21 g. A homemade crossbow-type device was used for archery. The boom velocity values were 15–16 m/s. In most cases, the glass cracked from the impact of an arrow (with a metal head with a diameter of 11 mm) into the glass. At the same time, it cracked

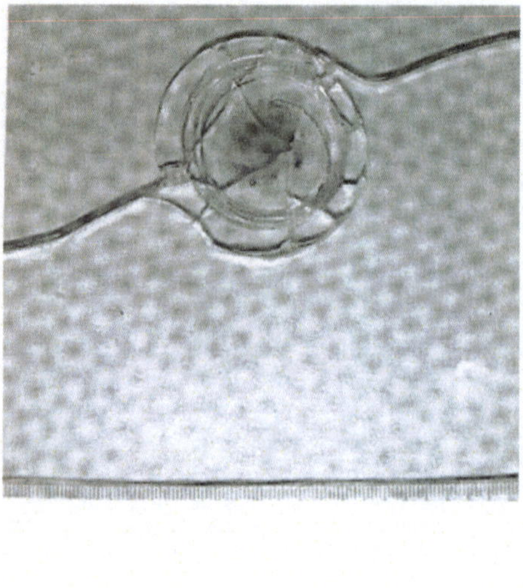

Fig. 4.54 The result of heating a glass plate with a nichrome spiral with a diameter of 4 cm at a power of 260 watts. The left crack appeared after 11 s; the annular crack, the right crack, and radial cracks appeared after 5 min of heating and subsequent cooling. The spiral is soldered into the glass

along several radial cracks, starting from the point of impact, and the central glass part of 8–10 cm in size was broken into small pieces. In rare cases (about one in twenty) there were two or three radial cracks, but the central part of the glass was also destroyed.

In a number of works, it has been suggested that round holes in glasses without cracking can appear when soft objects (for example, a ball) hit the glasses. Our experiments have not confirmed this assumption. Thus, the nature of damage to the glass when hitting the glass with an arrow with a flat rubber head with a diameter of 32 mm and a weight of 28 g remained the same as when hitting the glass with an object with a small surface. At the same time, radial cracks still appeared in the glass, and the central round part with a diameter of 8 cm also crumbled into fragments, but of large size.

In experiments with the impact of a tin bullet from an air pistol on glass, the bullet mass (0.46 g) was 30–40 times less than that of an arrow, and its speed (120–170 m/s) was about an order of magnitude greater. Radial cracks in this case did not have time to reach the edges of the glass, and the central part around the impact site with a diameter of 4 cm was dispersed into powder.

Discussion

Simulation of the interaction of BL with glasses has shown that the methods described above for local heating of glass or impact of a flying object on glass do not lead to results completely similar to the effect of BL on glass.

Heating the glass with a wire spiral usually allows you to keep the central circular disk intact, but the rest of the glass cracks due to thermal stresses. Mechanical stresses in the radial direction during rapid heating of glass by a laser beam (Kolosovsky 1981) are not so strong, since the central region of the molten glass is deformed, it relieves stress, and the glass around the heating area remains intact. However, at the same time, the central disk greatly changes its shape (a hole appears in it, it deforms, etc.). As we said above, this was not observed in the case of the appearance of glass discs as a result of the impact of the BL.

The impact of a solid object on the glass causes the glass to remain intact around the hole, and this probability increases with a decrease in the mass of the object and an increase in its speed. However, the central part always crumbles into small fragments.

As the most resembling traces of BL, we should consider the results presented in Fig. 4.54 when the stresses in the glass caused by heating were removed due to the propagation of an accidental crack.

The results obtained allow us to draw certain conclusions about the properties of BL. A: nothing should prevent the BL from approaching the window glass, most likely, on the contrary, there should be forces that attract it to the glass. B: it should have a minimum energy reserve—of about 20–100 kJ, which can be spent on heating the glass. C: this energy should be transferred to the glass for a short period of time (5–10 s), while the power released by the BL should be at least 2–20 kW. D: the BL should be able to selectively heat one of the glass surfaces, and often not where the BL is located, but opposite. And finally, E: the BL must be able to generate a force pulse that causes the disc to fall out of the glass before cracking the glass left in the frame.

References

Abakumov VI, Bychkov VL, Bikmukhametova AR, Chernikov VA, Golubkov GV, Safronenkov DA, Zaitsev SV (2018) Production of long-lived heterogeneous structures using plasma. J Phys Conf Ser 1112:012022

Abrahamson J, Dinniss J (2000) Ball lightning caused by oxidation of nanoparticle networks from normal lightning strikes on soil. Nature 403:519–521

Abrahamson J (2002) Ball lightning from atmospheric discharges via metal nanosphere oxidation: from soils, wood or metals. Phil Trans Roy Soc Lond 360(1790):61–88

Aleksandrov AF, Isaev KSh, Chernikov VA (1990) Irradiation and chemical composition of erosion plasma leaking to air. Teplophys High temper 28(N3):833–842

Aleksandrov AF, Bychkov VL, Bychkov DV (2011) Electrohydrodynamic peculiarities of corona discharge interaction with a liquid surface. Moscow Univ Phys Bullet 66:390–397

Aleksandrov AF, Chernikov VA, Chuvashev SN, Ershov AP, Shibkov VM, Timofeev IB (1999) Long-lived plasma formations in air. AIAA-99-4977. 9-th Int Space Planes and Hypers Sys Technol Conf 1–4 November, Norfolk, Virginia

Aleksandrov VYa, Borodin IP, Kirichenko EV, Podmoshensky IV (1982a) Rapid coagulation of submicron aerosols into filamentous bulk structures. ZTF 52 N 4:818–820

Aleksandrov VYa, Golubev EM, Podmoshensky IV (1982b) Aerosol nature of ball lightning. ZTF 52 N 10:1987–1992

Alexeff I, Parameswaran SM, Thiayagarajan M, Grace M (2004) An experimental study of ball lightning. IEEE Trans Plasma Sci 32:1378–1382

Alexeff I, Parameswaran SM, Thiayagarajan M, Grace (2005) An observation of synthetic ball lightning. IEEE Trans Plasma Science 33:498–499

Anderson RW (1958) Absence of diffusion in creation random lattices. Phys Rev 109(5):1492–1505

Avramenko RF, Bakhtin BI, Nikolaeva VI, Poskacheeva LP, Shirokov NN (1990a) Experimental study of plasma formations initiated by erosive discharge. ZTF 60(12):73–78

Avramenko RF, Bakhtin BI, Nikolaeva VI, Poskacheeva LP, Shirokov NN (1990b) Investigation of a laboratory analogue of ball lightning. In the collection Ball lightning. 1 Prof Smirnov BM ed. IVAN Publishers, Moscow, p 17

Avramenko RF, Bakhtin BI, Nikolaeva VI, Poskacheeva LP (1994) Energy-intensive plasma formations initiated by erosive discharge are a laboratory analogue of ball lightning. In: Avramenko RF, Bychkov VL, Klimova AI, Sinkevich OA (eds) The Collection of articles ball lightning in the laboratory. Khimia Publishers, Moscow, pp 15–56

Barry JD (1980) Ball Lightning and Bead Lightning. Plenum Press, New York

Brovkin VG, Kolesnichenko YuF, Khmara DV (1994) The principle of structure formation in an initiated microwave discharge and a low-threshold ball discharge. In: Avramenko RF, Bychkov VL, Klimov AI, Sinkevich OA (eds) Ball lightning in the laboratory. Khimia publishers, Moscow, pp 119–135

Bychkov VL (1994) Polymer ball lightning model. Phys Scr 50:591–599

Bychkov VL (2010) Artificial and natural fireballs as combustion objects. IEEE Trans Plasma Sci 38:3289–3290

Bychkov VL (2014) Ball lightning with a cover filled by a vapor. IEEE Trans Plasma Sci 42:3912–3915

Bychkov VL, Anpilov SV, Savenkova NP (2014) Gas-dynamic modeling of the "Plasmoid" created by the Gatchina discharge. Khim Fiz 33(2):58–63

Bychkov VL, Anpilov SV, Savenkova NP, Stelmashuk V, Hoffer P (2018) On modeling of "plasmoid" created by electric discharge. Int Interdiscip Conf "Euler Readings MRSU 2017" IOP Publishing IOP Conf Series: Jour Phys: Conf Series 996 012012 doi:https://doi.org/10.1088/1742-6596/996/1/012012

Bychkov VL, Bychkov AV, Timofeev IB (2004) Experimental production of long-lived luminous organic polymeric objects. Tech Phys 49:128–133

Bychkov VL, Chernikov VA, Osokin AA, Stepanov AI, Stepanov IG (2015) Modeling of artificial ball lightning with a help of capillary discharge. IEEE Trans Plasma Sci 43:4043–4047. https://doi.org/10.1109/TPS.2015.2478441

Bychkov VL, Gridin AYu, Klimov AI (1999) Investigation of the structure and physical properties of polymer plasmoids in the atmosphere. Khim Fiz 18(7):66–72

Bychkov VL, Gridin AYu, Klimov AI (1994) On the nature of artificial ball lightning Investigation of the structure and physical properties of polymer plasmoids in the atmosphere. Teplofiz Vysok Temper 32:190–194

Bychkov VL, Klimov AI, Babaev VG, Lukianov AE (1996) SEM and REM investigation of the framework of polymer plasmoid. Izvestia Russ Acad Sci 60(2):112–115

Cawood W, Patterson HS (1931) A curious phenomenon shown by highly charged aerosols. Nature 128:637

Corum KL, Corum JF (1989) Tesla's production of electric fireballs. Tesla Coil Builder's Association News. 8(3):13–181

Chalmers JA (1949) Atmospheric electricity. Clarendon, Oxford

Chukanov KB (2008) Ball Lightning—source of free energy. In Proc. 10th Int Symp Ball Lightning (ISBL08) and 3rd Int Symp on Unconvent Plasmas (ISUP08). Kaliningrad, Russia, pp 22–26

Dijkhuis GC (1982) Threshold current for fireball generation. J Appl Phys 53(5):3516–3519

Dijkhuis GC (1981) A model for ball lightning. Nature 284:150–151

DikhtyarV, Jerby E (2006) Fireball ejection from a molten hot spot to air by localized microwaves. Phys Rev Lett 96:045002-1-045002-4

Dimitrov SK, Zhdanov SK, Kirko DL (1994) In: Avramenko RF et al. (Ed) Appearance of luminescent formations at interaction of the capillary Influencedischarge torch with water and liquid In Ball lightning in laboratory. Khimia publishers, Moscow, pp 79–86

Emelin SE, Semenov VS, Bychkov VL, Belisheva NK, Kovshyk AP (1997) Some objects formed in the interaction of electrical discharges with metals and polymers. Tech Phys 42:269–277

Emelin S, Bychkov V, Astafiev A, Kovshik A, Pirozersky A (2012) Plasma combustion nature of artificial ball lightning. IEEE Trans Plasma Sci 40:3162–3165

Emelin SE, Pirozerskyi AL, Egorov AI, Stepanov SI, Shabanov GD, BychkovVL (2002) Modeling of ball lightning with a help of electric discharge through a surface of weak water solution. In Mater 9-th Rus Conf Cold Nuclei Transmut Chem Elem and Ball Lightning. 2001. Dagomys, city of Sochi. NITs FTP "Ersion" Publishers, Moscow, pp 240–248

Emelin SE, Pirozersky AL, Vassiliev NN (2006) The dust-gas fireball as a special form of the electric erosive discharge afterglow. Proc 9th Int Symp Ball Lightning (ISBL06). Eindhoven, Netherlands, pp 51–61

Emelin SE, Astafiev AM, Pirozerski AL (2008) Investigation of space-time structure of the discharge with an electrolyte anode and face-type, air half-space directed cathode (Gatchina's discharge). In Proc 10th Int Symp Ball Lightning (ISBL08) and 3rd Int Symp Unconvent Plasmas (ISUP08). Kaliningrad, Russia, pp 42–45

Emelin SE, Pirozerski AL (2006) Some questions of power-consuming plasma-chemical ball lightning. Khim Fiz 25(3):83–89

Ershov AP, Timofeev IB, Chuvashev SN, Bytskevich SP (1990) Evolution of the structure and parameters of a plasma jet during pulsed injection into the atmosphere. Teplofiz Vysok Temper 28:583–589

Ershov AP, Rozanov VV, Sysoev NN, Timofeev IB, Chuvashev SN, Shibkov VM (1994) Physical hydrodynamics. Is. 4, Preprint physical dep. MSU, N.8? Moscow

Evstigneev NM, Ryabkov OI, Shakhparonov IM, Lavrov SI (2011) Some relations on a conductor in the form of a Mobius sheet. Nat Tech Sci N 6:50–58

Ford RA (2002) Homemade lightning. McGraw Hill, New York

Furov LV (2019) On determining the shape of a ball lightning. Mater Conf Wave hydrodynamics of conducting fluid. Long-lived plasma formations and poorly studied forms of natural electrical discharges in the atmosphere. Filigran Publishers, Yaroslavl, pp 111–115

Golka RK Jr (1994) Laboratory-produced ball lightning. J Geoph Res 99(ND5):10679–10681

Golubnichiy PI, Gromenko VM, Krutov JM, Snizhko AP (2006a) Plasma of electric discharge in water as a raw material for laboratory analogue of a ball lightning. In: Proceedings of the 9th International Symposium on Ball Lightning (ISBL-06). Eindhoven, the Netherlands, pp 60–66

Golubnichiy PI, Krutov JM, Nikitin EV (2006b) Long-living luminous objects, formed from the products of disintegration of water plasma. In: Proceedings of the 9th International Symposium on Ball Lightning (ISBL-06). Eindhoven, the Netherlands, pp 67–73

Gromyko AI (2004) On a nature of ball lightning. Fundam investigations 3:62–64

Imyanitov I, Tikhiy D (1980) Beyond boundary of science laws. Atomizdat Publishers, Moscow

Juettner B, Noak S, Versteegh A, Fussmann G (2007) Long-living plasmoids from a water discharge at atmospheric pressure. Proc 28th Int Conf on Phenom Ioniz Gases. Czech Republic, Prague, pp 2229–2234

Kamogawa M, Tanaka H, Ofuruton H, Ohtsuki YH (1999) Possibility of microwave localization to produce an experimental plasma fireball. In Proc Jap Acad 75. Ser B N10:275–280

Kamogawa M, Ofuruton H, Liu JY, Tanaka H, Ohtsuki YH (2004) Study of microwave localization mode for appearance of plasma fireballs. In Proc 8th Int Symp on Ball Lightning (ISBL04). Chung-li, Taiwan, pp 118–119

Kapitsa PL (1955) On nature of ball lightning. Doklady AN SSSR 101:245–248

Kapitsa PL (1969) Free plasma filament in high frequency field at high pressure. ZETF 57 N 6(12):1801–1866

Kirko DL, Savjolov AS, Egorov ID (2014a) Properties of luminous regions under capillary discharge interaction with metalic foils/proc 13-th Int Sympos Ball Lightning (ISBL-14) 22–28th June 2014, Kaliningrad, Russia, pp 34–36

Kirko DL, Savjolov AS, Egorov ID (2014b) Luminescence formation in liquid nitrogen under pulse discharge influence. Proc 13-th Int Sympos Ball Lightning (ISBL-14) 22–28th June 2014, Kaliningrad, Russia, pp 37–39

Klimov AI, Melnichenko AA, Sukovatkin NN (1994) Long living energy intensive excited formations in liquid nitrogen. In: Avramenko RF et al. (Ed) Ball lightning in laboratory. Khimia publishers, Moscow, pp 72–78

Konev EB (1977) Physical bases of the plant materials combustion. Nauka publishers, Novosibirsk

Kolosovsky OA (1981) Investigation of ball lightning trace on the window pane. Zhurnal Tech Fiz 51:856–858

Klimov AI (2003) Plasmoid generation in airflow and its physical properties. Materials 10-th Rus Conf on Cold Nucl transmut Chem Elem and Ball Lightning. Dagomys, city of Sochi, 2002, Moscow 2003, pp 229–235

Klimov A, Moralev I, Zhirnov K, Kutataliev V (2006) Materials 13-th Rus Conf on Cold Nucl transmut Chem Elem and Ball Lightning. Dagomys, city of Sochi, 2005, Moscow 2006, pp 298-306

Kopeikin VV (2011) Fireball - ball lightning. “Trovant” Publishers, Troisk

Kunin VN, Pleshivtsev VS, Furov LV (1997a) Experiments on the nature of ball lightning. Teplofiz Vysok Temper 35(63):866–870

Kunin VN, Pleshivtsev VS, Furov LV (2006a) A method of obtaining of independent long-lived plasma formations in free atmosphere. PMMA were carried out

Kunin VN, Furov LV (1990) Investigations of plasma toroidal vortices appearing at electric explosion of diaphragms in air. Izvestiya VUZov Fiz 33N(6):119

Kunin VN, Pleshivtsev VS, Furov LV (1997b) Experimental laboratory research of the ball lightning nature. In Proc 5th Int Symp Ball Lightning (ISBL97). Tsugawa-Town, Japan, pp 225–229

Kunin VN and Pleshivtsev VS (1999) On the possible mechanism of transformation of toroidal plasma vortex into the laboratory analog of ball lightning. In Proc 6th Int Symp Ball Lightning (ISBL99). Antwerp, Belgium, pp 233–235

Kunin VN, Kondakov VP, Pleshivtsev VS, Furov LV (2006b) Experimental results of anomalous decay of independent long-lived plasma formation in free atmosphere. Khim Fiz 25(3):101–103

Kunin VN, Furov LV (2006) On refraction mechanism of energy accumulation in a volume of long-lived plasma formation. In Mater 13-th Rus Conf Cold Nuclei Transmutat of Chem Elem and Ball Lightning. 2005. Dagomys, city of Sochi. NITs FTP "Ersion" Publishers, Moscow, pp 284–289

Kuz'min RN, Miskinova NA, Shvilkin BN (2003) Laboratory model of ball lightning. In: Materials of 10-th Rus Conf on Cold Nuclei Transmut Chem Elem and Ball Lightning. 2002. Dagomys, city of Sochi. NITs FTP "Ersion" Publishers, Moscow, pp 236–242

Kuz'min RN, Miskinova NA, Shvilkin BN (2006) Laboratory model of ball lightning. Khim Fiz 25 N 3:90–93

Landau LD, Lifshits EM (1982) Electrodynamics of continuous media, Nauka publishers, Moscow

Lazarouk SK, Dolbik AV, Labunov VA, Borisenko VE (2006) Spherical plasmoids formed upon the combustion and explosion of nanostructured hydrated silicon. JETP Lett 84:581–584

Leonov RA (1965) Ball lightning Enigma. Nauka Publishers, Moscow

Lushnikov AA, Negin AE, Pakhomov AV (1990a) Experimental observation of the aerosol-aerogel transition. Chem Phys Lett 175:138–142

Lushnikov AA, Negin AE, Pakhomov AV (1990b) Transition aerosol - aerogel. The collection of Ball lightning prof Smirnov, BM. IVTAN Publizshers, Moscow, pp 11–14

Mikhailov EF, Vlasenko SS, Kiselev AA (1995) Some features of the formation of fractal clusters during the relaxation of aerosol plasma. Fiz Plazmy 21:442–448

Mitsuk AV, Mitsuk VE (1994a) Laboratory ball lightning during discharge in the air. In: Avramenko RF, Bychkov VL, Klimov AI, Sinkevich OA (eds) Ball lightning in the laboratory. Khimia publishers, Moscow, pp 151–153

Mitsuk AV, Mitsuk VE (1994b) Investigation of parameters of an electrodeless microwave discharge of a diffusive type in air at atmospheric pressure Laboratory ball lightning during discharge in air. In: Avramenko RF, Bychkov VL, Klimov AI, Sinkevich OA (eds) Ball lightning in the laboratory. Khimia publishers, Moscow, pp 154–156

Nauer HZ (1953) Angew Phys 5 N 12: 441

Nikitin AI (2006) Electrodynamic model of ball lightning. Khim Fiz 25(3):38–62

Ofuruton H, Ohtsuki YH (1989) Experimental research on ball lightning. Science of Ball Lightning (Fire Ball). Ohtsuki Y-H (ed) World Science, Singapore, pp 310–324

Ofuruton H, Kamogawa M, Tanaka H, Ohtsuki YH (1999) Experiment of the localization of the electromagnetic wave in the waveguide. In Proc 6th Int Symp Ball Lightning (ISBL99). Antwerp, Belgium, pp 207–211

Ofuruton H, Kamogawa M, Tanaka H, Ohtsuki YH (2004) Experiments on microwave localization for ball lightning research. In Proc 8th Int Symp Ball Lightning (ISBL04). Chung-li, Taiwan, pp 120–122

Ofuruton H, Kondo N, Kamogawa M, Aoki M, Ohtsuki YH (1997) Experimental condition of artificial ball lightning by using microwave and discharge. In Proc 5th Int Symp Ball Lightning (ISBL97). Tsugawa-Town, Japan

Ofuruton H, Kamogawa M, Ohtsuki YH (2006) Experiments on ball lightning by using microwave localization. Proc 9th Int Symp Ball Lightning (ISBL06). Eindhoven, Netherlands, pp 157–158

Ofuruton H, Kamogawa M, Ohtsuki Y-H (2008) Experiments for artificial ball lightning by using microwave localization. In: Proc. 10th Int Symp on Ball Lightning (ISBL08) and 3rd Int. Symp on Unconventional Plasmas (ISUP08). Kaliningrad, Russia pp 99–100

Ohtsuki YH, Ofuruton H (1990) Experimental research on ball lightning. Il Nuovo Cimento 13C:761–768

Ohtsuki YH, Ofuruton H (1991) Plasma fireballs formed by microwave interference in air. Nature 350:139–141

Paiva GS, Pavão AC, de Vasconcelos EA, Mendes OJr, da Silva EF (2007) Production of ball-lightning-like luminous balls by electrical discharge in silicon. Phys Rev Lett 98:048501-1-048501-4

Paiva GS, Ferreira JV, Bastos CC, dos Santos MVP, Pavao AC (2010) Energy density calculations for ball-lightning-like luminous silicon balls. UFN 180(2):218–222

Pancova MB, Nesterova TV (2006) Example of unknown nature water solutions activation by plasmoid. In: Proc 2-nd Intern Sympos Unconvent Plasmas (ISUP-06). Eindhoven, the Netherlands, pp 117–122

Pankova MB, Leonov SB, Shipilin AV (1994) Modeling of the interaction of ball lightning with physical phenomena accompanying the flight of bodies in the atmosphere. In: Avramenko RF, Bychkov VL, Klimova AI, Sinkevich OA (eds) The collection of articles ball lightning in the laboratory. Khimia Publishers, Moscow, pp 95–118

Pankova MB, Nesterova TV, Dobrodeev LL (1999) Absorption of high-frequency waves in the propagation zone of a hydrocarbon plasma model of ball lightning. Khimich Fiz 18(7):101–106

Physical values (1991) Reference book. Grigorjev IS, Meilikhov EZ eds. Enrgoatomizdat Publishers, Moscow

Pirozersky AL, Emelin SE (2006) Long-living plasmoids generation by high-voltage discharge through thin conducting layers. Proc 9th Int Symp Ball Lightning (ISBL06). Eindhoven, Netherlands, pp 180–190

Pirozersky AL, Mikhailovsskii VYu, Lebedeva EL (2014) Artificial fireball generation via an erosive discharge with tin electrodes. Int J Appl engine research 9:3162–3165

Planté G (1891) Electrical phenomena in atmosphere. Paris

Powell JR, Filkenstein D (1970) Ball lightning. Amer Sci 58:262

Silberg PA (1965) A review of ball lightning. In Coroniti SC ed. Problems of Atmospheric and Space Electricity, pp 436–466. Proc. 3rd Int. Conf. on Atmosp and Space Electricity. 1963. Montreux, Switzerland. Elsevier, Amsterdam

Shabanov GD (2002) Optical properties of long-lived luminescent formations. Pis'ma ZTF 28(4):81–86

Shabanov GD (2019) About the possibility of creating a natural ball lightning by pulsed discharge of a new type in laboratory conditions. UFN 189(1):95–111

Shabanov GD, Zherebtsov OM (2003) Experimental modeling of ball lightning analog. In Mater 10-th Rus Conf Cold Nuclei Transmut of Chem Elem and Ball Lightning. 2002. Dagomys, city of Sochi. NITs FTP "Ersion" Publishers, Moscow, pp 285–301

Shabanov GD, Zherebtsov OM (2004a) Electric properties of independent luminous formations. In Mater 11-th Rus Conf Cold Nuclei Transmut Chem Elem and Ball Lightning. 2003. Dagomys, city of Sochi. NITs FTP "Ersion" Publishers, Moscow, pp 279–290

Shabanov GD, Zherebtsov OM (2004b) Electric discharge to air half-space. Optich Zhur 71(1):6–8

Shabanov GD, BYu, Sokolovskyi (2005) Macroscopic separation of charges in the pulsed electric field. Fizika Plasmy 31:560–566

Shabanov GD, Zherebtsov OM, Sokolovskyi BYu (2006a) Experimental modeling with a help of pulsed electric discharge to air half-space of linear lightning leader and ball lightning. In Mater of 13-th Rus Conf Cold Nuclei Transmut Chem Elem and Ball Lightning. 2005. Dagomys, city of Sochi. NITs FTP "Erzion" Publishers, Moscow, pp 246–265

Shabanov GD, Krivshich AG, Sokolovsky BYu, Zherebtsov OM (2008a) Ball lightning nature. In Mater 14-th Rus Conf Cold Nuclei Transmut Chem Elemand Ball Lightning. 2006. Dagomys, city of Sochi. NITs FTP "Erzion" Publishers, Moscow, pp 243–254

Shabanov GD, Krivshich AG, Sokolovsky BYu (2008b) In Mater 14-th Rus Conf Cold Nuclei Transmut Chem Elem and Ball Lightning. 2006. Dagomys, city of Sochi. NITs FTP "Erzion" Publishers, Moscow, pp 255–266

Shabanov GD, Krivshich AG, BYu, Sokolovsky, Zherebtsov OM (2006b) Performance of the laboratory of ball lightning. Proc 9th Int Symp on Ball Lightning (ISBL06). Eindhoven, Netherlands, pp 202–209

Shabanov GD, Zherebtsov OM, Sokolovsky BYu (2006c) Independent long-lived luminescent formations in open air. Experimental check of ball lightning formation by the linear lightning leader hypothesis. Khim Fiz 25 N.4:74–88

Shakhparonov IM (1994) The use of undirected contours in the generation of ball lightning in laboratory conditions. In: Avramenko RF, Bychkov VL, Klimova AI, Sinkevich OA (eds) Ball lightning in the laboratory. Khimia Publishers, Moscow, pp 15–56

Shibkov VM, Alexandrov AF, Kuzovnikov AA (1994) Freely localized microwave discharge in the air. Avramenko RF, Bychkov VL, Klimov AI, Sinkevich OA (eds) Khimia publishers, Moscow, pp 136–150

Singer S (1971) The nature of ball lightning. Plenum Press, New York

Slyusarev NM (1990) Ball lightning and wood. In the collection. Ball lightning prof. Smirnov BM ed. IVAN Publishers, Moscow, pp 18–29

Smirnov BM (1988a) Physics of fractal clusters. Nauka Publishers, Moscow

Smirnov BM (1988b) The problem of ball lightning. Nauka Publishers, Moscow

Stakhanov IP (1985) On physical nature of the ball lightning. Energoatomizdat Publishers, Moscow

Stelmashuk V, Hoffer P (2017) Experimental study of a long-living Plasmoid using high-speed filming. IEEE Trans Plasma Sci 45:3160–3165. https://doi.org/10.1109/TPS.2017.2770224

Stepanov SI (2014) Ultrasonic Probing of the Plasmoid. Zhtf 84:109–114

Stephan KD, Massey N (2008) Burning molten metallic spheres: one class of ball lightning? J Atmos Solar-Terr Phys 70:1589–1596

Tanaka K, Tanaka M (1997) Is ball lightning "Anderson localization"? In Proc 5th Int Symp on Ball Lightning (ISBL97). Tsugawa-Town, Japan. Pp 137–141

Tesla N (1978) Colorado Springs Notes. 1899–1900. Nolit Publishers, Beograd

Urutskoev LI, Lixonov VI, Tsinoev VG (2000) Experimental detection of "strange" radiation and transformation of chemical elements. Priklad Fiz VIMI Publishers, Moscow, N 4:83–100

Versteegh A, Behringer K, Fantz U, Fussmann G, Jüttner B, Noak S (2008) Long-living plasmoids from an atmospheric water discharge. Plasma Sour Sci Tech 17(2):02414–02421

Vinogradov GK, GZh, Imanbaev, Polak LS, Slovetsky DI (1983) Experimental study of the formation of polymer films in nonequilibrium fluorocarbon plasma. Khim Vysok Energ 17:372–377

Vlasov AN (1997) A ball lightning is a natural nuclear reactor. In Proc 5th Int Symp Ball Lightning (ISBL97). Tsugawa-Town, Japan, pp 75–79

Vlasov AN (1999) Bose-Einsten condensed electron catalyzed fusion as a possible mechanism of the ball lightning energy supply. In Proc 6th Int Symp Ball Lightning (ISBL99). Antwerp, Belgium, pp 133–138

Vlasov AN (2006a) Ball lightning as the current layer induced in a vortex. Proc 9th Int Symp Ball Lightning (ISBL06). Eindhoven, Netherlands, pp 243–249

Vlasov AN (2006b) Experimental modeling of ball lightning on a basis of electric explosion of a wirespiral rolled in torus. In Mater 13-th Rus Conf Cold Nuclei Transmut Chem Elem and Ball Lightning. 2005. Dagomys, city of Sochi. 2006b. NITs FTP "Ersion" Publishers, Moscow, pp 290–297

Vlasov AN (2008) Magnetohydrodynamical model of plasma object capable to be generated at impact of ordinary lightning. In Proc 10th Int Symp Ball Lightning (ISBL08) and 3rd Int Symp Unconvent Plasmas (ISUP08). Kaliningrad, Russia, pp 149–150

Vlasov AN, Kolesnikov SA (2008) Calculations of an installation parameters for obtaining of plasmoids at electric explosions of wire spirals rolled in torus. In Mater14-th Rus Conf Cold Nuclei Transmutation of Chem Elem and Ball Lightning. 2006. Dagomys, city of Sochi. NITs FTP "Ersion" Publishers, Moscow, pp 162–179

Vlasov A, Vlasov A (2006) Getting a fireball by means of electric explosion of wire spiral twisted into a ring. Proc 9th Int Symp Ball Lightning (ISBL06). Eindhoven, Netherlands, pp 234–242

Vlasov AN, Dubkov M, Byrobin MA, Manoshkin AB (2015) Formation of dense plasma by electric explosion of toroidally mounted spirals and creation of a submillisecond generator with peak current up to 400 kA. Intern XLII (Zvenigorod) Conf Plasma Phys and CNF Publishers, February 9–13, pp 260

Wojciechowski BV, Wojciechowski MB (1986) Drop-breakdown glow of ball lightning. DAN SSSR 287:331–334

Yasuda H (1985) Plasma polymerisation. Academic Press, New-York

Zhiltsov VA, Leitner JF, Manykin EA, Petrenko EA, Skovoroda AA, Handel P (1995) Spatially-localized discharge in the atmosphere. ZTEF 108 N 6(12):1966–1985

Chapter 5
About Theoretical Hypotheses of BL

Abstract This chapter is devoted to a brief discussion of the theoretical models of ball lightning that have appeared in recent decades. Material models and models of BL as chemical-thermal formations are given.

Keywords Theoretical models · Material models · Chemical-thermal formations

In this section, we give a brief overview of the theories (hypotheses) of BL that existed at the beginning of the twentieth century, and trace what happened to them over the past 20 years. There are quite a lot of review papers on BL theories (hypotheses) (Singer 1973; Barry 1980; Stakhanov 1996; Stenhoff 1999; Smirnov 1993; Bychkov et al. 2010), so our consideration does not include the creation of an encyclopedia of BL models and hypotheses. Our task is to cover mainly material and thermochemical models, which are not given enough attention in the review literature.

5.1 New BL Hypotheses

By the end of the twentieth century, thanks to the statistical analysis of the material on the observations of BL collected by Brand, Singer, Barry, Reilly, McNally, Imyanitov, Dmitriev, Stakhanov, Grigoriev, Kel, Smirnov, and Bychkov, a certain idea of the "average" BL was formed, the Hypotheses of BL are presented in Singer (1973), (Barry 1980; Stakhanov 1996; Stenhoff 1999; Smirnov 1993; Bychkov et al. 2010). It was represented as an object of 20 cm in size with a density equal to the density of air, but, however, capable of having a surface tension like a liquid. This object had no electric charge but was able to carry charges accumulated on objects during a thunderstorm over long distances. The energy of this object was no more than 20 kJ and was compared with the energy of 10 matches, and its energy density was about 30 J/cm^3. The lifetime of the object was about 20 s. Some rare cases of demonstration of the possibility of BL to have high energy, high power of optical and radio emission, and a large charge were excluded as statistically unreliable.

At the same time, the theory of BL with an external energy source was considered viable, the "start in life" of which was given by Kapitsa P.L. (1989). Ball lightning

V. L. Bychkov, *Natural and Artificial Ball Lightning in the Earth's Atmosphere*,
Springer Atmospheric Sciences, https://doi.org/10.1007/978-3-031-07861-3_5

was considered a plasma object formed in the air as a result of the action of powerful radio emission. The advantage of this theory was that it easily explained the cases of "free" passage of BL through glass. The disadvantage of this theory is the absence in nature of an energy reservoir capable of feeding the resulting electric discharge with power for a long time.

Properties of the "average" BL have an "aerogel" model of Smirnov B.M. (Smirnov 1988, 1993). BL is a framework of fractal filaments formed by the condensation of nanoparticles and clusters. In an electric field, the aggregates are connected in threads. The frame has the properties of a solid, liquid, and gas at the same time. Due to the surface tension, it takes the shape of a ball. Its surface remains cold, only individual points, heated to a temperature of 1800 K due to a chemical reaction, shine. Burning starts simultaneously from many points (their number reaches 1000) and spreads in the form of heat waves. The substance of the Stakhanov I.P. "cluster" BL (Stakhanov 1996) (is a mixture of positive and negative ions surrounded by shells of polarized water molecules. He assumed that the presence of shells significantly slows down the process of ion recombination, however, this question is open until now.

Bychkov V.L. (Bychkov 1994, 2002) proposed a polymer model of BL—a structure of polymer molecules, soot, and silicon with fractal structure and with reserve of energy in the form of electric field energy. This model appeared much earlier than the similar model by (Abrahamson and Dinnis 2000), (Abrahamson 2002). The BL as a whole can be electroneutral, but it is capable of removing and transferring electrical charges between various objects. In the strong field created by these charges, an electric discharge occurs near the surface of the BL, providing its glow. In principle, such a structure can also be unipolar. This work was a development of the Smirnov fractal model (Smirnov 1988, 1993) for the case of polymer or composite charged structures. Such a BL in the process of formation accumulates charges from the surrounding gas, strongly ionized in the conditions of lightning electricity. Taking into account the possibility of charge accumulation by polymer films and filaments, and high experimental values of charge density on their surface, it was found in (Bychkov 1994, 2002) that the energy density of such BLcan reach $W = 10^5$ J/m^3, which is much higher than assumed Smirnov and Stakhanov. This model can explain the appearance of BL during thunderstorms near trees that are struck by linear lightning discharges. In the unipolar case, under the action of charges on the structure, the BL can have an arbitrary tangle shape. However, this model cannot explain the existence of BL with an energy density up to $W = 10^{10}$-10^{11} J/m^3. At that time, such high energy densities were not seriously considered by a majority of investigators. In addition, it has difficulties in explaining the preservation of the structure at high concentrations of charges on the surface.

Norman G. E. (Norman 1999, 2000), (Manykin and Norman 1999) believed that BL is a sphere of nonideal plasma with energy densities of the order of $W = 3 \cdot 10^4$J/m^3. Manykin E. A., Ojovan M. I., and Poluektov P.P (Manykin et al. 1981, 1982, 1983, 1999) believed that the BL is "Rydberg" matter—a substance made of atoms with a large value of the principal quantum number in the excited state. Tangles of "Rydberg" matter with a density close to the density of air have the properties of a solid body and have a surface tension. The energy density stored in the "Rydberg" matter is about 1 J/cm ^{3}.

Dijkhuis G.C. (Dijkhuis 1980) suggested that electrons in strong confinement can combine into pairs obeying Bose-Einstein statistics, forming vortex tubes with a radius of 53×10^{-12} m. Positive ions rotate around the tubes, the radius of their orbit is 10^{-7} m. When the electron-ion tubes are stretched, the ions are accelerated. These tubes are stacked within a limited volume so that when adjacent tubes touch, the directions of the ion velocity vectors coincide.

Raniada A. F. and Trueba J. L. (Rañada and Trueba 1996), (Rañada et al. 1999), suggested that the BL may appear at the site of an air breakdown in a system of magnetic field lines free of forces (Force-free field). A magnetic field of this configuration can be obtained by starting a current that creates a second magnetic field along the lines of force of the first magnetic field. The current flows through closed threads. They occupy a small part of the volume of the visible BL. It itself is a ball with a volume of 0.07 cm^3, heated to a temperature of 18,000 K. The ball consists of a streamer with a thickness of 100 microns and a length of several meters, twisted into a ball.

Koloc P. (Koloc 1999a, b) proposed a BL configuration in the form of a "Plasmak", which is a development of the "Tokamak" configuration. According to Koloc, the BL is a plasma torus suspended inside a shell of high-temperature plasma. There are three types of currents flowing in this system: (1) current along the axis of the toroid (toroidal current), creating a poloidal magnetic field; (2) current along the inner surface of the plasma shell, induced by a toroidal current. Its direction is opposite to the direction of the toroidal current; (3) the poloidal current along with the torus generator (equivalent to the current through the coil wound on the toroidal chamber of the tokamak). The stability of the system is maintained by atmospheric pressure, which is transmitted from the shell to the core due to the interaction of magnetic fields.

Vlasov A.N. (Vlasov 2006, 2009) believed that the BL core is a toroidal ring of rotating plasma, inside which there is a current layer of relativistic electrons rotating around the annular axis of the toroid. This current layer creates a toroidal magnetic field. The pressure of this field compensates for the pressure of the atmosphere. There are positive ions between the electron current layer and the region occupied by the magnetic field. When deuterium ions enter the fast electron layer, a fusion reaction of nuclei occurs. The plasma vortex ring is surrounded by a halo of luminous air, which is perceived by the observer as a ball.

Oreshko A.G. (Oreshko 2006) believed that the BL consists of a negatively charged nucleus and an outer ring formed by positive charges that rotate around the nucleus. This ring, due to the Coulomb interaction of charges, causes the rotation of the inner electron core. The rotation of the charges generates a magnetic field of such a configuration that the induction of the magnetic field in the region of the CMM core is minimal. The glow of such a CMM occurs due to the time-stretched electrical breakdown of the region between the separated charges.

According to Shchedrin A.I. (Shchedrin 2018), the BL is an environment consisting of concentric regions of different signs. The inner region consists of electrons moving in orbit at speeds close to the speed of light. The outer region is the positive ions orbiting. The direction of rotation of electrons and ions coincides, so the directions of the magnetic field vectors generated by their motion are opposite. As a result, there is a precession of the magnetic moments of the currents of electrons and ions, the frequency of which is 50 MHz for ions and 150 GHz for electrons. Due to this, the BL emits radio waves. The reason for the visible glow of the BL is the synchrotron radiation of relativistic electrons and the interaction of ions with air in the outer shell.

Ofuruton H. et al. (Ofuruton et al. 1997, 2001) proponents of the BL model developed by Kapitsa suggest that in nature high-frequency field strengths can be formed due to interference of radio waves on objects randomly located in space. By placing ceramic plates inside the waveguide, they actually observed a 100-fold increase in radio emission power in some areas.

According to Handel P. (Handel 1975), excited water molecules can be a reservoir for storing the energy of electromagnetic radiation from lightning discharges. A huge cloud of such molecules serves as the working medium of an atmospheric maser, in the beam of which an air breakdown can occur and a BL can form.

Kawano S. (Kawano 2004, 2006) considered the BL as the corona discharge at the end of the leader of the linear lightning. The tip of the leader can glow directly, as well as a piece torn off from it, fed by the current of streamers.

Matsumoto T. (Matsumoto 1997, 1999a, b) found that when electric discharges occur in water, spherical or annular objects with a size of about 50 microns are formed. They behave like little BLs. According to Matsumoto, these micro-sized objects consist of highly compressed hydrogen clusters.

Lowke J. (Lowke 1996), (Lowke et al. 2012) believed that the BL is a corona-type discharge in the air, supported by an electric field of charges moving in the Ground after a linear lightning strike into it. The ion current of such a discharge flowing to the surface of the earth slowly neutralizes these charges. If the smoldering discharge switches to arc discharge mode, a BL explosion may occur.

Rabinovich M. (Rabinowitz 2001) believed that the BL is a small black hole that happened to be near the Earth. The source of its energy is the energy of gravity, which is radiated in the form of a gravitational field (gravitons) in the direction of another massive body, for example, the Earth. This radiation creates a reactive force that allows the black hole to levitate. For example, a black hole with a mass of 1/3 g will levitate if the power of this radiation is greater than 10^6 W. A small black hole becomes visible due to the excitation and ionization of ambient air molecules by the energy emitted by the hole of electrons and positrons.

Shmatov M. L. (Shmatov 2003) proposed a model of the BL in the form of a spherical skeleton of positively charged ions, relative to which electrons oscillate in the radial direction. The total kinetic energy of electrons in a ball with a radius of 80 cm can reach up to 3×10^9 J, and the lifetime of the BL, determined by the time of energy emission due to electron braking radiation, is from 1 to 5 s. It is assumed that the core of the BL has a small uncompensated positive charge, and the forces of Coulomb's repulsion of charges are compensated by the force of atmospheric pressure on its shell. Criticism of this work and the impossibility of explaining the observed lifetime of the BL is presented in (Bychkov 2008).

In the last decade, there has been a qualitative shift in the science of BL: along with the "usual" information about visual observation, videos have appeared. The result of this shift is difficult to overestimate. Firstly, a support was knocked out from under the feet of the "skeptics" who doubt the existence of the BL, and, secondly, a frame-by-frame analysis of the videos revealed subtle details of the structure and behavior of the BL that were inaccessible to the human eye. In addition, for the first time in the world, an analysis of the window glass was carried out, through which a BL "freely" passed in front of a witness (Bychkov et al. 2016). An area of glass with traces of BL exposure was discovered, which allowed us to draw conclusions about its material nature and the size of its core elements. A unique case of simultaneous video recording of a BL by three operators was analyzed, which made it possible to study its interaction with the electric field of a cloud in the presence of a strong wind (Nikitin et al. 2018a, b). The analysis of the motion of a pair of connected BL was carried out, which allowed to build a theory of this interaction. Based on the video, the process of dividing the BL into two parts was investigated, confirming the fact that it has an electric charge (Bychkov and Nikitin 2014). A snapshot of the burn mark left by the gold chain vaporized by the explosion of the BL allowed us to confirm the assumption about its ability to generate powerful radio emissions (Bychkov and Nikitin 2014).

For some unknown reason, these achievements, based on a solid base of reliable video materials, did not have a noticeable impact on the international team of BL researchers. On the contrary, his great interest was aroused by the construction of a "skimeron" made of rubidium atoms, the re-invention of the BL model in the form of a plasma shell filled with radio waves (Wu 2016) or a new solution of Maxwell's equations in the form of a tangle of a self-enclosed electromagnetic field (Cameron 2018).

Important consequences of this change in the quality of the BL observation results were (1) Proof of the possibility of the existence of a BL with an energy density above 10^{10} J/m^3; (2) Proof of the presence of an electric charge of 10^{-3}-10^{-1} C (moreover, in an explicit form, and not in a "hidden" form, as in Avramenko R.F. (Ball lightning in the laboratory 1994); (3) Its ability, along with light, to emit radio waves in the range of 1–10 cm, sometimes in the form of a powerful pulse (Nikitin et al 2018a, b).

The analysis of the BL models described above allows to come to the conclusion that none of them can explain the complexity of these three properties.

At the same time, models based on the material nature of the BL can cope with this task. Let us start with the justification of the possibility of retaining an electric charge in a small volume. From general considerations, it is clear that a homogeneous system (aerosol tangle, "Rydberg" matter, etc.) cannot hold a large charge: Coulomb forces cannot be resisted by ordinary surface tension. However, a heterogeneous system—a kind of charged nucleus placed in a spherical shell—can quite cope with this task.

5.2 High-Energy BL Hypotheses

Before we move on to high energy BL models, we will consider the questions that need to be answered for the development of these models.

There are several important questions. Why observation shows that BL energy density can reach 10^{10}–10^{11}J/m^3. What BL types can ensure such energy density? How to explain that BL electric charge can reach 10^{-3}–10^{-1} C, and there is no immediate leakage of the charge to the environment. At that, it is well known that the breakdown of electric field in the atmosphere is about $3 \cdot 10^6$V/m Smirnov BM (1988), (Atmosphera 1991). And there is a question: Why are there electric fields on the high voltage tip of the corona discharge $3 \cdot 10^8$ V/m without breakdown (Akishev et.al 1999)?

Let us estimate an energy density near a charged surface. For estimates, we use the simplest equation for electric field energy density (Terletsky and Rybakov 1980) $w = \varepsilon \cdot \varepsilon_0 E^2/2$, where ε_0-is the dielectric constant of vacuum, ε-of air. Inserting $E = 3 \cdot 10^6$ V/m into this equation, one gets $w = 20$ J/m^3 and $E = 3 \cdot 10^8$ V/m $w = 2 \cdot 10^5$ J/m. This shows that high electric field can not explain observed energy density of BL.

Therefore, in this section, we consider mechanisms ensuring BL's high-energy density. In case of the model (Nikitin 2006), it is ensured by moving ions inside BL, and in case of Bychkov (Nikitin et al. 2018a, b), the reactions of evaporation.

We consider that one of the most important features of such BL is the presence of the dielectric shell, which was observed in experiments with erosive discharges (Emelin et al 2012), (Bychkov et al. 2015), (Abakumov et al. 2018). Its presence allows to realize high-energy BL.

5.2.1 *Dipole Force (Nikitin 2006)*

Consider a sphere with an inner radius $R = 5$ cm and a wall thickness $a = 1$ cm, inside which a charge Q is placed. The charge carriers will gather on the inner surface of the sphere and will stretch it with a force $F_q = Q^2/4\pi\varepsilon_0 R^2$. (Here, $\varepsilon_0 = 8.854 \cdot 10^{-12}$ F/m is an electric constant). In the electric field of the charge Q, the matter of the shell will be polarized (we consider it to consist of a dielectric, for example, water) and the formed dipoles will tend to compress the shell with force

$$F_{sh} = 4\sigma a Q/\varepsilon_0(2R + a). \tag{5.2.1}$$

Here, σ is the charge density of the dipoles on the inner surface of the shell, for water $\sigma = 1$ C/m^2. For small values of the charge $Q > Q^2$, it can be expected that $Fsh > Fq$. A sphere of water molecules with $R = 5$ cm and $a = 1$ cm can hold a charge $Q = 1.14 \times 10^{-2}$ C inside itself. At $Q = 5 \times 10^{-3}$ C, the sphere can, in addition to the charge pressure, withstand the force F_k of additional pressure caused by the movement of charge carrier particles. The strength of this pressure $F_k = F_{sh} - F_q = 1.15 \cdot 10^8$ N. The area of the inner surface of the sphere $S = 4\pi R^2 = 3.14 \cdot 10^{-2}$ m^2; hence, the pressure exerted by the charge carriers on the shell, $P = F_k/S = 3.66 \cdot 10^9$ N/m^2.

According to the virial theorem, the kinetic energy density of the particles ρ_E is related to the pressure on the vessel wall by the ratio $\rho_E = 3P$. For our case, $\rho_E = 1.1 \cdot 10^{10}$ J/m^3, which corresponds to the results of measuring the energy density of BL that heated water or destroyed buildings. Of course, the pressure of 10^4 atmospheres seems fantastically large and it is unclear how relatively thin shells can withstand such pressure. This requires special research.

The appearance of such a force ensures the presence of a high charge of the BL, which allows BL to levitate in the electric field of the earth and explain observations with a current shock from the BL.

5.2.2 *Electric Field of a Highly Charged BL (Bychkov et al. 2020)*

Estimates of the electric field strength near charged fireballs having the parameters of the observed high-energy BL raise the question of why the discharge to the ground and its discharge are not realized. A similar question arises in the study of corona discharges. Such studies (Akishev et al. 1999) show that with strong fields near the tip of the charged electrode, fields of the order of 10^8 V/m are created, comparable to the fields at the surface of the BL, and the discharge begins to pulsate (pulsation of the Trichel), and there is no breakdown to the Ground.

Ignorance of these facts is often present when stating that the BL is an uncharged or weakly charged object. The presence of a strong electric field on the BL allows explaining its levitation at a high value of the BL's own weight.

We consider this situation from the point of view of electrostatics and plasma physics. So, let us consider a charged ball in an electric field created by a large proper charge of the ball. Moreover, this electric field on the surface of the ball is large and capable of ionizing air.

It is known from electrostatics that the electric field near a charged ball varies depending on the distance R from it, as

$$E = E_s(r_s/R)^2, \tag{5.2.2}$$

where E_s, R, respectively, are the electric field on the surface of the sphere and the distance to the sphere, r_s is the radius of the sphere. The magnitude of the electric field is related to the electric charge on the sphere Q and the radius of the sphere by the known ratio $Q = 4\pi\varepsilon_0 \cdot r_s^2 E_s$ or

$$E = \frac{Q}{4\pi \cdot \varepsilon\varepsilon_0 \cdot R^2}, \tag{5.2.3}$$

where—ε_0 is the permittivity of the vacuum and ε the dielectric constant of the air.

If the initial electric field is greater than the so-called field breakdown in which the speed of the ionization processes is compared to the rate of loss of electrons, the field on the sphere is associated with a field breakdown E_{br} (in the air $E_{br} \approx 2.8$-$3 \cdot 10^6$ V/m) ratio from (5.2.2)

$$r_{br} = r_s (E_s/E_{br})^{0.5}, \tag{5.2.4}$$

where r_{br} is the distance from the center of the sphere to the boundary of the ionization In the space around a charged sphere is formed by a plasma layer, which screens the electric field of a charged sphere.

At high values of the electric field strength in the air, as calculations (Ardelyan et al. 2014)

(Ardelyan et al. 2017) have shown that plasma is formed in the air, consisting mainly of NO^+ ions and electrons; therefore, the electron concentration for estimates can be determined on the basis of a simplified equation

$$\frac{dn_e}{dt} = q + \nu_i \cdot n_e - \alpha_{dr} \cdot n_e^2. \tag{5.2.5}$$

Here, t is the time, q– the rate of formation of electron-ion pairs in the air due to fast background particles, $\nu_i = \nu_i(E/N)$—the frequency of ionization of air molecules by electrons in an electric field (its value depends on the parameter E/N (N is the concentration of gas particles) can be found in (Ardelyan et al. 2013), (Smirnov 1987), $\alpha_{dr} = 2 \cdot 10^{-7} \, (T_g/T_e)^{0.5}, \; cm^3/s$—the coefficient of dissociative recombination of electrons and molecular ions, $T_g, \; T_e$ the temperature of gas and electrons, respectively. Note that we consider the case of high values of the parameter *E/N*, when the role of negative ions can be neglected (Ardelyan et al. 2017). Under the conditions of quasi-neutrality of plasma, the concentrations of electrons and positive ions are equal, $n_e = N_i$, N_i—the concentration of positive ions. At the same time in the plasma layer, $E \geq E_{br}$. For example, as follows from (Ardelyan et al. 2017), at E = 4 MV/m ($E/N = 150 \cdot 10^{-15}$ V cm^2), the typical time to increase the electron concentration to $n_e \sim 10^2$ cm^{-3} is $t \sim 3 \cdot 10^{-9}$ and less.

Let us determine how the plasma affects the electric field in the layer $r_s < r < r_{br}$. According to (Smirnov 1987), the potential of a charged sphere in a plasma is determined by the expression

$$\varphi = \frac{Q}{4\pi \cdot \varepsilon \cdot R} \cdot exp(-R/r_D) \tag{5.2.6}$$

where

$$r_D = (\varepsilon_0 \cdot T_i \cdot T_e / e^2 n \cdot (T_e + T_i))^{0.5}$$

—the radius of the Debye shielding in the case of a two-component plasma, e is the electron charge, n is the concentration of plasma particles, and T_i, T_e the temperatures of ions and electrons, respectively. If $T_i \gg T_e$, which is performed with strong electric fields in the plasma, then

$$r_D = (\varepsilon_0 \cdot T_i / e^2 n)^{0.5}. \tag{5.2.7}$$

The absolute value of the electric field, determined on the basis of the expression (5.2.6), has the form

$$E_p = \frac{Q}{4\pi \cdot \varepsilon_0 \cdot R^2} \cdot exp(-R/r_D)(1 + R/r_D). \tag{5.2.8}$$

Calculations (Ardelyan et al. 2017) show that for breakdown fields $n \approx 10^{16}\ \mathrm{cm}^{-3}$, then the Debye radius turns out to be $r_D \approx 1{,}2 \cdot 10^{-6}$ cm, (5.2.8) is simplified and takes the form:

$$E_p = \frac{Q}{4\pi \cdot \varepsilon_0 \cdot R^2} \cdot R/r_D \cdot exp(-R/r_D) = \frac{Q}{4\pi \cdot \varepsilon_0 \cdot R^2} \cdot F, \tag{5.2.9}$$

where

$$F = R/r_D \cdot exp(-R/r_D) = R \cdot \left(\frac{e^2 \cdot n}{\varepsilon_0 T_i}\right)^{0,5} \Big/ \exp R \cdot \left(\frac{e^2 \cdot n}{\varepsilon_0 T_i}\right)^{0,5}$$

Substituting in F the specific values of the parameters in the SI system, we get

$$F = R \cdot n^{0,5} \cdot 2.3 \cdot 10^{13} / \exp(R \cdot n^{0,5} \cdot 2.3 \cdot 10^{13}),$$

where R in m and n in m^{-3}. This value tends to zero; therefore, it can be assumed that the plasma field $E_p = 0$ is in the area $r_s < r < r_{br}$ of the radius of the sphere for all concentrations of plasma particles. Thus, plasma processes turn off the electric field. As noted above (Ardelyan et al. 2017), plasma with strong fields appears at $t \sim 10^{-9}$ s and less so that the electric field is turned off. Consequently, no breakdown phenomena occur in this spatial region.

When $r \geq r_{br}$ the electric field is again described by Coulomb's law $E = \frac{Q}{4\pi \cdot \varepsilon_0 \cdot R^2}$.

When $r = r_{br}$, $E = E_{br}$, the thickness of the layer in which electrons and ions are generated is of the order of the length of the plasma oscillations—the thickness of the layer. In this layer, in the stationary case, it follows from (5.2.3) that

$$0 = q + \nu_i \cdot n_e - \alpha_{dr} \cdot n_e^2$$

The specific value of the electron concentration in stationary conditions with an electric field $E = 3.3$ MV/m close to breakdown turns out to be of the order of $n_e \sim 5 \cdot 10^{15}$ cm^{-3} when $T_e \sim 3$ э B, and the Debye radius is $r_D = 1.8 \cdot 10^{-5}$ cm, and this layer can be noticeable as a halo over the BL in the air (Ardelyan et al. 2017).

5.2.3 The Lifetime of the Charged Ball Relative to Neutralization by Air Ions

The presence of a plasma layer affects the lifetime of the ball due to the movement of oppositely charged particles in the gas to the surface of the ball. Consider reducing the charge of a charged ball. To do this, we also consider two areas $r_s < r < r_{br}$ and $r > r_{br}$. Oppositely charged ions move against the sphere's field towards the boundary. It can be assumed that it is the sphere of radius r_{br} that attracts ions. The equation describing the decrease of ions from this region has the form:

$$\frac{dQ}{dt} = -4\pi R^2 \cdot e \cdot N_i w_{dr} = -4\pi R^2 \cdot e \cdot N_i\, b_i\, E$$
$$= -4\pi R^2 \cdot e \cdot N_i\, b_i \frac{Q}{4\pi \cdot R^2 \varepsilon_0} = -e \cdot N_i\, b_i\, Q/\varepsilon_0, \tag{5.2.10}$$

where w_{dr} is the drift velocity of the ions, b_i is their mobility. From (5.2.10), we can find the characteristic time of charge transfer across the boundary r_{br}

$$\tau_{dis} = \varepsilon\varepsilon_0/(b_i\, e \cdot N_i).$$

Accounting for that in air under normal conditions $N_i \approx 10^3$ cm^{-3}and $b_i \approx 2.8 \cdot 10^{-4}$ m^2/(V·s), we obtain the time of ion transfer through the boundary r_{br} $\tau_{dis} \approx 1.9 \cdot 10^2$ s. Crossing this boundary, the ions fall into an area where the electric field of the ball does not act. There they simply diffuse to the surface of the ball. The characteristic motion time of ion diffusion over the surface of the ball is of the order of (Raiser YuP 1992) $\tau_{dif} \approx (\pi \cdot (r_{br} - r_s))^2/D$, where D is the diffusion coefficient of neutral molecules for a mixture of N_2-O_2 $D \approx 0.2\ cm^2/s$ (Physical quantities 1991). For example, if $r_{br} - r_s \approx 10$ cm $\tau_{dif} \approx 5 \cdot 10^3$ s, .e., the lifetime of a charged ball relative to its neutralization by ions of the opposite sign increases significantly and varies depending on the difference, i.e., on the size of the ball and its charge.

The Effect of the Plasma Layer on the Movement of the Ball

Obviously, if the position of the ball above the Earth's surface h satisfies the condition $h > r_{br}$, then the plasma layer will not affect the movement of the ball. Otherwise, in the absence of an electric field on the surface of the ball, this will lead to a situation where gravity will be the only force acting on the ball, and the ball will move towards the Earth.

Therefore, the presence of a large charge on the surface of the ball, exceeding the breakdown field, significantly affects the ability of the ball to ionize the surrounding air. There is no breakdown of air in the area between the surface of the ball and the radius of the breakdown field. This increases the lifetime of the charged ball compared to neutralization by aeroions. The movement of the ball may have features when it approaches the surface of the Earth.

5.2.4 *High Energy Electrodynamic BL (Nikitin 2006), (Nikitin et al. 2018a, b)*

In this model, it is assumed that the BL has a shell and a core. The space inside the BL shell is filled with some "plasmoids" with an uncompensated electric charge. A "plasmoid" usually means a system of moving charges held by their own electric and magnetic fields (Bostik 1956). However, it has been proved that stabilization of the plasmoid only by internal forces is, in principle, impossible. This requires some kind of external force, for example, the force of gravity (Shafranov 1957). In this case, the compression force of the shell acts as this external force. According to formula (5.2.1), for a BL with an inner radius of the shell $R = 5{\cdot}10^{-2}$ m with a charge value $Q = 10^{-2}$ C, the compressive force of the shell $F_k = 2.97{\cdot}10^8$ N, and the pressure inside the shell $P_{in} = F_k/4\pi R^2 = 9.5{\cdot}10^9$ Pa, which is almost 10^5 times greater than the atmospheric pressure force.

Charge carriers in BL are plasmoids formed by relativistic electrons moving in closed orbits and ions rotating around them. Electrons are a source of synchrotron radiation. In rare cases, it has the form of a narrow beam, but more often the light of the core is re-emitted and scattered by the matte shell of the BL. The periodic movement of charges serves as a source of BL radio emission in the wavelength range of 1–10 cm. There is no infrared component in the synchrotron radiation spectrum, which explains the absence of a feeling of "heat" emanating from the BL when in close contact with it. The power of the BL radiation is maintained at a low level due to the uniform distribution of electrons along the orbit. When this uniformity is violated, the BL emits energy in the form of a powerful pulse in the optical and radio bands.

In the "quiet" mode, the power of the BL radiation fluctuates randomly, the radiation may disappear completely for a while, and the ball lightning may turn into a "black" one. The intensity of the electric field on the surface of the BL significantly exceeds the intensity of the breakdown of air. Due to this, a corona discharge occurs near the surface, which causes the formation of nitrogen dioxide and ozone and is detected as a glow of nitrogen dioxide and other gases.

In the model (Nikitin 2006), the configuration in the form of an energy-intensive core makes the spherical shape of the BL natural and makes clear the reason for the existence of surface tension at its shell. The study of the type of charge distribution between numerous small objects forming its core allows us to find a way to explain the cases of changing the shape of a ball lightning—its transformation into a "melon", a ribbon, and cases of its passage through small holes. The presence of an uncompensated charge easily explains the reason for its complex movement in the atmosphere. Thus, a high-energy BL with an outer diameter of 12 cm and a charge of $Q = 5 \cdot 10^{-3}$ C has a shell with a volume of 380 cm^3. If its density is equal to the density of water, then the mass of the BL is $m_{bl} = 0.38$ kg, and its weight is $P_{bl} = 3.74$ N. With a charge value of $Q = 5 \cdot 10^{-3}$ C, it can levitate under the influence of an electric field $E_{ex} = P_{bl}/Q = 750$ V/m (which is comparable to the intensity of the "good weather field" of 100 V/m). Therefore, there is no reason to postulate that the weight of the BL is equal to the weight of the air displaced by it (Barry 1980). The wind moving at a speed of v = 10 m/s presses the ball with a radius of $R = 6$ cm with a force $F_f = C_x \cdot (\rho_a v^2/2) \cdot \pi R^2 = 0.2$ N. (Here $\rho_a = 1.13$ kg/m^3 is the air density, and Cx = 0.3). Under the action of a field with a strength greater than 40 V/m, a BL of this size, with a charge of $5 \cdot 10^{-3}$ C, can fly against the wind. There are reports of BL, with a diameter of about 1 meter, flying next to the plane. At the speed of the aircraft v = 100 m/s, the drag force of the BL with a radius of 0.5 m is $F_f = 1330.6$ N. Let the BL be at a distance of $l = 1$ m from the metal wing of the aircraft. The force of its attraction to the wing $F_{at} = Q_{bl}^2/4\pi\varepsilon_0(2l)^2 = 22.48 \cdot 10^8 Q_{bl}^2$ F. Equating F_{at} and F_f we find $Q_{bl} = 7.69 \cdot 10^{-4}$C.

A charged BL constantly loses its charge. The limited size of its charge is the reason for the limited duration of its life. Simultaneously with the loss of charge, the BL loses energy for radiation. When the charge and energy loss rates are balanced, the size and luminosity of the BL remain unchanged. If the rate of charge loss becomes greater than the rate of energy loss, the BL may explode; otherwise, it will collapse. Near conductive objects, the charge will flow mainly in their direction. This will create a "cushion" of charges of the same sign as the BL charge. This explains the cases of BL movement over extended conductors and the earth's surface, as well as the existence of connected BL (pairs, triples, etc. (Nikitin et al. 2016).

Comment. We described a very attractive hypothesis of Nikitin A.I. (Nikitin 2006). This seemingly coherent theory can explain all possible observations of BL. However, the following can be noted. It can explain a lot if BL is a stationary object. It has implicit limitations. The time of formation of such a BL is limited by the time of existence of a vacuum cavity in which ions and electrons move in orbits. The vacuum must be very high so that these particles do not slow down. It is not clear from the model how this can be implemented. At the same time, the lifetime of the vacuum cavity is of the order $t \sim D_{bl}/a_s$, where D_{bl}, is the diameter of the vacuum cavity, in order of magnitude equal to the diameter of the BL, a_s is the speed of sound in the air. Substituting $D_{bl} = 20$ cm, $a_s = 300$ m/s, one gets a formation time of t ~ 10^{-3} s. So, in this short time, it is necessary to form a shell, the movement of ions and electrons in orbits, and synchronize them. At current development of technic, this seems unrealistic.

5.3 Chemical-Thermal Model (Bychkov and Abakumov 2018)

Let us consider the "chemical-thermal" model of BL. According to it, the BL is a unipolar charged object with a shell and a core. The core may consist of atomized particles of water vapor or atoms of matter that appear when the linear lightning strikes various materials, such as SiO_2, Al_2O_3, etc., or soil with or without water in it. For simplicity of analysis, we assume that the energy of linear lightning in this case is sufficient to convert the material that was affected into atoms of an ideal gas. This approach allows to consider the possibility of high-energy BL formation, although it leaves open the question of the properties of BL, the core of which includes vapors and droplets of the formed substance.

Electrons carried by linear lightning, when struck into materials at atmospheric pressure, quickly attach to neutral molecules, forming negative ions. Therefore, the core substance contains these ions, which determine the electrical properties of the BL. In this case, the shell is a thin film of a melt or solidified substance. The charge is transmitted to the BL from linear lightning, and the stability of the BL with respect to destruction, is provided by the polarization force arising due to the Coulomb effect of the unipolar charge, see Sect. 5.2. Inside such BL is a mixture of gas and negative ions of an atomized substance. In this model, components connected with the random movement of particles inside the shell are present inside the shell. At the same time, the energy of such a BL is determined mainly by the vaporization energy released during the destruction of the object, and, to a lesser extent, during the combustion of the BL substance, as well as the average energy of particles inside the core at a given temperature and pressure. The release of energy during burning is insignificant. Taking into account the combustion processes of BL material in the air requires complication of the model, but does not make a significant contribution to the value of the BL energy density.

BL energy. The most important issue in this approach is the question of BL energy. To find an answer to this question, we first determine the pressure of the gas inside the BL acting on its surface. Then, when we know the pressure and enthalpy of vaporization (or atomization energy—the energy of splitting a substance into atoms), that is, the energy density per atom of a substance or a water molecule, we determine the specific energy of such a BL. Let us consider two possible cases of BL formation.

First: the composition of the shell and the gas inside the BL differ little (this possibility was indicated by the results of experiments with the evaporation of basalt and steel). At the same time, the existence of the shell is determined by nonequilibrium processes in the plasma that occur when a linear lightning strikes the earth. In this case, nonequilibrium processes of shell formation lead to a change in its physical properties without affecting the chemical properties. For estimates, in this case, silicon is considered, the content of which in the earth's crust reaches 50%. The gas consists of SiO_2 silicon dioxide, and the shell consists of a modified SiO_2 melt and its alloy with soil components. At the same time, for estimates, we assume that the temperature of the gas inside the shell is close to 2000 K and is equal to the typical boiling point of SiO_2. We assume that in this case a shell in the form of molten SiO_2 or its film with a temperature lower than the temperature of its destruction can be formed, and inside it, there are atoms of atomized matter formed when the linear lightning strikes the material.

Second: the core of the BL consists of water vapor, and the shell consists of molten SiO_2 or its film. At the same time, for estimates, we also assume that the temperature of the gas inside the shell is about 2000 K, i.e., about the boiling point of SiO_2, at which its destruction can occur.

Let us write down the equation of the balance of forces acting on the shell of such a BL:

$$F_{surf} + F_a + F_{sh} - F_{el} - w\,4\pi(R+a)^2 = 0, \tag{5.3.1}$$

where $F_a = 4\pi(R+a)^2 \cdot P_a$ is the force of atmospheric pressure, and, as above, $P_a = 1.03 \cdot 10^5\,Pa$ is atmospheric pressure.

$$F_{sh} = \frac{\sigma\, a(2R+a)}{\varepsilon_0(R+a)} Q$$

the strength of the polarization interaction of charges (5.2.1) σ—the surface density of charges on the shell (see 5.2.1), a—the thickness of the shell of the ball lightning, R—its radius, Q—the charge of the BL, $\varepsilon_0 = 8.85 \cdot 10^{-12}$F/m—the electric constant.

$$F_{el} = \frac{1}{4\pi\varepsilon_0} \frac{Q^2}{(R+a)^2}$$

the Coulomb interaction force of charges of the same sign acting on the shell.

$$F_{surf} = 8\pi\alpha(R+a)$$

the surface tension force, α- the surface tension coefficient of the shell.

w is the pressure inside the ball lightning (when simplified, we do not take into account the volume of the shell)—determines the internal energy of the BL, which goes to collisions of gas particles with the BBL shell and leads to its stretching. We believe that it is the processes of atomization of matter, Coulomb repulsion, and polarizing compression of the shell that lead to the establishment of pressure inside the BL. In turn, atmospheric pressure and surface tension practically do not contribute. From (5.3.1) follows the equation for its definition

$$w = \frac{F_{surf} + F_a + F_{sh} - F_{el}}{4\pi(R+a)^2} \approx \frac{F_{sh} - F_{el}}{4\pi(R+a)^2}. \tag{5.3.2}$$

The concentration of particles inside the BL can be determined based on the ratio

$$w = P = N_a kT, \tag{5.3.3}$$

where $k = 1.38 \cdot 10^{-23} J/K$ is the Boltzmann constant and T is the temperature of the gas particles inside the BL.

The mass of the BL is defined as

$$M = m_a N_a V = \frac{4}{3}\pi R^3 m_a N_a, \tag{5.3.4}$$

where m_a is the mass of the gas particle and V is the volume of the BL. The energy density of the BL is determined based on the expression

$$E = N_a\left(m_a\lambda + \frac{3}{2}kT\right) = \frac{w}{kT}\left(m_a\lambda + \frac{3}{2}kT\right), \tag{5.3.5}$$

where λ is the enthalpy of vaporization (sublimation energy) or the energy density of recombination of water atoms, (J/kg). For the main material SiO_2 $\lambda \approx 9.55 \cdot 10^6$(J/kg), and for the main material $\lambda \approx 5.1 \cdot 10^7$ J/kg (determined on the basis of data on bond breaking in a water molecule (Physical quantities 1991)). In all the cases, we have considered $m_a\lambda >> \frac{3}{2}kT$, which is natural, because the sublimation energy is of the order of the dissociation energy, and that, in turn, is much greater than the thermal energy of gas particles.

Figure 5.1 shows typical dependences of the pressure w ((J/m^3) in dimension coincides with Pa) on the charge Q of the BL. The polarization compression force of the shell of the BL is proportional to the charge Q, and the Coulomb force leading to the destruction of the shell is proportional to Q^2, they act in different directions, therefore, there is a charge value at which these interactions are compared. With a large BL charge, its main energy goes to stretch the shell and the internal energy that can be released and perform work falls. It can be seen from Fig. 5.1 that the pressure $w(Q)$ varies widely depending on the magnitude of Q and has a maximum at given

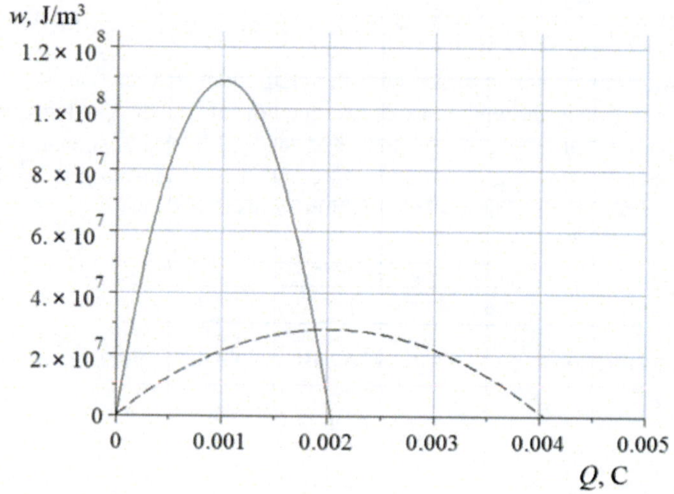

Fig. 5.1 Typical dependences of pressure w on charge Q of BL. The solid line corresponds to a BL with a radius of $R = 5$ cm, the dotted line corresponds to $R = 10$ cm. For both lines $\sigma = 1.5$ C/m^2, $a = 1$ mm

values of the thickness of the shell and the density of the polarization charges located on it.

In Tables 5.1 and 5.2, the parameters of the BL are presented at various values (σ, a, R, Q) calculated according to the formulas (5.3.2)–(5.3.5) presented above, at the maximum value of $w(Q)$, which is interesting for us (since we analyze the possibility of the existence of high-energy BL).

The data in Table 5.1 show that the energy density of the BL, determined by the sublimation of the material and the formation of gas inside the shell, varies widely, mainly depending on the shell material determining the surface charge density, the charge inside the BL, the thickness of the shell and the radius of the BL. The energy

Table 5.1 Model parameters for calculating the BL energy density

N	σ, C/m2	a, mm	R, cm	w, J/m^3	Q, C
1	1.5	1	5	$9.60.10^7$	0.001
2	1.5	1	10	$2.45.10^7$	0.002
3	1.5	1	50	$1.00.10^6$	0.01
4	4.5	1	5	$8,59.10^8$	0.003
5	4.5	1	10	$2.22.10^8$	0.006
6	4.5	1	50	$9.06.10^6$	0.03
7	4.5	10	5	$5.28.10^{10}$	0.03
8	4.5	10	10	$1.72.10^{10}$	0.06
9	4.5	10	50	$8.52.10^8$	0.3

Table 5.2 The parameters of the model and the resulting energy density of the BL

N	n, m^{-3} SiO_2, H_2O	M, kg SiO_2	M, kg H_2O	E, J/m^3, SiO_2	E, J/m^3, H_2O
1	$3.46.10^{27}$	0.19	0.06	$3.19.10^{9}$	$5.10.10^{9}$
2	$8.91.10^{26}$	0.37	0.11	$8.16.10^{8}$	$1.31.10^{9}$
3	$3.64.10^{25}$	1.83	0.55	$3.32.10^{7}$	$5.30.10^{7}$
4	$3.11.10^{28}$	1.67	0.5	$2.86.10^{10}$	$4.57.10^{10}$
5	$8.02.10^{27}$	3.33	1.0	$7.38.10^{9}$	$1.18.10^{10}$
6	$3.29.10^{26}$	16.6	5.0	$3.01.10^{8}$	$4.82.10^{8}$
7	$1.93.10^{30}$	166	50	$1.76.10^{12}$	$2.82.10^{12}$
8	$6.24.10^{29}$	333	100	$5.73.10^{11}$	$9.18.10^{11}$
9	$3.12.10^{28}$	1650	493	$2.83.10^{10}$	$4.52.10^{10}$

density is weakly dependent on the type of material inside the shell. Within the framework of this model, the BL can have energy comparable to the observed one, but less than in anomalous cases.

BL radiation. The question of the cause of radiation in this model does not have an unambiguous answer, since the BL has a complex internal and surface structure. The reason for the generation of BL radiation in the optical and radio ranges may be plasma created under the action of the electric field of the charge, the high temperature of the core, the movement of electric charges forming its core, or processes that cause plasma pulsation near its shell. BL radiation can be of two types. It can be in equilibrium when the shell of the BL, filled with hot steam, glows. In this case, heat is transferred to the shell by vapor particles located inside the shell. BL radiation can also be nonequilibrium when plasma appears near its surface under the action of a large charge of BL because of ionization processes in a strong electric field. The lifetime of nonequilibrium radiation is determined by the lifetime of the plasma on its surface, that is, the time of reduction of the BL charge. The parameters of nonequilibrium radiation require special analysis, which requires careful detailing of this model, which is beyond the scope of this chapter.

The parameters of the equilibrium radiation can be estimated as follows. We assume that the heating of the shell by steam is compensated by thermal radiation from the surface of the shell, and then we can use the following expression approximate with respect to the heat capacity of a real gas (Smirnov 1987):

$$\frac{3}{2}Nk_B\frac{4}{3}\pi R^3\frac{dT}{dt} = -\sigma_{SB}T^4 4\pi R^2, \tag{5.3.6}$$

where k_B is the Boltzmann constant, and σ_{SB} is the Stefan-Boltzmann constant. The solution of this equation for determining the cooling time to temperature $T1$, if the initial temperature of the gas is much greater than the final temperature of the gas $T >> T_1$, has the form:

$$t_x = \frac{N\,k_B R}{6\,\sigma_{SB} T_1^3} \tag{5.3.7}$$

For example, we substitute in (5.3.7) the following values of the average parameters of the BL from Table 5.1: $N = 8{\cdot}10^{27}$ m^{-3}, $R = 0.1$ м, $T_1 = 500$ K, we get $t_x \approx 260$ s. This value is comparable to the observed lifetime of the BL.

In erosive discharges, as shown above, luminous spherical objects up to 2 cm in diameter were obtained based on polymer materials, which glowed for 1–2 s. At the same time, it seemed that the intensity of their radiation did not change during their lifetime, and then these objects instantly went out. The analysis of the shooting video showed that during 0.04 s their smooth extinction occurred, which was not registered by the observer. Therefore, the constancy of the luminescence intensity noted by the BL observers is most likely an optical effect because the eye does not distinguish a decrease in the luminescence intensity when the BL is cooled from 2000 K to ~800 K. The subsequent cooling of the BL and the corresponding decrease in the intensity of the glow occurs quickly. During this time of extinction of the BL, the eye does not have time to adjust to a low intensity of illumination, since the physiological adaptation time of the eye when observing bright objects is $t \sim 10$ c.

According to (5.3.7), the exposure time of the BL radiation increases with the increase in the radius of the ball, so the observed increase in the lifetime of the BL with the increase in its radius is easily explained. The color of the BL in this model is determined by the material of which its shell consists. Being driven by the electric field of the earth, the BL can approach the windowpane and "penetrate" into the room through a crack or crack. In addition, it is able, having spent part of its energy, to heat the glass to the melting point and form a hole in it, see Chaps. 3, 4. However, this "material" BL is unable to pass through the glass without leaving at least some traces in it.

We have described some of the features of this model. Currently, the tasks of turning the BL from a ribbon into a ball, as well as folding the ribbon into a ball, remain unresolved within the framework of this model, the question of the magnetic properties of such a BL and others is open.

5.4 Ball Lightning as the Formation of a Physical Vacuum (Zaitsev and Bychkov 2021)

The opinion about the BL as an object of unusual nature begins with the well-known words of P.L. Kapitsa that the BL is a window into another world. The question of the possibility of describing BL from the point of view of physical vacuum (PV) according to the methodology of mathematical modeling was raised in (Zaitsev and Bychkov 2021). In the methodology of mathematical modeling (Zaitsev and Bychkov 2021), a mathematical model of an object is first constructed based on general laws and interpretation of experimental facts. Then it is studied theoretically and experimentally. The model is accepted as adequate if the consequences from it correspond to all well-established experimental facts. This methodology is used explicitly or implicitly in other areas of science, especially in the study of unusual and rarely occurring phenomena such as BL.

Model of BL must first explain the totality of the observed unique properties of this object, an unbiased description of which can be found in Chap. 2, a) the possession of anomalously large energy density of the order of 10^{11} J/m^3 and above; the appearance in the clouds and above the ground in the absence of thunderstorm activity, as if from "nothing"; typical diameter up to several meters, turning the ball into a ribbon, folding the ribbon into a ball; the ability to pass without leaving traces through glass at its size compared to the size of a BL, as well as through other bodies and several others.

It is clear that understanding the nature of BL and mastering the technique of its artificial creation is in all respects an extremely attractive area of research. However, modern physics, abandoning the concept of a PV medium, comes to a dead-end when interpreting the listed properties. This even leads to the denial of the existence of the BL phenomenon itself, that is, the denial of the objectivity of its observation in nature. This situation is a sign of the significant role of the PV movement in the phenomenon under consideration.

Let us proceed to the construction of the BL model as a formation from PV. The application of the consistent theory of BL in the model of PV hydrodynamics was proposed in (Zaitsev and Bychkov 2021), we follow this work below.

We first determine the value of the PV density, which we will need when determining the energy density of the BL.

To do this, consider the expression for the electric energy density W in electrodynamics

$$W = \varepsilon\varepsilon_0 \cdot \mathbf{E}^2/2, \tag{5.4.1}$$

where $\varepsilon\varepsilon_0$ is the dielectric constant. In hydrodynamics, $\mathbf{E} = d\mathbf{v}/dt$ and $W \sim \tilde{C} \cdot v^2/2$, where $\tilde{C}$ is some constant. For dimensional reasons, the energy $\tilde{C}$ should have dimension $[\tilde{C}] = \text{kg/m}^3$. Therefore, in hydrodynamics, a standard expression for the energy density of a liquid is obtained

$$W = \rho \cdot v^2/2. \tag{5.4.2}$$

The analysis done in (Zaitsev and Bychkov 2021) showed that the perturbation velocity in PV is equal to the c—speed of sound in it. Assuming that this velocity corresponds to the only known velocity in electrodynamics—the speed of light in the undisturbed state of PV, with an undisturbed value of the dielectric constant, it is possible to obtain a relationship between the density of the fluid and the known electrical characteristics, namely

$$W = \varepsilon \cdot \mathbf{E}^2/2 = \rho \cdot c_L^2/2. \tag{5.4.3}$$

Modern studies with lasers Akhmanov (1991) show that the maximum values of electric fields achieved in experiments in which there is no change in the electrodynamic constant are of the order of intra-atomic electric fields, $E \approx 10^{11}$ V/m. Substituting this value and the known data for ε and c_L in (5.4.3), we get $\rho \approx 2 \cdot 10^{-6}$ kg/m^3.

The expression relating to the speed of sound of PV allows to estimate the pressure in it. As is known, in the adiabatic case, when there is no heat exchange in the system, then Loitsyansky (1950) goes into the formula

$$\frac{\gamma \cdot p}{\rho} = c_L^2,$$

where γ is the ratio of heat capacities in the fluid—PV. Assuming $\gamma = 5/3$, as for a monatomic gas, we obtain an estimate for the pressure in the PV $p \approx 1.1 \cdot 10^{11}$ Pa.

Thus, we have obtained PV parameters at which the appearance of an object is realized, which can be called BL PV.

Let us imagine the BL as a large-scale PV vortex compared to a typical magnetic field. Suppose that at the boundary of the BL, the linear velocity of rotation reaches the speed of light. Since the speed of light is the characteristic speed of free propagation of disturbances in PV, it follows from the analogy with gas hydrodynamics that a boundary layer should be formed near the boundary of such a vortex, which is difficult to penetrate for structural media carriers, in this case for Newtonians, which were introduced by D.I. Mendeleev (see detail in (Zaitsev and Bychkov 2021). The size of Newtonians is extremely small even compared to the size of a proton, so the boundary layer is all the more difficult to penetrate for large PV formations: electrons, atoms, molecules, etc. From the law of conservation of momentum (inertia properties), objects in a vortex tend to cling to the boundary, like solid particles during separation in a centrifuge. As a result, the boundary layer is reinforced by the substance trapped in the BL during its formation. Some kind of shell is created that prevents the collapse of the PV vortex.

The flow of physical vacuum inside the BL shell can be quite complex, as, for example, the flow of gas inside the vessel. To obtain the simplest quantitative estimates, we represent the BL in the form of a PV stream rotating with angular velocity ω around the z-axis of Cartesian coordinates with the origin in the center of the BL, see Fig. 5.2.

The shape of the equilibrium figure of a rotating volume depends on the forces acting inside and outside it, as well as forces in the boundary layer, see, for example (Loitsyansky 1950). In the model of a compressible medium, the pressure required to maintain a steady flow in the vortex and the shape of the free boundary of the vortex can provide, for example, an appropriate distribution of the density of the medium.

BL is usually observed in a spherical shape. Therefore, we study the BL model in the form of a ball of radius r_0.

We introduce spherical coordinates (r,θ,φ) with unit basis vectors $\mathbf{i}_r, \mathbf{i}_\theta, \mathbf{i}_\varphi$, Fig. 5.1.

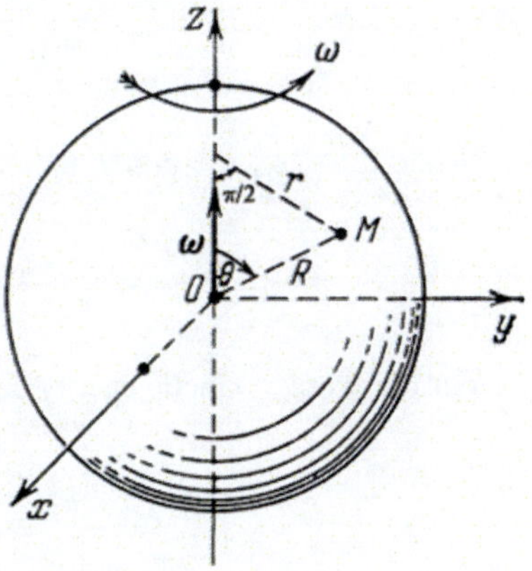

Fig. 5.2 Spherical BL

The linear speed of rotation PV in the simplest BL model is expressed in terms of angular velocity and distance to the axis of rotation (Fig. 5.2)

$$\mathbf{u} = \omega r \sin\theta \mathbf{i}_\varphi.$$

The condition for reaching the speed of light at the boundary r_0 gives the ratio

$$c = \omega r_0 \sin\theta.$$

From here,

$$\omega = \frac{c}{r_0 \sin\theta}.$$

PV has extremely low viscosity and self-diffusion (Zaitsev and Bychkov 2021), so the PV layers can move almost independently of each other. In this BL model, each layer rotates at a fixed θ with its constant angular velocity $\omega = \omega(\theta)$.

Substituting ω into $\mathbf{u}$, we find

$$\mathbf{u} = c\frac{r}{r_0}\mathbf{i}_\varphi.$$

According to the formulas from (Zaitsev and Bychkov 2021), the volumetric density of the kinetic energy PV in BL is

$$\mathcal{A} = \rho_m \mathbf{u}^2 = \rho_m c^2 \frac{r^2}{r_0^2}.$$

The kinetic energy stored in the entire BL is equal to the integral over its volume

$$K_{BL} = \int_V \rho_m c^2 \frac{r^2}{r_0^2} dV.$$

When $\rho_m \approx \rho_{m,*} = const$

$$K_{BL} = \rho_{m,*} c^2 \frac{1}{r_0^2} 2\pi \int_0^{r_0} \int_0^{\pi} r^4 \sin\theta d\theta = \rho_{m,*} c^2 \frac{4}{5} \pi r_0^3.$$

For the average kinetic energy density in BL we have

$$\overline{A}_{BL} = \frac{K_{BL}}{\frac{4}{3}\pi r_0^3} = \frac{3}{5} \rho_{m,*} c^2.$$

It is important to emphasize that in the considered BL model, its average energy density turned out to be a universal characteristic determined only by the PV parameters and independent of the radius of the BL and the angular velocity of rotation of the PV in it.

For $\rho_{m,*} \approx \rho_{m,0}$, we have

$$\overline{A}_{BL} \approx 1.1 \cdot 10^{11}, \mathrm{J/m^3}.$$

Taking into account the fact that not all of its kinetic energy can be transferred to the object when the BL is exposed to objects, the resulting value of $\overline{A}_{BL}$ is close to the anomalous energy density value observed in BL. Such correspondence in the methodology of mathematical modeling serves as one of the confirmations of the correctness of the model.

In (Zaitsev and Bychkov 2021), this approach is applied to the explanation of the nature of luminous balls in N. Tesla's experiments. They are considered PV vortices appearing in an RF installation.

5.5 Conclusion

The book reflects, from the author's point of view, the next stage of research on these objects. It follows that over the past 30 years there has been significant progress in understanding these phenomena.

Therefore, there was an understanding that in order to get new information about observational properties, it is not enough to use the classical analysis of the probability of observing certain properties of BL. Currently, when modern statistical methods of analysis are developed, it is possible to obtain information about the complex properties of an object by simultaneously analyzing a large set of observations, with information about a number of parameters of the BL. In-depth analysis requires the use of modern programs and computers capable of processing large amounts of data based on modern methods of statistical analysis. At the same time, the problem arises in interpreting the huge amount of information received. One can only hope that there will be enthusiasts along the way.

Recently, many books and articles have appeared describing anomalous phenomena, an unbiased analysis of which shows that most of them relate to LLO or BL in different thunderstorms, geographical, meteorological, and geophysical conditions. Combining such information in one book shows that the phenomena of LLO and BL are not at all rare, as previously thought.

Many photos and videos appeared confirming the reality of BL and LLO, while the problem of separating reliable objects from illusions and high-quality fakes was highlighted. Solving this problem is very important and requires additional painstaking work on mastering, analyzing, and applying new methods of video and photo shooting.

Many data on objects similar in manifestations to BL and LLO, but related to other physical phenomena, such as biological and various kinds of lights require additional statistical analysis and laboratory modeling. Experimental work on LLO in liquids, over liquids, and soil appeared. Determining the relationship of their results with phenomena such as crop circles and deep holes in the ground will help to remove questions about their mystery.

A huge number of experiments lead to artificial LLOs with many properties similar to those of natural BLs. The parameters by which they can be distinguished from real BL are low average energy density, lifetime, and modest size. The development of experiments encounters difficulties in realization of high-power voltage and current sources necessary for the creation of such objects, both from the engineering and financial side.

In theoretical studies, works have appeared describing the BL as a high-energy formation with a large electric charge, not necessarily with a density of matter like that of the surrounding air. There are difficulties in describing the BL as an object that changes shape and has abnormal electrical and magnetic properties. In general, one can see the directions that allow us to advance the description of such properties.

The works showing the connection of the BL with the physical vacuum require the accumulation of information about the super-anomalous properties of natural objects and, in this regard, the analysis and repetition of the works of N.Tesla, proceeding from the ideas about the fundamental role of ether (physical vacuum now) in nature.

Despite the practical lack of funding, the development of the science of LLO and BL continues with the work of enthusiasts, students, and concerned researchers. Their results are discussed with great interest by scientific and near-scientific communities because human interest in the mysteries of nature cannot be stopped.

References

Abrahamson J, Dinniss J (2000) Ball lightning caused by oxidation of nanoparticle networks from normal lightning strikes on soil. Nature 403:519–521

Abrahamson J (2002) Ball lightning from atmospheric discharges via metal nanosphere oxidation: from soils, wood or metals. Phil Trans Roy Soc Lond 360(1790):61–88

Abakumov VI, Bychkov VL, Bikmukhametova AR, Chernikov VA, Golubkov GV, Safronenkov DA, Zaitsev SV (2018) Production of long-lived heterogeneous structures using plasma. J Phys Conf Ser 1112:012022

Akhmanov SA (1991) In: Akhmanov SA (Ed.) Modern problems of laser physics. VINITI publishers

Akishev YuS, Grushin ME, Kochetov IV, Napartovich AP, Trushkin NI (1999) Establishment of trichel impulses in the negative corona in the air. Fizika Plasmy 25(11):1–6

Ardelyan NV, Bychkov VL, Kochetov IV, Kosmachevskii KV (2014) Prebreakdown air ionization in the atmosphere. In: Bychkov V, Golubkov G, Nikitin A (eds) The atmosphere and ionosphere: elementary processes, monitoring and ball lightning. Springer, Cham Heidelberg, pp 69–111

Ardelyan NV, Bychkov VL, Kosmachevskii KV (2017) On electron attachment and detachment processes in dry air at low and moderate constant electric field. IEEE Trans Plasma Sci 45:3118–3124. https://doi.org/10.1109/TPS.2017.2773021

Ardelyan NV, Bychkov VL, Bychkov DV, Kosmachevskii KV (2013) Electron-beam and non-self-maintained driven plasmas PAC. Plasma assisted combustion, gasification and pollution control 1, Matveev. Outskirts press. Denver, Colorado, pp 183–372

Atmosphere (1991) Reference book Sedunov YuS ed. Hydrometeoizdat Publishers, Leningrad

Ball Lightning in the Laboratory (1994) Avramenko RF, Bychkov VL, Klimov AI, Sinkevich OA (eds) Khimia publishers, Moscow

Barry JD (1980) Ball lightning and bead lightning. Plenum, New York

Bostik WH (1956) Experimental study of ionized matter projected across a magnetic field. Phys Rev 104(2):292–299

Bychkov VL (1994) Polymer ball lightning model. Phys Scr 50:591–599

Bychkov VL (2008) Response to "Comment on the paper by A. I. Nikitin "will it be Possible to solve the problem of ball lightning in the 21st century?". Khim Fis 27 N 2:84–86

Bychkov VL, Nikitin AI, Ivanenko IP, Nikitina TF, Velichko AM, Nosikov IA (2016) Ball lightning passage through a glass without breaking it. J Atmos Solar-Terr Phys 150–151:69–76

Bychkov VL (2002) Polymer composite ball lightning. Phil Trans R Soc Lond 360(1790):37–60

Bychkov VL, Abakumov VI (2018) Ball lightning as chemical-thermal electrically charged object. In Proc 14th Int Symp Ball Lightning (ISBL-18) 3–9 June 2018, Zelenogradsk, Kaliningrad region, Russia, pp 29–38

Bychkov VL, Nikitin AI (2014) Ball lightning: a new step in understanding. In The Atmosphere and Ionosphere. Elementary Processes, Monitoring, and Ball Lightning. Bychkov VL, Golubkov GV, Nikitin AI eds Springer, Cham, Heidelberg, pp 201–367

Bychkov VL, Chernikov VA, Osokin AA, Stepanov AI, Stepanov IG (2015) Modeling of artificial ball lightning with a help of capillary discharge. IEEE Trans Plasma Sci 43:4043–4047. https://doi.org/10.1109/TPS.2015.2478441

Bychkov VL, Esakov I I, Shestakov DV (2020) On electric field distribution the near a charged sphere. High-Energy Ball Lightning and its Big Charge. In Atmosphere, Ionosphere, Safety AIS Borshevkina OP, Golubkov MG, Karpov IV, eds. I. Kant university Publishers, Kaliningrad, pp 186–188

Bychkov VL, Nikitin AI, Dijkhuis GC (2010) Ball lightning investigations. In The Atmosphere and Ionosphere. Dynamics, processes and monitoring. Bychkov VL, Golubkov GV, Nikitin AI eds. Springer, Dordrecht, pp 201–373

Cameron RP (2018) Monochromatic knots and other unusual electromagnetic disturbances light localized in 3D. J Phys Commun 2:015024

Dijkhuis GC (1980) A model for ball lightning. Nature 248:150–151

Emelin S, Bychkov V, Astafiev A, Kovshik A, Pirozersky A (2012) Plasma combustion nature of artificial ball lightning. IEEE Trans Plasma Sci 40:3162–3165

Handel P (1975) Maser theory of ball lightning. Bull Am Phys Soc Ser. Ii 20(1):26

Kawano S (2004, 2006) Is a spatial stem a ball lightning? In Proc 8th Int Symp Ball Lightning (ISBL04). Chung-li, Taiwan, pp 79–82, Proc ISBL-06, pp 91–94

Koloc PM (1999a a) Formed PLASMAK artificial ball lightning results. In Proc 6th Int Symp Ball Lightning (ISBL99). Antwerp, Belgium, pp. 219–226; Fusion Technology, 1989, 15 N 3:1136

Koloc PM (1999b) Comparison between PLASMAK BL model and formed PMKs. In Proc 6th Int Symp Ball Lightning (ISBL99). Antwerp, Belgium, pp 227–232

Loitsyansky LG (1950) Mechanics of liquid and gas. State publishers of Technic-Theoretic literature, Moscow- Leningrad

Lowke JJ (1996) A theory of ball lightning as an electric discharge. J Phys D (appl Phys) 29(5):1237–1244

Lowke JJ, Smith D, Nelson KE, Crompton RW, Murphy AB (2012) Birth of ball lightning. J Geoph Res 117:D19107. https://doi.org/10.1029/2012JD017921

Manykin EA, Norman GE (1999) Ball lightning as supercooled plasma condensed phase. In Proc 6th Int Symp Ball Lightning (ISBL99). Antwerp, Belgium, pp 120–125

Manykin EA, Ozhovan MI, Poluektov PP (1981) On collective electronic state in a system of strongly-excited atoms. Doklady AN SSSR 260:1096–1098

Manykin EA, Ozhovan MI, Poluektov PP (1983) Theory of the condensed state in the system of excited atoms. ZhETF 84(2):442–453

Manykin EA, Ozhovan MI, Poluektov PP (1982) To the question of ball lightning nature. ZhTF 52:1474–1476

Manykin EA, Ozhovan MI, Poluektov PP (1999) Rydberg matter and ball lightning. In Proc 6th Int Symp Ball Lightning (ISBL99) Antwerp, Belgium, pp 114–119

Matsumoto T (1997) Ball lightning during underwater spark discharges and the Matsumae earthquakes. In Proc 5th Int Symp Ball Lightning (ISBL97). Tsugawa-Town, Japan, pp 193–201

Matsumoto T (1999a) Micro ball lightning during underwater spark discharges. In Proc 6th Int Symp Ball Lightning (ISBL99). Antwerp, Belgium, pp 249–254

Matsumoto T (1999b) Transport of micro ball lightning. In Proc 6th Int Symp Ball Lightning (ISBL99). Antwerp, Belgium, pp 255–262

Nikitin AI (2006) Electrodynamic model of ball lightning. Khim Fiz 25(3):38–62

Nikitin AI, Velichko AM, Nikitina TF, Stepanov IG (2018a) Analysis of the unique case of ball lightning observation in Mitino, the northwest district of Moscow. J Atmos Solar-Terr Phys 179:97–104

Nikitin AI, Bychkov VL, Nikitina TF., Velichko AM, Abakumov VI (2018b) Sources and components of ball lightning theory. Int Interdis Conf "Euler Readings MRSU 2017" IOP Publishing IOP Conf Ser: J Phys: Conf Ser 996:012011 doi:https://doi.org/10.1088/1742-6596/996/1/012011

Nikitin AI, Bychkov VL, Nikitina TF, Velichko AM (2014) High-energy ball lightning observations. IEEE Trans Plasma Sci 42:3906–391147

Nikitin AI, Velichko AM, Nikitina TF, Stepanov AI, Stepanov IG (2016) Observation and analysis of ball lightning group flights. Proc 5th Int Conf "Atmosphere, Ionosphere, Safety" (AIS -2016). Kaliningrad, pp 274–278

Norman GE (1999) Ball lightning as overcooled non-ideal plasma. Khim Fiz 18(7):78–86

Norman GE (2000) Ball lightning as a supercooled nonideal plasma. Chem Phys Rep18 N 7:1335–1352

Ofuruton H, Kondo N, Kamogawa M, Aoki M, Ohtsuki YH (1997) Experimental condition of artificial ball lightning by using microwave and discharge. In Proc 5th Int Symp Ball Lightning (ISBL97). Tsugawa-Town, Japan, pp 215–217

Ofuruton H, Kondo N, Kamogawa M, Aoki M, Ohtsuki YH (2001) Experimental condition for ball lightning creation by using air gap discharge embedded in microwave field. J Geoph Res 106(D12):12367–12369

Oreshko AG (2006) Generation of laboratory ball lightning. J Phys Conf Series 44:172–132

Physical quantities (1991) Handbook. Grigoriev IS, Meilikhov EZ (ed), Energoatomizdat publishers, Moscow
Rabinowitz M (2001) Little black hole ball lightning models agrees with observations. Abs 7-th Int Symp Ball Lightning July 26–29 Univ. of Missouri. St Louis. 2001. Missouri, p 21
Raiser YuP (1992) Physics of gas discharge. Nauka Publishers, Moscow
Rañada AF, Trueba JL (1996) Ball lightning as an electromagnetic knot? Nature 383:32
Rañada AF, Soler M, Trueba JL (1999) Ball lightnings and force-free magnetic knots. In Proc 6th Int Symp Ball Lightning (ISBL99). Antwerp, Belgium, pp 102–107
Shafranov VD (1957) On equilibrium magnetohydrodynamic configurations. ZETF 33 3(9):710–722
Singer S (1973) The nature of ball lightning. Mir Publishers, Moscow
Shmatov ML (2003) New model and estimation of the danger of ball lightning. J Plasma Phys 69(6):507–527
Shchedrin AI (2018) Relativistic model of BL. Hotline-Telecom Publishers, Moscow
Smirnov BM (1987) Introduction to plasma physics. Nauka Publishers, Moscow
Smirnov BM (1988) The problem of ball lightning. Nauka Publishers, Moscow
Smirnov BM (1993) The problem of ball lightning. Phys Rep 224:151–263
Stakhanov IP (1996) On the physical nature of ball lightning. Nauchnyi mir Publishers, Moscow
Stenhoff M (1999) Ball Lightning. An Unsolved Problem in Atmospheric Physics. Kluwer Publishers, New York
Terletsky YaP, Rybakov YuP (1980) Electrodynamics. Vysshaya shkola Publishers, Moscow
Vlasov AN (2009) Ball lightning—induction discharge in a vortex ring? Nauka I Zhizn N7:28–34
Vlasov AN (2006) Ball lightning as the current layer induced in a vortex. Proc 9th Int Symp Ball Lightning (ISBL06). Eindhoven, Netherlands, pp 243–249
Wu HC (2016) Relativistic-microwave theory of BL. Sci Rep 6:26263. https://doi.org/10.1038/srep2826311
Zaitsev FS, Bychkov VL (2021) Mathematical modeling of electromagnetic and gravitational phenomena by the methology of continuous media mechanics. MAKS PRESS Publishers https://doi.org/10.29003/m2011.978-5-317-06604-8

Conclusion

The book reflects, from the author's point of view, the next stage of research on these objects. It follows that over the past 30 years there has been significant progress in understanding these phenomena.

Therefore, there was an understanding that in order to get new information about observational properties, it is not enough to use the classical analysis of the probability of observing certain properties of BL. Currently, when modern statistical methods of analysis are developed, it is possible to obtain information about the complex properties of an object by simultaneously analyzing a large set of observations, with information about a number of parameters of the BL. In-depth analysis requires the use of modern programs and computers capable of processing large amounts of data based on modern methods of statistical analysis. At the same time, the problem arises in interpreting the huge amount of information received. One can only hope that there will be enthusiasts along the way.

Recently, many books and articles have appeared describing anomalous phenomena, an unbiased analysis of which shows that most of them relate to LLO or BL in different thunderstorms, and geographical, meteorological, and geophysical conditions. Combining such information in one book shows that the phenomena of LLO and BL are not at all rare, as previously thought.

Many photos and videos appeared confirming the reality of BL and LLO, while the problem of separating reliable objects from illusions and high-quality fakes was highlighted. Solving this problem is very important and requires additional painstaking work on mastering, analyzing, and applying new methods of video and photo shooting.

Many data on objects similar in manifestations to BL and LLO, but related to other physical phenomena, such as biological and various kinds of lights require additional statistical analysis and laboratory modeling. Experimental work on LLO in liquids, over liquids, and soil appeared. Determining the relationship of their results with phenomena such as crop circles and deep holes in the Ground will help to remove questions about their mystery.

V. L. Bychkov, *Natural and Artificial Ball Lightning in the Earth's Atmosphere*, Springer Atmospheric Sciences, https://doi.org/10.1007/978-3-031-07861-3

A huge number of experiments lead to artificial LLOs with many properties similar to those of natural BLs. The parameters by which they can be distinguished from real BL are low-average energy density, lifetime, and modest size. The development of experiments encounters difficulties in the realization of high-power voltage and current sources necessary for creation of such objects, both from the engineering and financial side.

In theoretical studies, works have appeared describing the BL as a high-energy formation with a large electric charge, not necessarily with a density of matter like that of the surrounding air. There are difficulties in describing the BL as an object that changes shape and has abnormal electrical and magnetic properties. In general, one can see the directions that allow us to advance the description of such properties.

The works showing the connection of the BL with the physical vacuum require the accumulation of information about the super-anomalous properties of natural objects and, in this regard, the analysis and repetition of the works of N.Tesla, proceeding from the ideas about the fundamental role of ether (physical vacuum now) in nature.

Despite the practical lack of funding, the development of the science of LLO and BL continues with the work of enthusiasts, students, and concerned researchers. Their results are discussed with great interest by scientific and near-scientific communities because human interest in the mysteries of nature cannot be stopped.

Index

V. L. Bychkov, *Natural and Artificial Ball Lightning in the Earth's Atmosphere*, Springer Atmospheric Sciences, https://doi.org/10.1007/978-3-031-07861-3